Joseph

Evolutionary Biology

VOLUME 2

Evolutionary
Biology

VOLUME 2

EDITORS:

THEODOSIUS DOBZHANSKY, *Rockefeller University*

MAX K. HECHT, *Queens College*

WILLIAM C. STEERE, *New York Botanical Garden*

Appleton-Century-Crofts   New York   1968

Division of Meredith Corporation

Contributors

THOMAS C. BARR, JR., *Department of Zoology and Institute of Speleology, University of Kentucky, Lexington*

J. DE LEY, *Laboratory of Microbiology, Faculty of Sciences, State University, Gent, Belgium*

THEODOSIUS DOBZHANSKY, *The Rockefeller University, New York*

RICHARD M. EAKIN, *Department of Zoology, University of California, Berkeley*

IRVING I. GOTTESMAN, *Department of Psychology and Behavioral Genetics, University of Minnesota, Minneapolis*

CHARLES L. REMINGTON, *Department of Biology, Yale University, New Haven, Connecticut*

J. P. SCOTT, *Department of Psychology, Bowling Green State University, Bowling Green, Ohio*

HERMAN T. SPIETH, *Department of Zoology, University of California, Davis*

We are pleased to present in this second volume of *Evolutionary Biology* a continuation of the scope and aims set forth by the Editors in Volume 1. We trust that the readers will find the expectations they now hold fulfilled by the quality of the present and future volumes. For all readers, we are glad to repeat the Editor's preface to Volume 1, below.

The Publisher

Preface to Volume 1

This is the first volume of a serial publication that reflects our conviction that evolutionary biology represents a unifying principle in the life sciences. We have conceived this serial as a forum in which critical reviews and commentaries, as well as original papers and even controversial views, can be brought together to cover a broad range of interest with provocative discussions.

Evolutionary Biology will provide research workers and students with an exceptional opportunity to read expert presentations of developments in areas of their field in which they are not specialists, and as specialists they will see how others assess these developments. An important feature is that contributions are not necessarily limited in length, subject, and other restrictions that usually prevail in basic research journals.

The eight chapters of the present volume, and those of the future volumes, combine to present a vast and impressive panorama of conceptualizations in the multiple disciplines of evolutionary and population biology, using data derived from studies of man, other animals, plants, and microorganisms, dealt with in terms of anthropology, biochemistry, developmental biology, ecology, genetics, molecular biology, paleontology, physiology, and so forth. The very fact of the association of these ideas in one volume will reinforce their mutual significance, so that information gained from one chapter may give the reader greater enlightenment and insight from the others.

In summary, *Evolutionary Biology* will serve to integrate a large and complex area of science that has previously been characterized more by its tendency toward divisiveness than toward synthesis.

Th. Dobzhansky
M. K. Hecht
W. C. Steere

Contents

Contributors   v

Preface   vii

1 **On Some Fundamental Concepts of Darwinian Biology**
   Theodosius Dobzhansky   1
 Introduction   1
 Vitalism, Mechanism, and Compositionism   3
 Adaptedness and Adaptation   5
 Adaptedness to Survive and to Reproduce   7
 Adaptability   9
 Evolutionary Plasticity   11
 The Problem of Quantification of Adaptedness   12
 Darwinian Fitness   15
 Varieties of Natural Selection   18
 Darwinian Fitness and Adaptedness   21
 Evolutionary Opportunism and Adaptive Radiation   23
 Progressive Evolution   27
 References   32

2 **Cave Ecology and the Evolution of Troglobites**
   Thomas C. Barr, Jr.   35
 Introduction   36
 Animal Life in Caves   43
 The Cave Ecosystem   47
 Regressive Evolution in Cave Animals   69
 Speciation and Adaptation in Troglobites   80
 References   96

3 **Molecular Biology and Bacterial Phylogeny**
   J. De Ley   103
 Introduction   104
 The Evolution of Microorganisms in the Precambrian   105
 Microbial Fossils   114
 Bacteria as "Living Fossils"   117
 The Relationship Between Bacteria and Bluegreen Algae   119
 DNA Nucleotide Composition of Microorganisms   122
 Evolutionary Changes in Bacterial DNA   128
 DNA Homology   138
 Phenotypic Aspects of Bacterial Evolution   148
 Summary   151
 References   153

4   **Evolutionary Implications of Sexual Behavior in Drosophila**
       Herman T. Spieth   157
    Introduction   157
    Types of Behavior   158
    Drosophila Preview   159
    Ontogeny of Mating Behavior   160
    Basic Courtship Patterns   160
    Adaptive Nature of Mating Behavior   165
    Specialized Mating Behavior of Hawaiian Species   168
    Summary of Mating Behavior   170
    Nature and Function of the Stimuli   171
    Mating Behavior of Other Diptera   175
    Significance of Drosophilid Mating Behavior   176
    Lek Behavior   178
    Sexual Isolation   181
    Intraspecific Variations   182
    Interspecific Behavior   185
    Origin of Sexual Isolation   190
    Conclusion   191
    References   191

5   **Evolution of Photoreceptors**
       Richard M. Eakin   194
    Introduction   194
    Ciliary Photoreceptors   196
    Rhabdomeric Photoreceptors   201
    Ciliary Line of Evolution   205
    Rhabdomeric Line of Evolution   216
    Other Taxa   233
    Protostomes Versus Deuterostomes   234
    References   235

6   **Evolution and Domestication of the Dog**
       J. P. Scott   243
    Introduction   244
    Taxonomy and Distribution of the Genus *Canis*   244
    Origin of the Dog   245
    Further Evidence for the Wolf as Ancestor to the Dog   246
    Time of Domestication   249
    The Place of Domestication   250
    The Process of Domestication   252
    Evolution of the Wolf   259
    Evolution of the Dog   260

Evolution of Behavior in Dogs and Wolves   264
Theoretical Considerations   268
References   274

7 **A Sampler of Human Behavioral Genetics**
    Irving I. Gottesman   276
Introduction   276
The Nature of Psychological Traits   278
The Twin Method   287
Intellectual Abilities   289
Schizophrenia   302
Summary   313
References   314

8 **Suture-Zones of Hybrid Interaction Between Recently
    Joined Biotas**
    Charles L. Remington   321
Introduction   322
Tabulation of Suture-Zone Examples   323
The Individual Nearctic Zones   329
Suture-Zones Outside North America   368
Stabilization After Early Suturing   371
Hybridization Outside Suture-Zones   372
Discussion   374
Acknowledgments   383
Tables   385
Appendix   411
References   415

Author Index   429

Subject Index   441

1

On Some Fundamental
Concepts of Darwinian Biology

THEODOSIUS DOBZHANSKY

The Rockefeller University, New York

Introduction 1
Vitalism, Mechanism, and Compositionism 3
Adaptedness and Adaptation 5
Adaptedness to Survive and to Reproduce 7
Adaptability 9
Evolutionary Plasticity 11
The Problem of Quantification of Adaptedness 12
Darwinian Fitness 15
Varieties of Natural Selection 18
Darwinian Fitness and Adaptedness 21
Evolutionary Opportunism and Adaptive Radiation 23
Progressive Evolution 27
References 32

Introduction

There are two approaches to the study of the structures, functions, and interrelations of living beings—the Cartesian or reductionist and the Darwinian or compositionist. This does not mean that some biological sciences are reductionist and others compositionist, or that there are Cartesian and Darwinian phenomena. Biological phenomena do have, however, Cartesian and Darwinian aspects. Some biologists view their subject more from the reductionist and others from the compositionist side, and some are more adept at using Cartesian and others Darwinian methodologies.

Descartes considered living bodies, including human bodies, to be automata, i.e., machines describable in physical, and eventually, in mathematical terms. They may, accordingly, be studied by means of the famous Cartesian method. Like any complex phenomena, the phenomena of life must be divided into the simplest components amenable to study. These components must eventually be described in mechanical, physical, and chemical terms. The hope is that, in the fullness of time, when the chemical study of the components of the living processes has advanced far enough, the more complex biological phenomena will be seen clearly as patterns of simpler chemical and physicochemical ones. Reduction is "The explanation of a theory or a set of experimental laws established in one area of inquiry, by a theory usually though not invariably formulated in some other domain" (Nagel, 1961). Biological explanations and laws will thus be shown to be special cases of the more powerful chemical explanations.

Whether reductionist explanations are by themselves sufficient may, however, be questioned. Warren Weaver (1964) wrote that "A person usually considers a statement as having been explained if, after the explanation, he feels intellectually comfortable about it." Many biologists do not derive enough intellectual comfort from reductionism alone. Dubos (1965) feels that "In the most common and probably the most important phenomena of life, the constituent parts are so interdependent that they lose their character, their meaning, and indeed, their very existence, when dissected from the functioning whole. In order to deal with problems of organized complexity, it is therefore essential to investigate situations in which several interrelated systems function in an integrated manner."

To feel "intellectually comfortable" about our understanding of biological phenomena, Darwinian, or compositionist, explanations are required. The matter can best be stated in Simpson's words (1964): "In biology then, a second kind of explanation must be added to the first or reductionist explanation made in terms of physical, chemical, and mechanical principles. This second form of explanation, which can be called compositionist in contrast with reductionist, is in terms of adaptive usefulness of structures and processes to the whole organism and to the species of which it is a part, and still further, in terms of ecological function in the communities in which the species occurs."

Biologists need not choose between Cartesian and Darwinian explanations. They are not only compatible but equally necessary, for the good reason that they are complementary. One of the most encouraging characteristics of biology today is its increasing unification and integration. The integration proceeds in the face of the tremendous growth of the store of available data and techniques. Reductionism and compositionism are not divisive, but unifying. Stebbins (1966) finds, in turn, two compositionist unifying concepts, which he describes splendidly as follows: "One is the concept of organization. This tells us that at every level, from the molecule through the supra-molecular organelle, the cell, the tissue, the organism, the individual, and up to the population or the society, the properties of life depend only to a small degree upon the substances of which living matter is composed. To a much greater degree living things owe their nature to the way in which the components are organized into orderly patterns, which are far more permanent than the substances themselves. The other unifying concept of biology is that of the continuity of life through heredity and evolution. This tells us that organisms resemble each other because they have received from some common ancestor hereditary elements, chiefly the chromosomes of their nuclei, which are alike both in respect to the substances which they contain and the way in which these substances are organized. When related kinds of organisms differ from each other, this means that in the separate lines of descent from their common ancestor changes in the hereditary elements have taken place, and these changes have become established in whole populations."

Vitalism, Mechanism, and Compositionism

Biological thinking is at present more fruitful and interesting than it was in the past, when the banal and wearisome vitalism-mechanism controversy was holding the attention of many biologists. In his great book *The Structure of Science* Nagel (1961) wrote: "It is a mistake to suppose that the sole alternative to vitalism is mechanism. There are sectors of biological inquiry in which physicochemical explanations play little or no role at present, and a number of biological theories have been successfully exploited which are not physicochemical in character. . . . Thus there is a genuine alternative in biology to both vitalism and mechanism—namely, the development of systems of explanation that employ concepts and assert relations neither defined in nor derived from the physical sciences." Both biology and philosophy would profit if this very lucid statement were to become more widely known and better understood than it is in actuality. Darwinian or compositionist biology furnishes a system of explanation which satisfies Nagel's demands.

The old-fashioned thoroughgoing vitalism of Harvey, Wolff, Bichat, Driesch, or Bergson has been a dead issue in biology for at least half a century. Mechanism has triumphed in biology certainly not because all life processes have been exhaustively described in physical and chemical terms. No reasonable mechanist seriously plans to have such a feat accomplished in the near, or even in the remote future, although he affirms that this is possible in principle. Vitalism has been rejected, rightly in my opinion, because it has turned out to be unnecessary and unprofitable as a guide to discovery. Mechanism, on the contrary, performs this function extremely well.

Reductionism and mechanism in biology are often hedged with reservations. One of these is that the now available knowledge of physics and chemistry may not be sufficient for a wholly satisfactory account of all biological phenomena. However, more complete knowledge and more advanced methods will eventually be powerful enough for the purpose. Cybernetics and computer technologies are often mentioned as examples of such powerful methods that have only recently become available and that are likely to be greatly improved. It is, of course, reasonable to expect that there will be progress both in biological and in physical sciences; but I do not regard this appeal to a conjectural future a persuasive argument in favor of the self-sufficiency of the reductionist methodology in affording a complete account of biological phenomena.

An eloquent statement of a modern form of the Cartesian creed in biology is given by Asimov (1960) in the following sentences: "Modern science has all but wiped out the borderline between life and nonlife. Nowadays the question 'What is life?' is asked by physicists as often as by biologists. In fact, biology and physics are merged in a new branch of science called biophysics—the study of the physical forces and phenomena involved in living processes. . . . And it is to biochemistry ('life chemistry') that biologists today are looking for basic answers to the secrets of reproduction, heredity, evolution, birth, growth, disease, aging, and death." And further: "A machine, we have seen, can calculate, remember, associate, compare and recognize. Can it also reason? The answer again is yes."

Nothing is more indicative of the present climate of opinion among biologists than that the few remaining adherents of vitalism do not willingly admit being vitalists. Sinnott, the most distinguished among them, has expounded his views in a series of biological and philosophical books and articles (1953, 1963, 1966), but "vitalism" is a word conspicuous by its rarity in his writings. Sinnott's chief biological argument stems from the truly remarkable and indisputable fact that the development of the organism seems directed to a goal, which is the production of a viable and healthy adult body. Starting from what looks like a simple beginning in a fertilized egg, processes of amazing complexity and precision succeed each other to produce in the end a "norm" of a body which is only comparable to a work of art. "Even when the growing organism is disturbed in various ways, this norm serves as the end toward which development moves. It is significant that this may take place not only over a single course but over others, depending on the character of the disturbance. Unity lies in the end, not the beginning, of development. This development toward a normal end result suggests a process that is purposive and hence resembles a mental one." To Sinnott this is more than a resemblance, and "there is something in life that produces harmony and pattern in a material system and that keeps it moving toward a definite end. . . . What this something is, which thus becomes the source of body, mind, and spirit, is the final problem."

Sinnott's perplexity arises because he sees only a part of the problem. Unable to produce a satisfactory Cartesian explanation, he is driven to embrace vitalism. It would be indeed nothing short of a miracle if atoms and molecules were to assemble by chance and form a living body as complex as any animal or plant. These bodies are, however, the most recent links in the chain of living beings extending from the past down to the origin of life. They are products not only of the individual development, which we observe on our time level, but also of the evolutionary development, which took many generations and perhaps as long as two billion years. Sinnott's predicament comes from his failure to take evolution seriously. He is not an antievolutionist, to be sure, but he fails to see that organic development is not only a Cartesian, but also a Darwinian, compositionist as well as reductionist, problem.

A hen is said to be an egg's way to produce another egg. An egg develops into a hen because it is a descendant of countless other eggs, hens, and, deeper down in time, of organisms which resembled neither the present eggs nor the hens. The goal-directedness of the ontogenetic development is a consequence of the nature of the phylogenetic development. The ontogeny develops by means of phylogeny. The ontogeny looks purposeful, in point of fact more purposeful than the assembly of an automobile in a modern factory. The operation of the assembly line is a product of "mind," or, if you wish, of "spirit." It is, however, futile to look for minds and spirits in a developing egg or an embryo. The reverse is true—the human mind and spirit are not the causes but the results of the ontogenetic and phylogenetic developments.

Do all ontogenies and phylogenies achieve the production of bodies able to live and reproduce? By no means; all die eventually, many prematurely, before they are able to reproduce, and extinction is the end of most phylogenetic lines. The countless and diverse organisms alive today are descendants of a minority of the inhabitants of the past, smaller and smaller minorities as we look further and further into the history of life. But some phylogenetic lines did survive and inherit the earth. They are living now, because they are adapted to the environments in which they live.

Adaptedness and Adaptation

When words are borrowed from everyday language to serve as technical terms, misunderstanding is liable to result. "Adaptation" is plagued with ambiguity, for it is used also in contexts which are biologically irrelevant.

Pieces of furniture, or implements, or machines are said to be "adapted" for certain purposes. Biological adaptation is concerned with survival and/or reproduction; it is found only in living bodies; a cadaver is no longer adapted, although certain organisms are adapted to live on cadavers. Adaptedness first arose with the origin of life, since this life did not become extinct; the origin of life was, however, not an adaptation. Some critics claim that adaptedness is a tautology, because what lives must be adapted to live. This is not so; no organism is adapted in the abstract, it can only be adapted to certain environments. Man is not adapted to feed on pasturage, while horses and cows are so adapted; palms and bananas have no adaptedness to live in Canadian forests, while larches and spruces do have such an adaptedness; certain microorganisms grow in laboratory media and others do not.

It is a curious, and to some orderly minds disconcerting, fact that many, even most, basic concepts of biology prove intractable to exact definition. There are no fully satisfactory definitions of what are life, gene, individual, species, mind, self-awareness. There is no generally accepted definition of adaptation or adaptedness. The situation is not as bad, however, as it may seem to a nonbiologist. Only rarely does one meet a situation where it is difficult to decide whether something is or is not alive, whether one is observing a single individual, or two, or several. Doubts about individuals or populations belonging to a single species or to different ones are more frequent; historically, these difficulties proved a blessing in disguise, because they suggested and eventually proved that species are not created but evolving entities. What is uncertain, indeed, is whether animals other than *Homo sapiens* have minds and self-awareness; strictly speaking, each of us is aware only of his own self.

Here is a sample of definitions, or descriptive statements, given by some recent authors. According to Simpson (1953), "An adaptation is a characteristic of an organism advantageous to it or to the conspecific group in which it lives, while adaptation or the process of adaptation is the acquisition within a population of such individual adaptations." To Mayr (1963): "The statement that every species is adapted to its environment is a self-evident platitude. In continental areas without physical barriers the border of the species range indicates the line beyond which the species is no longer adapted." To Grant (1963): "The reason why each kind of organism is restricted in nature to its own habitat and is not normally found elsewhere is that it is specialized or adapted for making a living under the particular set of conditions. . . . An organism is more than a bundle of separate adaptations; it is a coordinated complex of adaptations." To Wallace and Srb (1964): "We say that animals and plants are adapted to their environments or to their modes of life; we mean that even casual observation reveals these organisms to possess particular characteristics that enable them to survive under the special environmental conditions in which they are found." Bock and Wahlert (1965) use a more formal language, defining the ". . . degree of evolutionary adaptation, the state of being, as the minimal amount of energy required by the organism to maintain successfully the synerg if a single biological role of a faculty is considered, or to maintain successfully its niche if the whole organism is considered."

The statement that an organism is adapted to certain environments means then that this organism can survive and reproduce in these environments. An adaptive trait is a structural or functional characteristic, or more generally, an aspect of the developmental pattern of the organism, which enables or enhances the probability of this organism surviving and reproducing (Dobzhansky, 1956). Adaptedness is a state of being adapted: adaptation refers to the process of becoming adapted; adaptability means that the organism or population concerned can remain or can become physiologically or genetically adapted in a certain range of environments.

Adaptedness to Survive and to Reproduce

Explicitly or implicitly, all the definitions quoted above refer both to individual survival and to perpetuation of a strain or a population. This is as it should be: an organism must survive in order to reproduce and must reproduce in order to survive in the next generation. Yet difficulties arise. The viability, longevity, fecundity, and fertility are in general positively correlated, but not perfectly so. Virgin females of *Drosophila* live longer on the average than those inseminated and actively ovipositing.

The adaptedness of individuals, and that of populations such as species, may occasionally work at cross-purposes. The honeybee workers are richly provided with adaptive structural and behavioral features that enable them to collect food and to build nests; they are, however, underdeveloped sterile females, and their stingers have reversed barbs which cause a bee that stings an enemy to commit suicide. This suicide is probably adaptive as far as the defense of the colony and of the reproductive individuals ("queens") is concerned. Neither solitary bees nor most colonial species have this ambivalent adaptive trait.

The statement that an individual is alive does not tell us either the quality or the degree of its adaptedness. The same is true of the adaptedness of populations and of species. Substantive research is required to discover to which environments the organism is adapted, by what means, and to what extent. Perfect individual adaptedness would enable the individual to live forever. This sounds paradoxical, but it is not as fantastic as it sounds. In some species of trees, e.g., the California redwood, an individual tree may stand and produce viable seed for many centuries. Moreover, since the redwoods are capable of stump-sprouting, an individual tree may live indefinitely, as long as the external environment remains propitious. And yet, the individual near-immortality does not guarantee a perdurability of the species; in fact, the California redwood is a relict species in danger of extinction. By contrast, certain insect species in which the individual is very short-lived seem to be thriving.

The conflict of the individual and the species adaptedness is resolvable. Reproduction is a necessary, though not sufficient, condition for the perpetuation of species; survival of individuals is a necessary, though again not sufficient, condition for reproduction. Individual adaptedness for survival is generally highest before and during the age of reproduction; it dwindles toward the close of the reproductive age and thereafter. Long-lived organisms generally have extended reproductive periods. Very old redwood trees still bear cones with seed.

One may ask why adaptedness does not lapse immediately after the last offspring is produced, and why so many common and ecologically successful organisms are short-lived. Except where the postreproductive individuals consume scarce food, or otherwise deprive the young of the wherewithal for living, the species would gain no advantage from killing them off quickly. A rapid extermination would require a very special set of mechanisms that would, at the close of the reproductive period, immediately destroy the very same individual adaptedness, which until then, was so essential for the perpetuation of the species. In reality, resistance to diseases and environmental hazards declines gradually. Some organ systems in senile individuals stubbornly continue to operate normally until the very end.

In medical writings the reaction of the body to disease is sometimes referred to as "adaptation" (see Dubos, 1965). This apparently contradictory usage may be brought in line with that customary in biology if it is understood to mean that a disease is a manifestation of an incompletely successful struggle of the organism against environmental insults and internal disharmonies; in other words, a disease may be construed as an adaptive reaction of the organism which has at least temporarily miscarried.

The problem of the evolution of senescence and longevity has often been discussed (see Comfort, 1956; Brues and Sacher, 1965; and references therein). The most plausible view is that longevity is set in evolution as a compromise between high reproductive fitness in youth and the adaptive stability of the body. Evolution is opportunistic and immediate advantage is more potent than eventual gain. Genetic variants which enhance both the individual adaptedness and fecundity in youth are perpetuated even when these variants lower the adaptedness in older ages. It would be difficult, for physiological reasons, to achieve an organization of the body that would not wear out with age. Moreover, indefinite prolongation of life and reproductive capacity would lead to overpopulation. It would also interfere with further evolution, since evolution involves replacement of older forms of life by newer and presumably better adapted ones. Whether natural selection can act to prevent overpopulation and favor evolutionary plasticity is, however, a difficult problem. A negative answer has been urged recently by Williams (1966), who takes the extreme view, denying that any form of natural selection acts on populations rather than on individuals.

A writer of science fiction could, however, imagine a world with absolutely constant environments, populated by perfectly adapted organisms that have achieved individual immortality. These organisms would, however, have to dispense with the joys of parenthood.

Adaptability

If a species could inhabit a single and perfectly constant environment, evolution could conceivably arrive at a genotype optimally adapted to that environment. The evolution would then come to a halt. In reality, not only every species but probably every individual has confrontations with many environments, because environments vary in space and in time. Adaptedness in a narrow range of environments is overspecialization; an overspecialized organism may be highly successful for a time, but it risks death or extinction if the environment changes. Hence the importance of adaptability.

Physiological and genetic adaptabilities must be distinguished. Every genotype has a "norm of reaction," which is the array of phenotypes it can produce over the range of existing and possible environments. Any trait, favorable or unfavorable, healthy or pathological, morphological, physiological, or behavioral, which an individual may exhibit, is evidently a product of the norm of reaction of the genotype and of the biography of the individual concerned. The adaptedness of a genotype is a function of its norm of reaction and of the range of environments in which it occurs. Now, physiological adaptability depends upon a norm of reaction yielding a certain adaptedness in the environments which an individual, a population, or a species meet. Genetic adaptability occurs by changing the norm of reaction to produce a better adaptedness in some or all, old or new environments.

A genotype which reacts to some or to all environments that it meets by giving rise to traits adaptive in these environments may be said to possess a high adaptability. This high adaptability may depend upon physiological homeostasis or upon developmental homeostasis (Cannon, 1932; Dubos, 1965). The two are closely related concepts. Physiological homeostasis is generally thought to refer to what Claude Bernard called the fixity of the *milieu interieur.* The body rapidly makes such adjustments to environmental changes so that the orderly course of its physiological processes continues as before the change intervened. The maintenance of a constant body temperature despite temperature variations in the environment is an obvious example. Acclimatization to changes in the oxygen tensions at different elevations by means of changes in the blood composition is another.

Developmental homeostasis is a modification of the development pattern, such that life continues and the orderly course of the essential physiological processes is not interfered with. When food is scarce, the larvae of *Drosophila* and of many other insects do not die, but produce adults much smaller than those coming from well-fed larvae. Some plants are strikingly dissimilar when grown in water or on land, in shade or exposed to sun, when they can climb a support or when there is no support (Grant, 1963; Bradshaw, 1965). The behavioral plasticity of the human species is a striking example of developmental homeostasis. Man can develop generally any one of a great variety of skills and proficiencies by being properly trained. This is man's most potent instrument for adaptation to human environments.

Homeostasis does not mean absence of change; it is not a kind of biological obstinacy. Our internal body temperature does not change much when we are shivering or when we are perspiring. This temperature constancy is due to the physiological mechanisms that bring about the shivering or perspiration, mechanisms working quite differently in cold and in hot environments. What homeostatic adjustment brings about is first of all the conservation of life; the organism stays alive and continues to develop to reach the reproductive state. Wallace and Dobzhansky (1953), Dobzhansky and Levene (1955), and others found that the egg-to-adult viability of *Drosophila* homozygous for certain chromosomes found in natural populations varies greatly with environmental changes, such as temperature, food, crowding, and so forth. The viability of the heterozygotes for the same chromosomes tends to be much more uniform, and on the average higher, than in the homozygotes. The heterozygotes are, then, on the average more homeostatic, and thus more adaptable, than are the homozygotes. Is this a physiological or a developmental homeostasis? Presumably a mixture of both, or a situation intermediate between the two. The flies raised in the different environments in these experiments sometimes do and sometimes do not differ visibly from each other, except for the all-important fact that some of them reach the adult stage and others die before reaching it.

Evolutionary Plasticity

Genetic adaptability is also referred to as evolutionary plasticity, and this latter name is perhaps preferable to avoid ambiguity. Physiological adaptability is a property of the norm of reaction of carriers of certain genotypes. Evolutionary plasticity refers evidently to populations or to lines of descent. Individuals develop, populations evolve. Evolutionary plasticity depends on the presence in the population of genetic variance in adaptive traits that are available for natural or artificial selection to work on. The effectiveness of artificial selection practiced by animal and plant breeders is greater or lesser depending on the amount of genetic variance present in the strain subject to the selection. Prolonged selection for a trait, such as yield or size or qualtiy, sooner or later reaches a ceiling, a plateau, when the additive genetic variance is depleted and further selection no longer results in advances in the direction desired. The "ceiling" can be broken through only by new mutations or by the origin of new combinations of linked genes stored in the chromosomes of the population undergoing selection (Lerner, 1958, Falconer, 1960). A rather flippant but valid illustration of the importance of genetic variance is that if one wished to transform men into angels by selection, it would be easier to select for an angelic disposition than for a pair of wings.

In general, adaptedness can be achieved either by individual adaptability or by genetic adaptability. A genotype which confers upon its carriers an ability to react to the environments in which they find themselves by adaptive homeostatic reactions may achieve a high degree of adaptedness. A theoretically thinkable, but unrealistic, limiting case would be a genotype giving maximum adaptedness in all environments which its carriers could encounter. In contrast to this, genetic adaptability can produce different genotypes, each suited best for different environments or groups of environments. A limiting, but again unrealistic, situation would be a specialized genotype yielding maximum adaptedness in only one environment and a population containing as many such genotypes as there are environments. These are, then, the two ways of becoming adapted, and in reality different combinations of both are utilized in the evolutionary changes. Homeostatic genotypes are being selected for, but also a diversity of genotypes is being created to fit the available ecological opportunities.

The question which suggests itself is under what conditions is individual adaptability, or genetic adaptability, preferable to secure an adaptedness to the environments in which a population or a species live? This is a question of evolutionary "strategy" (Lewontin, 1961; Slobodkin, 1964), which has thus far been approached only as an exercise of mathematico-genetical theorizing. It is, after all, not self-evident that evolution will always follow the strategy which our ingenuity will show to be the best; as pointed out above, evolution is basically opportunistic, and opportunism is not always the best strategy.

Levins (1962, 1963, 1964) has devised most interesting mathematical models. His principal conclusions are, very briefly, as follows. Assume that a population or a species inhabits a territory with two accessible ecological niches. Is it preferable for this population to be genetically uniform or diversified? Some genotypes confer on their carriers wider capacities for homeostatic adjustment than do other genotypes. On the other hand, the difference between the phenotypic expressions optimal in the two niches may be small or large in relation to the amplitude of the homeostatic response of the genotype. If the optimal phenotypes are well within the reaction norm of the genotype, then the best strategy is for the population to be monomorphic for the genotype in question. The amplitude of the homeostatic response may, however, be smaller than the difference between the phenotypic expressions optimal in the two ecological niches. Then it is advantageous for the species to be either polytypic or polymorphic. Polymorphism is preferable if the different ecological niches recur regularly in time, for example seasonally, or alternate mosaic-fashion in space. A polymorphic population with two or more genotypes fitting these niches is then optimal. On the other hand, the two niches may be geographically separate, each being restricted to a part of the distribution area of the species. Then two geographic races, monomorphic for the genotypes best adapted in the respective territories, are preferable. Migration and gene flow are desirable in fluctuating environments; they permit the populations to become adjusted to widespread and long-term environmental fluctuations, at the same time damping the response to local and ephemeral environmental oscillations.

The Problem of Quantification of Adaptedness

Anything alive is adapted to be alive. As we have seen above, this tautology is not the whole story. In the first place, adaptedness exists only in relation to an environment or environments. Deprived of clothing and exposed to an arctic blizzard, I would rapidly cease being alive, because I am not adapted to such an environment. There is no individual in the species *Homo sapiens* so adapted. But polar bears are so adapted, and they survive arctic blizzards. Moreover, it is intuitively clear that one can be better or less well adapted to some environment. The problem of devising a meaningful measure of adaptedness has proved, however, rather intractable.

For individual adaptedness, the probability of survival, of reaching the reproductive stage of the life cycle, and, in the case of man, the degree of the sense of well-being (which is also hard to measure), are thinkable criteria. For populations, a statistic has been proposed named variously the Malthusian parameter, intrinsic rate of natural increase, or innate capacity for increase (Lotka, 1925; Fisher, 1930; Andrewartha and Birch, 1954). The last-named authors give the following "approximate definition" of this statistic: "We define r_m, the innate capacity for increase, as the maximal rate of increase attained at any particular combination of temperature, moisture, quality of food, and so on, when the quantity of food, space, and other animals of the same kind are kept at an optimum and other organisms of different kind are excluded from the experiment."

The value r_m is measured when the population has reached a stable age distribution, i.e., a constant age schedule of births and deaths. The basic equation then is:

$$\int_0^\infty e^{-r_m x} l_x m_x \, \delta_x = 1$$

where e is the base of natural logarithms, l_x the probability at birth of being alive at the age x, m_x the number of female offspring produced in unit time by a female aged x, and 0 to ∞ the life span. In practice, especially in experiments dealing with insects such as *Drosophila* flies, the age-specific mortality and fecundity rates are not measured in the same individuals, and a modified equation is as follows (Andrewartha and Birch, 1954; Birch et al., 1963):

$$a \sum_0^t e^{-r_m (x + F)} l_x m_x = 1$$

where a is the proportion of zygotes surviving to the adult stage, t the observed maximum life span, l_x the probability of survival of adult from the time of hatching to age x, m_x one half of the number of eggs laid by a female of age x counted from the beginning of the adult stage, F the duration of the development from egg to adult, and r_m the intrinsic rate of increase. An interesting derived value is the finite rate of increase, λ, which is simply:

$$\lambda = \text{antilog}_e r_m$$

which can be expressed, for example, as the rate of increase per day, per week, or other time interval. Small differences in the values of r_m and λ can produce enormous differences in the numbers of animals, such as *Drosophila* flies, within a fairly small number of generations.

The experiments needed to estimate these values are so laborious that they have been performed for very few organisms only, of which *Drosophila* flies and certain grain beetles, e.g., *Calandra, Rhizoperta,* and some species of *Ptinidae*. (See Andrewartha and Birch, 1964.) The values obtained are, of course, dependent on the environment in which the experiments are made. Ohba (1967) obtained values of r_m ranging from 0.256 to 0.275 for seven experimental populations of *Drosophila pseudoobscura* at 25°C when they were kept on an enriched culture medium, and r_m values of only 0.153 to 0.164 on a less favorable medium. This corresponds to finite rates of increase per 20 days (λ^{20}) of 1.292 to 1.316 on the enriched medium, and 1.167 to 1.178 on the poor medium. Dobzhansky, Lewontin, and Pavlovsky (1964) obtained for the same species, under more stringent conditions, r_m values between 0.154 and 0.222 at 25°C, and between 0.096 and 0.114 at 16°C. For *Drosophila serrata* and *Drosophila birchii* the r_m values varied from 0.135 to 0.249 at 25°C, and from 0.051 to 0.156 at 20°C. These values correspond to finite rates of increase per 20 days in the range 1.145 to 1.282 at 25°C, and 1.053 to 1.170 at 20°C (Birch et al., 1963).

As a help to visualize the biological meaning of these statistics it may be useful to consider how they are modified by various factors. The effects of making the environment more rigorous or more favorable are evident from the figures given— r_m's are greater in more favorable environments. The superiority at higher temperatures is due to a greater speed of development, which more than compensates for a decreased total fecundity. Moreover, an earlier beginning of the reproductive period and a greater fecundity at a younger age are more important than the longevity of the adults and fecundity at older ages. In the experiments of Dobzhansky, Lewontin, and Pavlovsky (1964) the estimates of the r_m's were made in each population twice, the second being one year to one year and a half later than the first. During this time interval the genetic adaptedness of the populations rose, and the r_m's also increased. It is, however, interesting that these increases occurred through increased fecundity, while the longevity not only did not increase but in fact somewhat deteriorated.

The usefulness as well as the defects of the r_m statistics as measures of adaptedness should now be made clear. They are only valid for the experimental environment in which they are assessed. To make such estimates for natural populations would be a formidable task. One would have to know the incidence of the various environments in which a population lives, the number of individuals in each environment, obtain r_m's for one environment after the other, and finally try to obtain a sort of weighted average. Furthermore, the r_m is a measure of the capacity of a population to increase in numbers when neither food nor space nor competitors are limiting. Yet the fate of a natural population may well depend on its ability to survive under some stringent conditions when it does not increase at all. For example, some *Drosophila* populations are decimated during severe winters or hot and dry summers, and the few survivors that are best "adapted" to hold on to life may be more important than the genotypes which develop rapidly, reproduce early, and thus have high r_m's under more favorable conditions. To be sure, the physiological characteristics giving higher r_m's under severe conditions may be quite different from those in favorable circumstances.

For organisms not closely related ecologically, such as an infusorium, *Drosophila*, a species of fish, and man, comparison of r_m's or of lambdas would tell us nothing about their adaptedness. This is not, however, a weakness of the method, because such comparisons are meaningless. Each of these organisms has a certain adaptedness in its own environment, but a zero adaptedness in the environments of the others. Whatever evolutionary progress may mean, it does not mean an ability of a higher organism, such as man, to reproduce faster than a fish or a *Drosophila* in either of their ecological niches. On the other hand, for genetically different populations of the same species, the r_m's usually, though again not always, coincide with other properties which it would seem reasonable to count under adaptedness. Beardmore, Dobzhansky, and Pavlovsky (1960) found that certain chromosomally polymorphic populations of *Drosophila pseudoobscura* can, in stringent environments, produce greater numbers of individuals, and a greater biomass, than do chromosomally monomorphic populations. In other words, the polymorphic populations have a superior efficiency in converting food into living bodies. Now, Dobzhansky, Lewontin, and Pavlovsky (1964) found higher r_m's in the polymorphic than in monomorphic populations under their more stringent environments, while Ohba (1967) found little or no difference under his more lenient environments.

The problem of devising measures of adaptedness is evidently in need of further study.

Darwinian Fitness

Natural selection is the process which tends to maintain or improve the genetic adaptedness in old environments, and contrive adaptedness to new environments. In Darwin's words (1859): "A struggle for existence inevitably follows from the high rate at which all organic beings tend to increase. . . . Hence, as more individuals are produced than can possibly survive, there must in every case be a struggle for existence, either one individual with another of the same species, or with the individuals of distinct species, or with the physical conditions of life. . . . Can it, then, be thought improbable, seeing that variations useful to man have undoubtedly occurred, that other variations useful in some way to each being in the great and complex battle of life, should sometimes occur in the course of thousands of generations? If such do occur, can we doubt (remembering that many more individuals are born than can possibly survive) that individuals having any advantage, however slight, over others, would have the best chance of surviving and of procreating their kind? On the other hand, we may feel sure that any variation in the least degree injurious would be rigidly destroyed. This preservation of favorable variations and the rejection of injurious variations, I call Natural Selection."

In a later edition of "The Origin of Species," Darwin saw fit to add: "I have called this principle, by which each slight variation, if useful, is preserved, by the term Natural Selection, in order to mark its relation to man's power of selection. But the expression often used by Mr. Herbert Spencer of the Survival of the Fittest is more accurate, as is sometimes equally convenient." As seen from the vantage point of our present knowledge, this is less accurate and is inconvenient. The superlative implies that there is a single "fittest" genotype, while in fact it is a great array of genotypes that constitutes the adaptive norm of a species. The emphasis on survival should not be allowed to obscure that differential reproduction is at least equally important. Natural selection could conceivably take place if all the progeny survived but different parents produced different numbers of progeny. This is not an entirely unrealistic contingency. There are species of insects which produce several generations per year, only few survive a cold or a dry season, increase rapidly in numbers during the favorable season, and are decimated, possibly at random, during the unfavorable season. In man, the pre-adult mortality is so greatly reduced in technologically advanced countries that 90 or more percent of the infants born survive; however, since the number of children per couple varies from zero to twenty and more, natural and artificial selections remain possible and in fact are probably operative.

For natural selection to occur there must exist two or more genotypes with different Darwinian fitness in the environment in which they occur. Darwinian fitness is sometimes referred to also as selective value or as adaptive value. The existence of these three terms, which are treated as synonyms and yet easily acquire different connotations, has caused some unnecessary disagreements, especially among philosophers writing about biological problems (e.g., Beckner, 1959, Grene, 1958, 1961, Goudge, 1961, Manier, 1965, Scriven, 1959, Smart, 1963). The dictionary meanings of "fit" are such as: adapted to an end, qualified, proper, right or becoming, preferred, ready, in fine physical condition, or good health. Hence it is advisable in biology to speak of "Darwinian fitness." The Darwinian fitness of a genotype can be defined operationally as the average contribution which the carriers of a genotype, or of a class of genotypes, make to the gene pool of the following generation relative to the contributions of other genotypes (Dobzhansky, 1962, and 1964).

Let a gene A_1 have a frequency p, and its alternative gene A_2 the frequency q in the gene pool. The values p and q are fractions, and $p + q = 1$. If the mating is random and the homo- and heterozygotes are equally viable, their frequencies, as shown by Hardy (1908) and by Weinberg (1908) will be:

$$p^2A_1A_1 + 2pqA_1A_2 + q^2A_2A_2 = 1 \tag{1}$$

Suppose now that the relative numbers of surviving progeny left by the carriers of the genotypes A_1A_1 and A_2A_2 are in the ratio 1 to $(1-s)$. The value s is called the selection coefficient; if for every 100 offspring of A_1A_1 only 90 surviving offspring are left by A_2A_2, then $s = 0.1$. The heterozygotes, A_1A_2, may leave as much progeny as A_1A_1 (if A_1 is dominant), or as much as A_2A_2 (if A_2 is dominant), or an intermediate number (if neither is dominant). The Hardy-Weinberg formula is then modified thus:

$$p^2A_1A_1 + 2pqA_1A_2 + q^2(1\text{-}s)A_2A_2 = 1 \tag{2}$$
$$p^2(1\text{-}s)A_1A_1 + 2pqA_1A_2 + q^2A_2A_2 = 1 \tag{3}$$
$$p^2A_1A_1 + 2pq(1\text{-}hs)A_1A_2 + q^2(1\text{-}s)A_2A_2 = 1 \tag{4}$$

The value, h, is the coefficient of dominance. The heterozygotes may exhibit the quality of hybrid vigor, heterosis, and be reproductively superior to both homozygotes, A_1A_1 and A_2A_2. We then have:

$$p^2(1\text{-}s_1)A_1A_1 + 2pqA_1A_2 + q^2(1\text{-}s_2)A_2A_2 = 1 \tag{5}$$

And finally (though this is rare except in species hybrids) the heterozygotes may be at a disadvantage compared to both homozygotes. The consequences of all these situations have been examined by mathematical geneticists; for our purposes it will be sufficient to state briefly the principal conclusions. The situation is simplest if the heterozygotes, A_1A_2, are equal in reproductive efficiency to one of the homozygotes, or intermediate between the two (formulas 2 to 4 above). Whichever gene, A_1 or A_2, confers a superior reproductive efficiency on its possessors will increase in frequency in the population. The increase will continue generation after generation; given enough time, i.e., enough generations, the more efficient gene will eliminate and supplant the less efficient one entirely. How rapid or slow will the selectional changes be, depends on the magnitudes of the selection coefficients. With the same selection coefficient, a favorable dominant gene will increase relatively rapidly at first, and slowly when it becomes frequent; with a favorable recessive gene the reverse will be true; if the heterozygotes are intermediate between the homozygotes (4), the selection is more uniformly efficient.

The most interesting, and at first sight paradoxical, outcome of selection will be if the heterozygote is superior to both homozygotes (5). Neither the gene A_1 nor A_2 is allowed to crowd the other out or to disappear entirely. Instead, a genetic equilibrium is reached, and the population attains the state of the so-called heterotic balanced polymorphism. All three genotypes continue to occur in the population, with frequencies dependent on the relative magnitudes of the selection coefficients s_1 and s_2 (5), so that at equilibrium the frequency of A_1 in the gene pool will be $p = s_2/(s_1 + s_2)$, and $q = s_1/(s_1 + s_2)$. This will be true even if the homozygotes are seriously incapacitated, inviable, or sterile ($s = 1$). The heterotic balancing natural selection may, thus, maintain hereditary defects and diseases in a population, provided that the genes responsible induce a high Darwinian fitness (heterosis) in the heterozygotes.

Varieties of Natural Selection

It should now be evident that natural selection is a common name for several by no means identical phenomena. Even before Darwin, biologists were aware of the fact that weak, malformed, and diseased individuals are weeded out (e.g., Blythe in 1835 and 1837, see Eiseley, 1959). This is normalizing selection. It is a conservative force which counteracts accumulation in populations of mutants which decrease the Darwinian fitness of their possessors, at least when these mutants become homozygous. Darwin was evidently familiar with this process, but since his attention was attracted to agencies inducing change rather than preserving the status quo, he gave the normalizing selection relatively less consideration.

Suppose that a gene A_1, the carriers of which have a high Darwinian fitness, mutates to the state A_2 which lowers the fitness. If A_2 is a dominant lethal or a sterility gene, the carriers of which always die or are unable to reproduce, then all the A_2 mutants are eliminated in the same generation in which they arise. A new crop of mutants will, of course, appear in the next generation. If, however, the selection is not so completely efficient, some mutant genes will escape its dragnet, and will be transmitted to the next generation. That next generation will contain all the newly arisen mutants, a part of the mutants that arose in the preceding generation, a smaller part of those having arisen two generations ago, and so forth. How great a "genetic load" of the as yet uneliminated mutants will a population accumulate, will depend principally on two factors—how often the mutation arises, and how much it lowers the Darwinian fitness.

Very simple formulas have been worked out to describe the situations that arise. Suppose that a deleterious mutation, $A_1 \rightarrow A_2$, occurs at a rate u per generation (the value of u for mutations of some human genes may be of the order of 10^{-5}). Suppose further that the mutant is discriminated against by a selection coefficient s (equal to 1 for completely lethal or sterile mutants). If, then, the mutant A_2 is dominant to the original state A_1, the frequency of A_2 in the gene pool will be u/s. If A_2 is recessive to A_1, its accumulated frequency will be much higher, namely $\sqrt{u/s}$. The reason deleterious recessive mutants are allowed to attain higher frequencies than equally deleterious dominants, is simply that a recessive mutant may be carried in many heterozygotes, in which it does not express itself, and is consequently protected, or sheltered, from the weeding-out action of natural selection. With mutants which are neither dominant nor recessive, the accumulation will be between u/s and $\sqrt{u/s}$.

Populations of presumably all living species, including human, carry genetic loads consisting of harmful mutant genes. This cannot be blamed entirely on culture, civilization, or on any other specifically human attributes. Populations of *Drosophila* flies and of other sexually reproducing organisms also carry genetic loads. The accumulation of the genetic loads is a necessary consequence of the occurrence of mutations, most of which are harmful, but not harmful enough to be always eliminated immediately after they are produced. Harmful mutations are accumulated until the numbers of the respective mutant genes arising by mutation become equal to the numbers eliminated by natural selection in the same population. The population is then said to be in the state of "genetic equilibrium." Muller (1950) termed the elimination of harmful mutants "genetic death." The genetic "death" is sometimes a cruel and sometimes a rather benign process. Death of a child from a severe hereditary disease, and a genetically conditioned failure to have one more child, are genetic "deaths." The higher the mutation rates and the more harmful mutants produced, the more frequent will be the genetic deaths. In populations which have reached genetic equilibria, the total number of genetic deaths will be equal to the total number of mutations which are subject to the normalizing natural selection.

A very different form of natural selection is the heterotic balancing selection. It occurs when the Darwinian fitness of a heterozygote exceeds the fitness of both homozygotes [a situation mentioned in Eq. (5)]. The heterotic balancing selection also leads to a genetic equilibrium, but not to an equilibrium like that between mutation and the normalizing selection. The balanced polymorphism that is established is due to the selection favoring the heterozygotes against the homozygotes. In a sexually reproducing population the heterozygotes tend, however, to produce a fresh crop of homozygotes in every generation. The maintenance of the sickle-cell anemia in human populations is an example of this form of selection. The homozygote for the sickle-cell gene is lethal; its Darwinian fitness is zero, and it is discriminated against by a selection coefficient $s_2 = 1$. But, as shown first by Allison, the heterozygote has a higher fitness than the "normal" homozygote in countries in which falciparum malaria is endemic, on account of the relatively high resistance to that disease.

In malarial environments, a population which contains some sickle-cell anemia genes has an advantage over a population free of these genes. The former population is in less danger from the ravages of malaria, although it "pays" for this advantage by sacrificing in every generation some individuals who die of the anemia. This advantage is removed in countries in which falciparum malaria is rare or absent. Populations native in these countries have, as expected, few or no sickle-cell genes. The lethal disease caused by homozygosis for the sickle-cell anemia gene certainly brings about some "genetic deaths"; it is a part of the genetic load of the population. But this balanced genetic load, because of the disadvantage of being homozygous for certain genes, is very different from the mutational load controlled by the normalizing selection. The former is maintained by the heterotic balancing selection, while the latter is maintained by recurrent mutation.

Another form of balancing natural selection is the diversifying selection.[1] In many discussions and mathematical analyses of selection, the simplifying assumptions are adopted, that the environment in which a population lives is uniform, and that the selection advantages and disadvantages of different genotypes are independent of their frequencies in the population. This simplification, however, flies in the face of reality. Many animals can subsist on a variety of foods, many plants grow on different soils, humans have to fill many different employments, functions, professions, and social roles. It is most likely that some genotypes will be fitter in some than in other environments. Diversifying selection will then favor different genotypes in different subenvironments, or ecological niches, which the population encounters. The beautiful experiments of Thoday and his colleagues (Thoday, 1959, and 1965; Thoday, Gibson, and Spickett, 1963) on artificial diversifying selection in laboratory populations of *Drosophila* have shown that it makes the population polymorphic, consisting of a variety of genotypes. The balanced polymorphism produced by the diversifying selec-

---

[1] This has also been called "disruptive" selection which is a most unsuitable name. Far from "disrupting" the adaptedness of a population, diversifying selection is one of the really constructive evolutionary agencies.

tion differs from that maintained by the balancing heterotic selection, because the former can be maintained without the heterozygotes being superior in fitness to the homozygotes.

There is another aspect of diversifying selection which demands attention. Ideally, the fitness of a population facing a diversity of environments would be maximized if every genotype were placed in the environment to which it is best adapted. This is not always realized. A plant must survive in whatever location a seed has fallen. A mobile animal is to a certain extent capable of moving to an environment more in accord with its individual requirements. However, even so highly mobile a species as man contains numerous individuals which for various reasons have not found environmental niches best suited to their particular needs. The population carries, then, a genetic load due to environmental misplacement of some (perhaps many) of its genotypes.

A special form of selection results in mammals from the incompatibility of certain maternal genotypes with those of their unborn children. The best studied case in man is that of a rhesus-positive fetus in a rhesus-negative mother. This selection should, theoretically, make the entire population either rhesus-positive or rhesus-negative. It does not appear to be doing this, for reasons that have not been clarified. Still another kind of selection is that due to so-called meiotic drives, disturbances of the Mendelian segregation mechanisms, which make sex cells carrying certain genes more or less frequent than they should be on a random basis. No proven case in man is as yet on record, but the possibility should be kept in mind.

The last to be mentioned, but in the long run most important form of selection, is directional selection. This is the natural selection which Darwin regarded as the main driving force of evolution. Suppose that the climate becomes warmer or cooler, that there appears a new source of food, a new predator, or disease, or that there occur some other prominent environmental changes. Some genotypes will, then, become more favorable, and others less favorable. Directional natural selection will operate to reconstruct the gene pool of the population in accord with the demands of the new environment. The assumption that the environment of any species is constant is a dangerous one; probably few or no populations achieve a genetic stability for any length of time; directional selection is likely to be always operating to some extent.

Darwinian Fitness and Adaptedness

Like adaptedness, Darwinian fitness is a function of both the genotype and the environment. A high adaptedness, as well as high Darwinian fitness in a certain locality or a geographic region at the present time, may become lower or higher in the future or in a different locality. Health, hardiness, and vigor enhance individual adaptedness (or "fitness" in the vernacular—here is a danger of a semantic muddle). Yet these qualities will be reflected in a high Darwinian fitness only if they lead to a high rate of transmission of the genes to the following generations. The point is that while adaptedness may be treated as an absolute measure, Darwinian fitness is a relative one.

Consider a genetically uniform array of individuals, such as a bacterial clone. It may be well or poorly adapted to the environment of a laboratory culture; its rate of growth in the log phase may be high or low on a certain nutrient medium; its ability to survive in the stationary phase may also be more or less satisfactory. This tells us nothing about its Darwinian fitness. This fitness will become manifest only when another genotype is introduced into the culture or arises by mutation, so that the two genotypes now compete in the same environment, side by side. It may turn out that a high adaptedness of one genotype may go together with a relatively low Darwinian fitness, if this genotype is exposed to competition with a still better adapted genotype.

Attempts to develop methods for estimation of the adaptedness in absolute terms have not, as we have seen above, been thus far entirely successful. It is nevertheless important that this is possible in principle. The statistic r_m, the "innate capacity for increase," is an example of a method which approaches, though hardly achieves, this goal. The point which must now be emphasized is that the r_m values obtained for populations are not necessarily proportional to their Darwinian fitness. The chromosomal polymorphisms found in natural populations of many species of *Drosophila* are instructive in this connection. Dobzhansky, Lewontin, and Pavlovsky (1964) found that, in *Drosophila pseudoobscura,* in populations of *Drosophila* polymorphic for certain chromosomal inversions, the Darwinian fitnesses of the homokaryotypes are lower than that of the heterokaryotype. Nevertheless, experimental chromosomally monomorphic as well as polymorphic populations can be maintained indefinitely in laboratory population cages. The monomorphic ones build numerically smaller populations than the polymorphic ones, given the same amount of food (Beardmore, Dobzhansky, and Pavlovsky, 1960). In nature, polymorphic colonies would outbreed the monomorphic ones, and eventually replace them. This can happen, however, only if the monomorphic and polymorphic colonies exchange migrants; if they are completely isolated, for example on oceanic islands, the replacement need not occur.

Another interesting situation is the chromosomal "sex-ratio" variant in *Drosophila.* A male carrying this genetic variant in its X-chromosome produces daughters and few or no sons when crossed to any female. Since daughters do and sons do not inherit X-chromosomes from their fathers, a "sex-ratio" male transmits his X-chromosome to his entire progeny, while a normal male transmits his X to only half of the progeny. Other things being equal, the "sex-ratio" genotype would have a higher Darwinian fitness in a population than the normal one (in reality this is not the case, since females homozygous for the "sex-ratio" have a low fitness; Wallace, 1948). Uncontrolled spread of the "sex-ratio" could result in a disaster and extinction of the population; the population might come to consist of females and no males, which in an organism incapable of parthenogenesis would spell extinction. However, as long as the shortage of the males is not acute enough to interfere with the reproduction, a population containing "sex-ratio" in high frequency might have a high Darwinian fitness. A collapse might occur rather suddenly.

Dunn (1956 and 1960) has studied a remarkable series of gene alleles at the t locus in the house mouse. These genetic variants have been observed to arise by mutation in the laboratory and they are widespread also in natural populations. The homozygotes for these mutant alleles are lethal or sterile. The most remarkable property of the t alleles is, however, that they distort the segregation mechanism in the male in such a way that more than 50 percent, and up to 95 percent, of the spermatozoa (at least of functioning spermatozoa) carry the mutant rather than the normal allele. The segregation in heterozygous females is normal. The t-alleles will, consequently, be favored by natural selection and will increase in frequency in populations, despite their clearly adverse effect on the adaptedness. Lewontin and Dunn (1960) have proposed a theoretical model to explain how the spread of these adaptively highly ambivalent genes is controlled in natural populations. The house mouse occurs in generally discrete colonies, the migration between which is more or less rare. In some of the colonies the t-alleles are lost because of chance, but more often they become so frequent that the adaptedness of the colony is lowered to the point where it becomes extinct. The place where this colony lived becomes repopulated by migrants from elsewhere, some of which may not carry the t-alleles at all.

Evolutionary Opportunism and Adaptive Radiation

As defined above, better adaptedness and Darwinian fitness are properties of individuals, genotypes, populations, or species at a given time level and in a given environment. It has been pointed out that adaptedness must be distinguished from adaptability. Darwinian fitness must not be confused with genetic adaptability or evolutionary plasticity. In some exceptional cases, like those of the "sex-ratio" condition in *Drosophila* and the t-alleles in mice, natural selection may lead to a loss of adaptedness and to extinction. Even leaving these unusual conditions out of account, evolutionary changes leading to high Darwinian fitness and adaptedness may mean a narrow specialization to life in a given environment, or in a limited range of environments, a loss of evolutionary plasticity, and extinction. Natural selection tends to maximize the Darwinian fitness in the environments which exist here and now. It has no information about the future, and cannot plan evolutionary "strategies." Yet present expediency may mean future disadvantage, and a gain now an eventual loss. Although some writers regard such names unduly anthropomorphic, it is reasonable to label this rather striking quality of organic evolution evolutionary opportunism.

Yet despite the opportunism of the evolutionary process, it has led not to extirpation of all life, but on the contrary to its expansion and aggrandizement. This is because life has "radiated," spread into an enormous variety of ecological niches, tried out a multitude of ways of exploiting the inorganic as well as the organic environments, and thus created a legion of "trial parties," most of which ended in extinction, but some of which "discovered" new ways of life and thus inherited the earth. Teilhard de Chardin (1959) has overstated this when he wrote that evolution is "pervading everything so as to try everything, and trying everything so as to find everything." Only a minuscule fraction of the potentially possible gene combinations have ever been realized, and consequently far from everything has been tried and discovered. Nevertheless, the existing organic diversity is impressive. Grant (1963) gives the following estimates of the numbers of species of organisms described so far:

Animals		Plants and Monerans	
Chordates	39,500	Angiosperms	286,000
Echinoderms	4,000	Gymnosperms	640
Arthropods	923,000	Ferns and allied	10,000
Annelids	7,000	Bryophytes	23,000
Molluscs	80,000	Algae	8,700
Nematodes	10,000	Fungi	40,400
Flatworms	6,000	Protozoans, diatoms	30,000
Coelenterates	9,000	Blue-green algae	1,400
Sponges	4,500	Bacteria	1,630
Miscellaneous	7,300	Viruses	200

The grand total is 1,492,000 described species. This is likely to be perhaps only one half of the species actually in existence. The total number of species that ever lived in the history of the earth is even harder to estimate; Simpson (1960) gives 50 million to 4 billion as the outside limits, with about 500 million as a reasonable guess.

It must however be noted that the numbers of contemporaneous (synchronic) and noncontemporaneous (allochronic) species are not really describing the same biological phenomenon. Simpson (1953) and Rensch (1960) have distinguished two types of evolutionary processes: (1) splitting of phylogenetic lines or cladogenesis, and (2) phyletic evolution or anagenesis, which is change of a population in time without subdivision. In Simpson's words: "When an ancestral species splits into two or more descendant species by segregation of existing variation, phyletic evolution is not an essential part of the picture; and when a single population undergoes extensive, cumulative change, there is no splitting. Other cases show, again, that the distinction is not absolute, that we are simply looking at two intimately connected parts of the whole."

The importance of cladogenesis follows from the simple consideration that life on earth arose presumably only once, or at any rate that the primordial organisms were all of the same or very few kinds. The immense organic diversity observed at present is evidently a consequence of the adaptive radiation, phylogenetic splitting, and branching. Since the time of the appearance of sexual reproduction, the crucial phenomenon of cladogenesis has been the multiplication of species. The achievement of the species rank makes the evolutionary divergence irreversible. Races or subspecies may differentiate or hybridize and fuse; full species can go through transformations in time, some or all may become extinct without issue, but they are unlikely ever to converge and to become one species again. The concept of cladogenesis includes, of course, not species splitting alone but also further divergence, making the species more and more distinct, thus originating genera, families, and other higher categories. The differentiation of species is crucial because it involves the origin of reproductive isolation between the subdivisions of what was formerly an inclusive Mendelian population. There is no known biological phenomenon to mark the generic or familial status; the higher categories are in this sense arbitrary; they are creations of the classifier made for his convenience. Species of sexual organisms are, on the contrary, tangible phenomena of nature.

The destiny of species in time varies from preservation with little change, to transformation, to extinction. Extinction is in the long run the most probable fate; according to Simpson (1953, 1960), not a single species known in fossil condition in Cambrian times has living descendants, and not more than one percent of the Mesozoic Tetrapods have descendants alive on our time level. Nevertheless, the total organic diversity, as measured by the number of species, has increased with time. This is evidently because some species not only survived but ramified into many derived species by adaptive radiation. In Simpson's words (1953): "It is strikingly noticeable from the fossil record and from its results in the world around us that some time after a rather distinctive new adaptive type has developed it often becomes highly diversified. This may follow soon after the origin of such a type or may be long delayed, but it is more likely than not to occur sooner or later if the type does survive and become well established."

Spectacular examples of adaptive radiation occur on oceanic islands, which are populated by accidental introductions of small numbers of species from other islands or continents. These strays often encounter on the islands where they happen to land an abundance of unoccupied ecological niches, and undergo "explosive" differentiation and multiplication of species. The faunas and floras of oceanic islands are often unbalanced, in the sense that many otherwise ubiquitous groups are not represented at all, while a small number of groups are represented by quite disproportionately large numbers of species (see, for example, Zimmerman, 1948, Fosberg, 1963). The Darwin finches on Galapagos Islands, the Drepaniid birds and Drosophilid flies on Hawaii are illustrations of such insular radiations.

Drosophilidae are one of the medium-sized families of flies, Diptera. Hardy (1965) quotes the following statistics concerning the numbers of species in the genus *Drosophila:* the world fauna, excluding Hawaii—750 species; Hawaii—243 species (further additions have, according to a private communication from Professor Hardy, been made to the Hawaiian species). Thus, of the approximately 1000 species of *Drosophila* in the world, about a quarter occur in the Hawaiian islands, which have an aggregate area of some 6423 square miles only. Another genus of the family Drosophilidae, *Scaptomyza,* has about 114 species in Hawaii and about 70 species in the rest of the world. The second largest family of flies in Hawaii are Dolichopodidae, with 188 known species. Other families, which elsewhere in the world outnumber Drosophilidae and Dolichopodidae, are represented in Hawaii by few or by no native species. The only plausible explanation of these phenomena is that one or two species of Drosophilidae reached Hawaii, a group of oceanic islands of volcanic origin, before other flies got there by chance dispersal. A remarkable adaptive radiation of species of Drosophilidae ensued, occupying numerous adaptive niches which are elsewhere occupied by representatives of other families. One of the smaller genera of Hawaiian Drosophilidae (*Titanochaeta,* 10 species) appears to have adopted a way of life most unusual for the family—they are parasites in the egg cocoons of spiders!

In phyletic evolution, species undergo extensive and cumulative changes in time, yet each remains a single species. Eventually the descended form becomes so different from the ancestral one that it has to be given another species name. It is evident that if a paleontologist would have a complete series of transitional forms from all time levels, the division of such series into allochronic species would be quite arbitrary. In reality complete series are rare. Paleontologists make virtue out of necessity, and use the gaps in the record for species boundaries. There is only one reasonable rule to follow, which is by no means always followed. This is to cut the series of fossil variants into sections, such that the extremes are about as different as are contemporaneous species in the same group of organisms, usually the now living species.

Human evolution, since at least the mid-Pleistocene or even earlier, is a splendid example of anagenesis. Although divergence of opinions and differences in nomenclature in this field are notorious, there is an approach to a consensus as far as the sequence of stages of human ancestry is concerned. This can be represented by the following series:

$$africanus \rightarrow habilis \rightarrow erectus \rightarrow sapiens$$

Some authors put each of these species in a separate genus; others recognize only a single genus—*Homo.* Some would regard *habilis* a race of *africanus* rather than a separate species. A "Neanderthal" stage, or species, may be inserted between *erectus* and *sapiens.* Another species, *robustus,* and the race *boisei* existed contemporaneously with *africanus,* and *habilis.* Some authors place *africanus* and *robustus* in the same genus, *Australopithecus;* others make them *Homo africanus* and *Paranthropus robustus* (Robinson, 1967).

What is reasonably certain despite all these doubts and disputes is that there were two hominid species in lower Pleistocene times, *africanus* and *robustus;* that *robustus* died out while *africanus* lived on and we are his descendants; and that the evolution passed through a series of anagenetic stages, culminating on our time level in *sapiens.* This does not mean that cladogenesis was entirely absent in human evolution; however, since the separation of *africanus* and *robustus* very early in Pleistocene or even in Pliocene, the cladogenesis occurred on the racial, subspecific, level only. We may have inherited some genes from every race of *africanus* and *erectus,* but we did not inherit any genes from *robustus* after its specific separation from *africanus.*

Progressive Evolution

Any organism, from the simplest to the most complex, is a beautifully engineered contrivance. In a sensitive and perceptive observer, the contemplation of the structure and the functioning of a living body evokes the same feelings as the contemplation of a work of art. And just as some works of art are perceived to be more admirable and perfect than others, so some organisms seem more advanced than others. A mammal, a bird, or a flowering plant seem higher, superior, more developed than an amoeba, an alga, or a bacterium. The evolution of life on earth started probably with beings at least as simple as microorganisms or viruses. It has now reached the level of forms as advanced as man. The evolution of the living world has, therefore, been on the whole progressive.

Starting with the classics of evolutionism and down to modern authors, attempts have been made to translate these intuitive insights into communicable judgments of what constitutes progress and advancement. All these attempts had, at best, only indifferent success. Teilhard de Chardin (1959) saw in evolution a "privileged axis" leading to man. This is too anthropocentric, since it makes the whole plant kingdom, all the phyla but the chordates, and all the classes but the mammals unprogressive by definition. This is not to deny that man is most progressive, in fact the supreme product of evolution. An objection has been made that another being, say a fish, would consider itself the supreme product. Simpson's answer (1949) is that any fish would be amazed to find that there are men who doubt the evolutionary superiority of the human species, and besides a fish who could consider such problems would have to be a man. Nevertheless, there are surely many kinds of progress, in addition to that in the line leading to man.

The fundamental thesis of the biological theory of evolution is that evolution is, at least in the main, an adaptive response of life to the challenges of the environment, mediated by natural selection. It does not follow, however, that the adaptedness always increases as the evolution progresses. Man is not necessarily better adapted to his environment than flies, or corals, or bacteria are to theirs. As pointed out above, such comparisons are operationally meaningless, since there is no way to compare the adaptedness of different organisms in different environments. Moreover, adaptedness must be distinguished from adaptability. Evolutionary changes which lead to a high adaptedness in a given environment may mean a narrow specialization for that environment, loss of evolutionary plasticity, and eventual extinction. This point has been lucidly analyzed especially by Thoday (1953, 1958; see also Manier, 1965).

Thoday defines the "fitness" of a "unit of evolution" (i.e., a strain, Mendelian population, or a species) as "its probability of leaving descendants after a given long period of time. Biological progress is increase in such fitness." Therefore, "The fit are those who fit their existing environments and whose descendants will fit future environments." Fitness in Thoday's sense is not at all the same thing as Darwinian fitness or adaptedness. Perhaps a better name for this variable is durability. It has several components, which he designates as adaptation (our adaptedness): variability which comprises genetic flexibility (evolutionary plasticity), phenotypic flexibility (physiological and developmental plasticity and homeostasis), and stability of the environment. This last is not an inherent property of the organism, and therefore should not, in my opinion, be included. This is not to deny that the stability of the ecological niche to which the organism is adapted is important for its preservation, but it is a conservative rather than a dynamic agent. "Progress" may then, according to Thoday, "be analyzed into a number of components all affecting either the ability of the unit to retain its adaptation and its ability to accommodate to environmental change or the amount of environment change it may meet. Fitness so considered is compounded of stability and variability, and increase in fitness becomes a resolution of the antagonism between stability and variability."

In his discussion of the "fitness" (the durability) of a unit of evolution, however, Thoday stresses a "long period," such as 10^8 years. This makes the operational usefulness of this concept rather low. Suppose that a zoologist or a botanist becomes well acquainted with the animals or plants which inhabited the earth in Cambrian, or Jurassic, or even Eocene times. Could he then pick out the forms which had a high durability and have descendants now living? This is most doubtful. Which form now living do we consider the most likely candidate to endure? Man is the most plausible choice; he could, if he wished, control not only his environment but his evolution as well. And yet, can one exclude the possibility of mankind succumbing to some suicidal folly? We know only too well that man is capable of stupid behavior.

In the light of natural selection, many otherwise paradoxical facts become intelligible. Together with remarkable adaptations, organisms often show astonishing imperfections which are necessarily incompatible with evolutionary progress. Can anything be adaptively more preposterous than the painfulness and hazard of childbirth in man? The puzzle becomes less puzzling if it is realized that natural selection operates not with fit and unfit traits in isolation, but with phenotypes as wholes and with genotypes that produce the phenotypes. Difficult childbirth may have arisen as a part of the constellation of changes which gave man his erect posture, and hands able to make and to use tools. The constellation as a whole is certainly adaptive, and its Darwinian fitness is high because its defects are presumably compensated for by its advantages.

Reproductive capacity is an important component of Darwinian fitness; does it follow that natural selection must work to increase indefinitely the number of the progeny produced? Lack (1954) and Cody (1966) have shown that in at least some species of birds the mean number of surviving young per nest is greatest in nests with a certain modal number of eggs. With too many eggs the parents are unable to take care of their progeny, and the net progeny size diminishes instead of increasing. Increases of fecundity would, then, be discriminated against by natural selection.

It has rightly been pointed out, both by philosophers and by biologists, that the theory of evolution has little predictive power. At the present level of our knowledge, long-term predictions of evolutionary events are extremely hazardous. The following may, however, serve as examples. Consider these three zoological species: the grizzly bear (*Ursus horribilis*), the Norway rat (*Rattus norvegicus*), and man (*Homo sapiens*). The adaptedness of the bear is restricted to a rather narrow range of the existing environments, that of a rat to a wider one, and man's range is the widest, because he is able to tailor environments according to his needs. There is not much point in trying to compare the Darwinian fitnesses of the three species, since they rarely compete for the same ecological niches. What, however, is their probable durabilities? It is not an implausible conjecture that the grizzly has the greatest and man the smallest probability of becoming extinct in, say, one thousand years (a much smaller number than that mentioned by Thoday!). This conjecture is based evidently on an extrapolation into the future of the environmental changes which have been taking place on earth over the last several centuries.

The grizzlies will either be exterminated or will continue in small numbers in nature preserves. The perspectives of the rat to endure are brighter, since at least until now this species has shown both a remarkable adaptedness and an adaptability. There is, however, a possibility that man may discover a way so to modify the rat's habitats that the toleration limits of the rat species will be exceeded. In fact, a drug denoted McN-1025 is highly toxic to rats but not to other mammals (Roszkowski, Poos, and Mohrbacher, 1964). Man as a species is most likely to have a greater durability than either the grizzly or the rat. This prediction has, however, an unknown margin of uncertainty; it is not quite inconceivable that some virus disease may not be controllable and may destroy the species, or that our species will commit suicide by means of an atomic war, population explosion, or like madness.

The durability of a unit of evolution, which enables it to have living descendants in the time to come, is evidently an important, though not precisely measurable parameter. But it is not the whole story as far as the problem of progressive evolution is concerned. Many relatively "low" or "primitive" microorganisms and viruses still survive and prosper. Some of them may even be products of regressive evolution, i.e., descendants of higher and less primitive ancestors. How is one to assess the primordial life on earth? It had the greatest number and variety of living descendants. Yet it was the base rather than the summit of the evolutionary progress.

Huxley (1942), Schmalhausen (1949), Simpson (1949), and Rensch (1960) are among modern authors who attempted to characterize progressive evolution. They recognize that, in Simpson's words, "evolution is not invariably accompanied by progress, nor does it really seem to be characterized by progress as an essential feature. Progress has occurred within it but is not of its essence. . . . Within the framework of the evolutionary history of life there have been not one but many different sorts of progress." Rensch lists the following six criteria: increasing complexity, rationalization of structures and functions, special complexity and rationalization of central nervous systems and of sense organs, increasing plasticity of structures and functions, open-ended improvement permitting further improvements, and increasing independence from or control of environment. These six criteria are obviously compatible, but the coincidence of all six is not necessary for an evolutionary change or an evolutionary line to be considered progressive.

Among Rensch's six criteria of progress, that dealing with the development of nervous systems and sense organs deserves special emphasis. This is important not because, or at least not only because man happens to stand at the summit of this particular kind of progress, but because it is probably most closely connected with the quality emphasized by Thoday, and has been referred to above as the durability of a unit of evolution in time. The reason for this, to quote Simpson (1949) again, is: "The progressive trend is to gather more and different kinds of information about the environment in which the organisms do, in fact, exist and to develop apparatus for appropriate adjustments in accordance with this information. On lowest levels this involves only diffuse sensibility and reactability to such signals from the environment as do have an effect on any protoplasm: some types of radiation, motion and bodily contact, temperature, and chemical effects. At higher levels it involves very complex and specialized sensory organs and equally or more complex nervous systems and other coordinating mechanisms."

Life reproduces itself by transforming materials selectively withdrawn from the environment into body constituents. The inception of biological evolution was the appearance of self-reproducing systems, i.e., of living organisms. This event was a turning point in the history of the Cosmos, which deserves the name of evolutionary transcendance. The transcendance became possible because the inorganic evolution has brought about conditions propitious for it, on at least one minor planet in the Universe. For about two billion years the biological evolution went on, and then reached another point of transcendance, which was the appearance of man. During these two billion years, the biological evolution was often, though not always and not everywhere, progressive. The setbacks, the regressive episodes, are explained by the opportunistic character of the principal impulse which brings about evolutionary changes—natural selection. Regressive evolution added to the diversity of life on earth, and did not interfere with the progressive evolution of other forms, or of the living world as a whole.

Though neither planned, guided, predestined or predetermined (except in the Laplacian sense of universal deterministic causality), the biological evolution gave rise to man. Although some rudiments of these abilities are observable on the animal level, man is the only being capable of abstract and symbolic thought, use of symbolic language, making and using tools, and creating culture transmitted not by genes but acquired in each generation de novo by learning and transmitted to the following generation by training. And finally man has achieved self-awareness and death-awareness. He has thus transcended his biological nature.

References

AMADON, D. 1964. The evolution of low reproductive rates in birds. Evolution, 18:105–110.

ANDREWARTHA, H. G., and L. C. BIRCH. 1954. The Distribution and Abundance of Animals. Chicago, Univ. Chicago Press.

ASIMOV, I. 1960. Intelligent Man's Guide to Science. New York, Basic Books.

BEARDMORE, J. A., TH. DOBZHANSKY, and O. PAVLOVSKY. 1960. An attempt to compare the fitness of polymorphic and monomorphic experimental populations of *Drosophila pseudoobscura*. Heredity, (London), 14:19–33.

BECKNER, M. 1959. The Biological Way of Thought. New York, Columbia Univ. Press.

BIRCH, L. C., TH. DOBZHANSKY, P. O. ELLIOTT, and R. C. LEWONTIN. 1963. Relative fitness of geographic races of *Drosophila serrata*. Evolution, 17:72–83.

BOCK, W. J., and G. V. WAHLERT. 1965. Adaptation and the form-function complex. Evolution, 19:269–299.

BRADSHAW, A. D. 1965. Evolutionary significance of phenotypic plasticity in plants. Advances Genet., 13:115–155.

BRUES, A. M., and G. H. SACHER, eds. 1965. Aging and Levels of Biological Organization. Chicago and London, Univ. Chicago Press.

CANNON, W. B. 1932. The Wisdom of the Body. New York, Norton.

CODY, M. L. 1966. A general theory of clutch size. Evolution, 20:174–184.

COMFORT, A. 1956. The Biology of Senescence. New York, Holt, Rinehart & Winston, Inc.

DARWIN, C. 1964 (1859). On the Origin of Species. (A facsimile of the 1st edition, introduction by Ernst Mayr.) Cambridge, Harvard Univ. Press.

DOBZHANSKY, TH. 1956. What is an adaptive trait? Amer. Naturalist, 90:337–347.

———. 1962. Mankind Evolving. New Haven, Yale Univ. Press.

———. 1964. Biology, molecular and organismic. Amer. Zool., 4:443.452.

DOBZHANSKY, TH., and H. LEVENE. 1955. Development homeostasis in natural populations of *Drosophila pseudoobscura*. Genetics, 40:797–808.

DOBZHANSKY, TH., R. C. LEWONTIN, and O. PAVLOVSKY. 1964. The capacity for increase in chromosomally polymorphic and monomorphic populations of *Drosophila pseudoobscura*. Heredity (London), 19:597–614.

DUBOS, R. 1965. Man Adapting. New Haven, Yale Univ. Press.

DUNN, L. C. 1956. Analysis of a complex gene in the house mouse. Sympos. Quant. Biol., 21:187–195.

———. 1960. Variations in the transmission ratios of alleles through egg and sperm in *Mus musculus*. Amer. Naturalist, 94:385–393.

EISELEY, L. 1959. Charles Darwin, Edward Blyth, and the theory of natural selection. Proc. Amer. Philos. Soc., 103:94–158.

FALCONER, D. S. 1960. Introduction to Quantitative Genetics. New York, Ronald Press.

FISHER, R. A. 1930. The Genetical Theory of Natural Selection. Oxford, Clarendon.

FOSBERG, F. R., ed. 1963. Man's Place in the Island Ecosystems. Honolulu, Bishop Museum Press.

GRANT, V. 1963. The Origin of Adaptations. New York, Columbia Univ. Press.

GRENE, M. 1958. Two evolutionary theories. Brit. J. Philos. Sci., 9:110–127, 185–193.

———. 1961. Statistics and selection. Brit. J. Philos. Sci., 12:25–42.

GOUDGE, TH. A. 1961. The Ascent of Life. London, Allen & Unwin.

HARDY, G. H. 1908. Mendelian proportions in a mixed population. Science, 28:49–50.

HUXLEY, J. S. 1942. Evolution, the Modern Synthesis. New York, Harper & Row.

LACK, D. 1954. The evolution of reproductive rates. *In* Huxley, J. S., A. C. Hardy, and E. B. Ford, eds., Evolution as a Process, London, Allen & Unwin.

LERNER, I. M. 1958. The Genetic Basis of Selection. New York, John Wiley & Sons.

LEVINS, R. 1962. Theory of fitness in a heterogeneous environment. I. The fitness set and adaptive function. Amer. Naturalist, 96:361–373.

———. 1963. Developmental flexibility and niche selection. Amer. Naturalist, 97:75–90.

———. 1964. The theory of fitness in a heterogeneous environment. IV. The adaptive significance of gene flow. Evolution, 18:635–638.

LEWONTIN, R. C. 1961. Evolution and the theory of games. J. Theor. Biol., 1:382–403.

———, and L. C. DUNN. 1960. The evolutionary dynamics of a polymorphism in the house mouse. Genetics, 45:705–722.

LOTKA, A. J. 1925. Elements of Physical Biology. Baltimore, Williams & Wilkins.

MANIER, E. 1965. Genetics and the philosophy of biology. Proc. Amer. Catholic Philos. Ass., 124–133.

MAYR, E. 1963. Animal Species and Evolution. Cambridge, Harvard Univ. Press.

MULLER, H. J. 1950. Our load of mutations. Amer. J. Hum. Genet., 2:111–176.

NAGEL, E. 1961. The Structure of Science. New York, Harcourt, Brace.

OHBA, S. 1967. Chromosomal polymorphism and capacity for increase under near optimal conditions. Heredity (London), 22:169–186.

RENSCH, B. 1960. Evolution Above the Species Level. New York, Columbia Univ. Press.

ROBINSON, J. T. 1967. Variation and the taxonomy of early hominids. *In* Dobzhansky, Th., M. K. Hecht, and Wm. C. Steere, eds., Evolutionary Biol., vol. 1, pp. 69–100, New York, Appleton-Century-Crofts.

ROSZKOWSKI, A. P., G. I. POOS, and R. J. MOHRBACHER. 1964. Selective rat toxicant. Science, 144:412–413.

SCHMALHAUSEN, I. I. 1949. Factors of Evolution. Philadelphia, Blakiston.

SCRIVEN, M. 1959. Explanation and prediction in evolutionary theory. Science, 130:477–482.

SIMPSON, G. G. 1949. The Meaning of Evolution. New Haven, Yale Univ. Press.

———. 1953. The Major Features of Evolution. New York, Columbia Univ. Press.

———. 1960. The history of life. *In* Tax, S., ed., Evolution after Darwin, vol. 1, 117–180, Chicago, Univ. Chicago Press.

———. 1964. This View of Life. New York, Harcourt, Brace and World.

SINNOTT, E. W. 1953. The Biology of the Spirit. New York, Viking.

———. 1963. The Problem of Organic Form. New Haven, Yale Univ. Press.

———. 1966. The Bridge of Life. New York, Simon and Schuster.

SLOBODKIN, L. B. 1964. The strategy of evolution. Amer. Sci., 52:342–357.

SMART, J. C. 1963. Physics and biology. *In* Philosophy and Scientific Realism. New York, Humanities Press.

STEBBINS, G. L. 1966. Processes of Organic Evolution. Englewood Cliffs, N.J., Prentice-Hall.

TEILHARD DE CHARDIN, P. 1959. The Phenomenon of Man. New York, Harper.

THODAY, J. M. 1953. Components of fitness. Sympos. Soc. Exp. Biol., 7:96–113.

————. 1958. Natural selection and biological progress. *In* Barnett, S. A. ed., A Century of Darwin, London, Heinemann.

————. 1959. Effects of disruptive selection. I. Genetic flexibility. Heredity (London), 13:187–203.

————. 1965. Effects of selection for genetic diversity. *In* Goerts, S. J., ed., Genetics Today, vol. 3, 533–540, Oxford, Pergamon Press.

————, J. B. GIBSON, and S. G. SPICKETT. 1963. Regular responses to selection. Genet. Res., 5:1–19.

WALLACE, B. 1948. Studies on "sex-ratio" in *Drosophila pseudoobscura*. Evolution, 2:189–217.

————, and TH. DOBZHANSKY. 1953. The genetics of homeostasis in Drosophila. Proc. Nat. Acad. Sci., 39:162–171.

————, and A. M. SRB. 1964. Adaptation. Englewood Cliffs, N.J., Prentice-Hall.

WEAVER, W. 1964. Scientific explanation. Science, 143:1297–1300.

WEINBERG, W. 1908. Über den Nachweis der Vererbung bei Menschen. Jahresh. Ver. vaterl. Naturk. Württemberg, 64:368–392.

WILLIAMS, G. C. 1966. Adaptation and Natural Selection. Princeton, N.J., Princeton Univ. Press.

ZIMMERMAN, E. C. 1948. Insects of Hawaii. Honolulu, Univ. Hawaii Press.

2

Cave Ecology and the Evolution of Troglobites

THOMAS C. BARR, JR.

Department of Zoology and Institute of Speleology
University of Kentucky
Lexington

Introduction 36
    Extent of the cave biotope 36
    Biological significance of caves 38
    Origin of caves 39
    Development of biospeleology 42
Animal Life in Caves 43
    Classification of cavernicoles 43
    Regional development of cave faunas 45
    Taxonomic distribution of troglobites 47
The Cave Ecosystem 47
    Physical parameters 47
    Energy sources and food webs 51
    Competition 61
    Circadian rhythms in cavernicoles 65
    Effects of geologic structure 67
Regressive Evolution in Cave Animals 69
    Lamarckian theories 69
    Orthogenetic theories 71
    Darwinian theories 72
    Conclusions 79
Speciation and Adaptation in Troglobites 80
    Preadaptation 80
    Colonization of caves 82
    Genetic changes accompanying troglobite speciation 86

Adaptation to the cave environment 89
Reproductive adaptations of troglobites 94
Conclusions 96
References 96

Introduction

Obligatory cavernicoles, or troglobites, have traditionally been of special interest to evolutionary biologists for several reasons. The existence of animal life in caves and other subterranean spaces at first attracted attention because of its novelty; intensive biological exploration of caves began little more than a century ago. Although the discovery and description of the cave faunas of the world is far from complete, especially in the Western Hemisphere, so much descriptive information has been compiled that we can safely assert that, at least in unglaciated, temperate parts of the world, the occurrence of numerous species of troglobites in any major limestone region is a common and highly probable phenomenon.

Extent of the Cave Biotope

The prevalence of caves and the vast amount of space in caves and other solutional cavities available to subterranean animals have been widely underestimated. In the United States, the Cave Files Committee of the National Speleological Society has, in the past 25 years, assembled information on approximately 7,500 caves. About 20 percent of these caves are located in Kentucky and Tennessee, where from personal experience I estimate that scarcely one third of the accessible caves have been located and recorded. If this same ratio is applied, there may very well be more than 20,000 caves in the United States which are accessible to man, i.e., with one or more entrances. Since many cave entrances (Fig. 1) appear to be fortuitous (collapse of the roof, intersection of a cave passage by a surface valley, and so forth), there must be many caves with no entrances. Working with data on the caverns of West Virginia (Davies, 1949), Curl (1958) developed a stochastic model which predicts that there are about 10 times as many caves with no entrances in West Virginia as there are caves with one or more entrances. If the karst region of West Virginia is typical of most of the cave areas of the United States, it is quite feasible to estimate that there are 200,000 caves in this country. Ninety percent of them would be closed to human penetration, but the experience of biospeleologists in artificially opened caves shows that very few of them are closed to cavernicolous animals.

A few caves are extremely long, with many kilometers of penetrable passages. The classic example is the vast system of caves in Mammoth Cave National Park, Kentucky, where more than 175 km of passages are

Fig. 1. Colglazier Sink, a lateral opening to the main trunk channel of Blue Spring Cave, Indiana's largest known cave system, is one of several sinkholes which collect surface runoff during rains and feed the subterranean stream.

developed beneath a surface area of about 10,000 hectares. Several other large caves, more than 10 km in total passage length, are reported in Alabama, Arkansas, Indiana, Kentucky, Minnesota, Missouri, New Mexico, Tennessee, Texas, Virginia, and West Virginia. However, most caves are much smaller, with a mean length of about 150 m. This figure still yields a rough estimate of 30,000 km of cave passages large enough to be penetrated by man. The total amount of solutional space available for habitation by small cavernicolous animals must also include innumerable small crevices in addition to these 30,000 km of relatively large passages. One can only conclude that caves and associated solutional openings in limestone terranes are not restricted, exceptional habitats, but in fact constitute a significant segment of environmental space in any of the major karst regions of the world.

Biological Significance of Caves

Despite the availability of considerable space, the subterranean habitat is a marginal one. In the absence of light, cave ecosystems are almost entirely heterotrophic, dependent on transfer of food energy from the surface. Because food input is sporadic and seldom profuse, the total biomass of such ecosystems is minute. Cave communities are consequently strikingly simple in comparison with natural epigean communities, embracing a relatively small number of species. The study of ecosystem dynamics in a cave is thus mitigated by relative simplicity.

Cavernicoles generally manifest a number of conspicuous rudimentations in a rather dramatic way. The reduction of eyes and pigment in animals living in perpetual darkness has been eagerly seized upon by Lamarckians and Neolamarckians and has been repeatedly explained away by Darwinians.

So many troglobites are relicts of formerly widespread faunas that Charles Darwin (1859) drew attention to this fact in "Origin of Species," and René Jeannel entitled his semipopular book on biospeleology "Les Fossiles Vivants des Cavernes" (1943). The alternating glacial and interglacial climates of the Pleistocene were unquestionably influential in this situation, the progenitors of existing troglobites having survived in the more equable microclimate of caves while their respective species became extinct at the surface. Cave faunas apparently include both thermophilic and cryophilic components (Vandel, 1964), but the latter certainly predominate in caves of the temperate zones.

Discontinuous habitats have attracted the attention of evolutionary biologists whenever it has been discovered that a multiplicity of closely similar, often allopatric species occur in them. Islands were the most obvious discontinuous habitats and for this reason were the first to receive careful consideration. Animals inhabiting mountains exhibit similar patterns of speciation because alpine habitats are effectively "islands" surrounded by a "sea" of lowlands (Barr, 1962a). Limestone caverns are a third major class of discontinuous habitats, with numerous, closely similar species occupying different cave systems. Even though caves offer more extreme cases of geographic isolation than either islands or mountains, multiplication of species in troglobites has rarely received close scrutiny. One of the chief reasons for this is that intensive biological exploration of caves began rather late in comparison with islands and mountains. Another reason may have been the preoccupation of some investigators with the search for evidence to support Neolamarckism, orthogenesis, and continental drift.

Modern biospeleology has not yet fully emerged from the descriptive phase, but the literature shows an increasing number of ecological and physiological studies, summarized and discussed by Vandel (1964) and Poulson (1964). However, despite the recent publication of Vandel's (1964) treatise on biospeleology and Mayr's (1963) profound book on animal speciation, the principal stages of speciation in troglobites have not yet been clearly stated, and the persistent problem of regressive evolution in cave animals still falls a bit short of final resolution.

Origin of Caves

The origin of limestone caves results from solution of the limestone by carbonic acid, formed when atmospheric and soil carbon dioxide dissolves in water, as well as by organic acids from soil, nitric acid from atmospheric and bacterial fixation, and occasionally sulfurous and sulfuric acids from the action of bacteria. Enlargement by abrasion takes place when cobbles, pebbles, and sand derived from overlying strata are swept through the solutional channels by flood water, but the load of subterranean streams is generally small compared with surface streams, and solution is probably a far more important factor in cave development than abrasion.

There is general agreement that much solution takes place at or below the water table, in the phreatic zone (Davis, 1930, Bretz, 1942, Swinnerton, 1932, 1942), but primitive networks formed beneath the water table may be greatly modified and enlarged subsequent to regional uplift, when surface streams invade the networks (Fig. 2) and establish an underground channel for part of their flow (Malott 1932, 1937, Woodward, 1961). W. M. Davis' (1930) cyclical theory of cavern origin postulated a phreatic stage of solutional enlargement beneath the water table, followed by regional uplift and draining of the caves. Deposition of stalactites, stalagmites, and other travertine speleothems (i.e., cave "formations") would take place in

Fig. 2. Lower entrance to the Sinks and Rises, Jackson County, Kentucky. South Fork of Station Camp Creek sinks and flows 700 meters through a large cave, illustrating Malott's invasion theory of speleogenesis.

the latter part of the cycle, when the caves became air-filled (Fig. 3). Swinnerton (1932) and Piper (1932) both felt that most of the solution took place at or just below the water table, in the zone of greatest flow near perennial surface streams. Bretz (1942) drew attention to the extensive deposits of "red clay" in caves of the Ozark plateau and other regions, suggesting that they were laid down by phreatic ground water prior to regional uplift. Bretz's original ideas have been considerably modified to include the many stratified sand, gravel, and cobble deposits which occur in most caves.

Caves usually exhibit considerable ceiling breakdown, which results from failure of the limestone beds of the ceiling, especially in large roof spans (Davies, 1951). Cave entrances often develop in this manner when the breakdown extends to the surface. Vertical shafts, or "domepits," may develop during the air-filled stage, when acid surface streams (usually run-

Fig. 3. Rimstone pools, or gours, are secondary deposits of travertine which provide quiet, silted pools often occupied by aquatic cavernicoles.

ning off sandstone strata above the limestone) work down along vertical joints in the limestone.

The significance of two-stage cavern development for cave biology is the possibility of approximately dating the time at which caves of a particular karst region became available for colonization by cavernicoles. In most of the cave areas of central Tennessee and Kentucky, for example, many large caves with rich faunas are remnants of late Pliocene trunk channels which were drained at the close of the Pliocene or beginning of the Pleistocene (Barr, 1961). Patterns of solutional development also suggest the most likely routes for subterranean dispersal of cavernicoles, or possible barriers to dispersal. To cite another example, if we assume the correctness of Piper's (1932) view that maximum solution takes place near perennial surface streams, we would expect to find networks of caves paralleling surface streams crossing limestone plateaus. The geographic distribution of the cave beetles *Pseudanophthalmus ciliaris* and *P. orlindae* in the Red River Valley along the Kentucky-Tennessee border, and the presence of entirely different species in another drainage system 15 to 20 km to the north is readily explained on this basis (Barr, 1959). For further details on the cycle of erosion in a limestone terrane, the origin and development of caves, and the principles of limestone hydrology, the reader is referred to the reviews of Swinnerton (1942), Barr (1961, pp. 3–27), and Thornbury (1954, pp. 316–353); to the classic papers of Cvijič (1918), Davis (1930), Swinnerton (1932), Piper (1932), Malott (1932), Gardner (1935), and Bretz (1942); and to the symposiums edited by Moore (1960, 1966).

Development of Biospeleology

The term "Biospéléologie" was coined by Armand Viré (1904). A brief history of biospeleology has recently been given by Vandel (1964, Chap. 3); elsewhere I have outlined the development of the science in the United States (Barr, 1966). The first published account of a troglobite was a popular description of the Yugoslavian cave salamander *Proteus anguinus,* mentioned in a book by Valvasor (1689). But it was in the first half of the 19th century that descriptive biospeleology really began, initially in the caves of Carniola (Schmidt, 1832, Sturm, 1844, Schiödte, 1851, Schiner, 1854) and in Mammoth Cave, Kentucky (Tellkampf, 1844a, 1844b, De-Kay, 1842, Motschulsky, 1862). The first checklist of European cave arthropods was published in 1875 by Bedel and Simon, followed in 1894 by Hamann's "Europäische Höhlenfauna." The foundations of European biospeleology were laid primarily by Emil G. Racovitza (1907) and René Jeannel (see References for principal titles) during the first three decades of the 20th century. In North America, A. S. Packard, Jr. (1888, 1894),

E. D. Cope (1887), A. M. Banta (1907), and C. H. Eigenmann (1909) made major contributions to our knowledge of cave faunas of the United States. During the period between World Wars I and II, the collection of cave faunas in the New World progressed at a much slower rate than in Europe. More than half of the known species of troglobites in North American caves have been collected since 1950. At least another decade will elapse before the major descriptive effort in North American biospeleology is completed.

Animal Life in Caves

Classification of Cavernicoles

Cavernicoles are customarily classified [1] as (1) *troglobites,* obligatory species which are unable to survive except in caves or similar hypogean habitats (Fig. 4); (2) *troglophiles,* facultative species which commonly inhabit caves and complete their life cycles there, but also occur in sheltered, cool, moist, epigean microenvironments (Fig. 5); (3) *trogloxenes,* which frequent caves for shelter and a favorable microclimate, but must return periodically to the surface for food (Fig. 6); and (4) *accidentals.* The first two terms were initially employed by J. R. Schiner (1854), and the third was added by Racovitza (1907).

This is an ecological classification, and its difficulties of application stem from insufficient knowledge of the ecology of species inhabiting caves. Obligatory deep-soil animals, the *edaphobites* (Coiffait, 1958, Jeannel, 1965), occasionally occur in caves, but if their presence is sporadic or if they also occur in "microcaverns" (cavities about the roots of trees, fractures in igneous and clastic rocks, and so forth), then an inflexible application of the Schiner-Racovitza terminology can lead to confusion. A similar situation exists with respect to many small aquatic subterranean species, chiefly crustaceans. Some of these animals inhabit not only caves and solutional crevices in limestone but also minute spaces in noncarbonate, nongypsum rocks, occurring frequently in seeps, springs, and wells. For such species, Motaş and Tanasachi (1946) proposed the term "phreatobite," because of their relationship to phreatic water (ground water which lies below the water table). Karaman (1954), a student of the phreatic fauna, has written, "Es besteht kein Höhlenfauna," obviously an extreme

---

[1] These terms have been widely accepted, but unfortunately they have not always been used consistently. While there is general agreement that a troglobite is an obligatory cavernicole, "troglophile" and "trogloxene" have been applied rather loosely, the latter often being synonymous with "accidental." See Racovitza (1907), Barr (1960a, 1964), and Vandel (1964) for further discussion.

Fig. 4. (Legend on page 45)

Fig. 5. Meta menardii *Latreille, a common troglophilic spider of caves in Europe and eastern North America.*

position. If one accepts the proposition that a given subterranean species can be simultaneously a troglobite and an edaphobite, or a troglobite and a phreatobite, the terminological difficulty is minimized. And there are certainly many species which are restricted to solutional openings in limestones, dolomites, or gypsum, or to (in a few cases) lava tubes; such animals are troglobites in the strictest sense of the word.

Regional Development of Cave Faunas

The occurrence of most troglobites is controlled by the distribution of soluble rocks, chiefly limestones and dolomites, in which most caves have

Fig. 4. Eurycea (=Typhlomolge) rathbuni *Stejneger, a troglobitic salamander from caves and artesian springs near San Marcos, Hays County, Texas.*

Fig. 6. Pipistrellus subflavus F. Cuvier, *a trogloxene, one of the most common cave bats of the eastern United States.*

been excavated. The presence of troglobites in lava tubes, gypsum caves, and sea caves has been established, but is exceptional. The richest troglobitic faunas occur in southern Europe from the Pyrenees to the Caucasus; Northern Africa and Asia Minor; southeastern United States between the margin of glaciation and the Fall Line; the Edwards plateau of central Texas; the highlands of Mexico; Japan; and New Zealand. Tropical caves typically have very rich faunas (see Leleup, 1956), but most of the few troglobites known from such caves are aquatic (Vandel, 1964). Limestone regions of Cuba and Florida have a well-developed fauna of aquatic troglobites.

The cave faunas of Europe are far better known than those of other parts of the world. In North America faunal collections have been made in approximately 2,000 caves, although few of these have been carefully and repeatedly investigated. An estimated 350 species of troglobites have been described from North American caves, and these probably represent only 30 to 40 percent of the total.

Toxonomic Distribution of Troglobites

The taxonomic groups in which the great majority of troglobitic species occur are the Tricladida, Gastropoda, Chelonethida, Araneae, Copepoda, Isopoda, Amphipoda, Decapoda, Diplopoda, Collembola, Diplura, Coleoptera (especially Carabidae, Catopidae, Pselaphidae), Teleostei (one or a few species in each of 10 families), and Caudata (Plethodontidae, Proteidae). Minor taxonomic groups for which troglobitic (or presumed troglobitic) species have been described are the "Rhabdocoela," Nemertea, Nematoda, Polychaeta, Oligochaeta, Hirudinea, Pelecypoda, Scorpionida, Opiliones, Palpigradi, Pedipalpi, Acarina, Ostracoda, Bathynellacea, Thermosbaenacea, Speleogriphacea, Mysidacea, Pauropoda, Onychophora, Chilopoda, Thysanura, Blattodea, Orthoptera, and Diptera. For more detailed information on these groups the reader is referred to the review and bibliography in the recent book by Vandel (1964). In caves, as on the surface, there are more species of arthropods than in all the other groups put together, and a majority of the arthropods are beetles.

The Cave Ecosystem

Physical Parameters

The physical environment of a deep cave is characterized by several peculiar features, the most striking of which are the absence of light, the generally high and relatively constant humidity, and the relatively constant temperature. Photosynthesis cannot take place in absolute darkness, consequently primary producers are virtually absent and almost all food energy must be imported from the exterior. Chemosynthetic autotrophs, notably sulfur and iron bacteria, have been demonstrated in a number of caves (Dudich, 1932, Caumartin, 1963), but their energy contribution to the ecosystem is probably negligible, except perhaps quite locally and sporadically. The apparent importance of cave clay in the nutrition of amphipods, beetles, salamanders, and other troglobites (Gounot, 1960, Vandel, 1964) may be related to its bacterial content. Terrestrial cavernicoles are almost all more or less stenohygrobic, so that dry, dusty caves are as devoid of animal life as the most inhospitable desert. For these reasons, food and moisture are critical limiting factors.

The total biomass of a typical cave ecosystem in the temperate zone is extremely low. Almost all the energy entering such a system is attributable to two sources: (1) rotting vegetation, small epigean animals, and other organic matter washed underground by sinking streams, and (2) the feces and dead bodies of trogloxenes, animals which feed outside by night and return to the cave by day. Bats and cave crickets (rhaphidophorine Orthoptera) are ecologically the most important trogloxenes because of their widespread use of caves, their gregarious roosting habits, and the guano which they produce.

Few caves are entirely free of the effects of climatic events which take place at the surface. Because of the density gradient between cooler and warmer air, temperate zone caves commonly "breathe," with denser, cooler, cave air flowing out of entrances in summer and outside air flowing into entrances in winter. Because of the insulating effect of the surrounding rock and overlying soil mantle, the temperature of the deep cave approximates the annual mean temperature of the geographic region in which the cave is located. On days in spring and fall, when the air temperature at the surface significantly rises above and later falls below the cave temperature, two reversals in direction of air flow are easily observed at cave entrances. In the so-called "breathing caves," the direction of flow oscillates with a relatively constant period, reversing every few minutes without apparent relation to surface temperature. V. A. Schmidt (in litt.) suspects that such oscillations are apparently associated with vertical air columns of considerable height in domes, pits, and other vertical shafts, and that the dynamic properties of the system resemble those of a Helmholtz resonator. Deep elevator shafts in Mammoth Cave, Kentucky, and Ruby Falls, Tennessee, are associated with marked oscillations in air flow in cavern passages near the base of the shafts.

Sudden drops in regional barometric pressure may induce movement of colder air into caves through small crevices. Cournoyer (1955) described a drop in temperature from 12.5° to 7.0° C. and in relative humidity from 97 to 78 percent in the Lost Passage of Floyd Collins' Crystal Cave,

Kentucky, during a February rainstorm. The temperature returned to 12.5° in 2½ hr, but the relative humidity did not reach 97 percent until 17½ hr had elapsed. Conn's (1966) studies of winds at the mouth of Jewel Cave, South Dakota, clearly demonstrate the effects of external barometric pressure on air flow in caves. Cropley's (1965) investigations on temperature fluctuations in West Virginia caves illustrate the combined effects of barometric winds, air temperature gradients, and the cooling and warming effect of sinking streams.

The principal effect of air flow in summer is to extend the cool, humid microclimate of the deep cave closer to entrances. In winter the effects are far more severe. Surface air entering the cave is much drier than cave air. As it warms up to cave temperature, the surface air absorbs more moisture. The net result is that cave passages near entrances are somewhat cooler in winter, and have a rate of evaporation (Fig. 7) which may be one hundred to two hundred times higher than in summer (Barr and Kuehne, 1968). Since most troglobites and many troglophiles are both stenothermic and stenohygrobic, the ecological consequences of winter "inhalation" are profound. Although the twilight or crepuscular zone and the aphotic zone of the cave may have more or less fixed limits, a considerable length of a cave passage near an entrance is a fluctuating temperature zone (Eigenmann, 1909), the boundaries of which shift back and forth between

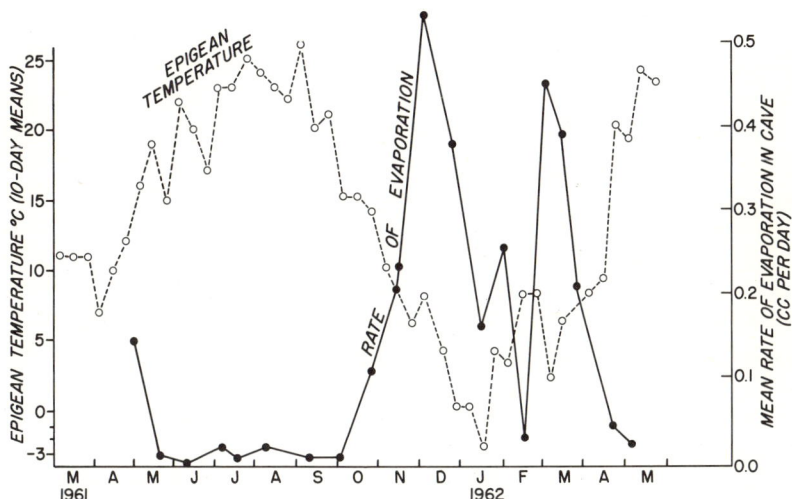

Fig. 7. Rate of evaporation at Moonlight Dome, Mammoth Cave, Kentucky (5 cm Livingston sphere evaporimeter), plotted with outside temperature.

summer and winter. The movements of cavernicoles coincide with the presence of moisture-laden air, nearer entrances in summer and farther into the cave in winter.

Aquatic environments in caves are subject to seasonal fluctuation in groundwater percolation. Still pools and slowly moving streams in the Mammoth Cave system of Kentucky contain dissolved oxygen in concentrations of about 7 to 16 parts per million (Barr and Kuehne, 1968). Oxygen content is slightly increased by dripping water or stream riffles. The normal pH range of such water is greater than 7.0 and less than 8.0, and free CO_2 is present only in minute traces.

In cave regions of the eastern United States, pool and stream levels normally reach a minimum in late October and November. Temporary pools are dry at this season, and some cave streams have little or no measurable flow. From the first heavy rains of winter, which usually fall between mid-December and mid-January, cave streams are subject to sudden and drastic flooding. Peak floods occur in February or early March. The levels of streams or pools near the water table in the lowest galleries of Mammoth Cave commonly fluctuate about a late winter mean 3 to 4 m higher than the October–November minimum. The maximum recorded level occurred March 2, 1962, when the cave water level rose nearly 15 m above the minimum.

Flooding is a common and widespread ecological feature of many caves in widely separated parts of the world. Hawes (1939) has described and assessed the biological significance of flooding in Yugoslavian caves. Gurnee, Thrailkill, and Nicholas (1966) report sudden, massive flooding of the subterranean course at the Río Camuy in Puerto Rico, with minimal stream levels occurring during the dry season, in February. Breder (1942) mentions heavy flooding in caves of San Luís Potosí, Mexico.

For aquatic cavernicoles, flooding means that they must often take refuge in crevices or beneath stones to avoid being swept out of the cave. A sudden and pronounced temperature change frequently results, oxygen concentration almost always rises, and the pH is altered, usually increasing slightly (but the direction of the change depends on the source of the flood water). Because many terrestrial animals are concentrated along the banks of the streams, they are subject to being swept away by floods, and in small, sewerlike passages they may be completely inundated. American trechine beetles in mud crevices and under stones react to splashing water by rapidly crawling away from the edge of the stream, yet are often very abundant in passages which flood to the ceiling and from which they could not readily escape. Vandel (in litt.) reports that *Aphaenops* in caves of the Pyrenees scurry into burrows to escape being washed away by flood waters. Some American species can survive total immersion in water for several hours (Barr and Peck, 1965). Hawes (1939) suggested that flooding serves to trigger reproductive cycles in aquatic cavernicoles (see also Poulson, 1964).

Energy Sources and Food Webs

Despite some obvious disadvantages, flooding streams bring into many cave ecosystems an annual supply of leaves, branches, logs, and other organic debris, providing food for detritus-feeding cavernicoles and initiating the short food chains of the cave community. Without floods such caves could be veritable biological deserts.

The guano of bats is a rich source of food for cavernicoles, although a typical guanobite assemblage includes many troglophiles and very few troglobites. In the "Bat Cave" of Carlsbad Caverns, New Mexico, immense guano deposits of *Tadarida brasiliensis mexicana* support tenebrionid beetles, muscoid flies, and tineid moths, which are in turn eaten by spiders, pseudoscorpions, and carabid beetles (Barr and Reddell, 1967). These species are all troglophiles. Parasites of the bats include fleas, mites, and streblid flies. The large *Tadarida* caves of central Texas support an even richer fauna of troglophiles and trogloxenes (Kohls and Jellison, 1948), and bats are regularly preyed upon by falcons as they leave one of the caves (Stager, 1941).

Few troglobites appear to be adapted to feed exclusively on bat guano, although some of them certainly utilize it when it is available. Some troglobite predators hunt their prey at the margin of guano piles, but few, if any of them seem to be specialized to feed on guanobites (see Jeannel, 1943; Vandel, 1964). Many species of troglobites have never been found near bat guano, even though it may be abundant in parts of the caves which they inhabit.

The guano of the cave cricket *Hadenoecus subterraneus* Scudder (Fig. 8) and related species in central Kentucky, central Tennessee, and northern Alabama, in contrast to the guano of bats, is extensively utilized by both troglophiles and troglobites. These crickets roost on cave ceilings during the day and emerge at night to feed on ants, millipedes, and other forest floor arthropods. Park and Reichle (1964) made an analysis of their food habits, based on examination of the contents of several hundred guts. Reichle, Palmer, and Park (1965) demonstrated a nocturnal rhythm in mean bihourly activity in *H. subterraneus*. In this study they concluded that approximately one third of the crickets fed outside the cave on any given night. Orientation in these crickets has not been experimentally investigated, but probably involves tactic responses to air currents as well as to temperature and humidity.

In the Mammoth Cave region of Kentucky, *Hadenoecus* guano accumulates on rock shelves and flowstone and may attain a thickness of 10 to

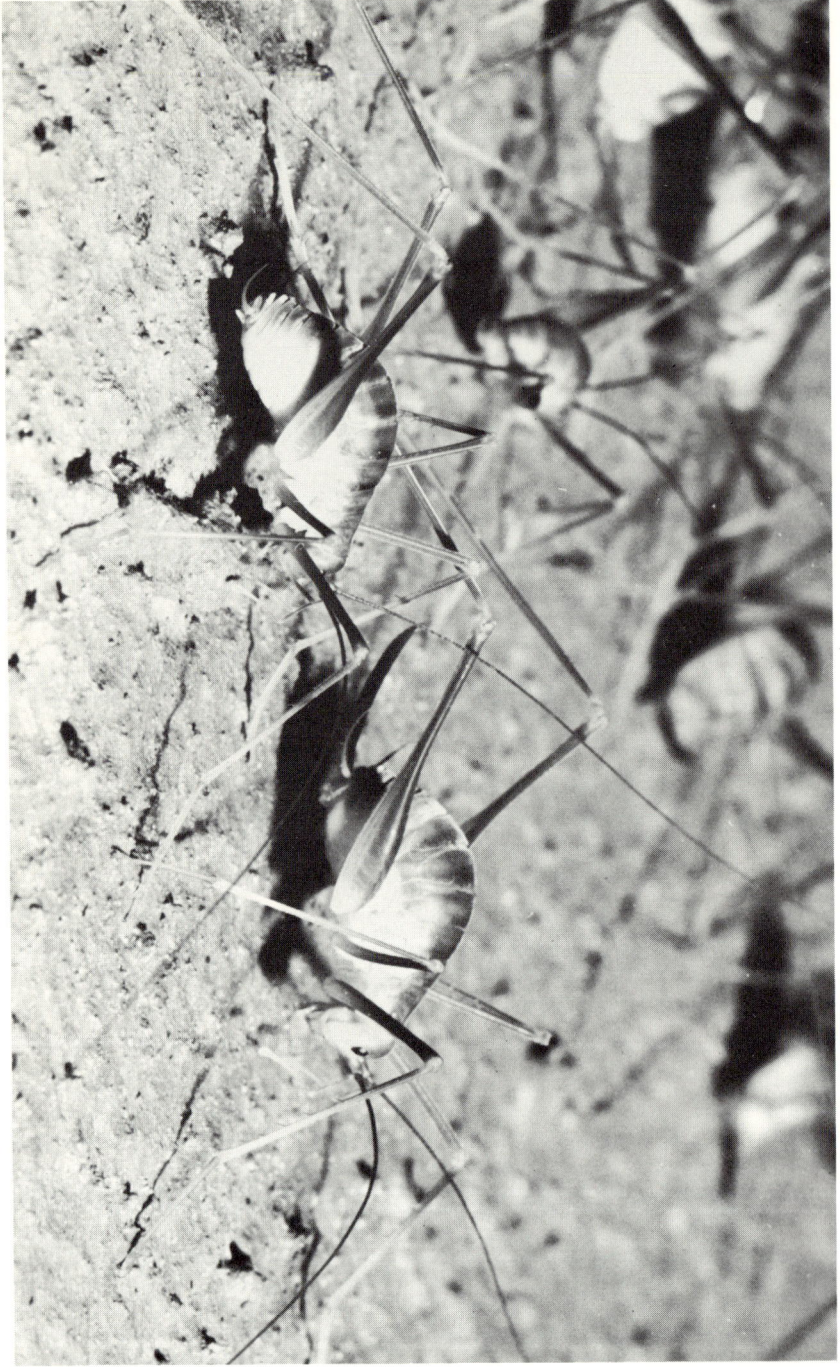

Fig. 8. Hadenoecus subterraneus Scudder, the common cave cricket of the Mammoth Cave region, on a cave ceiling.

12 mm. Bat guano is far less important in this region as a source of food for the cave community, partly because bat occupation of the caves occurs principally during hibernation (Fig. 9), when there is little guano deposition, and partly because bat colonies have been driven from the several tourist caves (including Mammoth Cave itself) in the area. Jegla and Hall (1962) showed that ancient bat skeletal material in Chief City, a huge chamber of Mammoth Cave remote from any existing natural entrance, pertained to the free-tailed bat (*Tadarida brasiliensis*), now extinct in Kentucky. Extensive deposits of dried guano associated with the bones were dated by the C^{14} method and were shown to be older than 34,000 years. It seems evident that Mammoth Cave was occupied by a large colony of *Tadarida* during the Sangamon (if not earlier) and that the ecology of at least part of the system must have resembled that of the Bat Cave in Carlsbad Caverns.

In Mammoth Cave National Park today, however, cavernicoles are dependent chiefly on *Hadenoecus* guano in the upper levels of the caves, and on organic debris washed into the lower levels by sinking streams. White Cave, near Mammoth Cave, has a rich fauna based on cricket guano. Detritus feeders include a troglobitic snail (*Carychium stygium* Call, Fig. 10), a blind catopid beetle (*Ptomaphagus hirtus* Tellkampf), a dipluran (*Plusiocampa cookei* Packard, Fig. 11), troglobitic millipedes (*Scoterpes copei* Packard [Fig. 12], *Antriadesmus fragilis* Loomis), terrestrial isopods (*Miktoniscus mammothensis* Muchmore), collembolans (*Tomocerus flavescens americanus* Folsom), and larvae of flies (*Bradysia* sp., *Amoebaleria defessa* Osten-Sacken). Predatory on these saprophiles are beetles (*Pseudanophthalmus menetriesii* Motschulsky, *Batrisodes henroti* Park [Fig. 20]), pseudoscorpions (*Kleptochthonius cerberus* Malcolm and Chamberlin [Fig. 13], *Pseudozaona mirabilis* Banks), a troglobitic harvestman (*Phalangodes armata* Tellkampf, Fig. 14), small blind spiders (*Anthrobia monmouthia* Tellkampf, *Phanetta subterranea* Emerton), large orb-weaving spiders (*Meta menardii* Latreille, Fig. 5), and an occasional salamander (*Eurycea lucifuga* Rafinesque). Two other important links in the food web concern the *Hadenoecus,* which eat cave beetles (Park and Reichle 1964), and eyeless carabids (*Neaphaenops tellkampfii* Erichson) which prey heavily on *Hadenoecus* eggs, which they extract from the damp silt of ledges and the cave floor with their elongate mandibles.

A remarkable case of parallelism in feeding behavior is seen in the anchomenine carabid *Rhadine subterranea* Van Dyke, a troglobitic beetle inhabiting caves north of Austin, Texas. *R. subterranea* feeds predominantly on the eggs of cave crickets (*Ceuthophilus cunicularis, C. conicau-*

Fig. 9. Hibernating bats, principally Myotis sodalis, in Long Cave, Mammoth Cave National Park.

dus, and related species), which it scoops from the silt using its elongate, flattened head as a shovel (Mitchell, 1965).

The young of the blind cavefishes *Typhlichthys subterraneus* Girard and *Amblyopsis spelaea* DeKay customarily inhabit deep, slowly moving subterranean streams, feeding principally on zooplankton (Poulson, 1963). Scott (1909) showed that the plankton of the Donaldson-Bronson-Twin cave system near Mitchell, Indiana—part of the old Indiana University "Cave Farm" where Carl Eigenmann investigated *Amblyopsis*—consisted almost entirely of epigean species which were washed into the system through sinkholes. Kofoid (1899) showed that this was also true for Echo

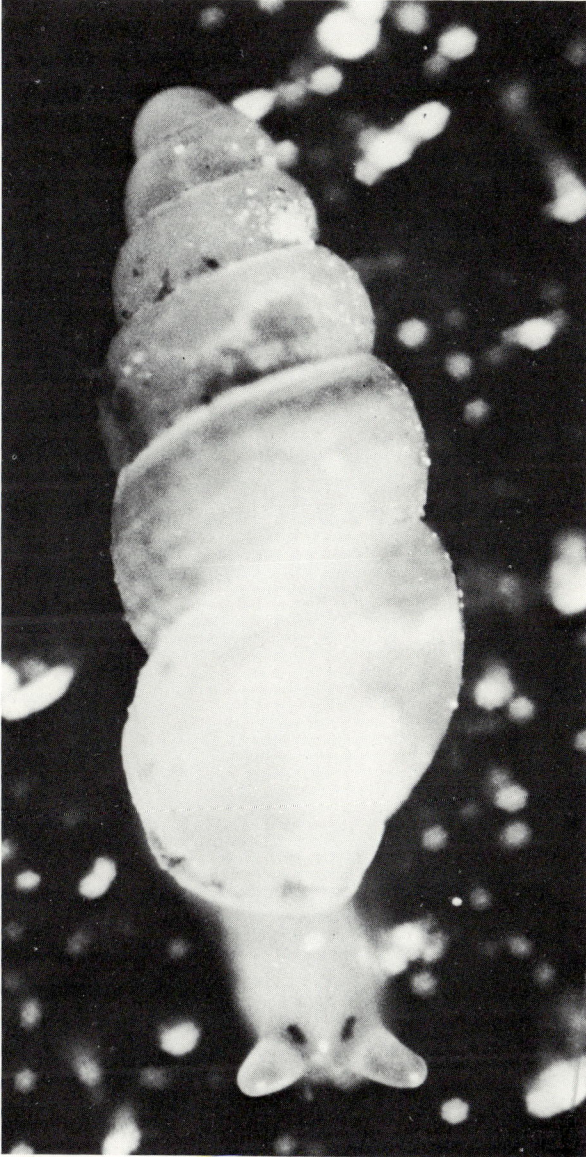

Fig. 10. Carychium stygium Call, a small (2 mm), terrestrial snail of the Mammoth Cave region; apparently a recent troglobite, it retains vestigial eyes.

Fig. 11. Plusiocampa cookei Packard represents a large genus of troglobitic diplurans (Campodeidae) in the Mammoth Cave region. It is a detritus feeder.

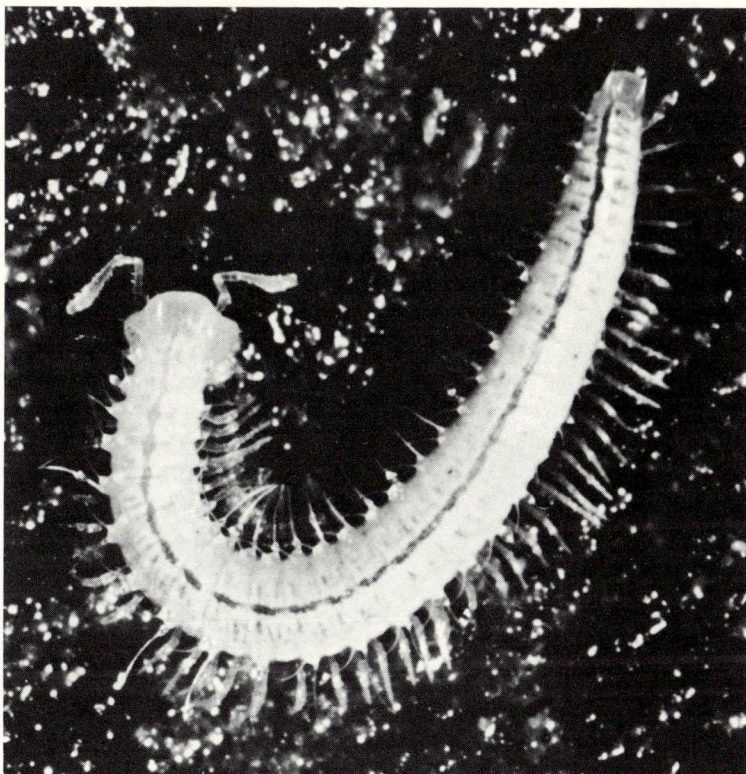

Fig. 12. Scoterpes copei Packard, a troglobitic millipede (Conotylidae), found on rotting wood or cricket guano in the Mammoth Cave region.

River in Mammoth Cave, where both species occur. Bolívar and Jeannel (1931) discovered subterranean species of copepods and ostracods—apparently phreatobites—in both of these systems, but they are greatly outnumbered by common epigean copepods and cladocerans. The two streams in the lowest levels of Mammoth Cave—the Echo and Styx rivers—empty into the canyon of Green River through large springs. During floods, the waters of Green River back up into the cave. Extreme flooding in late winter and early spring adds little to the plankton count, but a slight rise in midsummer carries large numbers of zooplankters into the cave river system. In late fall—the driest time of the year—a typical plankton tow of Echo River yields about 1 adult zooplankter and 2½ nauplii per 100 liters (Barr and Kuehne, 1968). From such observations it is apparent that the juvenile amblyopsids feed infrequently and are most likely to have food readily available in late spring and early summer.

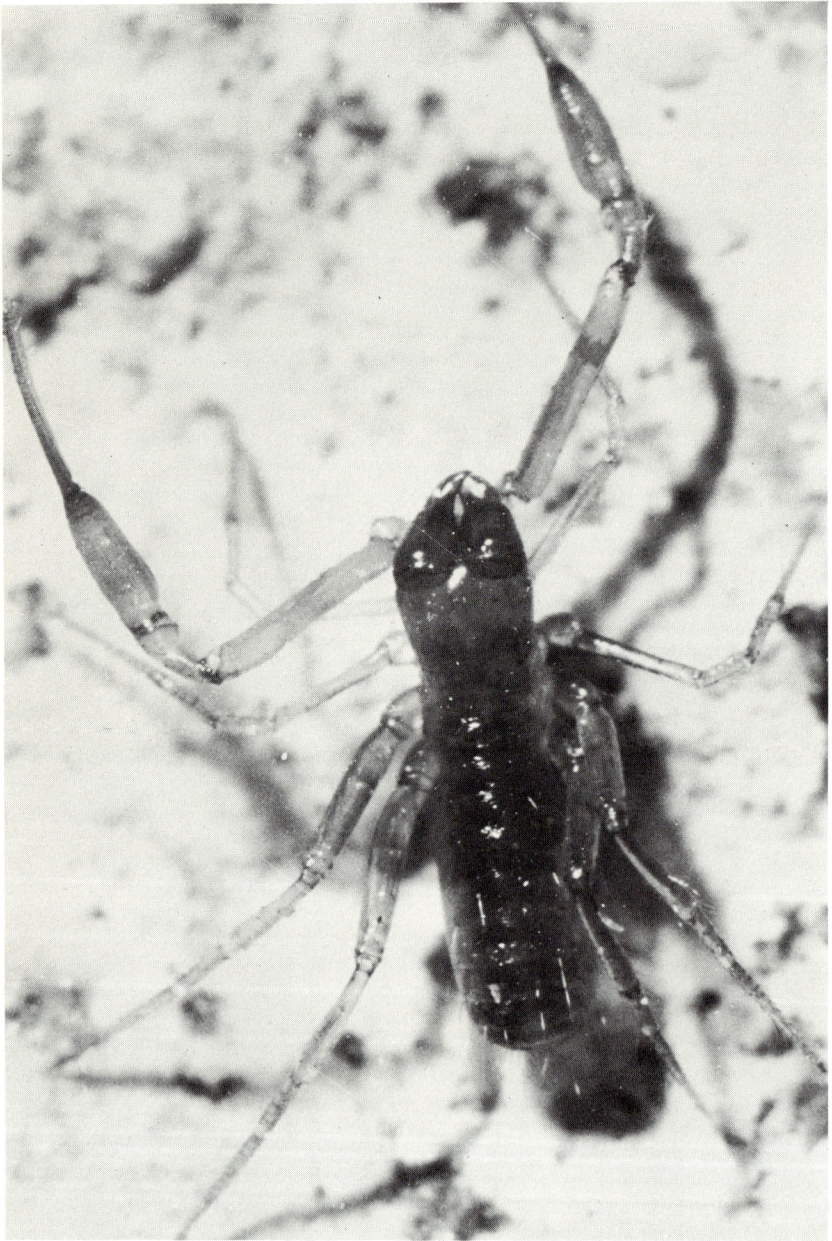

Fig. 13. Kleptochthonius cerberus Malcolm and Chamberlin, a troglobitic pseudo-scorpion from Mammoth Cave.

Fig. 14. Phalangodes armata Tellkampf, an eyeless opilionid from Mammoth Cave. The elongate second legs serve an antennalike function.

Detritus washed into caves through sinkholes is a very important food source. But, like guano, very abundant, fresh, detritus (logs, sticks, leaves) rarely attracts many troglobites, although it may be swarming with troglophiles and accidentals. Finely triturated or well-decomposed detritus can apparently be utilized much more efficiently. It is probable that troglobite detritus feeders use only decomposition products of wood and leaves, or perhaps the bacterial and fungal decomposers themselves. Troglobitic collembolans apparently feed rarely on decaying wood but commonly on the hyphae of polypores and other wood-rot fungi (see Christiansen, 1964). Isopods and amphipods on wet, rotting wood are probably feeding on the decomposers. Holsinger (1966) reports a huge population of asellid isopods and other small invertebrates in a polluted cave in southwestern Virginia. A septic tank drains periodically into this cave, the effluent flowing over dripstone and into small pools where the isopods congregate. Coliform bacteria were extremely abundant in this microhabitat and were probably eaten by isopods.

Palaemonias ganteri Hay (Fig. 15), a troglobitic atyid shrimp which is known only from Mammoth Cave National Park, occurs in shallow lakes in the Roaring River passage of Mammoth Cave. These lakes are situated about 3 m above low-water level and are replenished by the annual floods. The shrimp are bottom feeders, straining the fine bottom silt through their mouthparts. Microscopic examination of the silt reveals a moderately dense population of protozoans (*Paramecium, Peranema, Halteria, Phacus, Difflugia,* and so forth). The presumed food chain in the shrimp pools begins with fine bits of organic debris deposited by the flooding streams. The debris is decomposed by bacteria and fungi which supply food for protozoans, as well as isopods (*Asellus stygius* Packard) and amphipods (*Stygobromus exilis* Hubricht). The protozoans and other microorganisms are eaten by *Palaemonias*. Troglobitic crayfishes (*Orconectes pellucidus pellucidus* Tellkampf) and an occasional troglophilic crayfish (*Cambarus tenebrosus* Hay) are probably predatory on the smaller crustaceans. A few *Typhlichthys* inhabit the shrimp pools. They are probably stranded there by floodwaters, but it is not unlikely that they eat small isopods and amphipods.

Many species of troglobitic trechine beetles (*Pseudanophthalmus, Ameroduvalius*) appear to feed principally on minute tubificid and enchytraeid worms which burrow in the silt along cave streams. The beetles are commonly abundant among castings of the worms, and have been observed carrying partially eaten worms about in their mandibles. This food chain also begins with detritus and includes the decomposers, which constitute the food of the worms.

Fig. 15. Palaemonias ganteri Hay, a troglobitic atyid shrimp known only from Mammoth Cave and the Flint Ridge Cave system, Mammoth Cave National Park (Photo courtesy of R. W. Barbour).

Competition

Few troglobites appear to be strictly stenophagous. On the contrary, the limited data available indicate that most species take a wide range of food. Since food is a limiting factor in caves, this is by no means unexpected. In caves with a rich troglobitic fauna, the prevalence of euryphagy would normally lead to interspecific competition, but the coexistence of many species of detritus feeders and of predators in the same caves indicates at least partial niche exploitation.

The trechine beetles (Carabidae) offer a particularly appropriate case for investigation of competition. Although the majority of troglobitic trechines in eastern United States are allopatric, ranges of several species overlap in cave systems of the Mississippian plateaus of Indiana, Kentucky, and Tennessee, and in the Greenbrier River valley of West Virginia, where two, three, or four species commonly inhabit the same cave. The extreme situation occurs in the Mammoth Cave system, where *Neaphaenops tell-kampfii* (Fig. 16) coexists with *Pseudanophthalmus menetriesii, P. striatus, P. pubescens, P. audax,* and *inexpectatus* (Barr, 1967a). All six species have been found at the same spot, near the top of Mammoth Dome at a wet crevice which connects with the back part of White Cave via an opening too small for human penetration. Old, rotting timbers in this location provide the primary food source. *Neaphaenops* ranges widely through the system, including some of the drier passages, where it feeds on cave cricket eggs, but all the *Pseudanophthalmus* spp. are limited to wet areas.

P. *menetriesii* and P. *striatus* are so closely similar that Jeannel (Bolívar and Jeannel, 1931) found it difficult to believe that the two could possibly be sympatric. The genitalia of the males are virtually identical, a most unusual condition among trechine species. At one time I erroneously synonymized the two species (Barr, 1962b). However, P. *menetriesii* is abundant on rotten wood at all levels of the cave. P. *striatus* is extremely rare in the upper levels, but along the Styx and Echo rivers it outnumbers P. *menetriesii* 20 or 30 to 1. Apparently P. *striatus* is a tubificid feeder, and P. *menetriesii* feeds predominantly on small arthropods in rotting wood and other debris. Farther north, where the range of P. *menetriesii* overlaps that of P. *barberi,* the same sort of niche segregation is observed. It is interesting to note that P. *striatus* and P. *barberi,* although they belong to different species groups, are both primarily tubificid feeders. These two species are strictly allopatric despite lack of any obvious physical barrier separating their ranges. P. *menetriesii,* apparently not a regular tubificid feeder, has a range which overlaps that of both the other two species.

Although P. *striatus* and P. *menetriesii* have distinctly different feeding habits, the situation is less clear for P. *pubescens,* a species which belongs to a different species group. P. *pubescens* is found sporadically in Mammoth Cave, and is far more abundant in caves of the sinkhole plain to the south of the Mammoth Cave plateau. Normally P. *pubescens,* too, is a riparian tubificid feeder, but unlike P. *striatus,* occasionally moves into *menetriesii-* like habitats. In certain caves (e.g., Diamond Caverns, Walnut Hill Cave, and Cave City Cave) P. *menetriesii, P. pubescens,* and P. *striatus* are all relatively abundant in the same spot, and here their abundance distribution follows the broken stick model very closely (MacArthur, 1957). Scarcely enough is known of the life histories and feeding habits of any one of these three species to describe its niche in a wholly satisfactory manner. There are indications of partial niche segregation, but also of broad niche overlap.

The remaining two species of trechines in Mammoth Cave—P. *inexpectatus* and P. *audax*—are smaller in size, relatively rare, and seasonal in

Fig. 16. Neaphaenops tellkampfii Erichson, a common troglobitic beetle (Carabidae, Trechini) of the Pennyroyal plateau in Kentucky.

occurrence. At the Mammoth Dome—White Cave locality, *P. inexpectatus* is usually seen only in the winter following the first heavy seasonal rains (January, February, and March), and *P. audax* has been found only in late summer (August). All that can be said of these small species at the present time is that their ecological requirements appear to be quite different from those of the larger *Pseudanophthalmus*.

Among crayfishes, the strangely modified *Troglocambarus maclanei* (Fig. 17) (Hobbs, 1942) inhabits caves of north peninsular Florida with

Fig. 17. Troglocambarus maclanei *Hobbs, a small, unusually slender, troglobitic cray-fish from north-central Florida, crawls over ceilings of water-filled caverns (Photo courtesy of H. H. Hobbs, Jr.).*

the closely related *Procambarus lucifugus*. *Troglocambarus* rests and probably feeds on the ceiling of water-filled cave passages, thus exploiting a niche in which *Procambarus* offers no competition. Its slender, elongate appendages and small size are presumably direct adaptations to an unusual niche.

The blind cavefishes, *Amblyopsis spelaea* and *Typhlichthys subterraneus*, have almost entirely allopatric ranges, occurring together only in the subterranean streams of Mammoth Cave National Park. Even in Mammoth Cave they are not seen in the same pools. If the two species are confined in the same aquarium the fins of the *Amblyopsis* soon become frayed, possibly because of attacks by the *Typhlichthys*. Poulson (1963) has shown that *Amblyopsis* is in many ways better adapted to cave life than *Typhlichthys*, and for this reason he believes that the former has occupied caves for a much longer period of time. Conceivably the southward dispersal of *Amblyopsis* has been effectively blocked by competition with a more active, though ultimately less well-adapted species of similar habits (Woods and Inger, 1957). Both *Typhlichthys* and *Amblyopsis* may outcompete a troglophile amblyopsid, *Chologaster agassizi*, which never seems to occupy the same cave streams as its pale, blind relatives (Poulson, 1963; Barr, unpublished observations).

Circadian Rhythms in Cavernicoles

As might be expected, pronounced circadian rhythms can be demonstrated in trogloxenes, which remain inactive in caves during the day, but feed outside at night. Menaker (1959), for example, reported endogenous rhythms in body temperatures of hibernating bats, and Reichle, Palmer, and Park (1965) found a nocturnal activity pattern in *Hadenoecus subterraneus*, the common cave cricket of the Mammoth Cave area.

With troglobites, however, the few experiments which have been performed suggest absence of definitive diurnal activity rhythms. Park, Roberts, and Harris (1941) found no circadian rhythm in the activity of a cave crayfish, *Orconectes pellucidus*, although by averaging their data over 24-hr periods they excluded the possibility of discovering rhythms of longer period.[2] Specimens kept in constant darkness and at constant temperature

---

[2] Brown's (1961) paper reporting discovery of a 24-hr rhythm from reconsideration of these data is unconvincing. His graph is constructed from five points, each obtained by averaging activity percentages for three consecutive hours. One point is the lowest possible 3-hr average, another is the highest possible, and the other three are intermediate. Data for nine hours were ignored.. Poulson's (1964) recalculation of the same data led him to suggest a 26-hr rhythm, although the manner in which the data were presented does not permit critical analysis for periods longer than 24 hr. Clearly there is need for further experimentation with *O. pellucidus* if information is desired beyond what Park, Roberts, and Harris (1941) set out to determine.

showed an average hourly activity of about 4.2 percent of total activity over a 24-hr period, with minimum and maximum of 2.8 and 7.0 percent respectively, and a slight peak between 6 P.M. and 8 P.M. (Fig. 18). The curve of standard deviations of hourly activity generally closely parallels that of hourly activity; standard deviations are quite high, with a mean coefficient of variability which, from recalculation of the data, I estimate at 63 percent. Blume, Bünning, and Günzler (1962), working with the European cave amphipod *Niphargus puteanus,* found no diurnal activity rhythm in the number of times an amphipod swam past a beam of dark red light. The highly irregular activity periods observed by these investigators ranged from 10 to 57 hr. Ginet (1960) reports arrhythmicity in *Niphargus orcinus virei,* which during a 24-hr observation period is active for 5 to 6 hr (total activity time) and at repose for 18 to 19 hr.

Park (1956) studied frequency of movement of a cave pselaphid beetle, *Batrisodes valentinei.* Specimens were maintained on moist, sectored filter paper in Petri dishes, and their position in the sectors was noted at 2-hr intervals for several days. No rhythmic pattern was discovered. Deleurance-

Fig. 18. Activity of a troglobitic crayfish, Orconectes pellucidus *Tellkampf; average of ten 24-hr activity records in constant darkness at constant temperature (Redrawn from Park, Roberts, and Harris, 1941).*

Glaçon (1963a) investigated activity in the blind catopid beetle *Speonomus diecki,* using a modified Chauvin microactograph. During each 24-hr period the beetles showed five to ten brief activity periods (the number varied markedly with the ambient temperature) separated by rest periods of about the same duration. Males showed more and shorter periods of activity than females.

Repetition of some of these experiments and their extension to a wider taxonomic range of troglobites would certainly be desirable. If biological clocks are regulated by extrinsic geophysical forces (see Brown, 1961, 1965), distinct rhythms should be demonstrable in troglobites. These species have in most cases lived in perpetual darkness for hundreds of thousands of years. Some, at least, seem to have lost any vestige of a functional photoreceptor. Nevertheless, they are exposed to all of the same geophysical and extraterrestrial forces, except for most incident radiation (only attenuated cosmic rays penetrate caves), that are present in environments at the surface. If, on the other hand, circadian rhythms are endogenous, genetically controlled adaptations to the periodic environmental phenomenon of alternating light and darkness, we can expect to find that in cavernicoles such rhythms undergo regressive evolutionary changes just as eyes and pigment regress. We can expect to find rhythms of circadian periodicity present in trogloxenes and even in troglophiles, although in the latter the rhythms would probably be free-running in the absence of an LD (light-darkness) cycle to set the clock. In troglobites we can expect to find wide variability in periods of activity and considerable phenotypic plasticity in the response of troglobitic species to entrainment under an artificially imposed LD regime. Finally, there should exist species of troglobites whose activity shows no vestige of circadian rhythmicity.

Effects of Geologic Structure

The cave systems of the faulted and folded limestones in the Appalachian valley of southwestern Virginia and eastern Tennessee are ecologically very different from caves of the widely exposed, relatively undisturbed Mississippian limestones of the Interior Low Plateaus in southern Indiana, central Kentucky and Tennessee, and northern Alabama (Fig. 19) (Barr, 1967b). The differences are especially marked in the trechine beetles, which have many species in both regions (Barr, 1965). In the Appalachian valley, limestones are exposed in long, linear anticlinal valleys which are separated by synclinal ridges of sandstones, shales, and other clastics, restricting dispersal of troglobites through subterranean routes. In the Mississippian plateaus, caverniferous limestones are exposed over broad areas, and troglobite dispersal is correspondingly enhanced.

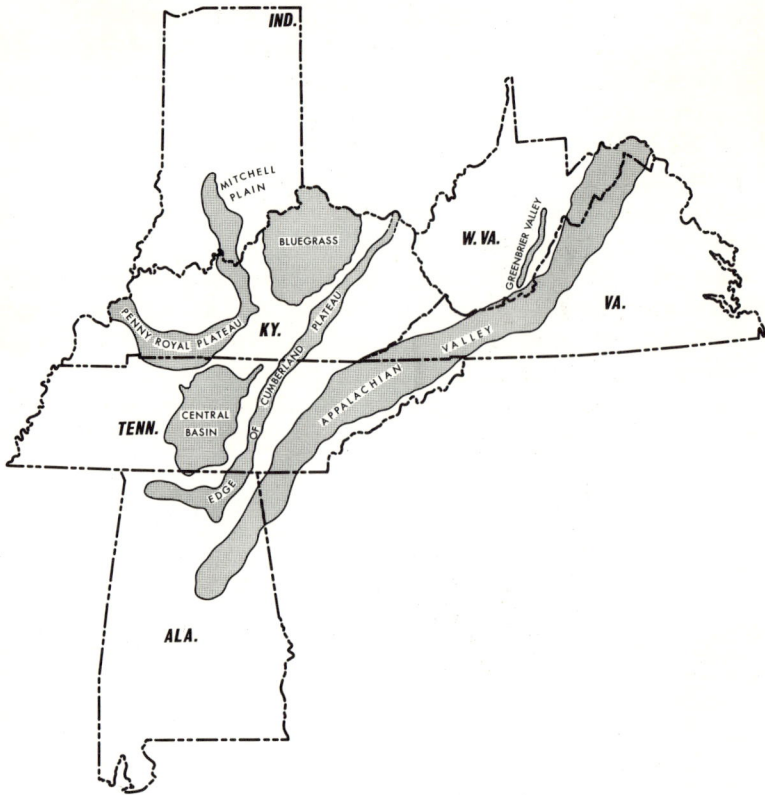

Fig. 19. Major cave regions of eastern United States. The greatest diversity of troglobitic faunas occurs in the Mitchell plain, Pennyroyal plateau, Cumberland plateau margin, and Greenbrier valley (From Barr, 1967b).

The ecological effects of reduced dispersal potential in cave areas of the Appalachian valley, compared to the Mississippian plateaus, are: (1) there are more species per unit area; (2) the species have much smaller geographic ranges; (3) sympatry of closely related species is very rare, although it is common in the Mississippian plateaus; (4) species diversity is much lower; (5) population density of troglobites is generally very low, and (6) modal size of trechines and certain other groups is lower. Species density, geographic distribution, and sympatry are readily explained by the greater isolation imposed in the Appalachian valley. Species diversity is related to the opportunity afforded (by extensive, open networks in the Mississippian plateau limestones) for dispersal into the same caves of many species of troglobites which have colonized caves at different times and in different localities by multiple invasion. This creates more complex cave communities in which drastic population fluctuations are relatively uncommon, unlike the simple communities of Appalachian caves. Greater modal size may result from broader niches, because the environment presents more elements from which to choose.

If we assume that the probability of successful colonization of a cave by a species of troglophile (see section on Speciation and Adaptation in Troglobites) is rather low, the sparse fauna of the Appalachian valley caves is not surprising. In very small limestone regions it is highly improbable that we would find any troglobites at all. Diverse troglobitic faunas may be anticipated in extensive karst regions, according to this theory, provided that, in the geologic past, the regions were exposed to alternating climatic or other major environmental changes (see pages 83 to 86). An increase in the number of highly local endemic species, relative rarity, and uncommon sympatry may be expected in regions of complex geologic structure which have produced many geographically isolated patches of caverniferous limestone.

Regressive Evolution in Cave Animals

Loss or degeneration of eyes and pigment in cavernicoles have been favorite examples for authors writing on the subject of regression of structures. Evolutionary modifications of this nature have been variously called regressive evolution (Lamarck, Weismann, and others), degenerative evolution (Eigenmann, 1909), rudimentation (Dobzhansky, 1951), and structural reduction (Brace, 1963, Prout, 1964). Of these terms, "structural reduction" is least desirable because not only structures but also physiological and behavioral patterns are subject to modification.

Literature references to cavernicole evolution are legion, although the number of papers by evolutionists who have actually visited a cave and studied cavernicoles in situ is manageably finite. The principal theories are listed in Table 1. The suggested explanations include almost all of the general theories that have been offered to explain regression, with the apparent exception of "meiotic drive," and at least two special theories primarily applicable to cavernicoles.

Lamarckian Theories

It is interesting to speculate whether Lamarck might not have become the first biospeleologist, had he not been professionally discredited by Cuvier, or had it not been for his failing eyesight. In the caves of southern France he would have found an as yet undiscovered, exceptionally rich troglobitic fauna which would have provided him with abundant illustrations of the effects of disuse. In any case, Lamarck made good use of the example of

Table 1. Theories of Regressive Evolution in Cavernicoles.

I. *Lamarckian*
  A. Classical Lamarckism (Lamarck, 1809)
  B. Neolamarckism (Packard, Cope, Racovitza, Jeannel)
II. *Orthogenetic*
  A. Orthogenesis (Jeannel)
  B. Organicism (Vandel, 1964)
III. *Darwinian*
  A. Direct selection of obscurely adaptive characters
  B. Material compensation (economy of growth energy)
  C. Escape theory (Lankester, 1893)
  D. Trap theory (Ludwig, 1942)
  E. Accumulation of random mutations
  F. Indirect effect of pleiotropy
  G. Genetic drift (sampling error)
  H. Negative allometry (Heuts, 1951)

eye degeneration in the only troglobite known to him, the Yugoslavian olm (*Proteus anguinus* Laurenti), a neotene salamander.

The *proteus,* an aquatic reptile, near the salamanders in its affinities, which inhabits the deep and dark cavities which are under the waters, has no more . . . than vestiges of the organ of sight; vestiges which are covered and hidden. . . . Here is a decisive consideration. . . . Light does not penetrate everywhere, consequently animals which habitually live in places where it does not reach lack the opportunity to exercise the organ of sight, if nature provided them with one. Animals which have a plan or organization in which eyes are necessary should have had them originally. Nevertheless, when one finds among them some who are deprived of the use of this organ and which have no more than hidden and covered vestiges of eyes, it becomes apparent that the decline and even the disappearance of the organ in question are the results, for this organ, of a constant lack of exercise.

Lamarck, 1809, I: 242

Charles Darwin himself fell back on an essentially Lamarckian view of troglobite eye loss:

It is well known that several animals, belonging to the most different classes, which inhabit the caves of Styria and Kentucky, are blind. . . . As it is difficult to imagine that eyes, though useless, could be in any way injurious to animals living in darkness, I attribute their loss wholly to disuse. . . . On my view we must suppose that American animals, having ordinary powers of vision, slowly migrated by successive generations from the outer world into the deeper and deeper recesses of the Kentucky caves, as did European animals into the caves of Europe. . . . By the time that an animal had reached, after numberless generations, the deepest recesses, disuse will on this view have more or less perfectly obliterated its eyes, and natural selection will often have effected other changes, such as an increase in the length of the antennae or palpi, as a compensation for blindness.

Darwin, 1859, ch. 5, pp. 137–138

A few pages later, in the same chapter, Darwin discussed at length the possible effects of compensation and economy of growth, relaxed selection, and correlation of parts, thus restating the theory of material compensation and anticipating the theories of accumulation of random mutations and indirect effect of pleiotropy (see subsequent discussion).

Failure of the selectionists to account for regressive aspects of evolution in a wholly satisfactory manner loomed large among factors which led to the rise of Neolamarckism in the last quarter of the 19th century. Biospeleology benefited enormously from the Neolamarckists (in basic descriptive data if not in theory), who discovered and collected large numbers of cavernicoles on both sides of the Atlantic. In America, E. D. Cope (e.g., 1887) and especially A. S. Packard, Jr. (1888, 1894) interpreted troglobite evolution in terms of inheritance of acquired characters, and were followed by Carl Eigenmann (1909) and A. M. Banta (1907) in the early twentieth century. In Europe, Emil G. Racovitza sought evidence for his Neolamarckian convictions among the rich and varied Mediterranean and Balkan cave faunas. His carefully prepared plan of attack, "Essai sur les Problèmes Biospéologiques" (1907) touches upon most of the significant aspects of cave biology and is still well worth reading. Racovitza's young protegé, René Jeannel, helped him launch the ambitious and fantastically productive "Biospeologica" program, which resulted in the collection and description of cavernicoles from over a thousand caves of Europe and Africa. Jeannel remained a Neolamarckist through most of his life (see Jeannel, 1950). Among thousands of pages of papers, chiefly on beetles, his chief contributions to biospeleology are his revision of the bathysciines (Catopidae) (1911), a book on the cavernicole fauna of France (1926), which includes many European troglobites, and his masterful, 1,800-page *Monographie des Trechinae* (1926–1930), a detailed treatment of a group of carabid beetles which are more widely prevalent and include more troglobitic species than any other comparable taxon.

Orthogenetic Theories

Classical orthogenesis has been evoked by various biospeleologists, notably Jeannel (1943, 1950) to explain regression in cavernicoles. Most trechine beetles are restricted to cool, humid environments. Some species have become edaphobites, with loss of melanin pigmentation and rudimentation of eyes. Many species are troglobites, with similar regressive modifications. According to Jeannel, loss of wings, pigment, and eyes, as well as the ability to tolerate warmer temperatures and low relative humidity, are all part of a *déroulement* inherent in the germ plasm of the trechines themselves. Cave beetles, by this interpretation, are not blind because they live in caves, but live in caves because they are blind and cannot survive elsewhere! Since Jeannel was also a proponent of Neolamarckism and consequently stressed the direct effect of the environment on the hereditary potential of an organism, he was led into more than one logical inconsistency.

A similar stress on *déroulement* is found in the recent book by Vandel (1964), whose principle of "organicism" rejects both environment and selection in favor of an unfolding pattern of phylogeny within organisms themselves. Strongly influenced by the paleontologist Schindewolf, Vandel points to a cycle of stages in the evolution of a phyletic line—rejuvenation, adaptive radiation, specialization, and phyletic senescence. A group of animals evolving thus passes through stages analogous to the life cycle of an organism—youth, maturity, and old age. Most troglobites belong to senescent groups, Vandel argues, and their rudimentations are the result of the terminal stage at which their respective phyletic lines have arrived. With senescence comes loss of the capacity for autoregulation and increasing restriction to specialized environments (e.g., caves).

Organicism in itself, of course, provides no causal explanation for the mechanism of regressive evolution, but instead postulates some unknown directive force generically similar, if not conceptually identical, to Bergson's *élan vitale*. Furthermore, for each supposedly "senescent" group of animals which includes troglobites among its species, another group can be cited which by no stretch of the imagination can be considered a "senescent" phyletic line. Most of the troglobitic crayfishes of the eastern United States belong to the large, widely distributed, actively speciating genera *Cambarus, Procambarus,* and *Orconectes.* The troglobitic catopid beetles of Europe (bathysciines) outnumber epigean species of the family, but the troglobitic catopids of Illinois, Kentucky, Tennessee, Alabama, and Georgia belong to the large, widely distributed genus *Ptomaphagus.* No one can dispute the existence of a large number of remarkable relicts in caves, but the equally numerous troglobites closely related to genera containing abundant, successful, epigean species which occur over broad geographic areas clearly show that phyletic senescence is not an indispensable prerequisite for existence in caves.

Darwinian Theories

These theories all have in common an acceptance of the validity and pervasiveness of natural selection in organic evolution, and cover a broad spectrum ranging from awkward apologisms to far more plausible schemes which take into account the discoveries of population genetics.

Direct selection against a character whose nonadaptiveness is only apparent has frequently been suggested as a cause of rapid regression of that character. Darwin, for example, speculated that the eyes of moles and other subterranean animals might be injured as they burrowed through the soil, and would consequently be reduced by selection. Others extended this idea to the supposition that animals might bump into the walls of the cave in the dark and thus injure their eyes (if they had them). The presence of many troglophiles in caves, as well as trogloxenes and even accidentals, with well-developed, undamaged eyes easily invalidates such an argument. Direct selection is usually very easy to suggest but difficult to demonstrate.

The principle of *material compensation* (economy of growth energy) arose from the natural philosophy of Goethe and from Geoffroy-St. Hilaire's "loi de balancement." It was given definitive form by Darwin ("Origin of Species," chap. 5), elaborated in considerable detail by Roux (1881), and briefly restated by Weismann (1889). Cope (1887) adopted the concept with but slight modification, calling it "growth force." The theory has recently been revived by Rensch (1959), whose arguments closely follow Darwin's, and by Poulson (1964). According to this theory, the different parts of an embryo compete with each other for growth substances (or for "energy"), and the more successful parts grow more rapidly and are correspondingly more highly developed. There are three main lines of evidence:

1. The anlagen of vestigial organs often appear in early stages, and it is assumed that the organs would in fact develop if the more rapid growth rates of other structures did not consume more energy and consequently prevent them from doing so.

2. Structures undergoing regression show increased variability because relaxation of selection favors preservation of those structures. "Variability" as applied to cavernicoles has often meant that eye and pigment development exhibit a wide norm of reaction, rather than a high degree of heterozygosity. Considerable phenotypic plasticity is exhibited in the pigment of certain crustaceans, which are pale when carotenoids are absent from their diet (as in caves) (Baldwin and Beatty, 1941; Beatty, 1941, 1949; Maguire, 1961), and in troglobitic salamanders, which develop a purplish color when reared in the light (for *Typhlotriton,* see Noble and Pope, 1928; Bishop, 1944; Stone, 1964a, 1964b; for *Proteus,* see Configliachi and Rusconi, 1819; Kammerer, 1912; Kosswig, 1937; Vandel and Bouillon, 1959; Vandel, 1961; Durand, 1964). Eye and pigment variability seen in cavernicolous forms of asellid isopods is apparently polygenic (de Lattin, 1939; Kosswig and Kosswig, 1940).

3. The unequal reduction of certain organs in males and females of the same species is taken as further evidence that ontogenetic competition of parts for body material is important in regression. Packard (1888) cited the curious case of sexual dimorphism in pselaphid beetles, apparently not known to Darwin or Rensch, in which some of the species have relatively large and well developed eyes in the males (in addition to antennal characters) but greatly reduced eyes in females. In cavernicolous *Batrisodes* spp. (Fig. 20) of eastern United States, recent taxonomic study has shown

Fig. 20. Batrisodes henroti Park, a pselaphid beetle from the Mammoth Cave system. The eyes of the male (illustrated here) are much larger than in the female.

that eye dimorphism is prevalent in almost all the species. *B. pannosus,* however, has two types of males; one has large eyes and the usual male antennal characters ("strong" form), the other has small eyes and no special antennal characters ("weak" form) (Park, 1951, 1956, 1960). Eyes and secondary sexual characters have been completely lost in the pselaphid genera *Arianops* (deep soil and cave species) and *Speleochus* (caves only).

As Darwin has summarized the theory, ". . . natural selection is continually trying to economize every part of the organization. If under changed conditions of life a structure, before useful, becomes less useful, its diminution will be favored, for it will profit the individual not to have its nutriment wasted in building up a useless structure." ("Origin of Species," chap. 5)

The *escape theory* of Lankester (1893) attributes the presence of eyeless animals in caves to a process of concentration of individuals which, having more or less degenerate eyes, were unable to find their way to the surface again when they accidentally fell or were washed into caves. The theory is primarily of historical interest because the accompanying polemics involved Herbert Spencer (1893) and others. In total darkness it is difficult to imagine how the presence or absence of eyes would be of any use at all to an animal attempting to find its way out of a cave. Most cave accidentals simply do not survive. It is furthermore extremely improbable that a sufficiently large enough number of individuals would be washed into the same cave at the same time to insure the successful colonization of the cave. When the additional probability that all or most of this group of castaways would have defective eyes is taken into account, the theory now appears even more untenable than it apparently did to Spencer. The unequal distribution of troglobites in a relatively small number of orders and tribes is very strong evidence that preadaptation is far more important than accident in the initial colonization process (see page 81).

A more sophisticated modification of the escape theory is Ludwig's (1942) *trap theory* (Fallentheorie), in which it is conjectured that paler individuals of a troglophilic species are more sensitive to light and consequently negatively phototactic. The paler individuals would tend to concentrate in caves. Janzer and Ludwig (1952) studied phototactic responses in the isopod *Asellus aquaticus,* reporting experimental evidence that the paler individuals are indeed more photophobic. However, their work also shows that during the breeding season both light and dark forms are photophilic. Presumably both light and dark isopods would interbreed at the mouths of caves and springs under these circumstances, consequently it is difficult to see how differential photophobia can act as a genetically effective mechanism in the evolution of troglobitic *Asellus*.

The theory that regression of eyes and pigment may result from *accumulation of random mutations* stems directly from proposals by Darwin (1859) and Weismann (1889) that natural selection might not affect such characters in the absence of light, where variants with more or less degenerate eyes and pigment would increase. Weismann named this process "panmixia," but pointed out that rudimentary structures would not completely disappear without the action of natural selection, citing Roux's (1881) version of the material compensation theory as the most likely manner in which selection would operate. In other (i.e., modern) words, mutation pressure for production of individuals with more or less degenerate eyes is no longer balanced by a corresponding component of selection pressure when a lucicole becomes a cavernicole. Essentially similar views have been advanced by Hubbs (1938), de Lattin (1939), and Brace (1963). Deamer (1964), using the Shannon information entropy equation, attributed regression to loss of information in the arrangement of base pairing in the DNA helix, which is merely another way of stating the theory.

This theory is attractive because of its inherent simplicity, but two major objections may be leveled against it in its unmodified form. The first is that it would take a rather long time for mutation pressure, acting alone, to fix a recessive allele within a population. Hubbs (1938) calculated that a recessive mutant allele which causes eye loss when homozygous would require 100,000 generations effectively to replace its normal allele, assuming a mutation rate of 10^{-3}, no increase in population size, and relaxation of selection favoring eye retention. Ludwig (1942), using a slightly different method and assuming a mutation rate of 10^{-5}, concluded that it would require a million or more generations. Since a number of genes are apparently involved in eye loss (see Kosswig, 1965) and their mutation rates will presumably vary (most rates probably much lower than 10^{-3}), the process would almost certainly take much longer. Many troglobites apparently colonized caves during the Pleistocene, some of them at least as late as the Sangamon (Riss-Würm) interglacial, about 300,000 years ago. The time available for regression via simple accumulation of random mutations might well be sufficient for fixation of at least a few recessives, but probably not for the sweeping genetic reorganization which most troglobites seem to have undergone.

A far more serious objection arises from the increasing amount of evidence that epigean and troglobitic species differ in a very large number of alleles, even when the cave species and its conjectured surface ancestor have not yet established isolating mechanisms. Because most troglobites are relicts, with no close relatives, the choice of populations which might be employed for genetic analysis of regressing characters is limited. Still further restrictions are imposed by the difficulty of rearing and maintaining cavernicoles under controlled conditions. Mexican fishes of the genus *Astyanax* (Characidae) and their troglobitic relatives (*Anoptichthys*) breed readily in the laboratory. Various pale, more or less eyeless cave populations, described as species by Hubbs and Innes (1936) and Álvarez (1946, 1947) produce fertile hybrids when crossed with *Astyanax mexicanus* Filippi, a species with eyes and pigment. The F_1 fishes are intermediate with respect to eyes and pigment degeneration, and great variability is exhibited in the F_2 (Sadoğlu, 1957), indicating multifactorial inheritance. Kosswig and Kosswig (1940) have shown that at least five loci are involved in pigment production in the isopod *Asellus aquaticus,* which has pigmented epigean and pale hypogean (*A. a. cavernicolus*) populations. Preliminary results of crossing lightly pigmented and unpigmented cave populations of the troglobitic collembolan *Pseudosinella hirsuta* (southeastern United States) suggest that inheritance of pigment in this species is polygenic (Christiansen, in litt.).

Even though the principal morphological differences cited between *Astyanax mexicanus* and *Anoptichthys jordani* are degeneration of eyes and pigment in the latter, Rasquin and Rosenbloom (1954) have documented numerous physiological, developmental, and reproductive disorders which result when *A. mexicanus* is maintained for long periods in the dark. In the same experimental group of fish, Atz (1953) demonstrated an inversion of the basophil-acidophil ratio in the transitional lobe of the pituitary, using the PAS and Gomori techniques. Poulson (1963) noted similar disorders in a dark-reared, swamp-dwelling amblyopsid fish, *Chologaster cornutus,* but found that the troglophilic *C. agassizi* did not exhibit endocrine imbalance in the dark. These endocrine disorders are possibly associated with disruption of the biological clocks of epigean species, and bear certain similarities to pathological effects observed in other animals kept in continuous darkness for prolonged periods (see review of Bünning, 1964, chap. 15). Vandel (1961) attributes eye and pigment regression in *Proteus* to hypothyroidism induced by failure of light stimulation of the pituitary. In any case, these data add to the view that the successful colonization of a cave by an epigean species may require a complete breakdown and overhauling of the epigenotype.

The *indirect effect of pleiotropy* as a causal mechanism for regression of characters was anticipated by Darwin in his discussion of correlation. In some unknown way, he felt that variation in an adaptively unimportant character could be influenced indirectly by natural selection, because it was "correlated" with another character with a high positive adaptive value. Many, if not all genes are pleiotropic. The net adaptive value of a pleiotropic allele will be the algebraic sum of its various phenotypic expressions (cf. Wright, 1929, 1964). In a lighted environment, a mutation causing reduction of eyes or pigment will be deleterious, because expression of the regressive character will be so disadvantageous that it will outweigh any possible advantages conferred by expression of its other effects. In an aphotic environment, however, the same reduction contributes little or nothing to fitness, and the adaptive value of the mutant allele will consequently be determined by its other effects. Pleiotropy and polygeny have been cited by Emerson (1949), Krekeler (1958), and Maguire (1961) as the most likely mechanism of relatively rapid regressive evolution in cavernicoles.

Since most cave populations are moderately small, it has occasionally been postulated that *genetic drift* would expedite the loss or fixation of alleles in cavernicolous species. Although it is easy to understand how sampling error in small populations would contribute to increasing homozygosity, which would in turn lead to stringent selection, it is difficult to see how genetic drift acting alone could produce a directed trend toward loss of eyes and pigment. In calculating the effect of drift in a hypothetical example, Ludwig (1942) concluded that this factor acting alone or even in combination with mutation pressure was inadequate to explain the origin of troglobite characters.

Negative allometry in growth of eyes and melanophores in the troglobitic barb, *Caecobarbus geertsi* Boulenger, from caves in the Congo, has been advanced by Heuts (1951, 1953a, 1953b) as an explanation for eye and pigment degeneration. Since eyes appear early in ontogeny and grow rapidly at first, Heuts believed that eye growth would be affected to an unusual degree by retardation of the overall body (absolute) growth rate. Melanophores, according to this author, appear early and do not increase in later life, a point disputed by Breder (1953). The low rate of growth in the Thysville caves inhabited by *Caecobarbus* was attributed to a small food supply. Breder (1953) pointed out that the caves in San Luís Potosí, Mexico, in which troglobitic characins have evolved (*Anoptichthys*), have large amounts of food in the form of flood detritus or bat guano, and that "practically identical depigmentation and eye loss has evidently here proceeded in the presence of abundant food."

Poikilothermic animals typically exhibit a broader norm of reaction in many developmental processes. The tendency of certain epigean species of fishes to have somewhat smaller eyes when reared at lower temperatures or on a reduced food supply is undoubtedly a manifestation of such phenotypic plasticity. The smaller eyes are functional and apparently normal in respects other than size (Breder, 1953), and it does not follow that negative allometry alone explains eye and pigment degeneration to the degree that these processes have occurred in troglobites. In any case, degeneration of these characters results from genetic control of development whether or not allometry is involved, and the negative allometry theory becomes merely a special case of the theory of accumulation of random mutations under relaxed selection.

Conceivably allometric growth may explain many of the exaggerated structures of troglobites, such as proportionately larger heads and longer fins in the more highly adapted cavefishes of the family Amblyopsidae (Poulson, 1963). Paralleling these morphological changes in amblyopsids are a decrease in metabolic rate and in absolute growth rate, presumably adaptations to a small and erratic food supply. The unusually elongate appendages of many troglobitic beetles are often most strikingly developed in the largest species of a group, a phenomenon at least partly the result of allometry. But careful biometric research on troglobites will probably eventually show that relative rates of growth for various structures exhibit patterns characteristic of the particular taxonomic category to which the troglobites belong, and we may expect different patterns in different groups.

Conclusions

What can be done with these numerous theories? In terms of population genetics, evolutionary change has occurred when there is a shift in the Hardy-Weinberg equilibrium, the causes of such shifts being normally reducible to selection pressure, mutation pressure, sampling error, and differential migration (Dobzhansky, 1951). As Ludwig (1942) and others have shown, neither mutation pressure nor sampling error, alone or in combination, appear very effective in the formulation of a satisfactory theory of regressive evolution in troglobites. The work of Janzer and Ludwig (1952) fails to establish the effectiveness of differential migration of paler individuals into caves, since both pigmented and unpigmented isopods are positively phototactic during the breeding season and would consequently interbreed, thus maintaining the equilibrium. We are left with the problem of discovering how selection may operate to expedite the degeneration and eventual disappearance of characters which are apparently neither advantageous nor disadvantageous. In the last analysis there is a choice between two theories, material compensation (developmental energy economy) and the indirect effect of pleiotropy. In my opinion, the pleiotropy theory is favored by presently available evidence. In the model of troglobite speciation presented on pages 86 to 88, the interaction of pleiotropy, polygeny, mutation pressure, sampling error, limited genetic potential, reduction of variability, and selection may all contribute to regression of eyes, pigment, and other characters. As discussed further on pages 89 to 96, troglobites show increasing levels of adaptation to the cave environment through energy economy in metabolism, behavior, and in producing fewer and larger eggs. It is not wholly inconceivable that energy economy may be selectively advantageous in development of the embryo.

For the future, it should ultimately be quite feasible to investigate both theories experimentally. How much energy is consumed by production of an eye, half an eye, or the mere rudiment of an eye? It is perhaps instructive to note that the Ozark cave salamander, *Typhlotriton spelaeus,* lives at the inner edge of the twilight zone during its larval stage, but moves to the recesses of the aphotic zone to undergo metamorphosis and live the remainder of its life in darkness. The larvae have functional eyes (Stone, 1964a), but at the time of metamorphosis a fold of skin grows over the eyes, which no longer function in the adult. Eyes are advantageous in the larval stage, but degenerate in the adult stage, where they are no longer advantageous. To what extent will heteroplastic transplants of eye anlagen from troglobite embryos develop into functional eyes in nontroglobite hosts? Stone (1964c) successfully interchanged right eyes between series of larvae of *T. spelaeus* and *Ambystoma punctatum* and found return of vision in both types of transplant, but did not carry his animals through to metamorphosis. What effect will removal of an eye anlage have on growth of other parts of embryos of cavernicoles? Will careful analyses of the genetics of troglobites reveal that one or more alleles contributing to eye degeneration also have phenotypic effects of obvious positive adaptive value? Definitive answers to these and similar questions will be critical to a final solution of the puzzle, which will quite probably draw on both embryology and genetics in its formulation.

Speciation and Adaptation in Troglobites

Preadaptation

The transition from life above ground to life in caves involves major morphological, physiological, and behavioral adjustments which few groups of animals have been able to achieve. Preadaptation to a cave existence has been stressed by most authors (for example, Jeannel, 1943; Vandel, 1964; Poulson, 1963; Barr, 1960a, 1965) cognizant of the peculiar ecological regimen of caves. The term was first employed by Davenport (1903). Cuénot's (1925) development of the "preadaptation theory" led to the curious notion that animals become randomly preadapted to all sorts of environments and can survive only in the environment to which they are preadapted. In modern usage, however, preadaptation has come to mean the development of characters which particularly suit a species or group of species for colonization of a certain habitat. In a broader view, especially from the time perspective of the paleontologist, preadaptation is "the existence of a prospective function prior to its realization" (Simpson, 1944). The concept is meaningless apart from specification of the mode of life to which a given phyletic line is supposedly preadapted. It implies only that the probability of successful adoption of a rather different sort of ecological niche is related to the morphological, physiological, and behavioral endowment of a phyletic line.

A rapid survey of the kinds of animals inhabiting caves strongly suggests that their presence there is not the result of purely random events. The troglobitic beetles of North America include representatives of four families, Carabidae (Agonini, Trechini), Catopidae (Ptomaphagini), Brathinidae, and Pselaphidae (Speleobamini, Bythini, Batrisini). Trechines, catopids, speleobamines, and batrisines are found not only in caves (principally in the Interior Low Plateaus and Appalachian valley), but also in moss, leaf litter, deep soil, and similar cool, moist, dark microhabitats at higher elevations in the mountains of east Tennessee, western North Carolina, and adjacent portions of Georgia, Virginia, Kentucky, and West Virginia. The tribe Speleobamini, in which certain basic characters of two major taxonomic categories of the Pselaphinae are united, includes two species, *Speleobama vana,* known only from a cave in Alabama, and *Prespelaea copelandi,* known only from leaf litter in the southern Appalachian mountains (Park, 1960). The troglobitic carabids of the Edwards plateau in Texas belong to the agonine genus *Rhadine* (Barr, 1960b), which includes a number of troglophilic species, species commonly inhabiting the burrows of mammals and owls (*Speotyto*), and species found under logs and in leaf litter at high elevations in western mountains. The staphylinoid family Brathinidae (in North America) includes three species of *Brathinus,* one of which, *B. nitidus,* is a common troglophile in eastern caves, and a single undescribed troglobitic species (R. L. Westcott, in litt.) from lava tubes of southeast Idaho.

Preadaptation in the amblyopsid fishes has been well documented by Poulson (1963). The family includes five species. *Chologaster cornutus* lives in swamps and rice fields in the southeast coastal plain of the United States, and its relative, *C. agassizi,* inhabits springs, sinkhole ponds, and cave streams in the Interior Low Plateaus. *Amblyopsis spelaea, A. rosae,* and *Typhlichthys subterraneus* are troglobitic species. Preadaptive characters in *Chologaster* include the ability to feed in darkness, negative phototaxis, positive rheotaxis, branchial incubation of the eggs, and the apparently associated anterior position of the cloacal opening. Rosen (1962) showed that the mechanism of opening the mouth in the amblyopsids would result "in a sudden very great increase in the orobranchial volume," presumably a trait preadaptive for feeding in a food-poor environment, since water and food organisms would be drawn into the mouth.

Colonization of Caves

The ancestors of troglobites have attained caves from three sorts of earlier habitats—soil, freshwater, and the ocean (Vandel, 1958). Most terrestrial troglobites were probably preadapted to cave life by earlier colonization of moss, leaf litter, or deep soil. The classic example is the case of the trechine carabid beetles, described in great detail by Jeannel (1926–1930; see also 1923). The trechines include (1) species with wings, eyes, and melanin pigment; (2) species without wings, eyes, and melanin; and (3) species without wings, but intermediate with respect to reduction of eyes and melanin. The eyeless species inhabit caves or deep soil, and the intermediate species occur in deeper humus layers, in wet talus piles, or sometimes in caves. Jeannel has interpreted the situation as the result of an orthogenetic trend leading inexorably to restriction to caves ("l'évolution souterraine"). However, the very large and widespread (Holarctic) genus *Trechus,* which certainly does not conform to such a trend, has few troglophilic and only one or two troglobitic species. Furthermore, the anilline carabids, which are all small, eyeless, depigmented edaphobites, are nearly worldwide in distribution, but very few of them are restricted to caves. Parallelism and convergence in the trechines is far more readily explained by preadaptation to a cool, moist environment, coupled with prolonged orthoselection under relatively stable ecological circumstances in deep soil or caves. Other groups of troglobites whose ancestors probably passed through a stage of living in humus, moss, or deep soil include catopid and pselaphid beetles, campodeid diplurans, collembolans, diplopods, pseudoscorpions, linyphiid spiders, and mites.

The freshwater groups which have invaded and successfully colonized caves include salamanders, fishes, various crustaceans, gastropods, and triclads, as well as a number of taxa of lesser numerical importance. A surprisingly large proportion of aquatic troglobites belong to predominantly marine groups, for example, desmoscolecid nematodes, polychaete annelids, cirolanid and sphaeromid isopods, a number of amphipods (*Niphargus, Ingolfiella*), a Mexican cave crab, and brotulid fishes (Vandel, 1964). Ancestors of these animals are thought to have entered interstitial crevices in limestone during a period of marine invasion and to have gradually been isolated in freshwater environments by later withdrawal of the sea. The tolerance of *Niphargus* (Amphipoda) and *Caecosphaeroma* (Isopoda) to brackish water, even though salt regulation increases the respiratory rate, is explained on this historical basis (Dresco-Derouet, 1952, 1959). At least two marine troglobites are known, *Heteromysis cotti,* a mysidacean;

and *Munidopsis polymorpha,* a galatheid. Both species inhabit the Jameo de Agua, a subterranean saltwater lake in the Canary Islands (Calman, 1904, 1932).

Since entrance into caves of animals ancestral to troglobites seems to have been non-accidental, but related to preadaptation, the evolution of troglobites almost certainly involves a preliminary, troglophile stage. de Lattin (1939) has written that troglophiles are troglobites "in statu nascendi." Troglophiles (see page 43) are facultative cavernicoles, able to exist either in caves or in similar dark, moist microhabitats outside of caves. Since their dispersal is not dependent upon underground channels through the limestone, their geographic distribution is not necessarily controlled by the extent of karst regions. Troglophiles do not differ taxonomically from members of the same species in epigean localities, and are not isolated in caves. Any shift in the gene pool of a cave population of a troglophile would be swamped periodically by occasional entrance of epigean individuals into the caves (see Packard, 1894). The mere fact that a given cave has been occupied by a troglophilic species consequently implies neither isolation nor incipient trogloby, unless sympatric speciation is postulated (see Mayr, 1942, 1963, for cogent arguments against sympatric speciation).

In marshalling evidence with which to construct a general theory of speciation of troglobites, three further observations must be included. (1) A large number of species (and higher categories) of troglobites are relicts, their presumed ancestors being extinct or surviving only in areas geographically remote from the cave region. (2) Troglobites are far more abundant in temperate zones than in the tropics, the relatively few tropical troglobites known being almost exclusively aquatic.[3] (3) Closely related species of troglobites are commonly found in different cave systems in the same geographic region, suggesting descent from a common ancestor which inhabited caves of the region as a troglophile. For example, trechine beetles of the *hubbardi* superspecies of *Pseudanophthalmus* inhabit caves of the northern Shenandoah Valley and upper Potomac basin (Barr, 1965).

---

[3] Poulson (1964) is mistaken when he states that absence of troglobites in the tropics is peculiar to trechines and pselaphids. Terrestrial troglobites are almost completely absent from tropical areas which have been carefully investigated (see Leleup, 1956), although trogloxenes and troglophiles are often exceptionally abundant (Vandel, 1964; Jeannel, 1965). There are unquestionably a number of widely scattered aquatic troglobites in tropical and subtropical areas, but even here the species diversity does not approach that of temperate areas. Jeannel (1965) attributed the absence of terrestrial troglobites in the tropics to the "isothermic" climate which has prevailed there for millions of years, in contrast to the "anisothermic" conditions of the temperate zones, where cold, wet periods alternated with hot, dry periods and the progenitors of troglobites survived only in cave refugia.

Pseudanophthalmus spp. of the *robustus* superspecies (*robustus, farrelli, beaklei*) occupy contiguous but mutually exclusive ranges in the eastern Highland Rim and margin of the Cumberland plateau in Tennessee and the edge of Kentucky (Barr, 1962c).

It is generally accepted that caves of the temperate zones have served as refugia for ancestors of troglobites against the climatic vicissitudes of the Pleistocene. These ancestral species were in many cases cryophilic, becoming widespread in taiga-like forests which bordered advancing continental glaciers (Jeannel, 1923), but becoming extinct or surviving in caves as the glaciers retreated and the climate of the cave regions became warmer and drier. Extinction of the cryophilic fauna would not have been immediate, since some of the species could survive for hundreds or possibly a few thousands of years in deep forested ravines, around the mouths of caves and springs, in cool, moist gorges near the bases of waterfalls, and in other, similar microhabitats. Holdhaus (1933) has documented much of the evidence that the present distribution of European cave faunas reflects the influence of Pleistocene glaciation (see also Vandel, 1958, 1964).

Trechus schwarzi is a carabid beetle with at least two subspecies in the high Appalachian mountains of western North Carolina, and a third subspecies (*T. s. cumberlandus*) in the Cumberland plateau of Tennessee and Kentucky (Barr, 1962a). *T. s. cumberlandus* inhabits ravines, gorges, sinkholes, and caves. It is a moderately successful, though sporadic troglophile. The present spotty distribution of this beetle has been interpreted as the gradual restriction of a much wider range during Wisconsin glaciation. Although *Trechus* belongs to the same tribe of Carabidae as *Pseudanophthalmus,* it is not a direct ancestor. The present distribution of *T. schwarzi* nevertheless suggests a way in which ancestral *Pseudanophthalmus* could have spread westward from the Appalachian mountains during a glacial maximum.

A similar pattern of dispersal, followed by restriction to the southern Appalachian mountains and to cool, moist microhabitats at lower elevations along the ridges of the Appalachian valley, is exhibited by the polytypic species *Trechus hydropicus,* which ranges northward from the high mountains of western North Carolina (*T. h. beutenmulleri* and other subspecies) to Virginia, easternmost Kentucky, West Virginia, and possibly western Maryland. Various populations of this species appear to be recently isolated, possibly since the Wisconsin. The present geographic range of *T. hydropicus,* like that of *T. schwarzi,* may also be repeating a pattern of dispersal of ancestral *Pseudanophthalmus* from the high mountains of Tennessee and North Carolina. There is at least one troglophile population of *T. h. canus,* in southwestern Virginia.

The southern Appalachians may thus have served as a source of colonizing species during different glacial maxima. The species of the *intermedius* group of *Pseudanophthalmus* occupy caves along the western margin of the Cumberland plateau, an area approximately equivalent to the present geographic range of *T. s. cumberlandus* (Barr, 1962c). If the ancestral stock of this species group dispersed in a *Trechus schwarzi* pattern, and if increasing adaptation to the cave environment is a true index of the length of time since the initial colonization of the caves, as Poulson (1963, 1964) maintains, then the *intermedius* group seems to include representatives of two distinct invasions, the *robustus* superspecies being the more recent. The two highly adapted species of the closely similar genus *Nelsonites* may even represent a third (earliest) invasion of the Cumberland plateau caves by the same phyletic stock.

The freshwater shrimps of the family Atyidae are a predominantly tropical and subtropical group which includes about 20 cavernicolous species throughout the world. *Palaemonias ganteri* from the Mammoth Cave system in Kentucky and *P. alabamae* from the huge underground lake in Shelta Cave, Alabama, should probably be regarded as thermophilic elements in the troglobitic fauna of the eastern United States.

The troglobitic crayfishes of the genus *Orconectes* inhabit caves of southern Indiana, the Pennyroyal plateau of Kentucky, and the margin of the Cumberland plateau from Kentucky to Alabama. These species are most closely related to *O. limosus*, an inhabitant of sluggish backwaters in northeastern streams. The single factor most significantly operative in extinction of the ancestor of these species in the Interior Low Plateaus was probably the regional uplift of the area which occurred at the close of the Pliocene or beginning of the Pleistocene. Prior to uplift, the major rivers of this region, notably the Kentucky and the Cumberland, flowed in floodplain valleys of low gradient and broad meanders. The ecological conditions favored by *O. limosus* would have been widespread. Following uplift the rivers would have been transformed into more rapidly flowing streams of steeper gradient, carrying heavy loads of silt, sand, and gravel. If an ancestral species of *Orconectes* was already capable of a troglophilic existence, it could have taken refuge and survived in the slowly-moving, quieter streams of caves at the headwaters of tributaries in limestone regions. Three or four separate colonizations are postulated to account for the known troglobitic species of *Orconectes* and their present distribution (Hobbs and Barr, in preparation).

Ancestral crayfishes of the genus *Procambarus* may have survived in submarine freshwater springs during inundation of northern Florida by the sea. Under these conditions of isolation, troglobitic species developed (Hobbs, 1942). The troglobitic isopods of the family Cirolanidae in North America (Bowman, 1964) may have been isolated in caves filled with brackish water during gradual withdrawal of the sea following

Cenozoic marine invasions of the Gulf and Atlantic coastal plains. Similar origins have been postulated for other troglobites of undoubted marine ancestry (see Vandel, 1958, 1964, for a more extended discussion).

Genetic Changes Accompanying Troglobite Speciation

We have seen that cave populations of troglophilic species, the troglobites "in statu nascendi," are no more isolated from their epigean populations than the peripheral populations of a wholly epigean species. Any temporary shift in gene frequency in the local cave population is quite likely to be swamped by occasional interbreeding with other individuals wandering into the cave from the surface. If the caves, with their troglophilic populations inside, were somehow sealed so completely that entrance by surface individuals was rendered impossible (as suggested by Mayr, 1963), then the food supply to the cave community would be cut off and the population would not survive. How, then, has the period of isolation required by the theory of geographic speciation been achieved?

The high proportion of relict species among troglobites is a very important clue. The ancestral species at some time in the past became extinct, at least in the region of the caves. If the extinction process included only surface populations of the species, and if the factors leading to extinction continued for some time outside the caves, the cave populations would be ecologically and genetically isolated. Drastic, long-term environmental changes during the Pleistocene have already been described—alternating cold, moist and warm, dry climates; brackish ground water being replaced by fresh water; and the transformation of a slow, sluggish stream into a rapidly-flowing one. The surviving cave populations could now adapt to the immediate conditions of existence in the caves, and changes in gene frequency could proceed unimpeded by gene flow from the exterior.

The process is depicted in the accompanying two-stage diagram (Fig. 21). In A, interbreeding occurs not only between individuals of the immediate cave population but also between the cave population and the surface populations. In B, following extinction of the surface members of the species, the cave populations are isolated and breed only within the caves. Exchange of genes between populations in different caves is possible only if there are interconnecting passages or crevices through which dispersal by cavernicoles can take place.

The newly isolated cave populations fulfill all the prerequisites for initiation of the genetic revolution (classically described by Mayr, 1963) which accompanies multiplication of species. The process is diagrammatically shown in Fig. 22. At A the level of variability in the troglophilic species is high. Following extinction of epigean populations of the species, the cave colony at B possesses only a small fraction of this total variability.

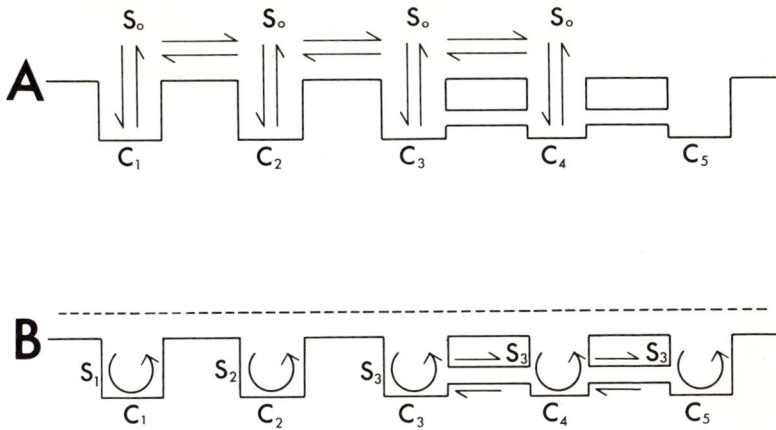

Fig. 21. Gene flow (A) in a troglophilic species S_0, and (B) in its descendants S_1–S_3 following extinction of S_0 in epigean environments. C_1–C_5 are caves; C_3, C_4, and C_5 are connected by subterranean channels, but C_1 and C_2 are isolated from C_3–C_5. C_5, originally unoccupied by S_0, is subsequently colonized by S_3 via subterranean channels (From Barr, 1967b).

Because the colony is of small to intermediate size, sampling error will increase homozygosity, leading to severe selection and an even greater decrease in variability. Like most peripheral isolates, the probability of survival of the colony at C must be rather low, otherwise the number of troglobitic species would be astronomical. If the colony survives, however, an extensive reorganization of the epigenotype takes place, adapting the colony more closely to the cave environment, and variability increases (D), attaining some new level E dependent on the abundance of the species throughout its geographic range.

Most of the more conspicuous regressive modifications of troglobites probably take place during this reconstruction of the epigenotype. With limited variability and strong selection, there may very well be a premium on the relatively few harmonious gene combinations available. This factor of limited genetic potential, coupled with pleiotropy and polygeny, sampling error, and severe selection, would lead to rather rapid evolution of characters of little or no selective advantage in caves. Loss or reduction of eyes, melanin and certain other pigments, wings (in insects), cuticular wax layers (in arthropods), a functional tracheal system (in insects), the ability to complete metamorphosis (in salamanders), and other characters would occur as incidental byproducts of the genetic revolution.

Certain problems arise in this view of troglobite speciation. Some of the regressive characters may have been acquired during the "period of preparation" (Vandel, 1964) for a cave existence, notably loss of wings in

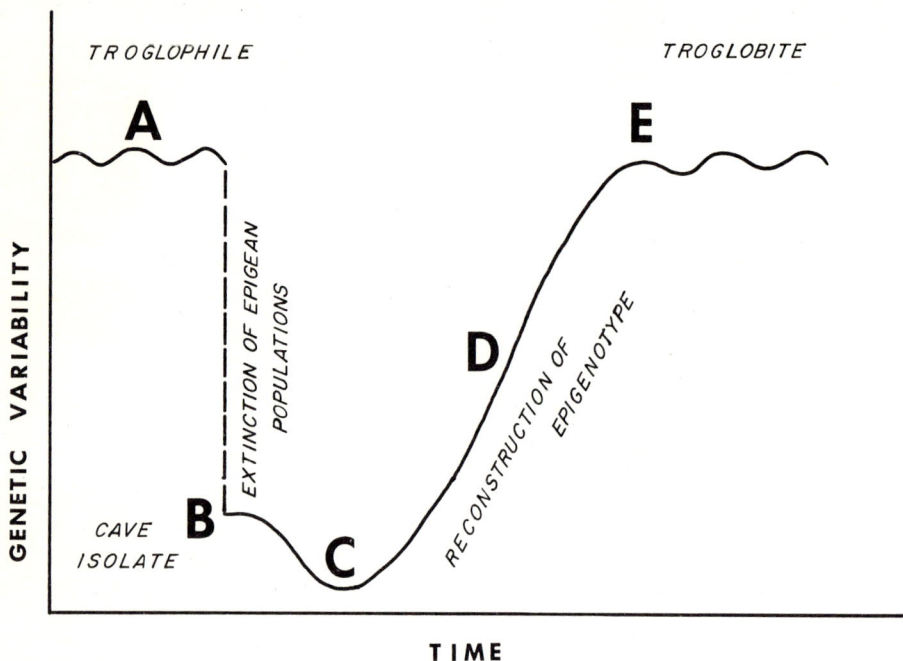

TROGLOPHILE   TROGLOBITE

A   E

GENETIC VARIABILITY

EXTINCTION OF EPIGEAN POPULATIONS

RECONSTRUCTION OF EPIGENOTYPE

D

CAVE   B
ISOLATE   C

TIME

Fig. 22. *Changes in genetic variability during troglobite speciation (Adapted from Mayr, 1963).*

insects, and perhaps even other characters in soil insects which eventually colonize caves. The same arguments presented above will apply whether the species is colonizing deep soil or caves. Aeppli (1952) has raised the possibility of species multiplication through autopolyploidy in cavernicolous flatworms of Central Europe, some species of which are apparently polyploid.

Two rather well described situations—each involving recent troglobites —may not exactly fit the process of troglobite speciation as described above, and further investigation would be most interesting. The first of these is the case of the cave isopod *Asellus aquaticus cavernicolus,* which includes a series of highly variable cave populations in Postojna Jama, Planina Jama, and other caves of the Piuka drainage in the Karst region of Yugoslavia. Variability of eyes and pigment has been described by de Lattin (1939) and the Kosswigs (1940). Apparently the cave populations are either a recent isolate of an epigean species, or are undergoing secondary hybridization (at least in Postojna Jama) with an epigean species.

The second interesting case is the group of characin fishes inhabiting caves of San Luís Potosí, Mexico (see page 77). These include several populations of troglobites (*Anoptichthys*) as well as an epigean species, *Astyanax mexicanus,* the supposed ancestor. In La Cueva Chica, *A. jordani* and *A. mexicanus* apparently hybridize near the exit of the cave stream into a surface river. If *A. mexicanus* is the ancestor of the troglobitic species, it has certainly not become extinct at the surface, and the nature of the isolation which must have occurred (barring a sympatric mechanism of speciation) has not been explained. These caves—la Cueva Chica, la Cueva de los Sabinos, el Sótano del Arroyo, el Sótano de la Tinaja, and la Cueva del Pachon—have all apparently developed by subterranean capture of surface streams on the west slope of the Sierra de El Abra, an anticlinal limestone ridge (Bonet, 1953). The cave streams emerge as Vauclusian springs at the base of the abrupt eastern fault escarpment of the Sierra, and—at least in their underground courses—do not appear to interconnect. Explored portions of the caves consist of horizontal galleries at successively lower levels, separated by abrupt waterfalls (J. R. Reddell, personal communication). If *Astyanax mexicanus* had lived in the west-slope streams prior to their capture, these fishes would have been carried underground along with the streams. Isolation could have taken place in the horizontal stretches, with high waterfalls and the siphons at the point of debouchement serving as barriers to prevent further invasions of *A. mexicanus* from the lower ends of the caves. Only in La Cueva Chica has erosional development of the cave permitted the ancestral species to come in contact with a cave population. The absence of isolating mechanisms and disruption of the extrinsic isolation has led to secondary hybridization between epigean and hypogean populations. Finally, Gordon and Rosen (1962) have described a peculiar, highly variable population of the fish *Poecilia sphenops* from a cave in the Arroyo del Solpho, near Tapijulapa, Tabasco, Mexico. Extreme variants are "almost colorless, have thick-lipped, terminal, horizontal mouths, and reduced eyes." These authors suggest that the population may represent intergrades between normal river fish and an undiscovered troglobitic species inhabiting the inaccessible upper reaches of the cave, but the possible effects of the cave water, described as saturated with calcium sulfate, on morphogenesis were not wholly excluded.

Adaptation to the Cave Environment

In most caves, adaptation to a decreased and erratic food supply has a high selective advantage. The reduction in respiratory intensity and a general decrease in activity present in many troglobites have been called "metabolic economy" by Poulson (1963, 1964). The standard and active metabolic rates of *Chologaster agassizi,* a troglophilic fish, are about two and a half times higher than the same rates in *Amblyopsis rosae,* a highly adapted troglobite (Poulson, 1963). Burbanck, Edwards, and Burbanck (1948) demonstrated that the troglobitic crayfish *Cambarus setosus* survived more than three times as long as the epigean *C. rusticus* when enclosed in sealed, one-liter jars of water under a layer of oil. Eberly (1960) found that oxygen consumption per gram of body weight in a troglophilic *Cambarus* (possibly *C. tenebrosus*) was 1.2 to 2.8 times that of a troglobitic *Orconectes* (probably *O. inermis*). Dresco-Derouet (1952, 1959) showed that oxygen consumption in the cave amphipod *Niphargus virei* is substantially lower than in the epigean *Gammarus pulex,* and obtained similar results comparing the troglobitic isopod *Caecosphaeroma burgundum* with a related marine species, *Sphaeroma serratum.*

The experiments of Schlagel and Breder (1947) showed that *Anoptichthys jordani* consumes slightly more oxygen per unit weight than its presumed ancestor, *Astyanax mexicanus.* However, this results from a behavioral difference associated with blindness, since both *A. jordani* and experimentally blinded *A. mexicanus* are more active than normal *mexicanus.* There is no apparent reduction of metabolic rate in the cave species, but all of the troglobitic fish populations in the San Luís Potosí caves have an abundant supply of food, either in the form of bat guano or logs and other debris washed into the caves by flooding surface streams (Breder, 1953; J. R. Reddell, personal communication).

Surprisingly little experimental evidence has been collected to support the frequent statements that troglobites have hypertrophied sensory receptors other than eyes to "compensate" for loss of vision. Light sensitivity has been reported in a number of troglobites with apparently nonfunctional eyes (Vandel, 1964). In caves of the eastern United States, negative phototaxis is most marked in troglobitic crayfishes (Park, Roberts, and Harris, 1941), which probably retain the function of the primitive light receptor in the sixth abdominal ganglion, and in the troglobitic salamander, *Gyrinophilus palleucus.* Trechine beetles are not uncommonly encountered in lighted areas in the bottom of sinkholes communicating with caves, provided that humidity and temperature conditions are not appreciably at variance with humidity and temperature of the deeper recesses, but the troglobitic species show no apparent phototaxis.

Behavioral and morphological data on amblyopsid fishes have been collected and evaluated by Poulson (1963, 1964). The neuromast cells in troglobitic amblyopsids do not sink into canals, but remain free and are surrounded by cupulae, a phenomenon which Poulson attributes to neoteny. In troglobitic species of this family, the neuromasts project farther out, have hypertrophied integrative centers, and are more sensitive to movements of a *Daphnia* than those of the troglophilic *Chologaster agassizi.* The hypertrophied cerebellums of troglobitic amblyopsids suggest better integration of sensory inputs from the lateral line, the otoliths, and the semi-circular canals.

A weak positive rheotaxis is exhibited in the troglobitic fishes *Typhlogarra widdowsoni* (Marshall and Thinés, 1958), *Typhlichthys,* and *Amblyopsis* (Poulson, 1963), and in the troglobitic salamanders *Haideotriton wallacei* (Carr, 1939), *Eurycea* (=*Typhlomolge*) *rathbuni* (Norman, 1900), and *Gyrinophilus palleucus* (Barr, unpublished). The large (to 33 mm) cave amphipod *Niphargus orcinus virei* is positively rheotactic but selects the weakest current in a gradient (Ginet, 1960). Positive rheotaxis would presumably prevent aquatic cavernicoles from accidentally swimming out of the caves, especially when combined with light sensitivity.

Disturbance in the water, particularly at the surface, is detected by several of the larger aquatic troglobites. Small disturbances result in a gradual approach to the source in *E. rathbuni* (Norman, 1900), *Haideotriton* (Carr, 1939), *Gyrinophilus palleucus* (Barr, unpublished), *Anoptichthys jordani* (J. R. Reddell, personal communication), and *Orconectes pellucidus* (Barr, unpublished). In all of these species, as well as in smaller crustaceans (*Asellus, Stygobromus, Crangonyx*), a major disturbance produces a flight reaction, with the escaping animal usually disappearing beneath a nearby rock or ledge. Gentle, repeated breaking of the surface tension of the water is sufficient to induce the approach of *O. pellucidus,* with a continuous waving of the antennae in the direction of the disturbance. If food is present at the source of disturbance, the chelae are suddenly flailed wildly upward and the food, whether it be a cave cricket, an isopod, or a bit of meat, is seized, dragged to the mouth, and slowly devoured (Barr, unpublished). Both *Orconectes* and *Anoptichthys* can be readily taken by hand if attracted by gentle disturbance at the water surface.

Although there have been no critical studies on the subject, most troglobites appear singularly insensitive to sounds and vibrations of low amplitude, although a sharp rap on an aquarium will usually produce a sudden response of brief duration. Many troglobites exhibit strongly thigmotactic behavior—the amphipods of the genus *Niphargus* (Ginet, 1960), most terrestrial troglobitic arthropods except when feeding, and amblyopsid fishes (Poulson, 1963). The long setae on the body and appendages of *Niphargus* are believed to be tactile. A similar function has been attributed to the "fouets," or whips, which are long setae on the elytra of cave trechine beetles. Jeannel (1926, 1943) has suggested that they may detect air currents, to which these insects are reputed to be highly sensitive. Certainly cold, dry air moving into the cave from the exterior would increase the rate of evaporation from the surface of an insect, and a negative response to airflow would appear to have selective value. Yet a number of individuals of three species of troglobitic trechines

(*Ameroduvalius jeanneli, Darlingtonea kentuckensis, Nelsonites jonesei*) have been observed feeding on tubificids and larvae of dipterans in a small "wind tunnel" in a Kentucky cave, where the airflow exceeded 40 m per min (Barr, unpublished). Humidity receptors in cave insects probably reside in the antennae, although unequivocal results of behavioral experiments have not yet been obtained. Little is known of chemoreception in troglobites, although a number of terrestrial species, notably beetles and millipedes, are attracted by cheese or decaying meat and presumably have well-developed chemoreceptors. Poulson (1963) believes that the non-random occurrence of *Amblyopsis spelaea* under rocks with greater concentrations of isopods is evidence of a well-developed olfactory sense in this species.

Chemical defense mechanisms in troglobitic arthropods have received little attention, but are apparently present in diplopods and beetles. The development of repugnatorial secretions in the trechines is highly variable. In the eastern United States many small species and at least one large species (*Pseudanophthalmus grandis*) emit objectionable odors when handled. Absence of such secretions may be only apparent in some species in the Mississippian plateaus; a faint odor was detectable in *Neaphaenops tellkampfii* only when a hundred or more individuals were collected into a large container during marking experiments (Barr, unpublished). Similarly, the presence of pheromones for location of females, conceivably advantageous in populations of low density, has not been investigated in troglobitic insects.

The long appendages of some species of troglobites (Fig. 23) have often been cited as "characteristic," although the selective advantages of elongation are not always clear. Poulson (1964) suggested that the long, spindly legs in the salamander *Eurycea rathbuni* held the animal off the substratum, minimized disturbance to the neuromasts, and was consequently of selective value in prey seeking. Banta (1910) attributed the greater sensitivity of the cave isopod *Asellus stygius* to tactile stimuli of its elongate, delicate legs and antennae. Longer antennae should be useful tactile structures in perpetual darkness. Certain cavernicolous mites (*Linopodes*) and opilionids (*Phalangodes*) walk on three pairs of legs and use the second, especially elongated pair as antennae, but these structures are also developed in non-troglobitic species. Slender bodies may be an adaptation to squeezing through narrow crevices or for attainment of large size with a maximum surface area for integumentary respiration. Allometric growth is quite probably involved in the appendage lengths in some of the larger, more bizarre species, but is insufficient in itself to account for all such modifications. Finally, many troglobites are known which are no more slender than their epigean relatives. Careful, comparative studies of the behavior, ecology, and physiology of both slender and robust species must

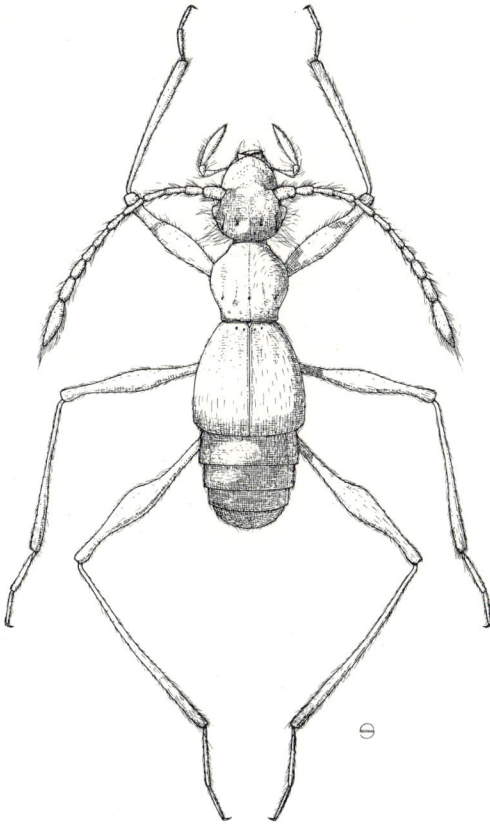

Fig. 23. Texamaurops reddelli *Barr and Steeves, a troglobitic pselaphid beetle from caves near Austin, Texas. Compare the slender, elongate appendages with Fig. 20 (From Barr and Steeves, 1963).*

be conducted before the adaptive value of elongation and slenderness can be intelligently assessed. Christiansen (1961), in a study combining morphology and behavior, was able to demonstrate that elongation of the unguis and empodial appendage of the foot in collembolans was related to the ability to walk on the surface of pools of water, an ability possessed only by species more highly adapted in a number of ways to life in caves. Elongation of antennae and legs, flattening of the terminal antennal segment, a thoracic "hump" resulting from larger muscles to extend the furcula, and larger body size are also characteristic of the more highly adapted troglobitic collembolans. These modifications probably result from adaptation to existence in open spaces in cave chambers, rather than the restricted spaces in soil and leaf litter usually inhabited by collembolans.

Reproductive Adaptations of Troglobites

Reproduction in troglobites may bear close resemblance to reproduction in closely related epigean species, or it may be modified in various ways in response to selection pressures of the cave environment. As a general rule, the number of eggs produced is lower and the eggs are larger and provided with more yolk in troglobitic invertebrates, as well as in many phreatobites and edaphobites. For example, Husson and Daum (1953) and Husson (1959) found that the troglobitic sphaeromid isopod *Caecosphaeroma burgundum* carries 4 to 15 eggs in its marsupium (average 7 to 8), each about 0.75 mm in diameter. In the marine species *Sphaeroma serratum* Hansen (1905) found 67 to 91 eggs with an average diameter of 0.39 mm. Vandel (1964, ch. 22) lists many other examples from other groups of invertebrates.

Deleurance-Glaçon (1963a, 1963b) has described stages of increasing adaptation in the life cycles of troglobitic bathysciine and trechine beetles. In the bathysciines, she demonstrated a series of reproductive patterns: (1) in the least specialized type, females lay a number of small, oligolecithal eggs, and there are three larval instars; (2) in an intermediate type there are only two larval instars, the larva molting in a temporary clay capsule at the end of the first instar and not feeding during the second instar; and (3) in the most advanced type, characteristic of many of the most specialized bathysciines, the female lays a single, very large egg, which hatches into a nonfeeding larva; the larva immediately begins building a clay capsule, in which it remains in a state of diapause for five to six months, then pupates. A similar telescoping of the life cycle was found in the few species of troglobitic trechines which Deleurance investigated, with the highly specialized members of the genus *Aphaenops* producing large eggs which hatched into nonfeeding larvae. In eastern North America most of the troglobitic trechines have six to eight ovarioles per ovary, produce a number of small, oligolecithal eggs, and apparently have larvae which feed actively. Only in the aphaenopsian genus *Nelsonites* are the ovarioles known to be reduced in number, and the eggs unusually large, for example 1.10 by 0.60 mm in a 7 mm female (Barr, unpublished).

Eigenmann (1909) demonstrated that the Cuban brotulid fishes *Lucifuga* and *Stygicola* are viviparous, but showed that the supposed vivipary in the amblyopsids was actually branchial incubation. Poulson (1963) reports a survival rate of 89 percent from egg to fry stage in *Amblyopsis spelaea* (Fig. 24). Population control, presumably desirable in an environment with low carrying capacity because of an erratic food supply, is achieved in *A. spelaea* by long life span, by aperiodic reproduction so that only a small proportion of mature females are found with eggs at any one time, and possibly by density-dependent cannibalism (Poulson, 1963). *Cholo-*

Fig. 24. Amblyopsis spelaea Dekay, *the northern cavefish, inhabits subterranean streams of the Mitchell plain, Indiana, and the northern half of the Pennyroyal plateau, Kentucky, as far south as Mammoth Cave.*

gaster agassizi, a troglophile, breeds every year, has a normal life span of one or two years, and has conspicuously fluctuating populations. Ginet's (1960) studies of the amphipod *Niphargus orcinus virei* suggest similar population control. The troglobitic trechines *Neaphaenops tellkampfii* and *Darlingtonea kentuckensis* may practice cannibalism in close confinement in captivity (Barr, unpublished). Most of these modifications of reproductive patterns may be viewed as adaptations to increased egg (or young) survival in caves, where food is in short supply and the environmental resistance is relatively low.

Seasonal reproduction in troglobites has been reported for amblyopsids (Poulson, 1963), Mexican cave characins (Breder, 1942), and *Niphargus* (Ginet, 1960), and apparently occurs in the atyid shrimp *Palaemonias ganteri* and in troglobitic crayfishes of the genus *Orconectes* (T. C. Jegla, personal communication; Barr, unpublished), based on occurrence of ovigerous females and/or form I males. Hawes (1939) and Poulson (1964) suggested that seasonal floods may trigger reproduction in aquatic cavernicoles. Terrestrial cavernicoles are subject to seasonal variations in air currents, temperature, relative humidity, rate of evaporation, and food supply, as well as seasonal inundation of their normal habitat if they are riparian feeders. Tenerals of central Kentucky cave trechines are relatively abundant in late summer and early fall, but rare at other times of the year, indicating seasonal emergence from the pupal stage, and—by extension—seasonal egg-laying. It would be unusual if there are not at least seasonal peaks in reproductive activity in most cavernicoles, terrestrial or aquatic, although this is a topic for which few data are available.

Conclusions

Troglobites have evolved from colonies of troglophiles which became isolated in caves through extinction of surface populations of the troglophiles. During reconstruction of the epigenotype which follows isolation, severe selection, sampling error, pleiotropy, and polygeny, as well as limited genetic potential interact to expedite adaptation to the cave environment. Loss of eyes, melanin pigment, and other characters of little or no selective value in caves occurs as a pleiotropic byproduct of selection for adaptively advantageous characters. Later adaptations in troglobite evolution include (in some species but not all) hypertrophy of sensory structures other than eyes, metabolic economy, increase in egg size and decrease in egg number, and population regulation via longer life span, infrequent reproduction, and perhaps density-dependent cannibalism. Reproduction is probably more or less seasonal in most troglobties.

References

AEPPLI, E. 1952. Naturliche Polyploidie bei den Planarien *Dendrocoelum lacteum* (Müller) und *Dendrocoelum infernale* (Steinmann). Z. indukt. Abst. Abst. Vererbungsl., 84:182–212.

ÁLVAREZ, J. 1946. Revisión del género *Anoptichthys* con descripción de una especie nueva (Pisc., Characidae). An. Esc. Nac. Ciencias Biol., 4:263–282.

———. 1947. Descripción de *Anoptichthys hubbsi,* caracinido ciego de la Cueva de los Sabinos, S. L. P. Rev. Soc. Méxicana Hist. Nat., 8:215–219.

ATZ, E. H. 1953. Experimental differentiation of basophil cell types in the transitional lobe of the pituitary of a teleost fish, *Astyanax mexicanus*. Bull. Bingham Oceanogr. Coll., 1953:94–116.

BALDWIN, E., and R. A. Beatty. 1941. The pigmentation of cavernicolous animals. J. Exp. Biol., 18:136–152.

BANTA, A. M. 1907. The Fauna of Mayfield's Cave. Carnegie Inst. Washington Publ. 67:1–114.

———. 1910. A comparison of the reactions of a species of surface isopod with those of a subterranean species. J. Exp. Zool., 8:243–310, 439–488.

BARR, T. C., Jr. 1959. New cave beetles (Carabidae, Trechini) from Tennessee and Kentucky. J. Tennessee Acad. Sci., 34:5–30.

———. 1960a. Introduction. Symposium: Speciation and raciation in cavernicoles. Amer. Midl. Nat., 64:1–9.

————. 1960b. The cavernicolous beetles of the subgenus *Rhadine,* genus *Agonum* (Coleoptera: Carabidae). Symposium: Speciation, and raciation in cavernicoles. Amer. Midl. Nat., 64:45–65.

————. 1961. Caves of Tennessee. Tennessee Dept. Conserv. and Comm., Div. Geol., Bull. 64:1–567.

————. 1962a. The genus *Trechus* (Coleoptera: Carabidae) in the southern Appalachians. Coleopterists' Bull., 16:65–92.

————. 1962b. The blind beetles of Mammoth Cave, Kentucky. Amer. Midl. Nat., 68:278–284.

————. 1962c. The *robustus* group of the genus *Pseudanophthalmus* (Coleoptera: Carabidae). Coleopterists' Bull., 16:109–118.

————. 1964. Non-troglobitic Carabidae (Coleoptera) from caves in the United States. Coleopterists' Bull., 18:1–4.

————. 1965. The *Pseudanophthalmus* of the Appalachian valley (Coleoptera: Carabidae). Amer. Midl. Nat., 73:41–72.

————. 1966. Evolution of cave biology in the United States, 1882–1965. Bull. Nat. Speleol. Soc., 28:15–21.

————. 1967a. Cave Carabidae (Coleoptera) of Mammoth Cave. Psyche, 73:284–287; 74:24–26.

————. 1967b. Observations on the ecology of caves. American Nat., 101. In press.

————, and R. A. KUEHNE. 1968. Ecological studies in the Mammoth Cave system of Kentucky. II. The ecosystem. Int. J. Speleol., 4. In press.

————, and S. B. PECK. 1965. Occurrence of a troglobitic *Pseudanophthalmus* outside a cave (Coleoptera: Carabidae). American Midl. Nat., 73:73–74.

————, and J. R. REDDELL. 1967. The arthropod fauna of the Carlsbad Caverns region, New Mexico. Southwestern Naturalist. In press.

BEATTY, R. A. 1941. The pigmentation of cavernicolous animals. II. Carotenoid pigments in the cave environment. J. Exp. Biol., 18:144–152.

————. 1949. The pigmentation of cavernicolous animals. III. The carotenoid pigments of some amphipod Crustacea. J. Exp. Biol., 26:125–130.

BEDEL, L., and E. SIMON. 1875. Liste générale des articulés cavernicoles de l'Europe. J. Zool., 4:1–69.

BISHOP, S. C. 1944. A new neotenic plethodontid salamander, with notes on related species. Copeia, 1944:1–4.

BLUME, J., E. BÜNNING, and E. GÜNZLER. 1962. Zur Aktivitätsperiodik bei Höhlentieren. Die Naturwissenschaften, 49:525.

BOLÍVAR, C., and R. JEANNEL. 1931. Campagne spéologique dans l'Amérique du Nord en 1928 (première série). Arch. zool. exp. et gén., 71:293–316.

BONET, F. 1953. Datos sobre las cavernas y otros fenómenos erosivos de las calizas de la Sierra de El Abra. Congr. Cien. Méxicana Mem. (V) 3, Cien. Fís. y Mat., Geología:238–266.

BOWMAN, T. E. 1964. *Antrolana lira,* a new genus and species of troglobitic cirolanid isopod from Madison Cave, Virginia. Int. J. Speleol., 1:229–236.

BRACE, C. LORING. 1963. Structural reduction in evolution. American Nat., 97:39–49.

BREDER, C. M., JR. 1942. Descriptive ecology of La Cueva Chica, with especial reference to the blind fish, *Anoptichthys.* Zoologica, 27:7–16.

————. 1953. Cave fish evolution. Evolution, 7:179–181.

BRETZ, J. H. 1942. Vadose and phreatic features of limestone caverns. J. Geol., 50:675–811.

BROWN, F. A. 1961. Diurnal rhythm in cave crayfish. Nature (London), 191:929–930.

————. 1965. A unified theory for biological rhythms. *In* Aschoff, J., ed., Circadian clocks, pp. 231–261. Amsterdam, North Holland Publ. Co.

BÜNNING, E. 1964. The Physiological Clock: Endogenous Diurnal Rhythms and Biological Chronometry. Berlin, Springer-Verlag.

BURBANCK, W. D., J. P. EDWARDS, and M. P. BURBANCK. 1948. Toleration of lowered oxygen tension by cave and stream crayfish. Ecology, 29:360–367.

CALMAN, W. T. 1904. On *Munidopsis polymorpha* Koelbel, a cave-dwelling marine crustacean from the Canary Islands. Ann. Mag. Nat Hist., ser. 7, 14.

———. 1932. A cave-dwelling Crustacean of the family Mysidaceae from the island of Lanzarote. Ann. Mag. Nat. Hist., ser. 10, 10:127–131.

CARR, A. F. 1939. *Haideotriton wallacei,* a new subterranean salamander from Georgia. Occ. Pap. Boston Soc. Nat. Hist., 8:333–336.

CAUMARTIN, V. 1963. Review of the microbiology of underground environments. Bull. Nat. Speleol. Soc., 25:1–14.

CHRISTIANSEN, K. A. 1961. Convergence and parallelism in cave Entomobryinae. Evolution, 15:288–301.

———. 1964. Bionomics of Collembola. Ann. Rev. Entom., 9:147–178.

COIFFAIT, H. 1958. Contribution à la connaissance des Coléoptères du sol. Vie et Milieu, suppl., 7:1–210.

CONFIGLIACHI, P. and M. RUSCONI. 1819. Del Proteo anguino di Laurenti. Pavia, 1819.

CONN, H. W. 1966. Barometric wind in Wind and Jewel caves, South Dakota. Bull. Nat. Speleol. Soc., 28:55–69.

COPE, E. D. 1887. The Origin of the Fittest: Essays on Evolution. New York, D. Appleton and Co.

COURNOYER, D. N. 1955. Appendix 6. *In* Brucker, and J. Lawrence, ed., The Caves Beyond. New York, Funk and Wagnalls.

CROPLEY, J. B. 1965. Influence of surface conditions on temperatures in large cave systems. Bull. Nat. Speleol. Soc., 27:1–10.

CUÉNOT, L. 1925. L'Adaptation. Paris, G. Doin.

CURL, R. L. 1958. A statistical theory of cave entrance evolution. Bull. Nat. Speleol. Soc., 20:9–22.

CVIJIČ, J. 1918. Hydrographie souterraine et l'évolution morphologique du karst. Rec. trav. inst. geogr. alpine (Grenoble), 6(4):1–56.

DARWIN, C. 1859. On the Origin of Species, 1st ed., facs., Cambridge, Mass., Harvard Univ. Press.

DAVENPORT, C. B. 1903. The animal ecology of the Cold Spring sand spit, with remarks on the theory of adaptation. Decennial Publ. Univ. Chicago, 10:1–22.

DAVIES, W. E. 1949. Caverns of West Virginia. West Virginia Geol. Surv., 19:1–353.

———. 1951. Mechanics of cavern breakdown. Bull. Nat. Speleol. Soc., 13:36–43.

DAVIS, W. M. 1930. Origin of limestone caverns. Geol. Soc. Amer. Bull., 41:475–628.

DEAMER, D. W. 1964. Entropy and cave animals. Ohio J. Sci., 64:221–223.

DEKAY, J. E. 1842. (Descr. *Amblyopsis spelaeus.*) Zoology of New York, or the New York fauna. Part IV. Fishes. Albany. P. 187.

DELEURANCE-GLAÇON, S. 1963a. Recherches sur les coléoptères troglobies de la sous-famille des Bathysciinae. Ann. Sci. Nat. (Zool.), sér. 12, 5(1):1–172.

———. 1963b. Contribution à l'étude des coléoptères cavernicoles de la sous-famille des Trechinae. Ann. Spéléol., 18:227–265.

DOBZHANSKY, T. 1951. Genetics and the Origin of Species. 3rd ed., New York, Columbia Univ. Press.

DRESCO-DEROUET, L. 1952. Influence des variations de salinité du milieu exterieur sur des crustacés cavernicoles et épigés. C. R. Acad. Sci. Paris, 234:473–475, 888–890.

———. 1959. Contribution à l'étude de la biologie de deux crustacés aquatiques cavernicoles, *Caecosphaeroma burgundum* et *Niphargus orcinus virei.* Vie et Milieu, 10:321–346.

DUDICH, E. 1932. Biologie der Aggteleker Tropfsteinhöhle "Baradla" in Ungarn. Speläol. Mon. (Wien), 13:1–246.

DURAND, J. P. 1964. Anatomie de l'orbite chez la larve de *Proteus anguinus.* Bull. Soc. Zool. (France), 88:278–298.

EBERLY, W. 1960. Competition and evolution in cave crayfishes of southern Indiana. Syst. Zool., 9:29–32.

EIGENMANN, C. H. 1909. Cave vertebrates of America: a study in degenerative evolution. Carnegie Inst. Washington Publ., 104:1–241.

EMERSON, A. E. 1949. Natural selection. In Allee, W. C., A. E. Emerson, O. Park, T. Park, and K. P. Schmidt. Principles of Animal Ecology. Philadelphia, W. B. Saunders.

GARDNER, J. H. 1935. Origin and development of limestone caverns. Geol. Soc. Amer. Bull., 46:1255–1274.

GINET, R. 1960. Écologie, ethologie, et biologie de Niphargus. Ann. Spéléol., 15:127–377.

GORDON, M. S., and D. E. ROSEN. 1962. A cavernicolous form of the poeciliid fish Poecilia sphenops from Tabasco, Mexico. Copeia, 1962, 360–368.

GOUNOT, A. M. 1960. Recherches sur le limon argileux souterrain et sur son rôle nutritif pour les Niphargus (Amphipoda gammaridés). Ann. Spéléol., 15:501–526.

GURNEE, R. H., J. V. THRAILKILL, and G. NICHOLAS. 1966. Discovery at the Rio Camuy. Explorers' J., 44:51–65.

HAMANN, O. 1896. Europäische Höhlenfauna: Eine Darstellung der in den Höhlen Europas Lebenden Tierwelt, mit Besonderer Berücksichtigung der Höhlenfauna Krains. Jena, Hermann Costenoble.

HANSEN, H. J. 1905. On the propagation, structure, and classification of the family Sphaeromidae. Quart. J. Micr. Sci., 49:69–135.

HAWES, R. S. 1939. The flood factor in the ecology of caves. J. Anim. Ecol., 8:1–5.

HEUTS, M. J. 1951. Ecology, variation, and adaptation of the blind African cave fish Caecobarbus geertsi Boulenger. Ann. Soc. Roy. Zool. Belgique, 82:155–230.

————. 1953a. Regressive evolution in cave animals. Sympos. Soc. Exp. Biol., 7(Evolution):290–309.

————. 1953b. Comment on "Cave fish evolution." Evolution, 7:391–392.

HOBBS, H. H., JR. 1942. The crayfishes of Florida. Univ. Florida Publ., Biol. Sci. Ser., 3(2):1–179.

————, and T. C. BARR, JR. The origins and affinities of the troglobitic crayfishes of North America. (Decapoda, Astacidae). II. The genus Orconectes. In preparation.

HOLDHAUS, K. 1933. Die europäische Höhlenfauna in ihren Beziehungen zur Eiszeit. Zoogeographica, 1:1–53.

HOLSINGER, J. R. 1966. A preliminary study on the effects of organic pollution of Banners Corner Cave, Virginia. Int. J. Speleol., 2:75–89.

HUBBS, C. L. 1938. Fishes from the caves of Yucatan. Carnegie Inst. Washington Publ. 491:261–295.

————, and W. T. INNES. 1936. The first known blind fish of the family Characidae: a new genus from Mexico. Occ. Pap. Mus. Zool. Univ. Michigan, 342:1–7.

HUSSON, R. 1959. Les crustacés pericaridés des eaux souterraines: considérations sur la biologie de ces cavernicoles. Bull. Soc. Zool. France, 84:219–231.

————, and J. DAUM. 1953. Sur la biologie de Caecosphaeroma burgundum. C.R. Acad. Sci. Paris, 236:2345–2347.

JANZER, W., and W. L. 1952. Versuche zur evolutorischen Entstehung der Höhlentiermerkmale. Z. indukt. Abst. Vererbungsl., 84:462–479.

JEANNEL, R. 1911. Révision des Bathysciinae—morphologie, distribution géographique, systématique. Arch. zool. exp. et gén., sér. 5, 7:1–641.

————. 1923. Sur l'évolution des coléoptères aveugles et le peuplement des grottes dans les monts Bihor en Transylvanie. C.R. Acad. Sci. Paris, 176:1670–1673.

————. 1926. Faune cavernicole de la France, avec une étude des conditions d'éxistence dans le domaine souterrain. Encycl. Entom., 7:1–334. Paris, P. Lechevalier.

————. 1926–1930. Monographie des Trechinae. Morphologie comparée et distribution géographique d'un groupe de Coléoptères. L'Abeille, 32:221–550; 33:1–592; 34:59–122; 35:1–808.

————. 1943. Les fossiles vivants des cavernes. Paris, Editions Gallimard. 321 pp.

————. 1950. La marche de l'évolution. Publ. Mus. Nat. Hist. Nat., Paris, no. 15.

————. 1965. La génèse du peuplement des milieux souterrains. Rév. d'écol. et biol. du sol, 2(1):1–22.

JEGLA, T. C., and J. S. HALL. 1962. A Pleistocene deposit of the free-tailed bat in Mammoth Cave, Kentucky. J. Mammal., 43:447–481.

KAMMERER, P. 1912. Experimente über Fortpflanzung, Farbe, Augen und Körperreduktion bei *Proteus anguinus* Laur. Arch. Entwicklungsmech., 33:349–461.

KARAMAN, S. 1954. Über unsere unterirdische Fauna. Acta Mus. Maced. Sci. Nat., vol. 1 (Cited in Vandel, 1964: 19–20).

KOFOID, C. A. 1899. The plankton of Echo River, Mammoth Cave. Trans. Amer. Micr. Soc., 21:113–126.

KOHLS, G. M., and W. L. JELLISON. 1948. Ectoparasites and other arthropods occurring in Texas bat caves. Bull. Nat. Speleol. Soc., 10:116–117.

KOSSWIG, C. 1937. Über Pigmentverlust während des Höhlenlebens. Zool. Anz., 117:37–43.

————. 1965. Génétique et évolution régressive. Rév. Quest. Sci., 26:227–257.

————, and L. KOSSWIG. 1940. Die Variabilität bei *Asellus aquaticus* unter besonderer Berücksichtigung der Variabilität in isolierten unter- und oberirdischen Populationen. Rev. Fac. Sci. (Istanbul) ser. B, 5:1–55.

KREKELER, C. H. 1958. Speciation in cave beetles of the genus *Pseudanophthalmus* (Coleoptera: Carabidae). Amer. Midl. Nat., 59:167–189.

LAMARCK, J. B. 1809. Philosophie Zoologique, vol. 1, facs., Weinheim, J. Cramer.

LANKESTER, E. R. 1893. Blind animals in caves. Nature (London) 47:389, 486.

LATTIN, G. DE. 1939. Über die Evolution der Höhlentiercharaktere. Sitzber. Ges. Naturf. Freunde (Berlin), 32:11–41.

LELEUP, N. 1956. La faune cavernicole du Congo Belge, et considérations sur les Coléoptères réliques d'Afrique intertropicale. Ann. Mus. Roy. Congo Belge (Tervuren, Belgique), sér.-in-oct., Sci. Zool., 46:1–171.

LUDWIG, W. 1942. Zur evolutorischen Erklärung der Höhlentiermerkmale durch Allelelimination. Biol. Zentralbl., 62:447–455.

MACARTHUR, R. H. 1957. On the relative abundance of bird species. Proc. Nat. Acad. Sci. U.S.A., 43:293–295.

MAGUIRE, B., JR. 1961. Regressive evolution in cave animals and its mechanism. Texas J. Sci., 13:363–370.

MALOTT, C. A. 1932. Lost River at Wesley Chapel Gulf, Orange County, Indiana. Indiana Acad. Sci. Proc., 41:285–316.

————. 1937. Invasion theory of cavern development (abstr.). Proc. Geol. Soc. America, 1936: 323.

MARSHALL, N. B., and G. L. THINÉS. 1958. Studies of the brain, sense organs, and light sensitivity of a blind cave fish (*Typhlogarra widdowsoni*) from Iraq. Proc. Zool. Soc. (London) 131:441–456.

MAYR, E. 1942. Systematics and the Origin of Species, from the Viewpoint of a Zoologist. New York, Columbia Univ. Press.

————. 1963. Animal Species and Evolution. Cambridge, Harvard Univ. Press.

MENAKER, M. 1959. Endogenous rhythms of body temperature in hibernating bats. Nature (London), 184:1251–1252.

MITCHELL, R. W. 1965. Ecological studies of the troglobitic carabid beetle *Rhadine subterranea*. Doct. dissert., Univ. Texas.

MOORE, G. W., ed. 1960. Origin of limestone caves: a symposium with discussion. Bull. Nat. Speleol. Soc., 22:3–84.

————, ed. 1966. Limestone hydrology: a symposium with discussion. Bull. Nat. Speleol. Soc., 28:109–166.

MOTAŞ, C., and J. TANASACHI. 1946. Acariens phreaticoles de Transylvanie. Notat. Biol. (Bucharest) 4. (Cited by Vandel, 1964:19).

MOTSCHULSKY, T. V. VON. 1862. Études entomologiques (11me Année). Dresden.

NOBLE, G. K., and C. H. POPE. 1928. The effect of light on the eyes, pigmentation, and behavior of the cave salamander, *Typhlotriton*. Anat. Rec., 41:21.

NORMAN, W. W. 1900. Remarks on the San Marcos salamander, *Typhlomolge rathbuni*. Amer. Nat., 34:179–183.

PACKARD, A. S., JR. 1888. The cave fauna of North America, with remarks on the anatomy of the brain and origin of the blind species. Mem. Nat. Acad. Sci. (U.S.A.), 4:1–156.

———. 1894. On the origin of the subterranean fauna of North America. Amer. Nat., 28:727–751.

PARK, O. 1951. Cavernicolous pselaphid beetles of Alabama and Tennessee, with observations on the taxonomy of the family. Geol. Surv. Alabama Mus. Pap. 31: 1–107.

———. 1956. New or little-known species of pselaphid beetles from southeastern United States. J. Tennessee Acad. Sci., 31:54–100.

———. 1960. Cavernicolous pselaphid beetles of the United States. Amer. Midl. Nat., 64:66–104.

———, and D. E. REICHLE. 1964. Observations on the ecology and behavior of the cave cricket, *Hadenoecus subterraneus* (Scudder). Bull. Nat. Speleol. Soc., 26:79 (abstr.).

———, T. W. ROBERTS, and S. J. HARRIS. 1941. Preliminary analysis of activity of the cave crayfish, *Cambarus pellucidus*. Amer. Nat., 75:154–171.

POULSON, T. L. 1963. Cave adaptation in amblyopsid fishes. Amer. Midl. Nat., 70:257–290.

———. 1964. Animals in aquatic environments: animals in caves. *In* D. B. Dill, ed. Handbook of Physiology, sect. 4, "Adaptation to the environment," ch. 47: 749–771. Washington, Amer. Physiol. Soc.

PROUT, T. 1964. Observations on structural reduction in evolution. Amer. Nat., 98:239–249.

RACOVITZA, E. G. 1907. Essai sur les problèmes biospéologiques. Arch. zool. exp. et gén., 36:371–488.

RASQUIN, P. and L. ROSENBLOOM. 1954. Endocrine imbalance and tissue hyperplasia in teleosts maintained in darkness. Bull. Amer. Mus. Nat. Hist., 104(4): 359–426, pl. 4–23.

REICHLE, D. E., J. D. PALMER, and O. PARK. 1965. Persistent rhythmic locomotor activity in the cave cricket, *Hadenoecus subterraneus,* and its ecological significance. Amer. Midl. Nat., 74:57–66.

RENSCH, B. 1959. Evolution Above the Species Level. New York, Columbia Univ. Press.

ROSEN, D. E. 1962. Comments on the relationships of the North American cave fishes of the family Amblyopsidae. Amer. Mus. Novit., no. 2109.

ROUX, W. 1881. Der Kampf der Theile in Organismus. Leipzig.

SADOĞLU, P. 1957. Mendelian inheritance in the hybrids between the Mexican blind cave fishes and their overground ancestor. Verh. Deutsch. Zool. Ges. Graz, 1957: 432–439.

SCHINER, J. R. 1854. Fauna der Adelsberger-, Lueger- und Magdalen-Grotte. Verh. zool.-bot. Ges. Wien, 3:1–40.

SCHIÖDTE, J. C. 1851. Bidrag til den underjordiske Fauna. Vidensk. Selsk. Skr. (Copenhagen), 5 Raekke, naturv. og math. Afd., 2 Bd.:1–39.

SCHLAGEL, S. R., and C. M. BREDER, JR. 1947. A study of the oxygen consumption of blind and eyed cave characins in light and darkness. Zoologica, 32:17–27.

SCHMIDT, F. J. 1832. *Leptodirus Hohenwartii,* n. g., n. sp. Illyrisches Blatt, Laibach, no. 3:9.

SCOTT, W. 1909. An ecological study of the plankton of Shawnee Cave, with notes on the cave environment. Biol. Bull., 17:386–407.

SIMPSON, G. G. 1944. Tempo and Mode in Evolution. New York, Columbia Univ. Press.

SPENCER, H. 1893. The inadequacy of natural selection. Contemporary Review, 63:153–167; 439–457; 743–761.

STAGER, K. E. 1941. A group of bat-eating duck hawks. The Condor, 43:137–139.

STONE, L. S. 1964a. The structure and visual function of the eye of larval and adult cave salamanders, *Typhlotriton spelaeus*. J. Exp. Zool., 156:201–218.

———. 1964b. Return of vision in transplanted larval eyes of cave salamanders. J. Exp. Zool., 156:219–228.

———. 1964c. Return of vision in larval eyes exchanged between *Amblystoma punctatum* and the cave salamander, *Typhlotriton spelaeus*. Invest. Ophthalmol., 3:555–565.

STURM, J. H. 1844. *Anophthalmus,* blind Laufkäfer, neue Gattung aus der familie der Caraben. Deutschl. Fauna in Abb. nach der Natur, Nürnberg, 5te Abt., 15:131, pl. 303.

SWINNERTON, A. C. 1932. Origin of limestone caverns. Geol. Soc. Amer. Bull., 43:663–694.

———. 1942. Hydrology of limestone terranes. *In* O. E. Meinzer, ed. Physics of the Earth, pt. 9, Hydrology: 656–677. New York, McGraw Hill.

TELLKAMPF, T. G. 1844a. Beschreibung einiger neuer in der Mammuth-Höhle in Kentucky aufgefundener Gattungen von Gliederthieren. Arch. Naturg., 10:318–322, pl. 7.

———. 1844b. Ueber den blinden Fisch der Mammuth-Höhle in Kentucky, mit Bemerkungen über einige andere in dieser Höhle lebenden Thiere. Müllers Arch. Anat. Physiol., 4:384–394, pl. 9.

THORNBURY, W. D. 1954. Karst topography. *In* Principles of Geomorphology, pp. 316–353. New York, John Wiley and Sons.

VALVASOR, J. W., JR. 1689. Die Ehre dess Hertzogthums Crain. Vol. 4, pp. 594–598 treats the olm (*Proteus*). Laybach, 4 vols.

VANDEL, A. 1958. La répartition des cavernicoles et la paléogéographie. Actes 2. Congr. Int. Spéléol., 2(3):31–43.

———. 1961. Eye and pigment regression of cave salamanders. Bull. Nat. Speleol. Soc., 23:71–74.

———. 1964. Biospéologie: La Biologie des Animaux Cavernicoles. Paris, Gauthier-Villars.

———, and MICHEL BOUILLON. 1959. Le Protée et son intéret biologique. Ann. Spéléol., 14:112–127.

VIRÉ, A. 1904. La Biospéléologie. C.R. Acad. Sci. Paris, 139:992–995.

WEISMANN, A. 1889. Essays Upon Heredity and Kindred Biological Problems. Oxford, Clarendon Press.

WOODS, LOREN P., and ROBERT F. INGER. 1957. The cave, spring, and swamp fishes of the family Amblyopsidae of central eastern United States. Amer. Midl. Nat., 58(1):232–258.

WOODWARD, H. P. 1961. A stream piracy theory of cave formation. Bull. Nat. Speleol. Soc., 23:39–58.

WRIGHT, S. 1929. Fisher's theory of dominance. Amer. Nat., 63:274–279.

———. 1964. Pleiotropy in the evolution of structural reduction and of dominance. Amer. Nat., 98:65–69.

3

Molecular Biology and Bacterial Phylogeny

J. DE LEY

Laboratory of Microbiology, Faculty of Sciences, State University
Gent, Belgium

Introduction 104
The Evolution of Microorganisms in the Precambrian 105
    The primitive earth and its atmosphere 105
    Life in the anoxygenic period 106
    The beginning of an oxygenic atmosphere 110
    The end of the Precambrian 112
    Geochemical evidence for the beginning of the oxygenic atmosphere .. 113
Microbial Fossils 114
    Indirect evidence 114
    Precambrian algae-like fossils 115
    Fossil bacteria 116
Bacteria as "Living Fossils" 117
The Relationship Between Bacteria and Bluegreen Algae 119
    Fossil bluegreen algae 121
    Phylogeny of bluegreen algae 121
DNA Nucleotide Composition of Microorganisms 122
Evolutionary Changes in Bacterial DNA 128
    The size of the bacterial genome 128
    Evolutionary drift in percent GC 129
    Drastic mutational changes in DNA base composition 134
    Possible dissimilarity between ancient and modern bacteria 135
DNA Homology 137
Phenotypic Aspects of Bacterial Evolution 148
Summary 151
References 153

ABBREVIATIONS

DNA	deoxyribonucleic acid
RNA	ribonucleic acid
A	adenine
T	thymine
G	guanine
C	cytosine
percent GC	average molar percent guanine + cytosine
UV	ultraviolet light
hμ	radiant energy
NAD and NADH	nicotinamide adenine dinucleotide and reduced form
NADP and NADPH	nicotinamide adenine dinucleotide phosphate and reduced form
ATP	adenosine triphosphate
G6P	glucose-6-phosphate
6PG	gluconate-6-phosphate

Introduction

The present day microorganisms are the result of complex evolutionary processes. If these events were known, a three- or multidimensional phylogenetic tree could be constructed with time as one of the axes. Its transverse section with the plane of the present day moment would provide us with a phylogenetic classification. Bacterial classification is mainly built on morphological, physiological, and biochemical features. When many of these data are compared by numerical (or Adansonian) analysis, they provide a fair picture of the degree of relatedness of many bacteria. However, it is nearly impossible to project this picture back into the past because numerical taxonomy covers at most some 20 percent of the bacterial genome, and orthodox bacterial taxonomy covers even less. Furthermore, paleontological, embryological, and comparative anatomical data are lacking for bacteria. Therefore, the evolutionary tree of bacteria remains largely invisible, if one uses only phenotypic features.

Molecular biology has provided methods to work directly with the bacterial genome, to determine its nucleotide composition, the compositional distribution of the four bases within the DA molecule and, in particular, to estimate quantitatively nucleotide similarities by DNA hybridization. As a result one starts to see very dimly phylogenetic relationships between some bacteria and there is hope that the hidden past of bacterial evolution is about to be unveiled. We can but take here the first hesitating steps on this long but promising path.

The Evolution of Microorganisms in the Precambrian

The earth is believed to have formed some 5 billion years ago. Primitive living systems probably originated some 3 or 4 billion years ago. At the beginning of the Cambrian, about 600 million years ago, the living world exploded into a tremendous variety of creatures, which set the foundation for the great diversity in form and function that one observes in the animals, plants, and microorganisms of today. Between the development of the first living forms and the Cambrian period there elapsed some 3 billion years and it is during this immense time span that the ancestors of our modern bacteria are believed to have originated. It is important, therefore, that we consider briefly some of the geological and atmospheric conditions that are presumed to have existed in Precambrian times.

The Primitive Earth and Its Atmosphere

During the formation of the earth nearly all of the inert gases escaped and thus the primitive atmosphere was lost. The Precambrian atmosphere or paleoatmosphere is believed to have had a volcanic origin, with the exception of oxygen. It is the general consensus that the paleoatmosphere contained rather large amounts of nitrogen and water vapor, probably some methane, ammonia, and CO_2, and possibly some ammonium nitrite. Hutchinson (1944) hypothesizes that the N_2 content of the atmosphere has increased rather uniformly through geological time. If such was the case, then the N_2 pressure of the paleoatmosphere at the time when the first microorganisms were developing, was possibly around 0.4 or 0.5 atm and increased to about 0.7 atm by the beginning of the Cambrian.

One can calculate that the amount of ammonia required to build up a biomass of the present size (which is largely composed of microorganisms) is surprisingly small. The total nitrogen in the present biomass is about 4×10^{16} gm. The total N_2 in the present atmosphere is about 4×10^{21} gm and it may have been about half that much at the time when microorganisms started to develop. It follows that a total of 0.002 percent ammonia in the early Precambrian atmosphere would have sufficed to allow the development of a biomass of the same size as there is today. Supposing that all this ammonia had been dissolved in the oceans (1.5×10^{21} liters) the concentration would have been only 3×10^{-5} gm/liter or 2×10^{-6} N. The pH of such a paleo-ocean could not have exceeded 9 and probably was near neutrality in view of the dissolved CO_2 and the volatility of ammonia. It is, therefore, unnecessary to assume that Precambrian oceans were strongly alkaline.

The most outstanding feature of the paleoatmosphere was its lack of free oxygen. The following arguments plead in favor of this hypothesis. (1) The earth's crust, which takes up oxygen primarily to oxidize ferrous to ferric iron and to some extent to oxidize manganous and sulphur compounds, is still suboxidized. (2) There is a greater abundance of fossilized carbon than atmospheric carbon, although both originated through photosynthesis. (3) The primitive organic building blocks for life would not have been stable in the presence of oxygen. (4) Nearly all inert gases escaped from the earth during its formation; others were trapped by chemical reactions. Berkner and Marshall (1965) have calculated that O_2 in the Precambrian atmosphere may have been regulated at less than 0.1 percent of the present concentration. The minute amounts of oxygen in the paleoatmosphere were derived from the photodissociation of water. This, however, must have been a self-limiting reaction since ozone shields water from further dissociation (self-regulated Urey-equilibrium).

More detailed information about the composition of the Precambrian atmosphere can be obtained from Dole (1965) and from several papers published in the Symposium on the Evolution of the Earth's Atmosphere (1965).

The composition of the Precambrian ocean is discussed by Dauvillier (1965) and by Holland (1965). No matter what the precise composition of the Precambrian ocean was, it is obvious that it must have contained sufficient minerals to support the growth of organisms. Since the experiments of Miller (1957), it is generally agreed that it also contained a variety of complex organic substances from which the primitive organisms were partially derived. There may have been as much as 1 percent dissolved organic material in the paleo-ocean.

Life in the Anoxygenic[1] Period

From the occurrence of the first living forms to the advent of the oxygen-producing algae a period of about 1 to 2 billion years elapsed. An important corollary to the lack of oxygen in the primitive atmosphere was the

---

[1] As pointed out by Rutten (1962, p. 62) one should use the term "anoxygenic" instead of "anaerobic" for organisms living in a Precambrian world, devoid of oxygen. Presently living anaerobic organisms (e.g., the Clostridia) are really likewise anoxygenic.

absence of an ozone layer. The earth was thus exposed to intense ultra-violet (UV) irradiation, down to wavelengths of about 200 mμ. Organisms built on the same principles as those existing today would have been very susceptible to the 200 to 400 mμ range irradiation. Berkner and Marshall (1965) pointed out that life could only have existed in the seas at depths below 10 m. Some life could also have existed in the soil in conditions where it was shielded from UV irradiation. Unavoidably, organisms must have been carried more closely to the surface by water currents and exposed from time to time to the potent mutagenic action of UV. Hence, we might expect a high mutation rate to have accelerated evolution at that time.

For reasons which are beyond the scope of this discussion, primitive microorganisms must have originated with a molecular structure rather similar to that of the present bacteria. We have to assume that they contained DNA, RNA, and catalytic proteins (enzymes) within a cell hull. These organisms were heterotrophic and used organic compounds from the sea as an energy source. This heterotrophic, anoxygenic world could not have existed very long, nor could it have developed very far. One reason for this was that the supply of organic energy sources from the sea would be exhausted. Furthermore, a hydrogen and/or electron acceptor was required; this would have had to be either an inorganic oxidant with a high redox potential (e.g., Fe^{+++}) or some organic redox system (e.g., triose phosphate dehydrogenase/ethanol dehydrogenase) such as those active in present day fermentations. However, most of the inorganic systems were in the reduced state already and the supply of hydrogen acceptor would also have become quickly exhausted. Had there not been continued mutation with the development of new forms, the living world would have been doomed.

This situation might have improved some by the formation of chemi-autotrophic microorganisms. Such organisms used CO_2 as their carbon source and thus helped by providing a reservoir of organic substrates for the heterotrophic population. Present day bacteria of this type use sulphate, nitrate, or carbonate as hydrogen acceptor and either H_2 or simple organic compounds as energy sources. Representatives of anoxygenic chemiautotrophic bacteria living today are the denitrifier *Micrococcus denitrificans,* the sulphate reducing bacteria *Desulfovibrio,* and the group of carbonate reducing bacteria such as *Methanobacterium, Methanosarcina,* and so forth. It should be noted, however, that no nitrates could have existed in early Precambrian times, since these compounds arose mainly through microbial oxidation of NH_3 and free oxygen was not available. It follows that the nitrifying bacteria (*Nitrosomonas* and *Nitrococcus*), chemiautotrophic nitrate reducers, and denitrifying bacteria must be of a more recent origin (less than 2 billion years old).

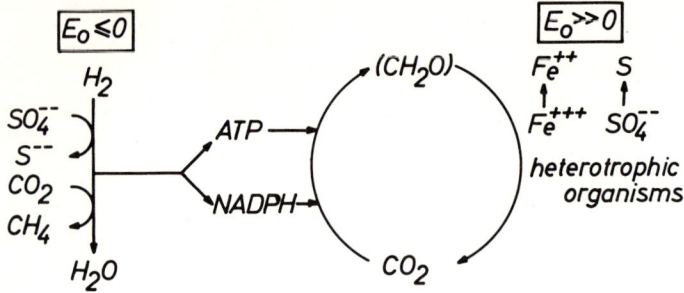

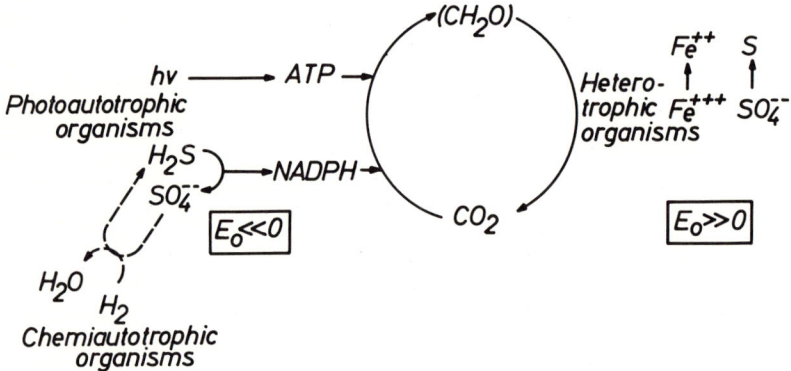

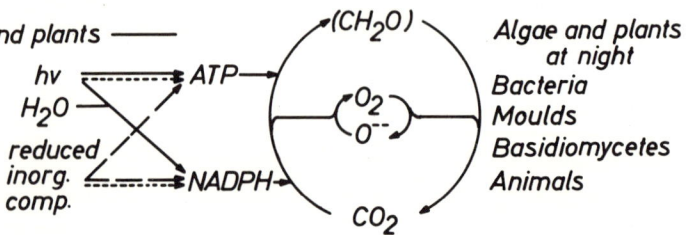

Fig. 1. (Legend on page 109)

Upon closer inspection, however, the intercalation of anoxygenic chemiautotrophs in the food cycle would be helpful only temporarily. For one, they would be dependent on the presence of a hydrogen acceptor, be it sulphate, carbonate, or something else. If they were to rely on H_2 as an energy source, they could not have obtained this gas readily from the atmosphere. Molecular hydrogen is only slightly soluble in water and for the bacteria to have inhabited the ocean surface would have meant exposure to lethal UV irradiation. These bacteria might have used H_2 produced by heterotrophic organisms, but that, again, would be of no ultimate advantage on a global scale. If they were to use organic compounds for that purpose it would in the end make no difference whether they or the heterotrophs used the organic low-entropy substances. Therefore it seems that the combination of anoxygenic chemiautotrophs and anoxygenic heterotrophs was of temporary importance. A possible carbon cycle for the anoxygenic period is shown in Fig. 1A. As we shall see below, the chemiautotrophic microorganisms may still have played another important role in the evolution of life.

Greater possibilities for evolutionary development and diversity opened up as the supply of organic substances available to the heterotrophic world became less limiting. This most likely was brought about by the formation of photoautotrophic bacteria. These bacteria differ from photosynthetic algae and plants in two main features. (1) They are unable to couple the dissociation of water and the concomitant liberation of O_2 from OH^- to the reduction of NADP to NADPH. (2) NADPH which is necessary for the Calvin cycle has to be provided by the oxidation of inorganic, or occasionally organic, substances. The Chlorobacteriaceae oxidize either H_2S, H_2, thiosulphate, or tetrathionate for that purpose. The Thiorhodaceae oxidize about the same compounds and the Athiorhodaceae use fatty acids for the production of NADPH. On the assumption that primitive photoautotrophic bacteria were in principle similar to those living now, a primitive anoxygenic world can be imagined as follows.

Photoautotrophic bacteria would convert CO_2 into cell material. Energy

---

Fig. 1. *The possible evolution of the carbon cycle.*

A. *The role of chemiautotrophic organisms in the anoxygenic precambrian period. A substrate with low redox potential E_0 (e.g. H_2) is oxidized with some hydrogen acceptor (e.g., SO_4 or CO_2) and provides the energy to drive the CO_2-incorporating Calvin cycle. Cell material (CH_2O) thus produced is oxidized by heterotrophic organisms, which need hydrogen acceptors (e.g. SO_4) with higher redox potential.*

B. *Joint action of photo- and chemiautotrophic organisms in the anoxygenic precambrian period. Both supply organic material for the growth of heterotrophic organisms. Chemiautotrophs may be used to recharge the hydrogen donor needed by the photoautotrophs.*

C. *Elementary principle of the present day carbon cycle. It may already have been active in this form at the beginning of the Cambrian.*

sources such as ATP were derived from light energy. The reducing compound (NADPH) was produced by the oxidation of reduced inorganic compounds such as H_2S, elemental S, organic substances, or some other system with a low redox potential. The organic compounds from the photoautotrophs served as a food source for the heterotrophs. The hydrogen acceptor which was required for the breakdown of those compounds might have been an internally produced compound such as those in present day fermentations, or possibly some inorganic compound (e.g., the reduction of Fe^{+++} to Fe^{++}, or of sulphate to elemental sulphur and to sulphide). In either case, a hydrogen accepting system with a high redox potential was required. If the original photoautotrophs were related to the Chlorobacteriaceae or the Thiorhodaceae, their activity was probably limited by the supply of oxidizable inorganic sulfur compounds. This supply, however, might likely have been recharged by the reverse reaction of sulphate and sulphur to sulphide. By carrying out these reactions, chemiautotrophic bacteria may have played an important role in the carbon cycle of this time as shown in Fig. 1B. Such a world, while allowing far more forms and greater evolutionary divergency, would still have been limited for two reasons. (1) Photoautotrophic bacteria were dependent on and could develop only in those regions where hydrogen donors were ample. (2) The heterotrophs would have had either a very limited biomass yield if an internally produced hydrogen acceptor was used, or would have likewise been tied to a certain habitat if an inorganic electron acceptor was required. However, the above hypothesis still remains attractive as a possible third step in the evolution of microorganisms since it leads logically to the next stage.

The Beginning of an Oxygenic Atmosphere

The fourth major step in the evolution of microorganisms was likely to have been the development of primitive anoxygenic photosynthetic (oxygen producing) "protoalgae." The transition from a bacteria-like photosynthetic pathway to a more plant-like pathway, would not have required the formation of many new enzymes. Although the difference between the two photosynthetic systems is not great, it is fundamental. In photosynthesizing plants and algae, water itself serves as the hydrogen donor and so primitive protoalgae with such a system would have been liberated from the problems of acquiring a foreign hydrogen donor such as H_2S and others. As a consequence, these microbes would have been less restricted to particular environments, a fact of great importance to the eventual distribution and evolution of life. The bottom of shallow pools, lakes, and protected seas, with 10 m of water above it no doubt provided optimal conditions for these primitive protoalgae. Berkner and Marshall (1965) have estimated that early photosynthetic "algae" were already in abundance at an early stage of the Precambrian era, since about 0.3 percent of the global area would have to be covered with photosynthetic organisms for the self-regulated Urey-equilibrium to have been upset and for an oxygenic atmosphere to have developed.

An important point concerning these primitive protoalgae is that they probably were unable to use oxygen for the breakdown of their polysaccharides and other reserve material. This is in contrast to present day algae and plants which both produce and consume oxygen. One reason for this might be that in existing algae and plants the sites where carbohydrates and ATP are synthesized are physically separated (paired lamellae and chloroplasts) from the sites where the energy is required. In the very primitive anoxygenic protoalgae this separation may not yet have existed. In these organisms, compounds such as ATP and glyceraldehyde-3-phosphate, the latter synthesized via the Calvin cycle, may have moved freely through the cell providing energy and substrate for further biosynthesis. According to our present classification the primitive photosynthetic protoalgae, were therefore neither algae as we know them today, nor bacteria, but more of an intermediate form.

Sooner or later a partition between the photosynthetic apparatus and the rest of the cell came about in the form of primitive chromatophores or paired lamella. From this stage on, an electron accepting system would have had to evolve in the cytoplasm. Several alternatives might have been possible. The most likely system would have been one similar to that found in algae today where O_2 acts as the terminal electron acceptor. In algae and plants, cytochromes have an important role in photosynthesis. These compounds also function in the anaerobic photoautotrophic bacteria. If cytochrome-like compounds had already existed in the Precambrian protoalgae, relatively few steps would have been required for the evolution of the system cytochrome—cytochrome oxidase—O_2.

Another speculation which can not be excluded is that the cytoplasmic electron-accepting system was different from oxygen. It is interesting to look for similar systems in present algae and plants. There are decided indications that they exist, as shown by the following examples. About 9 percent of the flowering plant species can germinate or grow with little or no oxygen (Siegel et al., 1965). It has also been suggested that winter rye lives anoxygenically by way of the internal redox reaction:

$R.CHNH_2.COOH + ADP + P_i + R'(SH)_2 \rightarrow R.CH_2.COOH + ATP + NH_3 + R'S_2$. We are not aware of attempts to grow algae in light under strict anoxygenic conditions nor of reports on anoxygenic heterotrophic growth of algae in spite of the fact that *Prototheca zopfii* and *Ochromonas malhamensis* can carry out a fermentation. There is a report (Nakamura, 1938) that *Oscillatoria,* a bluegreen alga, can use H_2S instead of H_2O as a hydrogen donor in photosynthesis; no O_2 is evolved but sulphur accumulates. Frenkel et al. (1950) studied a *Synechococcus* which uses H_2 for the reduction of CO_2 under anaerobic conditions.

In theory quite a variety of compounds could have taken over the function of oxygen: e.g., sulphate, carbonate, or nitrate with the concomitant production of sulphur, sulphide, methane, N_2, or ammonia. The reduction of ferric to ferrous iron was unlikely because the available iron was probably still mainly in the reduced state. Next to nothing is known about the reduction of SO_4^{--}, S, or CO_3^{--} in algae and plants.

Fermentation is another possible system that could have been utilized in the absence of an electron accepting oxygenic system. We have seen that those pathways may have been active in primitive protoalgae. Many algae can carry out fermentations of which there are three types: (1) lactic acid, (2) lactic acid, ethanol, and CO_2, and (3) acetic acid, ethanol, and CO_2, with the occasional evolution of H_2. Also plants, when subjected to anoxygenic conditions, start to ferment with the production of CO_2 and organic acids. For reviews we refer to Lewin (1962).

In summary, the reduction of sulphate or carbonate, or one of the systems of fermentation could have served as suitable electron acceptors, allowing primitive oxygen-producing photosynthetic protoalgae to live and grow in a completely anoxygenic atmosphere. On the other hand a far more efficient system, that of using O_2 as an electron acceptor, was soon to evolve if it had not already done so.

The End of the Precambrian

As the concentration of O_2 in the atmosphere rose to about 1 percent of its present level, UV irradiation was reduced, allowing living forms to inhabit all but the uppermost surface of the oceans. This must have been a very critical point in the evolution of life since this was approximately the oxygen pressure at which organisms change from an anoxygenic to an oxygenic way of life (the "Pasteur-point"). In view of the abundant supply of photosynthetic protoalgae during the precambrian era, the O_2 concentration in the shallow waters they inhabited may have been rather high, high enough anyway to allow for a natural selection of oxygenic or facultative oxygenic microorganisms and small metazoa. According to Berkner and Marshall (1965) this point was reached some 600 million years ago around the beginning of the Cambrian.

It would appear that by the end of the Precambrian, suitable conditions had existed for the formation of the main classes of bacteria: strictly anaerobic, facultative anaerobic and aerobic heterotrophs, anaerobic chemi-autotrophs, which later may have given rise to aerobic chemiautotrophs, and anaerobic photoautotrophs. If this is true then the stage was set for the evolutionary divergency that lead to the presently existing bacteria. The same was also true for the algae. From that moment on an explosive divergency set in. This first took place in the oceans. With the accumulation of O_2 in the atmosphere, the efficiency of the UV filtering ozone layer increased and the layer was slowly lifted up to its present position at 30 km altitude. The land surface was thus protected from lethal UV. This spreading of life to land is estimated to have occurred some 450 million years ago in the mid-Silurian period. In this case it might have coincided with the rise of the actinomycetes, morphologically the most complex of all heterotrophic soil bacteria.

One of the dangers of speculating about the evolution of microbial life is that it becomes tempting to think only in terms of presently existing forms. It is often tacitly assumed that primitive and modern bacteria are very similar. This would mean that there was not much evolutionary divergency since the Precambrian. Molecular biological evidence points to the contrary, as we shall see below. In the above discussion we have been concerned only with principles of biochemical pathways. If we had the imaginary possibility to study these primitive microorganisms, it is very likely that they would be considerably, if not totally, different from presently living bacteria.

Geochemical Evidence for the Beginning of the Oxygenic Atmosphere

Geochemical evidence indicates that the transition from an anoxygenic to an oxygenic precambrian atmosphere occurred between 1 and 2 billion years ago (Holland, 1962; Rutten, 1962). These findings are not necessarily in contradiction with Berkner and Marshall's (1965) hypothesis, since the concentration of oxygen that would have been required to oxidize the rocks is not known. Detritus of granites, uranium reefs, and iron formations from the precambrian, which date back to between 3 and 1.7 billion years, all contain ferrous or uranous components. This is regarded as evidence for an anoxygenic atmosphere at that time. The "red-beds" are sediments consisting primarily of quartz, clay, and a small amount of oxidized Fe^{+++}, mostly in the form of limonite $Fe_2O_3.nH_2O$. The latter mineral is formed during superficial oxidative weathering. Its existence serves as an indication of the presence of oxygen. Many red-beds were formed 450 to 200 million years ago and several others are still forming in deserts. There are also red-beds whose exact age is not known but which are estimated to be between 600 and 1,000 million years old. Older red-beds are not known.

Microbial Fossils

The wealth of literature on Precambrian fossils has been reviewed by Glaessner (1962) and by Rutten (1962). Only a very brief summary will be given here. There are very few fossils from the Early and Middle Precambrian. Organisms from this era were very small and rarely had hard cell constituents amenable to fossilization. Rocks bearing such fossils are often located at very great depths and are not infrequently transformed. Finds of undisturbed Precambrian fossils have been limited to old geological shields which have been preserved practically unchanged since their formation. Such shields exist in Canada, Scandinavia, Asia, Brazil, South Africa, and Australia.

There is a great variety of microfossils in late precambrian rocks which date back 600 to 1,000 million years. These fossil remnants include algae, radiolaria, and some metazoa, such as crustaceans. By this time, life had already evolved into a rich variation of forms far surpassing the morphological simplicity of bacteria.

Both direct and indirect evidence can be found in favor of the existence of living beings in the Early and Middle Precambrian.

Indirect Evidence

It consists of remains of presumed microbial activity. One example is the limestone of the Bulawayo area (Southern Rhodesia) which occurs in peculiar wavy layered structures. There are good reasons to believe that these are layers of biotic origin formed by lime-secreting organisms about 2.7 billion years ago. It is supposed by some investigators that algae were responsible for this phenomenon, since lime-secreting algae are known to have existed in later geological history. Algae, however, are not the only organisms capable of precipitating $CaCO_3$. Thus the find can be considered only as an argument for the activity of living organisms and not as evidence for the existence of algae at that time. Analogous limestone deposits exist in the Central Sahara. They are also thought to be of algal origin and are called "stromatoliths." The age of these deposits is not exactly known, but is believed, according to different authors, to be between 1 and 3 billion years old. A number of speculations have been made concerning the organisms responsible, but in reality next to nothing is known about them. Another indirect indicator of biological activity are the "red-beds" which were briefly discussed above.

Precambrian Alga-like Fossils

The work of Barghoorn and Tyler (1965) has revealed what are probably the oldest fossils on earth. The structurally preserved precambrian fossils were discovered in the Gunflint chert of the Northern Lake Superior region. The age of this formation is approximately 2 billion years, placing it in the lower third of the Middle Precambrian. Barghoorn and Tyler have observed a great variety of fossils in these rocks. Eight major groups have been selected for detailed description. The reader is referred to the original paper for the micrographs of these fossils. They were detected in thin sections of rock. The organisms range from single celled spheroidal bodies to filamentous branched and unbranched forms. The genus *Animikiea* contains multicellular unbranched, septated and sheeted filaments. In general morphology, this genus resembles organisms in the living genera *Oscillatoria* and *Lyngbya*. It is, however, premature to say that *Animikiea* was a bluegreen alga. Another genus of septated, multicellular filaments is *Gunflintia*, whose morphology is suggestive of green algae in the family Ulotrichaceae. Nonseptate filamentous organisms, possibly forming spore-like bodies, were grouped into the genus *Entosphaeroides*. Morphologically they are not unlike some bluegreen algae, *Chamaesiphon* or the iron bacteria *Crenothrix*. Structures with irregular weird bulbs were placed in the genus *Archaeorestis;* some of these forms vaguely resemble the green algae Vaucheriaceae. The genus *Huroniospora* consists of forms the size of yeast, having spherical to ellipsoidal bodies with sculptured walls. The genera *Eoastrion* (star-shaped), *Kakabekia* (tripartite organization, consisting of bulb, slender stipe, and umbrella-like extrusion) and *Eosphaera* (inner sphere and randomly attached tubercles, encompassed by an outer thick-walled spherical membrane) all contain forms which are completely distinct from presently existing microorganisms.

The unusual morphology of these latter fossil genera strongly suggest that they are indeed autochtonous and represent the true remains of the precambrian flora for two reasons. For one, it has been shown several times (Bien and Schwartz, 1965; Myers and McCready, 1966) that bacteria can penetrate rocks with surface water or air by way of fissures and pores in the rock. Myers and McCready believe that recent penetration of rocks with living organisms followed by petrification and opalization (which takes only 20 to 30 years under certain conditions) suffices to explain the presence of fossil-like microorganisms in many geological specimens. The unusual morphology of Barghoorn's organisms is a sure indication that this is not the origin of his fossils which are very likely as old as the rocks in which they are embedded. The second argument is derived from molecular biology. Because of the great number of possible nucleotide combinations in their chromosome, the number of possible microorganisms is astronomically high as we shall show below. It is a commonly known fact that organisms are ever changing in the course of evolution. Even bradytelic forms (the so called "living fossils") are distinctly different from their ancestors. Therefore it can be expected that microorganisms living at the dawn of time were morphologically quite different from presently living forms and it is not surprising that fossil forms are found which are morphologically not related to any known type.

There is, however, a reverse side to this picture. Analogies in morphology between fossils and modern microorganisms do not necessarily mean relatedness or analogy in function. Because *Animikiea* resembles the bluegreen algae *Oscillatoria* and *Lyngbya* it does not necessarily imply that it was analogous or similar. It might have been a bluegreen alga with completely different properties or even no bluegreen alga at all.

Barghoorn and Tyler's fossil-containing chert also contained a variety of organic compounds, e.g., alkanes and aromatic hydrocarbons, carbonyl, alcohol, and ester groups. Pyrolysis yielded a variety of hydrocarbons. There is evidence that the filamentous structures were really photosynthetic algae. The evidence consists of: (1) The structure of the algal domes containing the organisms is similar to algal domes of bluegreen algae existing today in shallow waters. (2) The ratio of $^{13}C/^{14}C$ is as expected from photosynthetic activity. The pyrite, organic matter, and the iron-bearing carbonates indicate reducing conditions; the algal domes and the granular chert suggest agitated water. *If* those organisms really possessed a plant photosynthetic system (which is not necessarily true, because the chemiautotroph *Beggiatoa* also has a similar morphology and a Calvin cycle) then the gunflint time of about 2 billion years ago represents the emergence of oxygen in the paleoatmosphere. This would have been the beginning of our fourth stage in which oxidizing microhabitats occurred within the largely anoxygenic atmosphere. The claim that *Kakabekia*-like organisms (Siegel and Giumarro, 1966) can be isolated from soils in ammonia-containing atmosphere, will need further confirmation.

Fossil Bacteria

On numerous occasions authors have described the occurrence of objects which are believed to be fossilized bacteria (reviewed in Myers and McCready, 1966; Kuznetsov et al. 1963; Beerstecher, 1954).

Possibly the best documented evidence is that presented by Schopf et al. (1965). In the same gunflint chert as mentioned above, electron microscopy revealed the presence of fossilized cocci and rod-like bacteria. Except for one thick-walled coccoid type, the other cells are indistinguishable from presently living bacteria. Very cautiously the authors point out that the fossils resemble the modern iron bacteria *Sphaerotilus natans, Sidero-capsa,* and *Siderococcus,* but that their actual relationship can not be known. Cloud (1965) has likewise observed fossils of microbial origin. The original paper should be consulted for the micrographs. These fossils resemble filamentous bacteria such as the iron precipitating *Sphaerotilus,* and possibly other iron bacteria such as *Gallionella* and *Metalogenium.* Observed also were other unidentified bacteria-like structures and larger forms resembling bluegreen algae.

Fossilized bacteria have been detected in all kinds of layers from the Precambrian up to recent formations. That these fossils so frequently resemble the iron and sulphur bacteria is striking. This implies that a great number of chemiautotrophic microorganisms, in particular bacteria, have been effective in oxidizing Fe^{++} to Fe^{+++}. Some 90 percent of the sulphur formation in the lakes of Cyrenaica in North Africa and lake Sernoye in the USSR are quoted as evidence. The sulphur deposits of the domes along the shores of the Gulf of Mexico also originated by biological activity, possibly by some *Desulfovibrio.* These bacteria reduce $^{32}SO_4^{--}$ more readily than $^{34}SO_4^{--}$ and the ratio $^{32}S/^{34}S$ in the several sulphur fractions (sulphate, sulphur, and sulphide) corresponds with this view. The age of these sulphur domes is about 800 million years. It may also be recalled that several microbiologists attribute the formation of peat, coal, and oil to the activity of microorganisms in ancient times.

What was said above concerning algae applies even more strictly to bacteria. It is impossible to draw taxonomic conclusions from the morphology of the fossils alone. Because the fossils resemble iron bacteria or blue-green algae does not mean that they actually *were* iron bacteria or blue-green algae. Summarizing, it can be said that there is good evidence that bacteria lived in the Precambrian, but as to their functions we can only speculate.

Bacteria as "Living Fossils"

Every now and then the notion arises that living bacteria, isolated from ancient geological formations, were actually living at the time when these rocks were formed, having been preserved in a state of latent life up to now. These bacteria would thus be real "living fossils," witnesses of times past. Lipman was the main advocate of this theory and published a series of papers between 1928 and 1937. The belief in surviving fossil bacteria has been revived more recently by Dombrowski (1960, 1963a and b) who isolated viable bacteria from a variety of salt deposits, dating from the Precambrian on.

The arguments in favor of this opinion are not convincing. They are as follows: (1) If bacteria can be preserved for several years by lyophilization for example, then they can as well remain vital for hundreds of million years. The fallacy of this argument will be refuted below.

(2) Preservation in dry salt is thought to be beneficial over long periods. There are some data indicating that preservation for 5 years in dry salt is not harmful. It is obvious that this conclusion can not be extrapolated over hundreds of million years.

(3) Dombrowski (1963a) produced photographs of a single viable bacterium embedded within salt crystals. It is claimed that these crystals have not changed since their formation in the Permian period. However, there is no irrefutable proof of this statement. In fact, the opposite is probably true. The presence of bacteria may be a proof that these crystals are indeed no longer primary.

(4) Strains isolated from Permian salts produce acids from several carbohydrates; those from Precambrian and Silur only rarely. Gram negative bacteria were not detected in the latter two types of rocks. This is taken as an argument that the enzymic outfit increases with evolution (Dombrowski, 1963a,b); however, the number of organisms isolated is far too small to be of statistical significance.

(5) The mere presence of these bacteria in rocks is considered as a proof of their geological age.

The criticism against this hypothesis is manifold.

(1) Accidental contamination of the rock samples with foreign bacteria by handling is easily possible.

(2) As contrasted with Dombrowski's results, Bien and Schwarz (1965) have never detected viable bacteria in the Zechsteinsalze.

(3) We know from hydrogeology that surface water migrates through rocks in every direction, down to some 4,000 m. It is obvious that this water contains great numbers of bacteria. The layers of Zechsteinsalze in Nauheim, from which Dombrowski isolated his "Permian bacteria," are in continuous contact with water, which emerges many miles further. The same bacteria "*Ps. halocrenaea*" can be isolated both from the salts and the well water. It has been demonstrated experimentally that bacteria can penetrate rocks (see Myers and McCready, 1966). Bien and Schwartz (1965) found Zechsteinsalze to be easily pervious to foreign bacteria through microcapillaries.

(4) The longevity of organisms was discussed by Sneath (1962, 1964) and De Ley, Kersters, and Park (1966) on the basis of observational and experimental data. They conclude that life can only rarely remain dormant for about 1,000 years as in seeds of the water-lily *Nelumbo*. Spore forming bacteria can still survive after storage for 300 years in dry soil, but human intestines and coprolites 1,000 to 3,500 years old are sterile. Revival of a dormant cell is impossible without the simultaneous presence of over 1,000

enzyme types and cofactors, many of which are very labile. Several of the latter compounds are so labile that they decompose noticeably even at $-20°$ C in the dry state. How then could bacteria remain viable for hundreds of million years? Even the more stable building blocks of proteins, the amino acids, decompose spontaneously during geological time spans, as demonstrated experimentally by Abelson (1959) and Vallentyne (1965).

(5) The case of *"Pseudomonas halocrenaea."* The relationship between modern pseudomonads and *"P. halocrenaea,"* an organism isolated from Permian salt deposits and claimed by Dombrowski (1960) to have lived 200 million years ago, was investigated by De Ley, Kersters, and Park (1966). The nucleotide sequence, nucleotide composition, and compositional distribution of the *halocrenaea*-DNA is indistinguishable from modern pseudomonad-DNA, in particular from *P. aeruginosa.* Eighty-eight different phenotypic tests showed that *P. halocrenaea* was merely *P. aeruginosa.* The explanation is obviously that *P. aeruginosa,* which is a common inhabitant in soil and water, infiltrated the water-permeable salt deposits rather recently. It is not necessary to invoke the theory that these bacteria have been incorporated in the salts when the Permian sea dried out 200 million years ago. If one does, one is faced with a number of formidable problems as pointed out by De Ley et al. (1966). It would mean that DNA from *P. aeruginosa,* and thus probably from many other bacteria, hardly changed by mutation over a period of 200 million years. This is in flagrant contrast to the observable rate of mutation of bacteria in laboratories and in hospitals. *P. aeruginosa* would be a case of bradytelic evolution, which is a rare phenomenon. It would mean that the rate of mutation of bacteria is nil as compared to higher organisms. Since the organism from Zechsteinsalze is very pathogenic for mice, this pathogenicity could only have arisen *after* the development of mammals and mice, which did not yet exist in Perm. All these inconsistencies can be overcome by the simple assumption that *P. aeruginosa* recently penetrated the water-permeable salt deposits with ground water. Summarizing, it may be said that no convincing arguments exist in favor of, and that many arguments are against, the theory about viable fossil bacteria.

The Relationship between Bacteria and Bluegreen Algae

More and more evidence is accumulating which points to a close relationship between bacteria and bluegreen algae. It would seem that ordinary heterotrophic, chemiautotrophic, and photoautotrophic bacteria, as well as bluegreen algae are all different evolutionary facets of a central group of primitive beings. These organisms display several properties which are encountered nowhere else (Stanier and van Niel, 1962; Echlin and Morris, 1965).

(1) They are procaryotic, having no separate nuclear apparatus so that their DNA is in direct contact with the cytoplasm. There is convincing evidence that in several bacteria—and possibly in most if not all—there is a single circular chromosome. Studies on the structure of DNA in bluegreen algae are woefully lacking.

(2) Nuclear division occurs by fission, not by mitosis.

(3) The photosynthetic bacteria and the bluegreen algae lack chloroplasts. Their photosynthetic apparatus occurs in very simple structures, such as paired lamellae or small spherical vesicles (chromatophores).

(4) These organisms lack mitochondria. In aerobic bacteria, electron transport and part of the Krebs cycle are mainly localized on the cytoplasmic membrane; a culmination of this situation is reached in the acetic acid bacteria (for reviews see De Ley 1960, 1964). The mesosomes of bacteria and lamellasomes of the Cyanophyta may be analogous.

(5) The cell wall of nearly all procaryotic organisms contains muramic acid and either diaminopimelic acid or lysine. Muramic acid has never been found in eucaryotic cells. The cell envelopes of bluegreen algae and Gram positive bacteria are similar as seen through the electron microscope. Ornithine biosynthesis and sensitivity to certain antibiotics are also similar in both groups. The primitive procaryotic ancestor of bacteria and bluegreen algae thus apparently had a naked, single chromosome, divided by fission, lacked chloroplasts and mitochondria, and had a cell wall with mucopeptides. Through evolution this ancestor diversified in several directions, which eventually lead to present day procaryotic protists which exhibit: (1) varied organization (single cells, multicellular, coenocytic); (2) different locomotion (mono- and multiflagellation; gliding movement; no motility); (3) a varied metabolism.

As oxygen became abundant, these procaryotic organisms developed a system for aerobic electron transport which became associated at least in part with the cytoplasmic membrane. It seems possible that the genes for the mitochondrion originated from those for the cytoplasmic membrane and that this event occurred with the rise of the algae proper. Mesosome-like structures (small vesicular structures, formed by folding of the cytoplasmic membrane) might have been intermediates in the evolution of the cytoplasmic membrane to the mitochondria. If we follow the current assumption that the protozoa evolved from primitive algae, this could explain why all eucaryotic organisms have mitochondria. The mitochondrion is about 20 times more efficient than the bacterial membrane in producing ATP. Thus the advent of mitochondria represented a great progress in cellular efficiency, which undoubtedly opened up new possibilities in cellular and multicellular development as well as biomass production in general.

Although there are many similarities between photosynthetic bacteria and bluegreen algae, there are also major differences as shown below.

bluegreen algae	photosynthetic bacteria
usually aerobic	usually anaerobic
no flagella	usually flagella
use H_2O in photosynthesis and produce O_2	do not use H_2O in photosynthesis and thus do not produce O_2
chlorophyll	bacteriochlorophyll
	Chlorobium chlorophyll

These differences are sometimes used as arguments for a lack of evolutionary relationship between these two groups. However, dissimilarities between organisms from the same origin are unavoidable, particularly when they started to evolve in separate ways some 1 or 2 billion years ago. From a molecular biological point of view, similarities are more significant than dissimilarities. Similarities mean genes, thus nucleotide sequences in common. A gene has about 1,000 nucleotide pairs, in this DNA region there are some $4^{1,000}$ possible nucleotide combinations, of which only a fraction constitutes the gene. The chance that the same gene will arise twice independently (convergence) in two organisms from a different evolutionary origin is very small indeed. The chances are obviously even smaller if several genes would have to arise. Furthermore continuous mutation and selection provoke changes in existing genes and lead to ever increasing dissimilarities.

Fossil Bluegreen Algae

Several fossils with a tubular structure are thought to be the empty sheets of bluegreen algae (for a review see Desikachary, 1959). However, as Fritsch (1965, p. 859) has pointed out, these fossils can not be used for phylogenetic considerations in view of their questionable nature and classification.

Phylogeny of Bluegreen Algae

Desikachary (1959) has reviewed the different hypotheses on the phylogeny of bluegreen algae. They are based mainly on morphological data.

The diversity of forms in the Cyanophyceae makes the task of reconstructing their possible evolutionary history a difficult one. In general, however, the diverse filamentous types are considered to have evolved along a plurality of lines from a common filamentous ancestry, which in turn was derived from coccoid-like forms (Fritsch 1965, p. 769).

The few data available on the percent GC of the bluegreen algae testify to their wide evolutionary divergency. *Lyngbya estuarii, Nostoc puncti-forme,* and *Aphanizomenon flos-aquae* have GC values of 64, 54, and 44 percent respectively (Serenkov, 1962). To have differences of 10 to 20 percent GC these algae would have to differ greatly in their genetic make-up. Yet on the basis of morphology they are placed in the same order Nostocales!

It is expected that studies on DNA nucleotide composition, DNA hybridizations, and other molecular biological approaches will bring significant contributions to this field. Until then, theories on evolution of Cyanophyceae can only be speculative.

DNA Nucleotide Composition of Microorganisms

With but a few exceptions, DNA from micro- and macroorganisms contains four kinds of organic bases [adenine (A), thymine (T), cytosine (C), and guanime (G)] as the major components. Because of the double-stranded nature of the DNA molecules the molar fraction of A is the same as T and of G is the same as C. Each type of DNA is characterized by its molar nucleotide composition and in practice this is represented as the overall molar percent $GC = 100 \ (G + C)/(A+T+G+C)$. The *sequence* of these bases constitutes the genetic information. Since many cistrons (units of functional information) differ from one group of organisms to another, the percent GC may also differ. Among vertebrates the difference in percent GC is imperceptibly small being typically about 42. In plants the range is somewhat wider (37 to 49 percent) and is wider yet among invertebrates. The percent GC content of algae, protozoa, and molds varies as much as 30 percent and is particularly wide in bacteria where it extends from 25 to 75 percent GC.

The explanation for this may be that vertebrates, which are of relatively recent origin and which in principle are built according to the same general plan, have many homologous cistrons which differ only slightly in nucleotide sequence. In fact Bolton et al. (1963–1964) have shown that some 20 percent of mammalian DNA from different organisms is nearly identical and that DNA homology in primates is as high as 95 percent. In spite of their simple and similar morphology, bacteria and other microorganisms have evolved over a much greater period of time and are—relatively speaking—in essence much more varied than the most widely varied groups of vertebrates. We shall see this confirmed below when we discuss interpretations from DNA homology.

The nucleotide composition has been studied rather extensively in bacteria, because of its usefulness in their classification. Classification itself is the grouping together of organisms with a certain level of relatedness. The degree of relatedness is the result of evolutionary processes. With phenotypical methods it was nearly impossible to determine the true degree of relatedness of bacteria. As a result, bacterial classification is confused and biased by the personal opinion of the investigators. It is likely that the more recently developed quantitative methods of numerical taxonomy, DNA nucleotide composition, and DNA homology will remedy this situation.

The classical picture of evolution as a flat-topped tree or bush steadily growing at the apex, applies also to bacteria. The intersection of the branches and twigs with a flat plane at the top is a fair picture of classification. Because of the lack of paleontological, embryological, and comparative anatomical data, the actual evolutionary tree of bacteria is unknown to us.

Percent GC values may give us our first clue. Extensive work in this direction will be discussed at length elsewhere (De Ley, to be published). A survey of the known percent GC values is summarized in Figs. 2, 3, 4, and 5. The following conclusions can be drawn from these data:

(1) Organisms with many morphological, physiological, and biochemical features in common (a genus), have also closely related percent GC values. This suggests that they have many cistrons and hence many nucleotide sequences in common and thus may have a common evolutionary origin. The percent GC range in several genera is variable and sometimes ranges over 20 percent GC. The percent GC values of the organisms in one genus are so close together that differences between species are usually not distinguishable. Furthermore, the entire concept of bacterial species is still unsettled. The only biological unit which has survived the close scrutiny of modern bacterial taxonomists is the genus, and they are the only groups which we shall consider here.

(2) There are several examples of organisms whose percent GC value falls completely out of the genus range. These exceptions are believed to be indicative of incorrect classification. In many cases the molecular biological conclusions have indeed been confirmed by phenotypic identifications. Several other cases remain to be resolved (e.g., *Cytophaga, Flavo-*

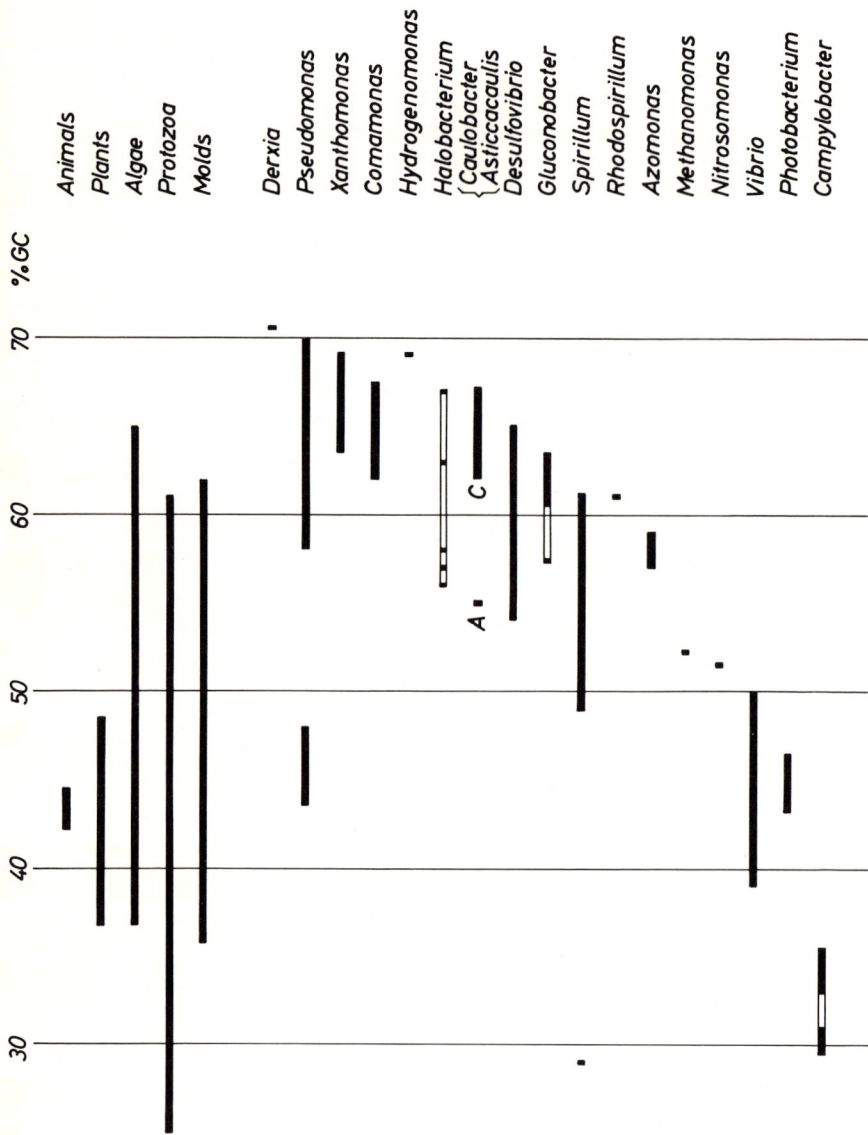

Fig. 2. DNA base composition of several genera of Gram-negative polarly flagellate bacteria.
A single dot (e.g. Derxia) means that only one strain of the genus has been studied. Longer horizontal bars mean that several or many strains have been used. Open spaces in bars mean that no typical strains have a percent GC in this range. Some genera are represented by two bars (e.g. Pseudomonas); strains with percent GC aberrant from the main group will have to be integrated in other genera. The percent GC ranges of animals, plants, algae, protozoa, and

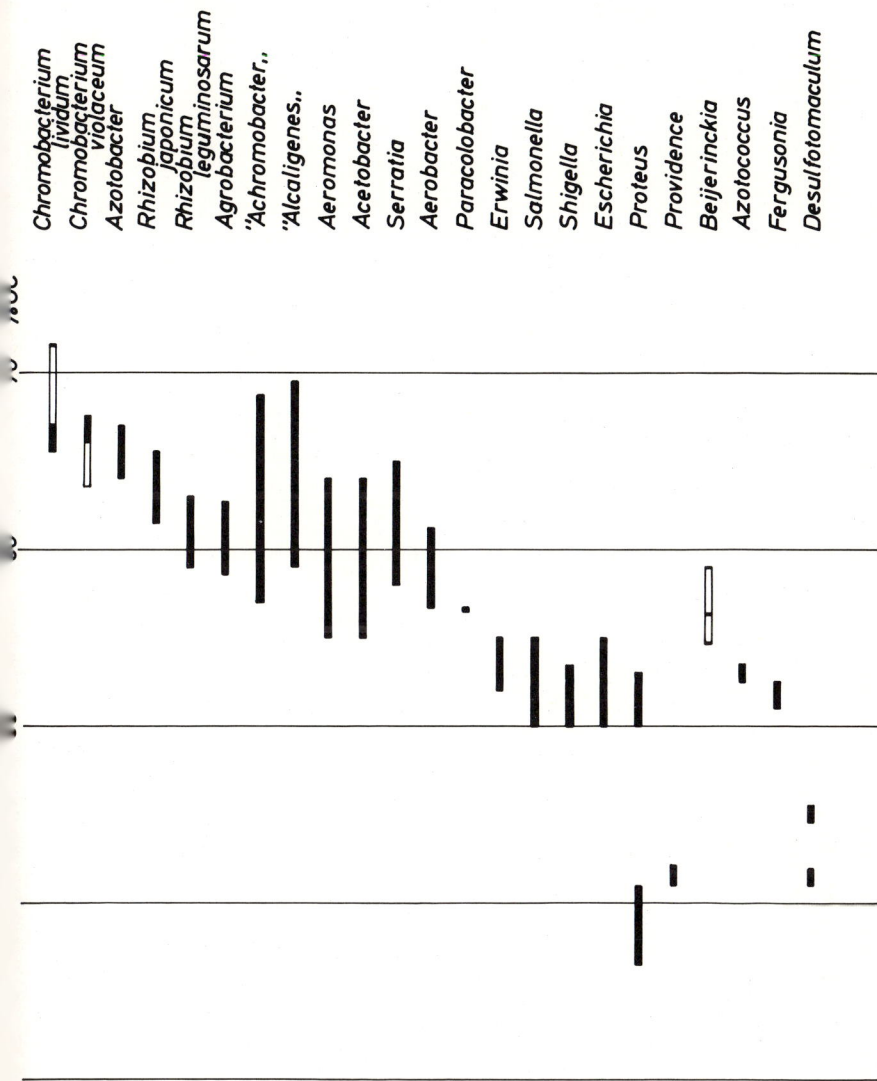

Fig. 3. DNA base composition of several genera of Gram-negative peritrichously flagellated bacteria. Explanation as in Fig. 2. The genera of motile "Achromobacter" and "Alcaligenes" are still ill defined. (De Ley, unpublished.)

125

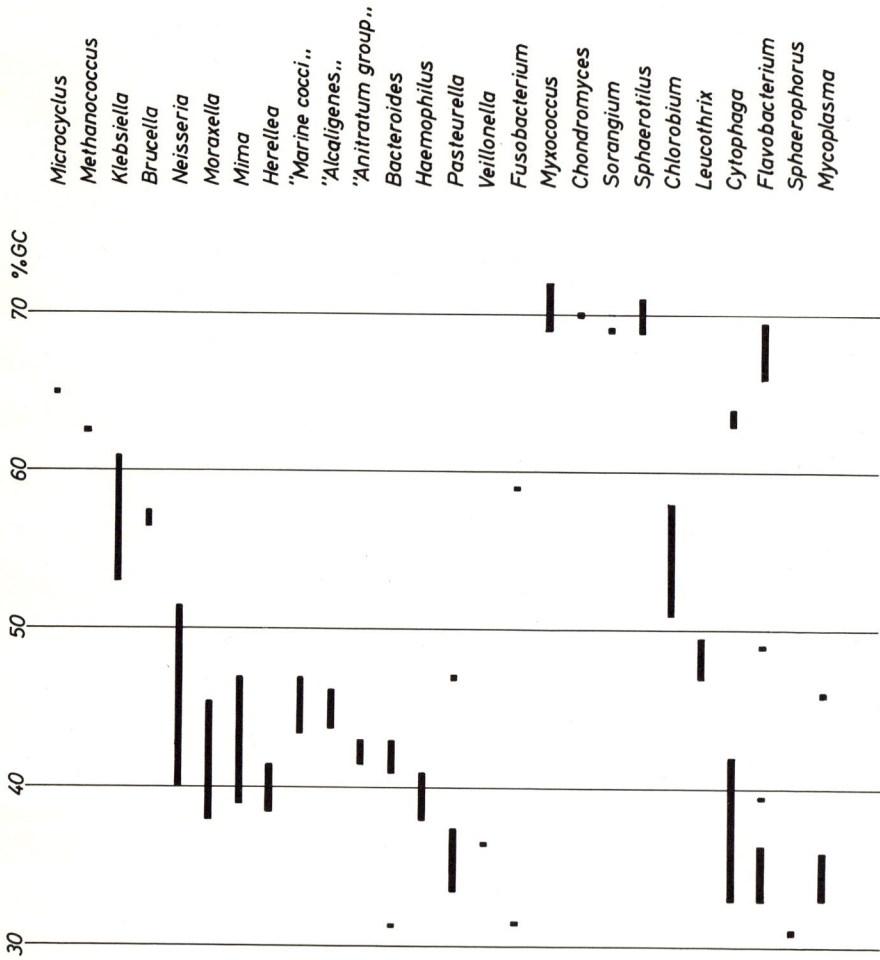

Fig. 4. DNA base composition of several genera, mostly Gram-negative nonflagellate bacteria, as well as some other genera. Explanation as in Fig. 2. The groups of "marine cocci," nonmotile "Alcaligenes" and

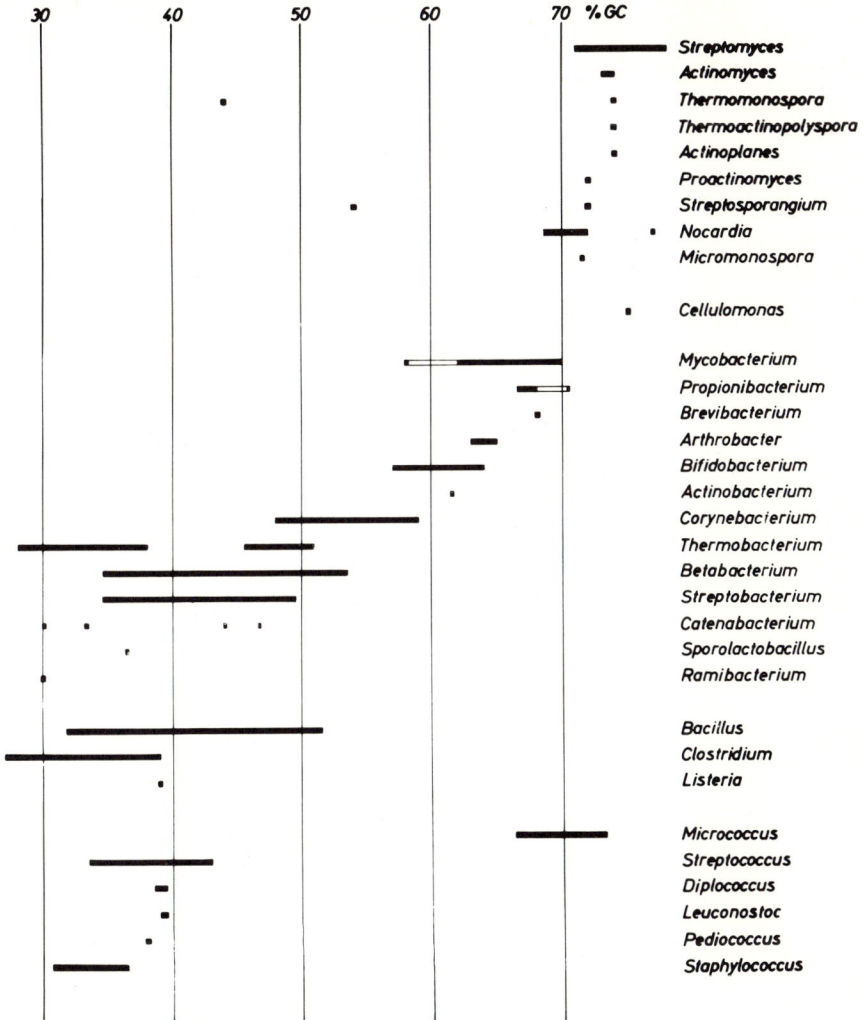

Fig. 5. DNA base composition of Gram-positive bacteria. Explanation as in Fig. 2. (De Ley, unpublished.)

bacterium, Pseudomonas, and Spirillum) but from molecular biological considerations it leaves no doubt that the existing exceptions will either be assigned to another genus, or that new genera will have to be created for them.

(3) Organisms or genera which have closely related percent GC values *may* have evolutionary ties; however, it need not be that way. Only DNA

homology studies can resolve these questions and have already partially done so (see below). As it now stands, the Fig. 2, 3, 4, and 5 show where relationship *may* be found. For example among the organisms with about 60 percent GC the following ones *may* be of the same evolutionary origin: *Rhizobium, Agrobacterium, Acetobacter, Serratia, Pseudomonas, Gluconobacter, Achromobacter, Alcaligenes, Aeromonas, Aerobacter, Klebsiella, Halobacterium, Desulfovibrio, Spirillum, Mycobacterium,* and *Bifidobacterium.*

(4) Organisms which are a certain GC distance apart can have no or very few nucleotide sequences in common. Thus they cannot be related and are phylogenetically far removed from each other. This can be calculated from the compositional distribution of DNA fragments from the chromosome, as determined by either density gradient centrifugation or thermal denaturation. Sueoka (1961) calculated that organisms which differ by 10 or more percent GC have only few chromosome fragments in common and are relatively unrelated in terms of evolution. Our own direct measurements on DNA homology and compositional distribution showed that 25 percent GC difference was more close to reality (De Ley, unpublished). This means that a *Clostridium* or a *Bacillus* with 30 percent GC has an almost completely different set of cistrons from organisms with a percent GC of 55 and higher.

(5) Some trends in evolution, which have been suspected on phenotypic grounds, appear to be corroborated. The outstanding example is found in the series of increasing morphological complexity of *Corynebacterium* (48 to 59 percent GC), *Arthrobacter* (63 to 65 percent GC), *Mycobacterium* (60 to 70 percent GC) and actinomycetes (68 to 78 percent GC).

Evolutionary Changes in Bacterial DNA

The primitive paleocells very likely had only a small genome. Evolution proceeded by increasing its size (increases in the total amount of DNA per haploid nucleus) and by mutational changes in the sequence of the four bases, A, T, G, and C, usually—but not always—leading to changes in overall DNA nucleotide composition. Each topic will be discussed separately.

The Size of the Bacterial Genome

The available information is collected in Table 1.

For bacteria the values vary over a 16-fold range, from 300,000 nucleotide pairs in *Mycoplasma* to about 5 million in *Escherichia*. The total DNA content in a haploid cell of higher organisms is about 1,000 times greater than in bacteria. Obviously great changes have occurred by addition and/or deletion.

Increase in the size of a genome is possible by several mechanisms. New genetic elements can exist either in an autonomous cytoplasmic state or in a state of integration with the bacterial chromosome.

(1) Kornberg (1965) showed that DNA polymerase can synthesize polymers in vitro without any template present. In spite of the fact that they are highly branched, it is conceivable that stretches are incorporated into chromosomal DNA, thus lengthening the chain. By the same mechanism of replication and slippage, nucleotides may be added to a chain of DNA primer. Those new additions have probably a nucleotide sequence which has nonsense information; mutation would change them into informational DNA.

(2) Genetic material can also be acquired from other microorganisms and can be introduced in a variety of ways: (a) By conjugation. Examples of this are the transfer of the episomal R factor (multidrug resistance) and the sex factor. Some of these transfers can occur between widely divergent genera, such as the Enterobacteriaceae and *Vibrio*. (b) By transduction through phage. (c) By transformation with free DNA.

(3) By gene or DNA nucleotide sequence duplication or insertion. These genes or DNA regions might then evolve independently by changes in their nucleotide sequence, ultimately leading to two different genes. The existence of iso-enzymes and the various hemoglobin chains in higher organisms support this hypothesis. Another powerful argument is the finding of the repetitive units in vertebrate DNA by Bolton and his colleagues (1964–1965).

It would thus be useful to know the molecular weight of DNA of a great number of bacterial genera. This would not only be valuable as an aid in classification but would allow a rough estimate of the range of the number of cistrons in the microbial world.

In higher organisms not all DNA is phenotypically expressed. In bacteria it appears that most, if not all of the DNA is genetically functional containing little or no nonsense DNA or unused genes. This conclusion was reached by De Ley et al. (1966) upon comparison of taximetric similarity of features and DNA homology of the same organisms (see also Fig. 13).

Evolution Drift in Percent GC

While the mean DNA base composition varies greatly among different bacteria, the heterogeneity in base composition of a DNA molecule from any one species is rather narrow. This, in terms of evolution of bacteria

Table 1. The Size of the Bacterial Genome.

Organism	Molecular weight in dalton	Nucleotide pairs per nucleoid	Estimated number of cistrons	Reference
Mycoplasma gallisepticum	ca. 0.2×10^9	ca. 0.3×10^6	300	Morowitz et al. (1962)
Haemophilus influenzae	0.72×10^9	1.2×10^6	1,200	Berns and Thomas (1965)
Aerobacter aerogenes	1.2×10^9	1.9×10^6	1,900	Caldwell and Hinshelwood (1950)
Pseudomonas campestris var. *pelargonii*	$(2.1 \pm 0.3) \times 10^9$	3.4×10^6	3,400	Park and De Ley (1967)
Pseudomonas fluorescens	$(2.5 \pm 0.7) \times 10^9$	4×10^6	4,000	Park and De Ley (1967)
Pseudomonas putida	$(2.7 \pm 0.3) \times 10^9$	4.4×10^6	4,400	Park and De Ley (1967)
Bacillus subtilis	1.3×10^9	2.1×10^6	2,100	Dennis and Wake (1966)
Bacillus subtilis	2.4×10^9	3.9×10^6	3,900	De Ley and Park (to be published)
Bacillus subtilis	2 to 4×10^9	3.4 to 6.5×10^6	3,000 to 6,000	Massie and Zimm (1965)
Escherichia coli	2.8×10^9	4.5×10^6	4,500	Cairns (1963)
Escherichia coli	$(3.1 \pm 0.2) \times 10^9$	4.9×10^6	4,900	Park and De Ley (1967)
Saccharomyces cerevisiae	43×10^9	70×10^6	70,000	Ogur, Minckler, and Mc Clary (1953)
Neurospora crassa	26×10^9	43×10^6	43,000	Horowitz and MacLeod (1960)
Aspergillus nidulans	25×10^9	41×10^6	41,000	Pontecorvo and Roper (1956)

The number of cistrons is calculated, assuming that there are about 1,000 nucleotide pairs per cistron. Data for a yeast and two molds are given for comparison.

with extensive differences in their percent GC content, must represent widely divergent organisms. Theoretical studies have sought to explain the origin of this phenomenon (Sueoka, 1962; Freeze, 1962). This theory is based on a rather uniform mutation and selection pressure affecting all base pairs equally in an organism with x percent GC. Nucleotides are converted at rates u and v

$$GC \underset{v}{\overset{u}{\rightleftharpoons}} AT$$

The rates of conversion u and v are probably smaller than 10^{-8} per nucleotide pair per generation.

From Sueoka's formula $\Delta p = v - (u + v) p_n$ in which $\Delta p = $ change of percent GC in one generation, $p_n = $ the GC content in mole fraction at a certain generation n, u and v as above, the following examples can be calculated.

The Clostridia, with a percent GC of about 25 will not change further when $v/u = 1/3$, *E. coli* (50 percent GC) when $v/u = 1$ and the Actonomycetes (75 percent GC) with $v/u = 3$.

It can also be calculated whether equilibrium has been reached. In other words, whether bacteria are evolutionary at rest. At equilibrium, (when no further changes occur in percent GC) the distribution of the base composition should be unimodal, which it is in practice. The theoretical variance σ is expected to be between 0.3 and 1.3 percent GC. In reality it is higher. Our experience shows that it is greater than 3.5 percent GC. This probably means that most bacteria today are approaching evolutionary equilibrium but have not yet completely reached it for present ecological conditions.

What would happen if an organism, say *E. coli* with 50 percent GC, in evolutionary equilibrium at a v/u ratio of 1, was obliged to live under new conditions of mutation and selection where the v/u ratio was, say 2? Figures 6 and 7 show that it would increase its percent GC to about 66. We can also make rough estimates as to how long this change in percent GC will take. Let us take an extreme case of an organism with 25 percent GC in evolutionary equilibrium. It is suddenly subjected to new conditions where the v/u ratio is 3. It will tend to its new equilibrium state of 75 percent GC. If we assume a generation time of 1 hr., and a value of about 10^{-9}/nucleotide pair/generation for v and u, then from Sueoka's formula:

$$n = \frac{1}{u + v} \ln \frac{\bar{p}_0 - \hat{p}}{p_n - \hat{p}}$$

in which n = number of generations, $\bar{p}_o$ = initial GC molar fraction (here 0.25), $\hat{p}$ = final GC molar fraction (here 0.75), p_n = intermediate molar fraction, we can calculate the following results.

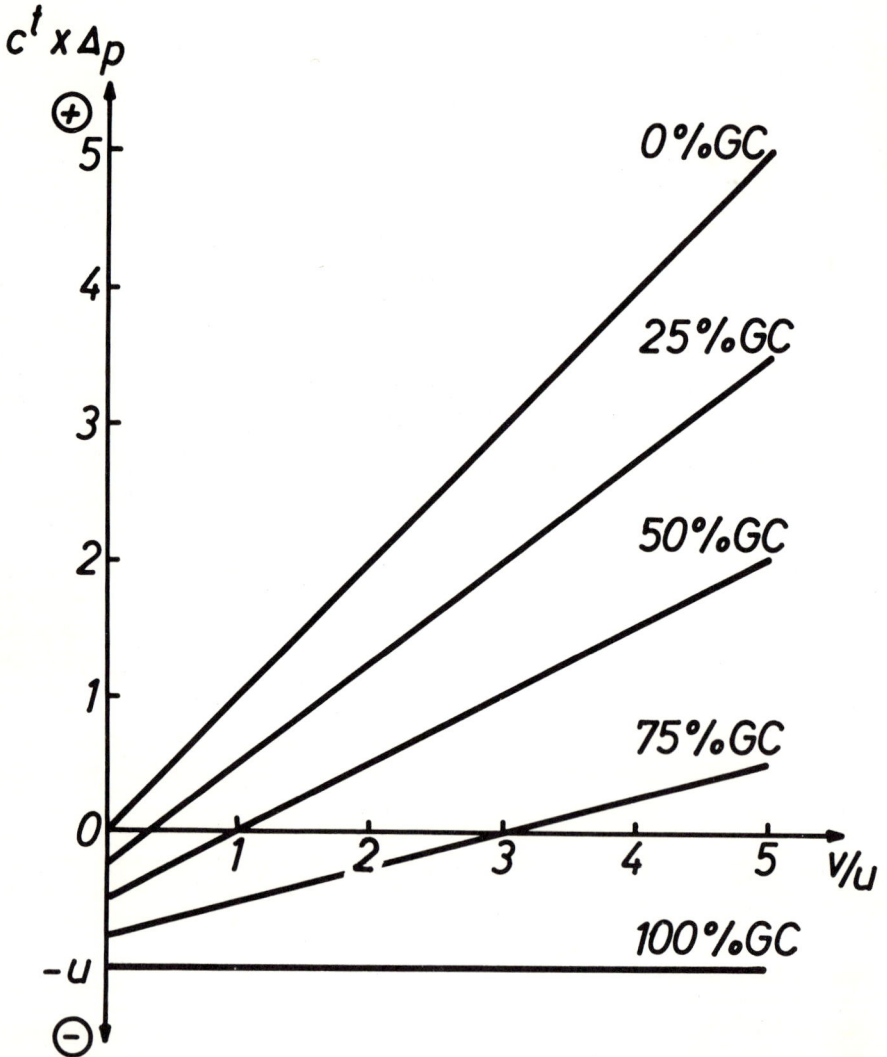

Fig. 6. Change in GC content (molar fraction) in one generation ($\triangle$p), either increasing (+) or decreasing (−), for different types of organisms (expressed as percent GC along the lines) as a function of the ratio of change of AT $\rightarrow$ GC (v) versus GC $\rightarrow$ AT (u). Data calculated from Sueoka's (1962) formula $\triangle p = v - (u + v)\, p$.

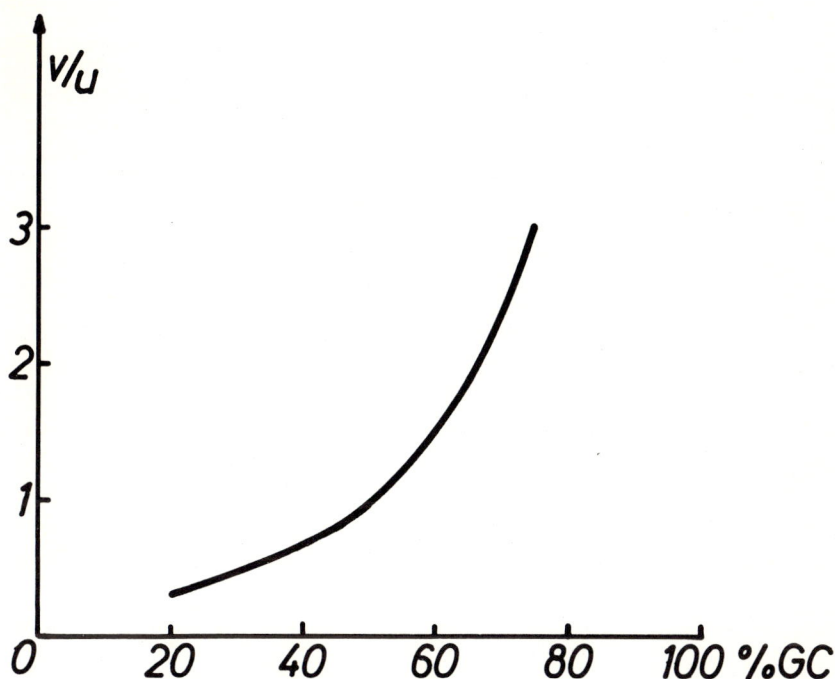

Fig. 7. The relation between the percent GC at equilibrium and the ratio of the rates of change $AT \rightarrow GC$ (v) versus $GC \rightarrow AT$ (u). Data calculated from Sueoka's (1962) formula $p = u / (u + v)$.

% GC	number of generations	Time needed in years $(n = 1 \text{ hr.})$	maximal nucleotide changes per year
25	0	0	
35	1×10^8	12,000	33
45	2.4×10^8	27,000	27
65	7.6×10^8	87,000	13
70	11×10^8	120,000	6
74	18×10^8	200,000	2
75	∞	∞	

The results are admittedly too low: (1) because they describe an idealized situation, and (2) because the values of u and v are difficult to estimate and are on the conservative side. These calculations are presented more as a matter of principle, however, than as an actual dating of changes. They illustrate the following points: (1) The mean percent GC value of bacterial DNA is very stable, as it takes at least several thousand years for a noticeable deviation from the genus range (10 percent GC). (2) The bacterial population may change considerably within relatively short geological periods, a conclusion which we will also reach on other grounds, as described below (page 37).

An experimental attempt was made (Park and De Ley, 1967) to determine the evolutionary drift in the GC content of the pseudomonads. By repetitive DNA hybridization of two pseudomonads (*P. fluorescens* and *P. putida*) and a xanthomonad (*P. campestris* var. *pelargonii*) we isolated a DNA stretch, consisting of about 50 percent of the chromosome, which was similar in the three organisms. Thus this common part had apparently not changed during the evolutionary divergency of the three organisms. This is in agreement with other observations that there are "hot spots" in DNA which mutate readily and other parts which are very resistant to mutational changes. The percent GC of the common part was found to be 62, whereas the percent GC of the total native DNA from the three organisms was also 62 percent. Unfortunately in this case it was not possible to determine the direction of the GC drift, because the three organisms have not yet diverged enough in their evolution. Similar experiments on a common DNA part from three organisms with a different percent GC might solve this problem.

Drastic Mutational Changes in DNA Base Composition

Natural mutation is presumed to have happened through a gradual change in base pairs and not by sudden gross alterations. Indeed, the percent GC of several mutants and their wild type parental strain is usually the same (see De Ley, 1964a). Several cases are now known, however, where mutagens provoke great changes in percent GC (for a summary, see Table 2). Substantial physiological and biochemical changes often accompany such mutations. In all cases up to now, the mutants have a higher percent GC and quite frequently they show impaired respiration, loss of enzyme systems, and so forth.

A few authors fear that the mere existence of these mutants endangers the value of bacterial classification systems. This is not so, since these mutants arise under highly artificial conditions which are not likely to be encountered in nature. The effect of UV suggests that these jump-mutations may have occurred in paleoconditions, when the UV radiation from the sun struck the surface of the seas and land undiminishedly.

Possible Dissimilarity between Ancient and Modern Bacteria

A bacterial chromosome harbors an enormous potential for diversification. An average bacterial DNA genome, e.g., of *Pseudomonas* contains a sequence of about 4×10^6 nucleotide pairs. It can theoretically yield $4^{4 \times 10^6}$ or $10^{24,000,000}$ different combinations. This is the total number of bacterial chromosomes that could in theory exist. Since the percent GC for bacteria is within the 25 to 75 percent range, only half of this number could yield cells. In addition many of the remaining possible DNA's would be nonsense DNA and would not yield viable cells. Still, the remaining number of possible viable individual bacteria must be astronomically high. One can roughly estimate the ratio of this potentiality present in modern bacteria. A generous rough estimate of the number of existing bacterial species amounts to about 10^4 and the number of varieties and strains within each species is probably less than 10^4. The total number of individual bacterial cells on earth may be estimated to be around 10^{27}. This is very small compared to the number theoretically possible, suggesting that an enormous number of bacteria could exist, each widely divergent from present day ones. (De Ley et al., 1966).

Another consideration to the same effect follows from the rate of mutation. Near the end of the Permian period, mammals and somewhat later birds started to branch off from the reptiles. It is known from the DNA hybridization experiments of Bolton et al. (1963–64) that some 80 percent of the reptilian chromosomal DNA changed by mutation into bird and mammalian DNA. One can calculate that some 2.8×10^9 nucleotide pairs out of 3.5×10^9 have changed or roughly an average of 10 nucleotide pairs per year. Under the same conditions a bacterial chromosome with its 10^7 base pairs, would be completely changed in about one million years. Obviously the rate of mutation and the effect of selection on bacteria and on higher organisms may be different and the estimate of 10^6 year for a complete turnover of the bacterial world may be quite wrong.

These calculations indicate that the bacterial world may be liable to considerable genetic change within relatively short geological periods and that two different geological periods may have widely varying bacterial flora.

Ecological conditions have changed greatly during geological times. For example, since the Precambrian the amount of UV irradiation reaching the earth's surface has decreased, the O_2 tension has increased, bacteria have been exposed to diversified terrestrial conditions, to considerable changes in temperature (ice ages), in magnetic fields, cosmic ray irradiation, and so

Table 2. DNA Base Composition of Parental Strains and Artificially Induced Mutants.

Organism	mutagen	% GC	Reference
Agrobacterium tumefaciens			
parental organism	—	61.8[a] }	De Ley (1964)
mutant M 39*	UV	65.5[a] }	
parental org.	—	57.7[b] }	Sébald (pers. comm.)
mutant M 39*	UV	60.3[b] }	
parental org.	—	58.8[d];58.3[b];62.3[a];58.2[c] }	van der Plaat (pers. comm.)
mutant M 39*	UV	61.8[d];63.8[b];67.0[a];63.3[c] }	
Staphylococcus aureus			
parental org.	—	32.4	Gause (1966)
mutant UV 2	UV	71.0	Gause (1966)
mutant UV 15	UV	69.2	Gause (1966)
mutant UV 16	UV	70.9	Gause (1966)
Bacterium paracoli			
parental org.	—	55	Gause (1966)
mutant 52-1	urethane	70	Gause (1966)
Bacillus subtilis			
parental org.	—	42	Gause (1966)
mutant FU 9	5-fluorouracil	64	Gause (1966)
mutant FU 12	5-fluorouracil	62.9	Gause (1966)
mutant sc-22	$CuSO_4$	65.1	Weed (1963)
mutant 4 G	$CuSO_4$	55 }	Duc-Nguyen and Weed (1964)
mutant 4 G-SC	$CuSO_4$	71 }	
mutant "opaque"	$CuSO_4$	71	Kelly and Weed (1965)

* The mutant M 39 was prepared by Dr. P. Manigault, Institut Pasteur, Paris, France. (a): from T_m; (b) from paper chromatography; (c) from CsCl centrifugation; (d) from E_{260}/E_{280}.

forth. Each new condition favored some groups of organisms and was detrimental to others.

Very likely all pathogenic and symbiotic bacteria arose from some non-pathogenic stock and adapted to their host. The evolution of metazoan parasites illustrates this point very clearly; it may be summarized by saying "parasites are living proofs of their own evolution" (de Beer, 1964, p. 41). Therefore it seems likely that, e.g., the rumen microorganisms arose after the advent of the ruminants, that the microorganisms in the gut of the termites arose after the advent of the termites, that *Rhizobium* originated after the leguminous plants came about, xanthomonads after the angiosperms and so forth. The number of examples can be greatly increased. It is also very likely that a number of metazoa, now extinct, suffered from all kinds of diseases due to microorganisms which no longer exist. Many paleopathological cases in fossils are indeed known (Tasnádi-Kubacska, 1962) of which only a few will be quoted. The oldest cases of caries are known in reptile teeth from the cretaceous. It is very likely that periostitis and osteomyelitis in the giant reptiles (Dinosaurs) of Perm and Carbon are caused by bacteria. Traces of tooth, jawbone, and vertebra injury, possibly caused by an actinomycosis or by tuberculosis, have been detected in several fossil animals.

The general conclusion is thus that microorganisms, in particular bacteria, can vary widely in their morphology, physiology, and biochemistry, that in theory they can vary rather rapidly, and that the changing ecological conditions favored these changes. Therefore it is quite possible that the presently living bacteria are quite different from those living in other geological periods, particularly in the Precambrian seas.

DNA Homology

Different organisms derived from a common ancestor may very well have several parts of their chromosomal DNA in common. This all depends on how much phylogenetic diversification has taken place. If they evolved very fast or for a very long time, all their genes may be different.

DNA hybridization provides methods for detecting similarities in nucleotide *sequences* between two or more organisms. It is not necessary to know the *absolute* nucleotide sequences, as we are only comparing different organisms. The principle of those methods consists of labelling the DNA from one organism with either heavy or radioactive isotopes, annealing it with ordinary DNA from other organisms, and determining what percent of the labelled DNA hybridizes with the ordinary DNA. Since hybridiza-

Table 3. DNA homology between several members of the Enterobacteriaceae and some other organisms.

Source of DNA	E. coli B	% [14] C-DNA bound relative to the homologous DNA		
		Proteus vulgaris	*E. coli* BB	*Aerobacter aerogenes*
Escherichia coli BB		14	100	49
Escherichia coli B	100			
Escherichia coli K12	101			
Salmonella typhimurium	71			60
Shigella dysenteriae	71			45
Aerobacter aerogenes 211	51			100
Aerobacter aerogenes 13048	45			105
Klebsiella pneumoniae	25			54
Proteus vulgaris	14	100	7	2
Providence		21	15	
Proteus morganii		8	5	
Aeromonas hydrophila				13
Serratia marcescens	7			11
Pseudomonas aeruginosa	1			2
Calf thymus	1			
Mouse liver	1			

Data from McCarthy and Bolton (1963).

tion occurs only between regions which have identical or similar nucleotide sequences, the results tell us what percent of the labelled DNA is similar to the DNA in question. If one knows also the molecular weight of the labelled DNA, one can easily calculate how many nucleotide sequences are involved; since it is fairly well established that an average cistron consists of about 1,000 nucleotide pairs, one can also make a fair estimate of the number of cistrons which are similar between two organisms. Before the advent of DNA hybridization techniques, there was no way of knowing whether different groups of bacteria had any genes in common, and if so, how many.

Early investigations involved hybridization between high molecular DNA from two bacteria, one containing ^{15}N and deuterium, and separating the hybrid by CsCl density gradient with an ultracentrifuge. This method was only qualitative. Hybridization can not be complete when high molecular DNA is used because slight imperfections in the nucleotide sequences hamper the annealing of homologous portions. Furthermore, this method is very expensive, time-consuming and not all organisms can be grown on

Table 4. DNA homology between two pseudomonads, a xanthomonad, and several other genera.

Source of DNA	% ^{14}C-DNA bound relative to the homologous DNA		
	Pseudomonas putida 520	*Pseudomonas fluorescens* 488	*Pseudomonas campestris* var. *pelargonii* P121
Serratia marcescens 293	31	36	
Gluconobacter oxydans 8131	18	20	23
Acetobacter aceti			
(*liquefaciens*) 9505	28	34	37
Rhizobium meliloti 1.5		38.5	
Rhizobium leguminosarum 6.2		56	
Escherichia coli B	17	30	19
Azotobacter chroococcum 9125	49		
Azotobacter beijerinckii B-2	41	46	
Azotobacter vinelandii C-1	50	40	
Azomonas macrocytogenes 9128	46	35	
Azomonas macrocytogenes 9129	57	36	
Azomonas insignis 9127	48	47	
Derxia gummosa III	16	12	
Beijerinckia derxii	29	28	
Beijerinckia fluminensis	19		
Azotococcus agilis K	21	19	
Azotococcus agilis S	33	22	
Azotococcus agilis 9	35	31	
Bacillus subtilis BQ2	11	6	

Several data are taken from De Ley et al. (1966) and De Ley and Park (1966), others are unpublished.

D$_2$0 and an inorganic ^{15}N source. Schildkraut et al. (1961) have shown that DNA of different strains of *E. coli* can hybridize; however, its extent could not be determined quantitatively. Other results demonstrated for the first time that different bacterial "species" have indeed DNA regions in common. DNA hybrids were prepared between *Bacillus subtilis* and either *B. natto* or *B. subtilis* var. *aterrimus* (Marmur, Seaman, and Levine, 1963). From their results it was calculated that *B. subtilis* DNA is about 80 percent homologous with *B. natto* DNA and 68 percent homologous with *B. subtilis* var. *aterrimus* DNA. Both *B. natto* and *B. subtilis aterrimus* are phenotypically nearly indistinguishable from *B. subtilis*. It was therefore gratifying to find that their DNA was so similar. DNA hybrids were also made between *Xanthomonas pelargonii* and eight different xanthomonads (Friedman and De Ley, 1965) and between *Acetobacter aceti* (mesoxydans) and several other acetic acid bacteria (De Ley and Friedman, 1964) using the same technique. The overall importance of these results is that they show that different bacterial nomenspecies frequently have long DNA regions in common. This technique was not pursued further in view of its quantitative shortcomings outlined above.

An excellent method was devised by McCarthy and Bolton (1963). Agar containing single stranded, high molecular weight DNA (ca. 5.10^6 dalton) from the organisms in question, is incubated with single stranded, low molecular weight ^{14}C-DNA (ca. 3.10^5 dalton) from a reference organism. The amount of ^{14}C-DNA which hybridizes with the ordinary DNA can be quantitatively measured so that the similarity with the reference strain can be determined to within some 5 percent.

DNA from 26 different strains of *Xanthomonas* hybridized for 75 to 100 percent with *X. pelargonii*. Quite frequently the DNA homology was over 90 percent. The chromosomal DNA from all these organisms is thus more than three fourths similar (De Ley et al., 1966). It may be recalled that there are many reasons for incorporating most if not all xanthomonads into one genetic species. An identical situation was encountered recently with *Agrobacterium*. Eleven different strains all showed over 70 percent DNA homology with the reference strain (Heberlein and De Ley, in preparation). *B. subtilis* and *B. natto* (a variety of *B. subtilis*) also hybridize for 70 to 100 percent (Takahaski, Saito, and Ikeda, 1966). In these cases—and possibly in many more to be investigated—a genetic species is thus a group of organisms which have at least 70 percent of their DNA in common.

Homology was also determined among pseudomonads and xanthomonads (De Ley and Friedman, 1965; De Ley et al., 1966). The results show that these organisms constitute one large group corresponding to a genus. DNA homology is over 45 percent, meaning that about half of the chromosomal DNA is similar in the different members of the genus. There is, however, not a very sharp borderline, as there are a few organisms, such as *P. iodinum* and *P. diminuta* which hybridize only to the extent of 30 to 45 percent, and which taxonomists have only reluctantly placed into this genus.

A very interesting case is presented by a few organisms such as *P. rubescens* and *P. pavonacea,* which are, morphologically speaking, pseudomonads, but have a completely different DNA, as their percent GC is very low; they share only some 5 percent of their cistrons with the pseudomonads proper. There are two alternatives for their origin. (1) They arose from the same stock as *Pseudomonas* but have moved away from it

by a considerable evolutionary change in percent GC. (2) They are examples of morphological convergence. The former interpretation seems
most probable in the light of comparative biochemical considerations.
Similar cases may exist in other groups. For example micrococci and
staphylococci are morphologically similar, yet differ greatly in percent GC.
Also in *Thermomonospora* and *Streptosporangium* there are representatives
with widely divergent percent GC values. In all these cases the genes
directing morphology seem to have remained nearly unchanged, while the
genes responsible for biochemical and physiological features have undergone considerable change.

Hybridizations within the genus *Bacillus* (Takahashi et al., 1966) point
out that this is genetically a very heterogeneous group, which may in the
future be fragmented into several genera, a conclusion which has also been
reached previously on phenotypic grounds (De Ley, 1962).

Quite a number of intergeneric DNA hybridizations have been carried
out. Some of these results are summarized in Tables 3 and 4. Others are
graphically represented in Figs. 8 to 12. Several conclusions can be drawn
from these data.

(1) Bacteria are not infrequently quite diversified, since they may
differ by some 95 percent of their chromosome (Fig. 8 and 12). This
shows that their evolution has been considerable. Relatively speaking they
are evolutionarily more diversified than the mammals which share some 20
percent of their DNA.

(2) In the Enterobacteriaceae evolutionary diversification has been
enormous. For example, *E. coli* is as different from *Providence, Proteus,*
and *Serratia* as it is from *Pseudomonas* or *Bacillus. Proteus morganii* and
Proteus vulgaris are two completely different entities, as predicted by their
percent GC. On the other hand *Escherichia* has some 70 percent homology
with *Shigella,* which may justify their future integration into one genus.

(3) Another interesting group consists of *Agrobacterium* and *Rhizobium* (Fig. 9). There are two subgroups. One is the crown-gall producer, with its nonpathogenic variety (the *A. tumefaciens—radiobacter*
group) and the other is composed of bacteria affecting roots either by
excessive root formation (*A. rhizogenes*) or by root nodule formation (the
rhizobia). Bacteria from both subgroups have about half of their chromosomal base sequences in common (Heberlein, De Ley, and Tijtgat, 1967).
It is obvious that they have had the same evolutionary origin. If the
reports are true that *A. tumefaciens* can infect conifers, then it is possible
that these bacteria arose as parasites on these plants. They may thus be
younger than 300.10^6 years. Since legume plants did not arise until later

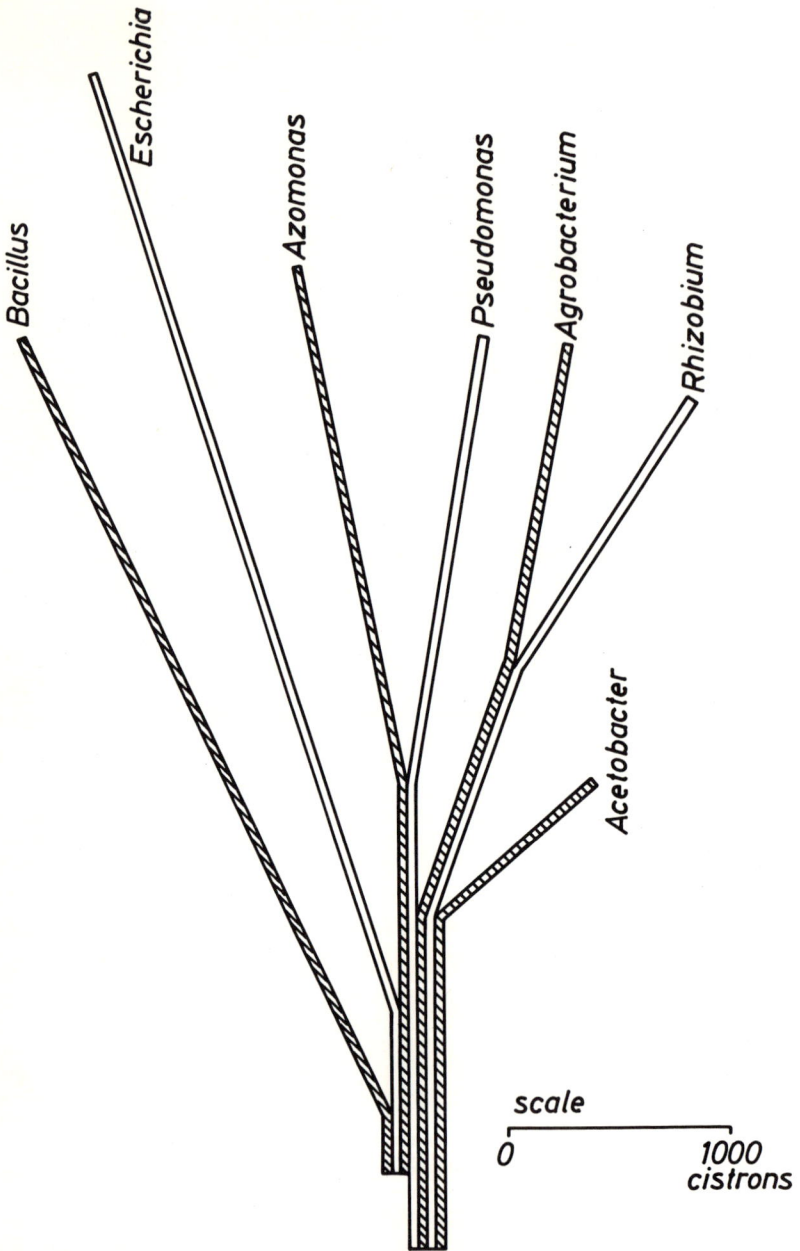

Fig. 8. DNA homology between seven organisms belonging to different genera.
Each bar represents a chromosome. Its length is proportional to the molecular weight
and thus also to the length of the chromosome. The degree of DNA homology between
each pair of organisms is represented by the adjoining parts. Loose parts are completely
different from all the rest. (De Ley and Park, to be published.)

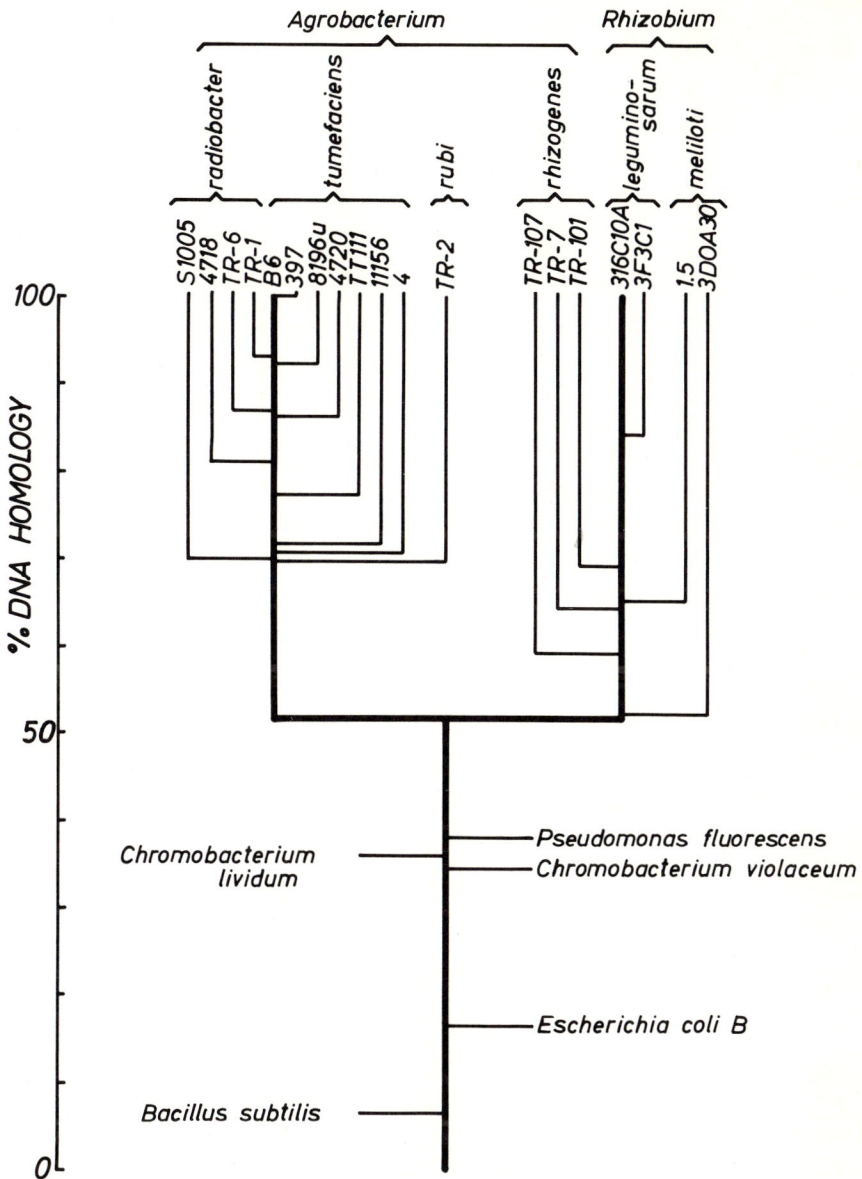

Fig. 9. DNA homology in the Agrobacterium-Rhizobium group.
At least half of the chromosomes are similar, indicating common ancestry. DNA from the different strains was hybridized with ^{14}C-DNA from both reference strains (heavy lines). Representatives from some other genera are included for comparison. (Simplified from Heberlein, De Ley, and Tijtgat, 1967)

(around 135.10^6 years ago), one may speculate that *Rhizobium* and *A. rhizogenes* are offshoots. One can imagine that the genes for hairy root and root nodule formation are only modifications of the crown-gall genes. The opinion that the chromobacteria belong in one family with the above *Agrobacterium—Rhizobium* group is illusory since they differ as much from it as e.g. *Pseudomonas*.

(4) The freeliving N_2 fixers (see Fig. 10, and De Ley and Park, 1966) such as "*P.*" *azotogensis, Azotobacter, Azomonas,* and so forth have widely diverse percent GC values and hybridize with the reference strains to different degrees. N_2 fixation, therefore, is a property found in widely diverse organisms. It may be recalled that N_2 fixation occurs in many other bacteria as well to a lesser degree and in several bluegreen algae. One explanation might be that N_2 fixation is a phylogenetically very old property, which originated very early and remained preserved during the evolutionary changes of the ancestral types. In agreement with our hypothesis is the suggestion by Fogg (1956) that the N_2 fixing bluegreen algae are remnants from the time of the onset of the oxygenic atmosphere, when the bloom of

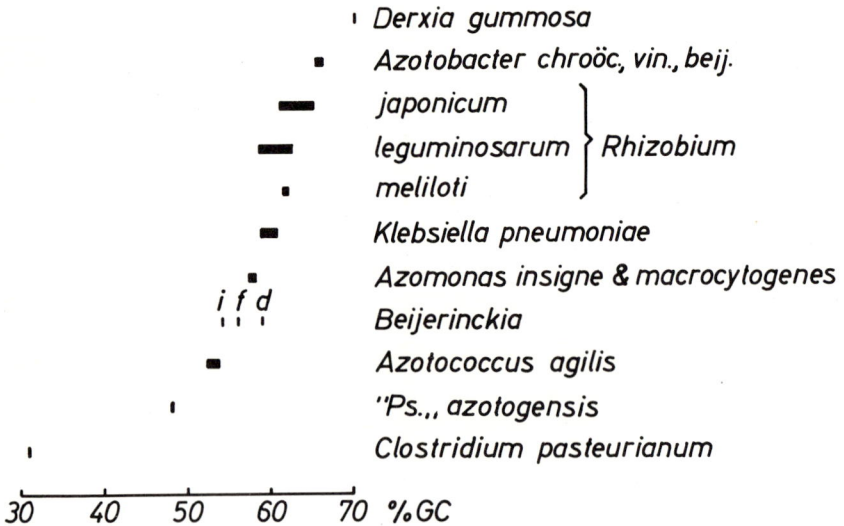

Fig. 10. DNA base composition of both free-living and symbiotic nitrogen fixing bacteria.
For Beijerinckia, *i* stands for indica, *f* for fluminensis, and *d* for derxii. "PS" azotogensis is not a real Pseudomonas, its taxonomic position has not yet been determined. (Adapted from De Ley and Park, 1966.)

oxygenic organisms depleted the soluble inorganic N and nitrogen fixing organisms became favored.

(5) Ancestral remnants in DNA. By far the most important result seems to us to be that all heterotrophic bacteria appear to be related and to be derived from a common ancestry. They share at least 5 percent, or some 200 cistrons, with many other bacteria (see Fig. 8 and 12). This region need not be the same in all bacteria, but in many cases it very likely is. This has been shown by cross hybridization of preselected DNA regions from two pseudomonads and a xanthomonad (Park and De Ley, 1967). The results are represented in Fig. 11. The common part between the three organisms consists of some 2.10^6 nucleotide pairs or about some 2,000 cistrons. They determine the features which are common and which these organisms inherited probably from the ancestor, such as cell shape, aerobic character, polar flagellation, and many soluble and particulate enzymes (G6P and 6PG-dehydrogenase, Entner-Doudoroff pathway, oxidase systems for many hexoses, pentoses, and so forth). In many other features the ancestor may have been completely different from all presently living pseudomonads. It was also established that the common part between two organisms is similar (less than 5 percent difference) but not perfectly identical. The *P. fluorescens* and *P. putida* chromosomes share with each other an additional 1,300 cistrons over that which they share with the xanthomonads. The remaining 650 cistrons would account for their species differentiation. The xanthomonads very likely share an additional 1,000 cistrons and the remaining 1,000 cistrons account for the species differentiation between the different xanthomonad strains. It is tempting to speculate that the xanthomonads diverged first or fastest from the ancestor and that the split between *P. fluorescens* and *P. putida* occurred only later. The above authors made some speculations on the rate of mutation, assuming that the xanthomonads arose only after the advent of their hosts the angiosperms, 135.10^6 years ago. For the stable common part the rate of mutation would be less than 1 base pair per 700 years and for the dissimilar region of the DNA it would be at least 1 to 10 base pairs per 700 years.

Similar results can be calculated from the results of Bolton et al. (1963–1964). In the chromosomal DNA from *Proteus vulgaris* there is a variable region of less than 20 percent of the total, which is similar to *Proteus morganii, E. coli,* and/or *Providence*. This region would have to be small considering that *P. vulgaris* has a percent GC of 39 which is far removed from the other organisms (40 to 50 percent GC).

(6) All available results have been graphically summarized in Fig. 12. This figure shows how far different bacterial genera are evolutionarily re-

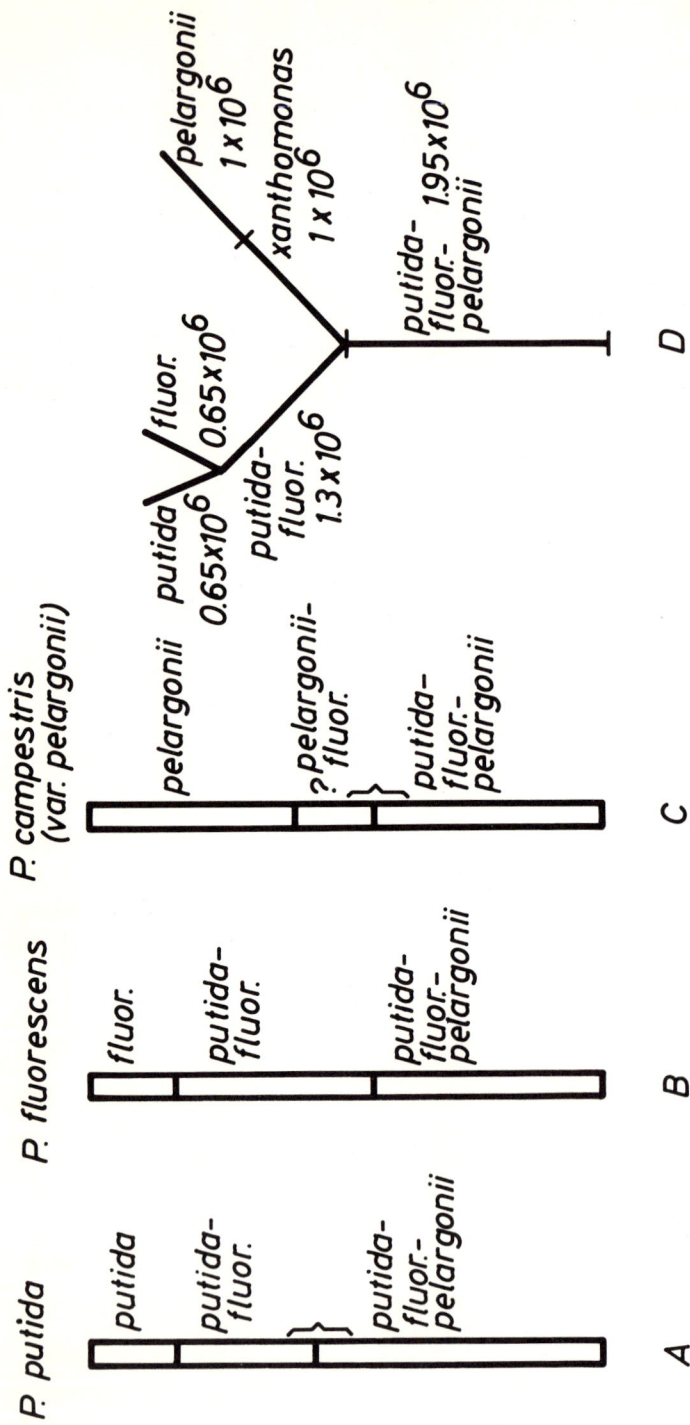

Fig. 11. Common and specific parts in the DNA from two pseudomonads (P. putida and P. fluorescens) and one xanthomonad (P. campestris var. pelargonii) expressed as percent of the total length. For illustrative purposes each part is depicted as a single stretch. (Common and specific parts are indicated by the names next to them. The numbers next to each part give the approximate number of nucleotide pairs involved. The results are an average of several determinations. (From Park and De Ley, 1967.)

146

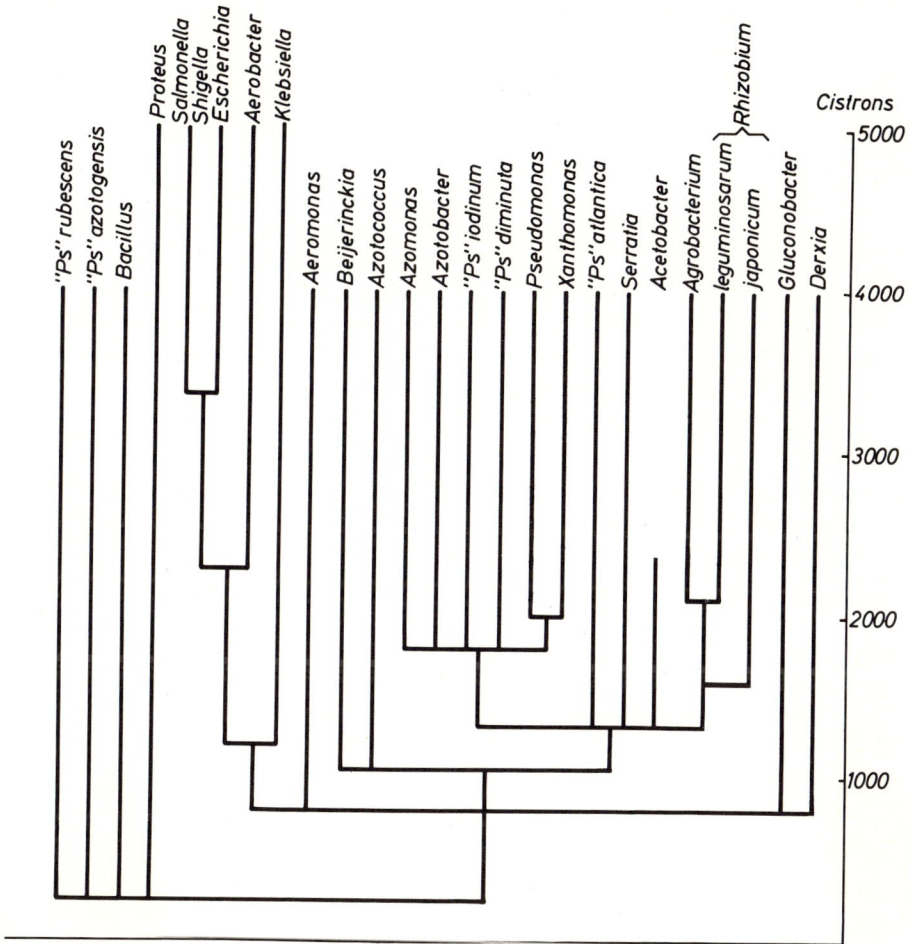

Fig. 12. DNA homology and tentative evolutionary relationship between several bacterial genera.

Data on the DNA homology between Enterobacteriaceae are taken from McCarthy and Bolton (1963), all the other results were obtained in the author's laboratory. Not all genera have been hybridized with each other, the reference strains being Bacillus subtilis, Escherichia coli, Azomonas macrocytogenes, Pseudomonas putida and P. fluorescens, Agrobacterium tumefaciens, Rhizobium leguminosarum and Acetobacter liquefaciens. The molecular weight of DNA is only known for Bacillus, Escherichia, Pseudomonas, Xanthomonas, and Acetobacter. In all other cases it is assumed to be roughly identical to Pseudomonas. (De Ley and Park, to be published.)

moved from others. For the present moment it is the closest picture of bacterial evolution which is possible.

Phenotypic Aspects of Bacterial Evolution

There have been some attempts to sketch an evolutionary tree for bacteria. Some are mainly based on morphology (Kluyver and van Niel, 1936), others on physiological features (Lwoff, 1943; Knight, 1945) or on a mixture of both (Bisset, 1962). These hypotheses had of necessity to be built on the limited number of features available at the time they were proposed. Molecular biology, numeral taxonomy, and comparative biochemistry are changing these pictures.

One of the several merits of numerical taxonomy is that it shows that each pleiston consists of a group of organisms with a core of some 70 to 80 percent of their features in common, whereas the remaining 20 to 30 percent may widely vary. It also shows that it is very rare that two strains of the same pleiston, or of the same species, are completely identical. This illustrates that each organism is subject to its own local evolutionary history. Numerical taxonomy makes no claims to elucidate the evolution of bacteria. Indirectly, however, it helps to understand the great lines of divergency and similarity, since the less similar two organisms are, the fewer cistrons they have in common and the farther removed they are from each other in terms of evolution. An evolutionary interpretation along these lines was attempted for the acetic acid bacteria (De Ley, 1961).

A comparison between DNA homology and phenotypic similarity should be instructive concerning the structure and expression of common genes. Present data are summarized in Fig. 13. For values between 60 to 100 percent, congruence is very good. Deviations from the theoretically expected diagonal are explained by the standard error on each method. The 100 to 60 percent region involves closely related organisms at the species level. The fact that the results obtained by these two methods are in good agreement means that all, or nearly all, DNA is phenotypically expressed and that the nucleotide sequences of homologous genes are for the most part still the same. Below 60 percent, DNA homology drops faster than phenetic similarity. This region concerns organisms from different genera. The explanation is partially due to the limited number of features which are used in numerical taxonomy, but also to different nucleotide sequences of homologous genes. All enzymes with a similar active center (similar amino acid sequence) but with a different amino acid sequence in the rest of the protein molecule, will score as the same phenotypic feature, but their cistrons won't hybridize. The lower part of the curve thus suggests that the

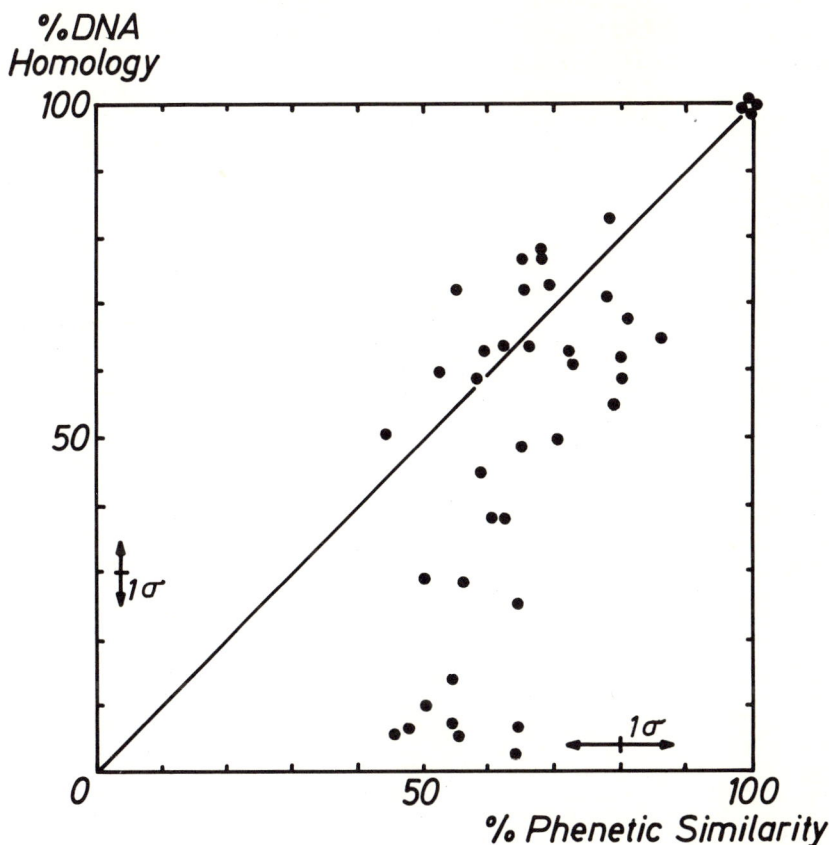

Fig. 13. Congruence between DNA homology and phenetic similarity. The data concern strains from Pseudomonas, Xanthomonas, Enterobacteriaceae, Agrobacterium, and Bacillus subtilis. All DNA data from the author's laboratory. Phenetic data from Lysenko (1961), Sneath and Cowan (1958), and Graham (1964). The arrows represent roughly the standard error of estimate. (Adapted from Sneath, personal communication.)

lower the phenetic similarity, the more homologous genes differ in their nucleotide sequence.

Comparative intermediary metabolism, in conjunction with numerical taxonomy and DNA homology, is expected to shed much light on bacterial phylogeny. The principle is that organisms which have a common enzymic pathway are also of a common phylogenetic origin. The broadest expression of this concept was already laid down as "the unity in biochemistry" by Kluyver and Donker. Several mechanisms, such as the glycolytic pathway and the hexose monophosphate shunt, are so universally distributed in most organisms, that they cannot be used for detailed phylogenetic interpretation, except maybe in those few organisms where some of the enzymes are lacking and thus indicating an evolutionary loss (e.g. the lack of aldolase in homofermentative lactic acid bacteria). This approach will be most useful in the study of the distribution of pathways which are limited to a few groups. This aspect has already been discussed on a previous occasion (De Ley, 1962). A new example came recently to light in this laboratory from the study of the occurrence of the Entner-Doudoroff pathway (Kersters and De Ley, unpublished). It was detected in *Pseudomonas, Xanthomonas, Chromobacterium, Agrobacterium, Rhizobium, Azotococcus, Azomonas, Klebsiella, Escherichia, Comamonas,* and *Aeromonas.* It is absent in several other genera tested. Comparison with Fig. 12 shows that these genera also share a rather high DNA homology. The interpretation is thus that the Entner-Doudoroff mechanism arose in the common ancestor of these genera, which would thus form a separate evolutionary branch. Another example is the comparative biosynthetic pathways of unsaturated fatty acids (Erwin and Bloch, 1964). It is expected that the study of the occurrence of the phosphoketolase pathway might clarify the relationship of several genera with the heterofermentative lactic acid bacteria. Also the study of the mechanism of the propionic acid fermentation (acrylic or methylmalonic acid pathways) may clarify the relationship between organisms which produce propionic acid. This approach is based on the hypothesis that an enzyme arises only once during evolution. This is indeed most likely according to the following considerations. Suppose an enzyme with a molecular weight of 90,000, consisting thus of some 600 amino acids; its cistrons contain 1,800 nucleotide pairs. If we consider the experimentally based fact that some 10 to 20 percent of the amino acids are immutable for enzyme activity, it follows that 180 to 360 nucleotide pairs need to have a well determined position. There are, however, 4^{180} to 4^{360} possibilities, so that the chance for a de novo formation with the right configuration is about $1/4^{270}$ or extremely small. It is still smaller if a pathway consists of several enzymes. One can, however, reason that if a certain cistron has been formed once it can be formed more than once, since an improbability becomes a possibility over a very long time. This too could be detected if an enzyme or enzyme system occurs in an organism or group of organisms entirely different from others.

In some cases, the mere common occurrence of certain chemicals in a group of organisms may also point to evolutionary relationships. An extreme case is the difference in the cell wall composition of Gram-positive and negative bacteria. This probably signifies a very old separation. Further evolutionary changes within each group yielded the present enormous variety of genera. It should be noted that the percent GC of both groups spans the entire range from 25 to 75 percent GC. Other examples are the composition of the cell wall as determined by Cummins (1962) and a number of other compounds, reviewed by Aaronson and Hutner (1966). This useful approach should, however, be complemented by the knowledge of the biosynthetic pathways of these compounds.

Summary

In the framework of present hypotheses on the geochemistry and geophysics of the Precambrian, a scheme is proposed on the possible evolution of microorganisms up to the Cambrian. On biochemical grounds several stages are postulated.

(1) Heterotrophic microorganisms were feeding first on organic substances present in the ocean.

(2) The second stage might have been a combination of chemiautotrophic and heterotrophic microorganisms. The former served as food source for the latter.

(3) A combination of photoautotrophic, chemiautotrophic, and heterotrophic microorganisms. Both former groups might have provided organic material as food for the heterotrophs.

(4) A combination of photosynthetic, oxygen-producing protoalgae and heterotrophs. The former organisms represented the start of the oxygenic atmosphere. They themselves were probably still anoxygenic. Geochemical and fossil evidence localizes the start of this period at about some 2 to 3 billion years ago. It appears that nitrifying and denitrifying bacteria could only have arisen later.

(5) The anoxygenic protoalgae probably gave rise to the more modern oxygenic algae by intracellular separation of paired lamellae and/or chloroplasts. The evolution of cytoplasmic membrane to mesosome to mitochondria is imagined to have stimulated the rise of the algae, protozoa, plants, and animals, because of the much greater energy coupling efficiency.

The intense UV irradiation during the Precambrian probably provoked considerable changes in DNA base composition of microorganisms, increasing the spectrum of evolutionary possibilities. Several examples of artificial mutants, provoked in this way, are given.

(6) As pointed out by Berkner and Marshall (1965), at the end of the Precambrian the oxygen level was about 0.002 atm, allowing the rise of all oxygenic organisms, leading to the invasion of land in the Middle-Silurian. The ancestors of modern Actinomycetes may date back that long. Plant and animal pathogenic and symbiotic microorganisms are supposed to have arisen only after the advent of the host.

Direct and indirect evidence for the existence of fossil microorganisms is briefly reviewed. Claims that some "fossil" bacteria are still viable are extremely unlikely.

Arguments are reviewed to consider all presently living bacteria, including the Actinomycetes and bluegreen algae, as diversified offshoots from the same evolutionary ancestry.

Many prehistoric microorganisms are expected to have been considerably different in many aspects (morphologically, physiologically, biochemically, and enzymically) from presently living ones. This follows from considerations on changing ecological conditions, DNA homology, the high number of possible nucleotide combinations in microbial DNA, and the rate of mutation (less than 1,000 nucleotide pairs per nucleoid per century).

The great range in DNA base composition in bacteria implies that they are evolutionarily greatly diversified. Several implications for relationships between groups of bacteria are discussed.

Changes in chromosomal DNA have occurred by both increase in the amount of DNA per nucleoid and mutational changes in DNA base composition. It is estimated that bacteria are genetically rather stable over several thousand years, but that the bacterial world may change considerably from one geological period to another.

DNA hybridization methods detect similarities in nucleotide sequences between different organisms. The degree of evolutionary divergency can thus be determined. Knowledge in this field is summarized. Bacteria belonging to the same genetic species (e.g. the xanthomonads and many *Agrobacteria*) have at least 70 percent of their DNA in common. In bacteria belonging to the same genus, DNA homology is over 45 percent (e.g., *Pseudomonas*). Considerable phylogenetic kinship has been detected in many cases, e.g., between *Agrobacterium* and *Rhizobium,* between *Pseudomonas* and *Xanthomonas,* and so forth. DNA regions which are similar between different organisms have to be considered as ancestral remnants. They can be isolated in sizeable amounts for further study.

Many genera of bacteria are evolutionarily very far removed from each other. Evolutionary divergency in the bacterial world is considerable. Several examples are given of genera which have at most 5 percent of their DNA in common. The great DNA dissimilarity between N_2 fixing microorganisms suggests that N_2 fixation is a very old property.

At the phenotypic level, the study of comparative intermediary metabolism is able to contribute considerably to the better understanding of phylogenetic relationships.

Acknowledgment

The author is indebted to the Nationaal Centrum voor Biochemie en Molekuulbiologie (Belgium) for research and personnel grants and to Dr. G. Heberlein for correcting the foreign flavor in the text.

References

AARONSON, S., and S. H. HUTNER. 1966. Biochemical markers and microbial phylogeny. Quart. Rev. Biol., 41:13–46.

ABELSON, P. H. 1959. Paleobiochemistry and organic geochemistry. Fortschr. Chem. Organ. Naturst., 17:379–403.

BARGHOORN, E. S., and S. A. TYLER. 1965. Microorganisms from the Gunflint chert. Science, 147:563–577.

BEERSTECHER, E. JR. 1954. Petroleum Microbiology, New York, Elsevier Press.

BERKNER, L. V., and L. C. MARSHALL. 1965. History of major atmospheric components. Proc. Nat. Acad. Sci. USA, 53:1215–1225.

BERNS, K. I., and C. A. THOMAS. 1965. Isolation of high molecular weight DNA from *Hemophilus influenzae*. J. Molec. Biol., 11:476–490.

BIEN, E., and W. SCHWARTZ. 1965. Geomikrobiologische Untersuchungen. VI. Ueber das Vorkommen konservierter toter und lebender Bakterienzellen in Salzgesteinen. Z. Allg. Mikrobiol., 5:185–205.

BISSET, K. A. 1962. The phylogenetic concept in bacterial taxomy. *In* Ainsworth, G. C., and P. H. A. Sneath, Microbial Classification. 12th Sympos. Soc. Gen. Microbiol., 361–373, Cambridge, University Press.

BOLTON, E. T., R. J. BRITTEN, T. J. BIJERS, D. B. COWIE, B. HOYER, Y. KATO, B. J. MCCARTHY, M. MIRANDA, and R. B. ROBERTS. 1963–64. Carnegie Inst. Wash. Year Book, 63:366–397.

———, R. J. BRITTEN, D. B. COWIE, R. B. ROBERTS, P. SZAFRANSKI, and M. J. WARING. 1964–65. Carnegie Inst. Wash. Year Book, 64:313–345.

CAIRNS, J. 1963. The chromosome of *Escherichia coli*. Cold Spring Harbor Sympos. Quant. Biol., 28:43–45.

CALDWELL, P. J., and C. HINSHELWOOD. 1950. Nucleic acid content of *Bacterium lactis aerogenes*. J. Chem. Soc., 1415–1418.

CLOUD, P. E. 1965. Significance of the Gunflint (Precambrian) Microflora. Science, 148:27–35.

CUMMINS, C. S. 1962. Chemical composition and antigenic structure of cell walls of *Corynebacterium, Mycobacterium, Nocardia, Actinomyces,* and *Arthrobacter*. J. Gen. Microbiol., 28:35–50.

DAUVILLIER, A. 1965. The Photochemical Origin of Life, New York, Academic Press.

DE BEER, G. 1964. Atlas of Evolution, London, Th. Nelson and Sons.

DE LEY, J. 1960. Comparative carbohydrate metabolism and localization of enzymes in *Pseudomonas* and related microorganisms. J. Appl. Bacteriol., 23:400–441.

———. 1961. Comparative carbohydrate metabolism and a proposal for a phylogenetic relationship of the acetic acid bacteria. J. Gen. Microbiol., 24:31–50.

———. 1962. Comparative biochemistry and enzymology in bacterial classification. *In* Ainsworth, G. C., and P. H. A. Sneath, Microbial Classification. 12th Sympos. Soc. Gen. Microbiol. 164–195, Cambridge, University Press.

———. 1964. *Pseudomonas* and related genera. Ann. Rev. Microbiol., 18:17–46.

———. 1964a. Effect of mutation on DNA composition of some bacteria. Antonie Leeuwenhoek, 30:281–288.

———, and S. FRIEDMAN. 1964. Deoxyribonucleic acid hybrids of acetic acid bacteria. J. Bacteriol., 88:937–945.

———, and S. FRIEDMAN. 1965. Similarity of *Xanthomonas* and *Pseudomonas* deoxyribonucleic acid. J. Bacteriol., 89:1306–1309.

———, and I. W. PARK. 1966. Molecular biological taxonomy of some freeliving nitrogen-fixing bacteria. Antonie Leeuwenhoek, 32:6–16.

———, K. KERSTERS, and I. W. PARK. 1966. Molecular biological and taxonomic studies on *Pseudomonas halocrenaea,* a bacterium from Permian salt deposits. Antonie Leeuwenhoek, 32:315–31.

————, I. W. PARK, R. TIJTGAT, and J. VAN ERMENGEM. 1966. DNA homology and taxonomy of *Pseudomonas* and *Xanthomonas.* J. Gen. Microbiol., 42:43–56.

DENNIS, E. S., and R. G. WAKE. 1966. Autoradiography of the *Bacillus subtilis* chromosome. J. Mol. Biol., 15:435–439.

DESIKACHARY, T. V. 1959. Cyanophyta. New Delhi, Indian Council of Agr. Research.

DOLE, M. 1965. Natural history of oxygen. *In* Oxygen, 5–27, Boston, Little Brown and Co.

DOMBROWSKI, H. 1960. Balneologische Untersuchungen der Nauheimer Quellen. II. *Pseudomonas halocrenaea.* (nova species). Z. Bakt. Parask. Infekt. Hyg., Orig., 178:83–90.

————. 1963a. Bacteria from paleozoic salt deposits. Ann. N.Y. Acad. Sci., 108:453–460.

————. 1963b. Organismes vivants du paléozoique. La Presse Médicale., 71:1,148–1,152.

DUC-NGUYEN, H., and L. L. WEED. 1964. D-Ornithine as a constituent of a bacterial cell wall. J. Biol. Chem., 239:3372–3376.

ECHLIN, P., and I. MORRIS. 1965. The relationship between blue-green algae and bacteria. Biol. Rev., 40:143–187.

ERWIN, J., and K. BLOCH. 1964. Biosynthesis of unsaturated fatty acids in microorganisms. Science, 143:1006–1012.

FOGG, G. E. 1956. The comparative physiology and biochemistry of the blue-green algae. Bact. Rev., 20:148–165.

FREEZE, E. 1962. On the evolution of the base composition of DNA. J. Theor. Biol., 3:82–101.

FRENKEL, A., H. GAFFRON, and E. H. BATTLEY. 1950. Photosynthesis and photoreproduction by the blue-green alga *Synechococcus elongata.* Näg. Biol. Bull., 99:157–162.

FRIEDMAN, S., and J. DE LEY. 1965. "Genetic species" concept in *Xanthomonas.* J. Bacteriol., 89:95–100.

FRITSCH, F. E. 1965. The structure and reproduction of the algae. Vol. 2. Cambridge, University Press.

GAUSE, G. F. 1966. Microbial Models of Cancer Cells. Amsterdam, North Holland Publishing Co.

GLAESSNER, M. F. 1962. Principles of Micropaleontology. London, Hafner Publishing Co.

GRAHAM, P. H. 1964. The application of computer techniques to the taxonomy of the root-nodule bacteria of legumes. J. Gen. Microbiol., 35:511–517.

HEBERLEIN, G. T., J. DE LEY, and R. TIJTGAT. 1967. Deoxyribonucleic acid homology and taxonomy of *Agrobacterium, Rhizobium,* and *Chromobacterium.* J. Bacteriol, 94:116–124.

HOLLAND, H. D. 1962. Model for the evolution of the earth's atmosphere. *In* A. E. J. Engel, Petrological Studies, 447–477. New York, Geol. Soc. of America.

————. 1965. The history of ocean water and its effect on the chemistry of the atmosphere. Proc. Nat. Acad. Sci. USA, 53:1173–1183.

HOROWITZ, N. H., and H. MAC LEOD. 1960. The DNA content of *Neurospora* nuclei. Microbial Genet. Bull. 17:6.

HUTCHINSON, G. E. 1944. Nitrogen in the bio-geochemistry of the atmosphere. Amer. Sci., 32:178–195.

KELLY, A. P., and L. L. WEED. 1965. (as quoted by Gause, G. F. 1966.) Microbial Models of Cancer Cells. Amsterdam, North Holland Publishing Co.

KLUYVER, A. J., and C. B. VAN NIEL. 1936. Prospects for a natural system of classification of bacteria. Zbl. Bakt. [Orig.], 94:369–403.

KNIGHT, B. C. 1945. Growth factors in microbiology. Some wider aspects of nutritional studies with micro-organisms. Vitamins and Hormones, 3:105–228.

KORNBERG, A. 1965. Synthesis in DNA-like polymers de novo or by reiterative replication. *In* Bryson, V., and H. J. Vogel, Evolving Genes and Proteins, 403–417, New York, Academic Press.

KUZNETSOV, S. I., M. V. IVANOV, and N. N. LYALIKOVA. 1963. Introduction to Geological Microbiology, New York, McGraw Hill.

LEWIN, R. A. 1962. Physiology and Biochemistry of Algae, New York, Academic Press.

LWOFF, A. 1943. L'évolution physiologique. Etude des pertes de fonctions chez les microorganismes, Paris, Hermann et Cie.

LYSENKO, O. 1961. *Pseudomonas*—an attempt at a general classification. J. Gen. Microbiol., 25:379–408.

MARMUR, J., E. SEAMAN, and J. LEVINE. 1963. Interspecific transformation in *Bacillus*. J. Bacteriol., 85:461–467.

MASSIE, H. R., and B. H. ZIMM. 1965. Molecular weight of DNA in the chromosomes of *E. coli* and *B. subtilis*. Proc. Nat. Acad. Sci. USA, 54:1636–1641.

MCCARTHY, B. J., and E. T. BOLTON. 1963. An approach to the measurement of genetic relatedness among organisms. Proc. Nat. Acad. Sci. USA, 50:156–162.

MILLER, S. L. 1957. The formation of organic compounds on the primitive earth. *In* Oparin, A. I., The Origin of Life on Earth, 73–85, Moscow, Publ. House Acad. Sci. USSR.

MOROWITZ, H. J., M. E. TOURTELOTTE, W. R. GUILD, E. CASTRO, C. WOESE, and R. C. CLEVERDON. 1962. The chemical composition and submicroscopic morphology of *Mycoplasma gallisepticum,* avian PPLO 5969. J. Molec. Biol., 4:93–103.

MYERS, G. E., and R. G. L. MC CREADY. 1966. Bacteria can penetrate rock. Canad. J. Microbiol., 12:477–484.

NAKAMURA, H. 1938. Ueber die Kohlensäureassimilation bei niederen Algen in Anwesenheit des Schwefelwasserstoffs. Acta Phytochim. (Japan), 10:271–281.

OGUR, M., S. MINCKLER, and D. O. MCCLARY. 1953. Deoxyribonucleic acid (DNA) and the budding cycle in the yeasts. J. Bact., 66:642–645.

PARK, I. W., and J. DE LEY. 1967. Ancestral remnants in the deoxyribonucleic acid from *Pseudomonas* and *Xanthomonas*. Antonie Leeuwenhoek, in press.

PONTECORVO, G., and J. A. ROPER. 1956. Resolving power of genetic analysis. Nature (London), 178:83.

RUTTEN, M. G. 1962. The geological aspects of the origin of life on earth. Amsterdam, Elsevier.

SCHILDKRAUT, C. L., J. MARMUR, and P. DOTY. 1961. The formation of hybrid DNA molecules and their use in studies of DNA homologies. J. Molec. Biol., 3:595–617.

SCHOPF, J. W., E. S. BARGHOORN, M. D. MASER, and R. O. GORDON. 1965. Electron microscopy of fossil bacteria two billion years old. Science, 149:1365–1367.

SERENKOV, G. P. 1962. Nucleic acids and the evolution in the algal group. Izs. Akad. Nauk SSSR, Ser. Biol., 27:857–868.

SIEGEL, S. M., and C. GIUMARRO. 1966. On the culture of a microorganism similar to the Precambrian microfossil *Kakabekia umbellata* Barghoorn in NH_3-rich atmospheres. Proc. Nat. Acad. Sci. USA, 55:349–353.

———, G. Renwick, O. Daly, C. Giumarro, G. Davis, and L. Halpern. 1965. The survival capabilities and the performance of earth organisms in simulated extraterrestrial environments. *In* Mamikunian, G., and M. H. Briggs, Exobiology, 119–178, London, Pergamon Press.

SNEATH, P. H. A. 1962. Longevity of microorganisms. Nature (London), 195:643–646.

———. 1964. The limits of life. Discovery, 25:20–24.

———, and S. T. COWAN. 1958. An electrotaxonomic survey of bacteria. J. Gen. Microbiol., 19:551–565.

STANIER, R. Y., and C. B. VAN NIEL. 1962. The concept of a bacterium. Arch. Mikrobiol., 42:17–35.

SUEOKA, N. 1961. Variation and heterogeneity of base composition of deoxyribonucleic acids; a compilation of old and new data. J. Molec. Biol., 3:31–40.
———. 1962. On the genetic basis of variation and heterogeneity of DNA base
composition. Proc. Nat. Acad. Sci. USA, 48:582–592.
Symposium on the evolution of the earth's atmosphere. 1965. Proc. Nat. Acad. Sci.
USA, 53:1169–1226.
TAKAHASHI, H., H. SAITO, and Y. IKEDA. 1966. Genetic relatedness of spore bearing
bacilli studied by the DNA agar method. J. Gen. Appl. Microbiol., 12:113–118.
TASNÁDI-KUBACSKA, A. 1962. Paläo-Pathologie. Jena, G. Fischer Verlag.
VALLENTYNE, J. R. 1965. Two aspects of geochemistry of amino acids. *In* Fox,
S. W. The Origin of Prebiological Systems, 105–120. New York, Academic Press.
WEED, L. L. 1963. Effects of copper on *Bacillus subtilis*. J. Bacteriol., 85:1003–1010.

4

Evolutionary Implications of
Sexual Behavior in *Drosophila*

HERMAN T. SPIETH

Department of Zoology
University of California
Davis

Introduction 157
Types of Behavior 158
Drosophila Preview 159
Ontogeny of Mating Behavior 160
Basic Courtship Patterns 160
Adaptive Nature of Mating Behavior 165
Specialized Mating Behavior of Hawaiian Species 168
Summary of Mating Behavior 170
Nature and Function of the Stimuli 171
Mating Behavior of Other Diptera 175
Significance of Drosophilid Mating Behavior 176
Lek Behavior 178
Sexual Isolation 181
Intraspecific Variations 182
Interspecific Behavior 185
Origin of Sexual Isolation 190
Conclusion 191
References 191

Introduction

Investigations by biologists studying diverse groups of animals have unequivocally shown that the behavior of organisms displays a structure which can be utilized for comparative analysis in a manner comparable to that which morphological structure provides. This is not surprising since behavior is the action of structure responding to stimuli, either internal or external, that have been received, perceived, and processed by the organism. Comparative behavior can thus serve as an effective tool for the analysis of the evolution of animals, particularly related groups of species.

The use of any particular approach to the study of the evolution of organisms presents the investigator with both advantages and handicaps. Behavior, unlike morphology, has the disadvantage that it is a transient activity performed by the creatures being studied and therefore can not be preserved for later review. It is, indeed, imperative that the behavioral elements of the organisms be perceived as they occur and recorded by the observer for future reference. One of the advantages of using behavior as a means for studying the evolution of a group of closely related organisms is that the investigator perforce must study living organisms, and by inclination often checks his laboratory findings against field studies, thus acquiring a "healthy respect for the all-pervading power of selection" (Tinbergen, 1960).

Types of Behavior

The life of an animal is one continuum of behavior, but the continuum can be separated for analysis into discrete units such as breathing, feeding, drinking, resting, sleeping, mating, escaping, and habitat selection. Each of these categories represents the responses of the individual to divers stimuli received from the external physical, the external biotic, or from its own internal environment.

For some types of behavior, the observer has difficulty ascertaining clues to the stimuli causing a response as, for example, in the selection of resting and/or sleeping sites by solitary insects. In other cases, the stimuli are determinable but groups of related species may display essentially similar, if not identical, responses to the same type of stimuli, as in the feeding behavior of many groups of cyclorrhaphous Diptera.

In comparison, mating behavior in various (but certainly not all) groups of bisexual animals has often proved amenable to analysis both in the laboratory and in the field. The basic minimal unit for such behaviors is not one individual but rather a pair, each of which serves as a producer of stimuli and also as a receiver and responder to the stimuli produced by the other individual. Those species in which the two sexes alternate sequentially in time in the production of a number of unique signals and responses have proved the most rewarding to behaviorists.

Species of the genus *Drosophila,* with a few unique and illuminating exceptions, engage in complex and elaborate mating activities which can be observed and recorded by the observer. In comparison, many if not most cyclorrhaphous Diptera, including other genera within the family Drosophilidae and especially their close relatives the *Scaptomyza,* present courtship behaviors which are not so effectively analyzable with presently available techniques of study (see subsequent section, p. 175).

Drosophila *Preview*

The genus *Drosophila* is a large, diverse group with world-wide distribution. To date over 1,000 species have been described (Hardy, 1965). Additional new species are constantly being described and the total size of the genus must be at least 2,000 species (Stone et al., 1960). A considerable percentage of these species can be successfully maintained in laboratory cultures. Many, if not most, are prolific, and the generation cycle is short, ranging from 10 to 30 days.

Ever since Morgan and his associates introduced them early in this century as prime subjects for the study of animal genetics, the members of the genus *Drosophila* have been intensively studied by specialists representing various subsections of biology. As a result, there has developed a massive and detailed amount of information which forms a basis for and elucidates the taxonomy, evolution, and phylogeny of the group. It is against this superb background of information that the student of drosophilid mating behavior, especially comparative behavior, can analyze and interpret his data.

Because drosophilids are small organisms, only 2 to 7 mm in body length (with the vast majority of species being in the 2 to 3.5 mm range), magnification is necessary to observe the details of their courtships. This has been accomplished in the laboratories by confining the flies in small transparent containers or cells of glass or plastic of various shapes and sizes of the investigator's choice, and then viewing them with the aid of a low power stereoscopic binocular microscope. Some observers use only one pair of individuals per observation, but most introduce a few to many specimens into the observation cell simultaneously.

Sturtevant (1915) initiated the study of drosophilid sexual activity by describing the behavior of *D. melanogaster* and some of its mutants. Subsequently, and especially during the past two decades, various authors have added to our fund of knowledge concerned with the description and analysis of the elements of the mating behavior of numerous species of the genus *Drosophila.* To date the courtship behavior of at least 200 species has been recorded in varying degrees of detail and completeness. Manning (1965) has recently utilized the accumulated fund of information to analyze the parameters of mating behavior within the genus *Drosophila* and to relate it to the behavior of other organisms. His elegant and imaginative contribution is a bench mark for the analysis of mating behavior and its evolution within the genus.

Ontogeny of Mating Behavior

The data are fragmentary concerning the ontogeny of the sexual cycle in various species of drosophilids. It appears that (1) *very* young females are ignored by mature males; (2) mature males will and do court young females before the latter are willing to engage in copulation; (3) young males probably court before they are able to achieve copulation, even with a receptive mature female; (4) sexual receptivity rises with time and then gradually declines in both sexes. With rare exceptions, a period of time, ranging from several hours in *D. melanogaster* to 10 days in the Hawaiian drosophilids, must elapse before the individuals will engage in courtship. *D. spinofemora* individuals, however, are sexually mature when they eclose. The majority of species become sexually active within the first two to three days after eclosion and most reach a peak of receptivity three to five days after they have become sexually active (Bastock and Manning, 1955; Spieth, 1952, 1958). It is significant to note that most investigations which have been conducted on mating behavior and sexual isolation have employed specimens that have been separated as to sex a few hours after eclosion, and then isolated either individually or in small numbers for several days before being utilized for study.

Basic Courtship Patterns

The complex mating behavior displayed by *Drosophila* involves a series of actions that can be observed and identified as units of behavior. For any given species these units are typically performed repeatedly and within limits of normal variation seemingly identically by all individuals of interbreeding populations. It is the identification and interrelationships of these units that the behaviorist uses for his analysis when he is studying a population of a given species, a number of intraspecific populations, or comparing the behavior of different species.

The nature and relationships of these units to each other can best be elucidated by consideration of specific mating behaviors displayed by three diverse examples: *D. melanogaster, D. picticornis,* and *D. virilis.*

Drosophila melanogaster Meigen. *D. melanogaster* is a cosmopolitan species whose present distribution is due to modern commerce and transportation. Its ancestral home is unknown, but presumably was in southeastern Asia. It is a small yellowish fly averaging 2 mm in body length.

When sexually mature males of *D. melanogaster* are introduced into the presence of females, the beginning of a courtship is indicated by the orientation of a male toward another individual that is walking nearby or that the male himself is passing at close proximity. Orientation involves the turning of the male toward the other individual. This is accompanied by a slight raising of the male's body. He then approaches, extends and elevates one of his fore legs, and strikes downward with a sharp tapping movement against the body of the other fly. Often, and especially if the latter is moving, the male may quickly repeat the tapping movement.

If the tapped fly is a female, the male then faces her body, extends his wing which is nearest her head and vibrates it rapidly. After a pulse of vibration, the wing is returned to its resting position. He may repeat the vibration before he circles to her rear and positions himself directly behind her with his face close to the tip of her abdomen. Having assumed this position, he curls the tip of his abdomen under and forward, and at the same time engages again in wing vibration. After one or two bursts of vibration he moves forward with a slight lunge, licks the female's genital area with a stabbing motion of his extended proboscis, the labial lobes of which are opened, and simultaneously extends both of his fore legs under her abdomen and quickly vibrates his fore tarsi against her venter. This short lunge is then immediately and invariably followed by the curling of the male's abdomen still farther forward and the rearing upward of his head and thorax as he thrusts the tip of his abdomen forward to make contact between the genitalia of the two individuals. At the same time he lifts his extended fore legs outward and upward above the female's abdomen, thus positioning them so that they can be used to spread the female's wings if she allows intromission. If this occurs, the male forces the relaxed wings of the female apart with his fore legs and head, grasps her abdomen with his middle legs, and completes his mount by grasping the dorsal surface of her wings with the tarsal claws of his fore legs. During copula, the female vibrates her wings intermittently (from every 30 sec up to every 2 min) and occasionally kicks with her hind legs. At the end of the copulatory period, which lasts for 17 to 20 min., the male pushes upward and backward, thus visibly elongating his abdomen, typically turns 180° and withdraws.

If a courting male lunges onto a nonreceptive female who does not spread her vaginal plates thus preventing the male from achieving intromission, he does not try to spread her wings or complete his mount but drops backward to the original courting position behind her. He may then, and usually does, engage in vibration and subsequent lunging, and attempt to mount the female. Such courtship may continue for prolonged periods, but normally it is interrupted by circling and scissoring. The male circles about the female, close to her and facing her as he moves around her. Occasionally while circling, he scissors both wings out and in, in increasing amplitude, until the wings reach 70° to 80° from the resting position. Sometimes he pauses in his circling in front of the female and extends his wings outward and then upward about 45°. The wings are held in this position for one to two sec and then returned to the resting position. As Manning (1959b) notes, scissoring often grades into vibration. After completion of his circling, the male again assumes his posture rearward of the female, but it is significant that he may at any time during the sequence (i.e., from tapping onward) break off courtship. He may then rest for a time, immediately seek another individual to court, or turn to some other activity such as preening, feeding, or flying away.

The nonreceptive female also displays a gamut of responses to the courting male. She may rapidly run or even fly away from him as soon as he begins to court her. Males usually chase running females, vibrating and attempting to lick, or alternatively they may circle in front in an effort to prevent the females' escape. If a female does not decamp, she may kick vigorously rearward with her hind legs while engaging in cleaning motions of her wings and genitalic area after a male has lunged unsuccessfully. She may also depress the tip of her abdomen downward close to the substrate, or she may engage in extruding. The latter involves a complex series of actions which constitutes an effective means of repelling the male. The female compresses the two vaginal plates against each other medially and at the same time displaces them posteriorly, thus exposing the articulating membranes connecting them with the more anterior sclerites. If she happens to be feeding, she will extrude and also elevate the tip of her abdomen. Young females appear not to be able to extrude and although they repel the males effectively, they usually do so by kicking, depressing, and decamping.

Males, particularly those that are sexually active, often orient upon other males and tap, frequently attempting to mount. The courted male countersignals by moving both wing vanes slightly laterally from the resting position and then vibrating them rapidly in small amplitude. He typically spins about to face his suitor, striking at him with one or both fore legs.

Drosophila picticornis Grimshaw. It is illuminating to compare the courtship of *D. melanogaster* with that of one of the Hawaiian drosophilids, for the latter are quite removed in certain specialized aspects from the main stream of drosophilid evolution (see subsequent discussion, p. 168).

D. picticornis is a blackish, moderately sized fly (3.2 to 3.6 mm body length), with heavily pigmented picture wings (Hardy, 1965). Under laboratory conditions the males of these flies do not reach their peak of sexual activity until they are 14 to 20 days old. When introduced to females, the males do not seem to include tapping as a necessary part of the courtship pattern. Rather, the male orients with respect to a fly, and attempts to position himself in front of the female. If she tries to avoid him, he will arc back and forth to stop her movement. When she stands still, he extends his proboscis toward her and repeatedly strikes it down against the substrate, periodically opening and closing the labellar lobes. At the same time the middle legs are extended and curled upward so that the tips of the tarsi are above the top of the thorax of the male and directed upward; from this position, the legs are swung sharply down against the substrate. After a period of proboscis striking and midleg action, the male circles quickly to the rear of the female, depresses his body slightly, and assumes a position directly behind her with his head under her wings and close to the tip of her abdomen. He then folds and elevates both front legs in a complex manner as follows: the fore trochanters are swung forward, each femora extending forward and upward so that the tips are alongside the anterior margins of the eyes and the tibia and tarsi form a straight line hanging downward. At the same time the tip of the abdomen is slightly curled upward so that the anal papilla is pointed almost directly upward. One other obscure (to the observer) but important action also occurs, i.e., the third antennal segment is extended, as a result of which the aristae which normally are directed forward are now pointing straight upward and, most important, are in contact with the under surface of the female's wings.

After assuming the courting stance, the male engages in two alternative types of action which are not exhibited with equal frequency. The most common action involves the extremely rapid back and forth vibration in small amplitude of the suspended fore legs. As pulses of this activity occur, a small bubble of fluid is periodically extruded and retracted from the male's anal papilla. Infrequently but never simultaneously with the fore leg vibration, bursts of wing movement occur. Both wings are involved in such action; first one and then the other is flicked alternately outward from the resting position with the vane flat to the substrate. At first the amplitude of movement is small, but with each subsequent outward flick the tempo and amplitude increase until finally both wings almost simultaneously reach 90° or a right angle to the body of the fly. At this time the wings are vibrated backward and forward at a rapid rate for a fraction of a second in an arc of about 20°. At the end of the burst, the wings return to the resting position.

After a period of courtship, usually prolonged, the female may accept. She signals acceptance by spreading her wings and extruding her ovipositor. Apparently the lateral movement of the wings transmits her acceptance signal via the male's aristae. Just as her wings start to move laterally, the male extrudes his proboscis and licks the female's vaginal plates while concurrently bringing his fore tarsi downward alongside the tip of her abdomen, vibrating them rapidly for a brief time, and then lunging forward and up-curling the tip of his abdomen forward to achieve intromission.

During copulation, which lasts for 15 to 17 min, the pair is quiet. Other individuals may brush against or even walk over them without causing a response movement on the part of the copulating pair. Only during the last one to three min of copula does any movement occur, when the female kicks with her hind legs at the point of genitalic union.

Nonreceptive females kick and preen with their hind legs, depress the tip of the abdomen, and decamp but never extrude. A male never attempts to mount a nonreceptive female, but occasionally tries to lick and vibrate his tarsi against her abdomen which always causes the nonreceptive female to react with vigorous kicking.

Males court each other freely and persistently; apparently there exists no countersignalling mechanism such as that displayed by *D. melanogaster*.

Drosophila virilis Sturtevant. On the basis of detailed study of evolution within the genus *Drosophila* (Patterson and Stone, 1952; Throckmorton, 1962), it is agreed that *D. melanogaster* belongs to a monophyletic evolutionary stem that is designated as the subgenus *Sophophora*. *D. picticornis* is a member of the subgenus *Drosophila* but, as is characteristic of the members of this subgenus which dwell in Hawaii, displays a mixture of primitive and highly specialized characteristics. It is instructive, therefore, to use a species such as *D. virilis* to compare with both the unique *D. picticornis* and the sophophoran *D. melanogaster*. Throckmorton (1962) considers *D. virilis* to have been "derived almost directly from the stem population of the subgenus" *Drosophila*.

D. virilis is a medium sized fly (3.2 mm). Its ancestral home is not known with certainty, but today it has wide distribution and is associated with human habitations.

The male, when courting, orients upon the female and then taps with his fore leg and immediately moves to her rear, depresses his body slightly and assumes a position directly behind the female with his head under the wing tips and close to the tip of her abdomen. He then extends his proboscis forward and licks the female's vaginal plate area, concurrently rubbing the ventral surface of her abdomen with his extended fore legs, moving them alternately and rapidly backwards and forwards. Both actions (licking and rubbing) continue for relatively long periods, with only occasional short pauses of interruption. Meanwhile one of the male's wings is extended from the resting position 10° to 15° and vibrated rapidly in a small amplitude of movement. The wing vibrations are intermittent in character and are not synchronized with the leg and proboscis actions.

A receptive female spreads her wings outward and upward about 45° and at the same time spreads her vaginal plates. The male then mounts and achieves intromission. During copula he rubs the sides of the female's abdomen with his mesothoracic legs in intermittent, rapid bursts of movement. The pair is quiet during copula, which lasts for two to four min, until almost the termination when the female kicks at the area of genitalic union with her hind legs.

A nonreceptive female displays several refusing actions such as decamping, kicking backwards with her hind legs, fluttering her wings rapidly in small amplitude, and depressing both the abdomen and wings so that the male can not assume the head-under-wings posture that is necessary for licking and rubbing to occur. If the male achieves the head-under-wing courting stance, he will occasionally lunge forward and attempt unsuccessfully to mount a nonreceptive female. Rarely, he will also circle about such a nonreceptive individual. Nonfecundated females do not extrude, but recently fecundated females do so, although such action does not seem to deter the males from vigorously courting them.

The mating behavior of the three species described above presents a fairly comprehensive overview of the elements constituting the sexual display performed by members of the genus *Drosophila*. To summarize briefly, a male engages in orienting, tapping, circling; in movements of the legs, wings, and abdomen; in kicking, lunging, and perhaps in countersignalling. If she is nonreceptive, a female kicks, decamps, depresses the tip of the abdomen and/or wings, and in most cases extrudes. A receptive female engages in wing and genitalic movements. Some species show a paucity of some of these elements, as is the case of *D. victoria* and *D. lebanonensis* (Spieth, 1952), while others such as the Hawaiian *D. petalopeza* and *D. adiastola* (Spieth, 1966c) exhibit complicated courting displays.

Whether the sexual display consists of few or many identifiable elements, two significant facts emerge. (1) Unique, clear-cut signals are produced by portions of the fly's anatomy, which at times other than during courtship are used for functions such as feeding, walking, flying, ovipositing, and defecating. Each of the courtship signals clearly involves complicated neuromuscular activities and, in the terminology of ethology, have become ritualized. (2) The similarity of the signalling actions is of such a nature that they can logically be considered as homologous actions in the same way that morphological structures are determined to be homologous.

Adaptive Nature of Mating Behavior

It is a biological truism that just as an organism's structure is adapted to the environment in which the species normally lives, so also is its total behavior. The question can therefore be raised as to how the complicated sexual behavior displayed by drosophilids is adapted to the existence of a "wild" individual living in its normal habitat.

Relatively few detailed observations of mating behavior have been recorded dealing with *Drosophila* in their normal habitats, but I have had the opportunity to make a number of such observations employing various types of magnification including a Questar telescope. The significant feature of these observations is that they have given assurance that the various elements of the behaviors observed in the laboratory situation also occur under natural conditions. In other words, we have reasonable assurance that what is observed in the confined and unnatural conditions of the laboratory also occurs in the normal life of the drosophilid populations in the field.

Drosophila as adults are microphags feeding upon liquids and small particles of food which they sponge from the surface of their food sources. Their larvae are microphags also and live in a variety of soft plant materials. Both adults and larvae, with rare exceptions such as *D. funebris,* depend upon plant materials for their foods. The adults make use of a broader spectrum of food types than that available to the larvae since the adult females often feed upon materials in which they do not oviposit as well as upon those materials in which they do oviposit.

Sturtevant (1942) and Patterson and Stone (1952) have divided the larvae on the basis of their feeding habits into the following types: (1) general scavengers, represented by a small number of species which now have cosmopolitan distribution; (2) fruit feeders; (3) sap feeders; (4) fungous feeders, and (5) phloem, flower, leaf, and pollen feeders. These categories, however, should not obscure the fact that in every case the larvae are actually eating microorganisms, i.e., yeasts, bacteria, pollen, and perhaps small fragments of decomposing material. The fermenting sap, the rotting fungi, and the fermenting fruits simply serve as media in which the microorganisms live. The adults imbibe both the liquid and microorganisms which the food materials produce.

Such feeding and breeding sites to which the adults are oriented by odor are usually physically rather small and restricted in nature, e.g., a rotting fruit, a slime flux on a tree, a decomposing fungus, a rotting leaf. Numerous individuals, both male and female, of various ages and various species congregate together upon such discrete food masses.

In the temperate and semitropical areas the adults visit the food and ovipositional sites for short periods only, early in the morning and late in the afternoon; during the rest of the time they remain hidden in secluded, probably humid areas of low light intensity. In the tropics and during cloudy periods in temperate regions, some individuals will visit food sites throughout the day, but the peak feeding activities of the bulk of the population still occur in the evening and morning. It should be noted that although *Drosophila* are polyphagous, each species does have preferences (Dobzhansky and da Cunha, 1955) and not all of the species present in a given area will be attracted to each of the available food substances.

When the individual drosophilid arrives at the feeding site, it is in close proximity not only to other members of its own species but also to individuals of other drosophilid species, as well as to a menagerie of quite different creatures such as large and small Diptera, Hymenoptera, or Coleoptera. Observation shows that the drosophilid females spend practically the entire period of each diurnal visit in feeding, whereas the males spend only short intervals in feeding and devote the remainder of their time to sexual investigation of other individuals.

All of the insects do move about on the food mass and as they move the sexually mature males constantly attempt to initiate courtships. Four elements are involved in stimulating a male to attempt to initiate courtship with another individual: (1) size; (2) overall facies; (3) movement, and (4) information received by his tapping action (Spieth, 1966a).

Large muscoid flies and small fungous gnats, ants, beetles, and other creatures are either ignored by the male drosophilid or he flees from them. Insects of approximately the same size as he, but of quite different general facies, are also ignored. The male orients upon and taps other insects that display approximately the same size, shape, and movement that are characteristic of other individuals of his own species. A significant point in regard to size is that considerable variation is found wthin a population of wild drosophilids of most, if not all, species—much more so than is observed in typical laboratory-reared populations. A male does investigate other flies that are considerably larger or smaller than most of his own females, if the movement and general facies of the other individual are appropriate.

If the male taps an individual of another genus or another distantly related drosophilid, he then turns away. If, however, he has tapped a female of his own species, he immediately proceeds with courtship. If the female is nonreceptive, and in the vast majority of instances she is not only nonreceptive but is also engaged in feeding, she typically responds with one of the following three reactions: (1) an immediate turning of her body in such a way that the male is not able to assume the courting position, e.g., she will turn and face the male; (2) decamping by running or flying away; (3) remaining stationary but engaging in repelling actions such as depressing the tip of her abdomen, kicking, or elevating the tip of her abdomen and extruding. All of the actions, and especially those of depressing, elevating and/or extruding, allow the female to refuse the male and at the same time to continue with her feeding activities without interruption during her short diurnal visit to the food site.

A male that is repulsed by a nonreceptive female turns from her after a variable but usually quite short period of time, and seeks another individual to court. A small percentage of females are receptive and some copulations do occur on the food site, but the number is so low that it is clear that most of the males during any given feeding session are not successful even though most of them have engaged in numerous attempts at courting.

Because of the paucity of field studies, the data given above have been derived mainly from the scavenger, fruit, sap, and phloem feeders. Brncic (1966) has observed adults of *D. flavopilosa* Frey (a species whose larvae feed upon the pollen of the fleshy flowers of *Cestrum parqui* L'Her.) copulating on the *Cestrum parqui* plants during the early morning and evening.

Specialized Mating Behavior of Hawaiian Species

The abundant and diverse endemic Hawaiian species of *Drosophila* differ from the basic pattern of feeding and courtship described above, a difference due to the evolution of a behavioral pattern which spatially separates courtship from feeding and oviposition (Spieth, 1966c).

A number of Hawaiian species breed in fungi but the majority use rotting leaves, phloem, flowers, and fruits of a number of shrubs and trees, but especially the leaves of *Cheirodendron, Tetraplasandra,* and *Pterotropia* (Araliaceae) and the flowers, fruit, leaves, and phloem of *Clermontia* and *Cyanea* (Lobeliaceae). The adults also use the same sources for food, but some have been observed feeding upon other naturally occurring fermenting substances such as tree fern stumps. Only a limited number of the species are attracted to baits made of rotting fruits such as bananas, tomatoes, and so forth. Extensive field observations of the feeding behavior of the adults on their natural foods indicate that the flies do not court when on a feeding site but, rather, both sexes restrict their activities solely to feeding, plus ovipositing in the case of the females. When on a food site the individuals are quiet and cryptic in their behavior. While feeding, they are tolerant of the immediate presence of other individuals, but as soon as they leave the food site and fly by a rapid, darting flight to the surrounding vegetation, the flies and especially the males become highly intolerant of the nearby presence of other individuals.

Like all other drosophilids, the Hawaiian males spend much less time feeding than do the females, and since they do not court on the food site they leave quickly and move into the adjacent dense vegetation where they sit on the surfaces of leaves, usually the under sides, on fern frond stems, and on tree trunks. Males, presumably sexually mature, often accumulate in particular areas such as on the trunk of a single tree, the leaves of one branch of a shrub, the stems of the fronds of a single tree fern, or even on the individual pinnae of a single fern frond (see Fig. 1, p. 179).

Each male defends a limited area such as a single leaf, a fern pinna, a section of a fern frond stem, or a portion of a tree trunk. Concurrently he apparently advertises his presence visually or otherwise. Thus males of some species (e.g., *D. crucigera, D. grimshawi,* and *D. engyochracea*) repeatedly drag the tip of their abdomens over the substrate and in doing so deposit a thin film of liquid (see Fig. 2, p. 180); others assume a ritualized posture and extrude and retract a bubble of fluid from their anal papillae (see Fig. 3, p. 180). Defense of territory against intruders, typically other males but occasionally nonreceptive females, usually involves physical contact between the individuals and many of the species exhibit ritualized postures such as curling and slashing which involve wing, leg, and body movements (Spieth, 1966c). The most vigorous fly, usually the largest, appears to be the victor in such contests, the original possessor of the area often being displaced by the intruder. The mere assumption of the aggressive posture, without actual physical contact, frequently is sufficient to determine the outcome of the contest. If the two individuals are of disparate size, the assumption of the aggressive posture usually causes the smaller to flee. Under field conditions the contestants apparently do not physically injure each other, but under the confined conditions of the laboratory at least wing vanes are broken in many individuals.

Presumably the sexually receptive females are attracted to the advertising males and seek them out, a conclusion confirmed by laboratory observations. Significantly the females of the Hawaiian species lack the ability to engage in the unique repelling actions of extruding and elevating the tip of the abdomen which females of many other species utilize when males court them while they are feeding.

Both in the field and in the laboratory, males have been repeatedly observed to assume first an aggressive posture before an intruding female, and then to court her. If she is nonreceptive, the male turns to aggression and drives her away.

Field and laboratory studies also indicate that males of most Hawaiian *Drosophila* species patrol and defend a small but definite courting territory. These sites are not randomly distributed in the rain forests where the flies live, but each species has special preferences, apparently determined by factors such as light, humidity, temperature, and spatial conditions. Invariably they are in close proximity to where the individuals feed and the females oviposit. As mentioned above, each leaf on a single branch of a given shrub may have a male on its under surface, or the frond stems of a particular tree fern may have a male patrolling and defending a two to four ft section of each stem. Experienced collectors locate these spots and can return repeatedly over extended periods, even many months, and each time find male occupants.

Accompanying this behavior has been the evolutionary development of sexual dimorphism in the males. With all the species whose behavior has been studied, more than 50, the male dimorphic structures have been importantly involved in courtship and appear on various parts of the body, but mainly on the appendages. Fore leg modifications are the most common, but in a few species the middle legs are involved. Mouthpart modifications, involving the proboscis, are found in numerous species and may involve excessive enlargement of the proboscis or the development of complex, heavily sclerotized structures made up of the modified labellar hairs which form a complex grasping organ. Dimorphic wing modifications consisting of clouds or patches of pigmentation appear in various species, and in a few others extraordinary abdominal modifications are present. Antennal dimorphism is displayed by one species group on the basis of which Hardy (1965) has established the genus *Antopocerus*.

It is to be noted that certain species groups from other parts of the world (e.g., the *Drosophila quinaria* species group) have females that are similar to those of the Hawaiian species in that they also lack the ability to extrude and elevate the abdomen as repelling actions. Further, the courtships of the males of the *quinaria* group are more similar to those of the Hawaiian species than to any other species group studied so far. Field studies have not been made on the *quinaria* species group and it is possible that they parallel the Hawaiian species in having courtship spatially separated from feeding and ovipositing.

Summary of Mating Behavior

The field studies in combination with laboratory findings show that (1) drosophilids are gregarious when they feed, oviposit, and also, excepting the Hawaiian species and perhaps a limited number of other species, when they engage in courtship; (2) a limited amount of time is spent upon the food-ovipositing-mating site; (3) when they are not feeding, they individually scatter into the surrounding area where courtships, if they occur, must be accidental and infrequent; (4) the males are promiscuous and they initiate the courtships; (5) the courtship behavior is such that the males engage in a series of ritualized movements; (6) the females determine whether copulation shall or shall not occur; (7) the females display both repelling and accepting actions; (8) the courtship behavior is such that the female can continue with feeding and even ovipositing while refusing the male's courting overtures, and (9) the Hawaiian species have superimposed upon the basic drosophilid mating behavior a true lek type of behavior in which each male selects and defends a courting arena near the food and ovipositional site. Here the males advertise their presence and the females are attracted to them. Concurrently with this behavior, male sexual dimor-

phism and extremely complex, diverse, and unique courtship patterns have
evolved.

Nature and Function of the Stimuli

The nature and function of the stimuli passing between the drosophilids
engaged in courting were first investigated by Sturtevant (1915). Subsequently, various investigators (see especially Manning, 1965) have continued to study and analyze the elements involved. Because of the small
size of the flies, it has not yet been possible to employ techniques enabling
one to record the action potentials of the insects' nervous system. Therefore more indirect methods have been used involving: (1) normal, apparently healthy males and females, either wild type or mutant, upon whom
observations are made as they engage in courtship and respond to each
other's actions, and (2) surgically mutilated individuals.

With the latter technique, wings, parts of legs, antennae, proboscis, labial
palps, or even the entire head can be removed from individuals of both
or either sex whose subsequent behavior is then studied. Comparison of
the behaviors and the courtship success of such specimens gives clues as to
the nature of the stimuli and reception of the stimuli involved. Some investigators have used genetic mutants such as *antennaeless* and *white-eyed*
flies. Since most genes have multiple effects upon the phenotype, such
procedures compound the uncertainty as to how the effects observed have
been caused, i.e., whether due to the obvious deficiency or to some other
unknown effect of the mutation.

The sense receptors of an insect are considerably different from those of
many other organisms, especially vertebrates, both in distribution over the
body of the insect and in structure. Nevertheless it is possible to categorize
the sensory modalities in comparable terms. For instance, both contact
and distance stimuli are involved in drosophilid courtship. The former
appear to consist of chemical and mechanical stimuli; the latter of visual,
air borne chemical and air borne mechanical stimuli. Diptera are known to
be well supplied with chemosensory receptors on various parts of their
bodies, but especially on the labellar lobes and palps, the tarsi, the oviposi-
tors, and the wings (Dethier, 1963). Stimuli received by chemosensory
receptors, but not from the mechanoreceptors, are transmitted directly by
a single neuron to the brain. Further mechano- and chemoreceptors appear
to be combined often in the same receptor unit (Hodgson, 1965). As
shown in the courtship behaviors described above, those parts of the insect's body which are more frequently involved in the mating sequences are
exactly those which are most abundantly supplied with chemoreceptors,

but as yet we cannot separate the contact mechanical effects from the chemical effects.

The compound eye constitutes the major visual organ of the drosophilids. Many species, but not all, have the ability to mate in the dark, indicating that stimuli other than vision per se are sufficient to enable the sexual process to be completed. However, direct observations on courtships of species that are not light dependent show that what they do under lighted conditions depends typically upon vision for the initiation of orientation by the males. Philip et al. (1944) showed that *D. subobscura* is light dependent and can not mate in the absence of light. Subsequently a number of other species have been determined to perform in a similar fashion (Manning, 1965; Grossfield, 1966). Visual stimuli thus serve an important role in the mating behavior, especially in the initiation of courtships by the males.

The existence of air borne chemical stimuli, released by one or both sexes during sexual activities, is more difficult to isolate than are those types considered above. Experiments (Flügge, 1934) show that drosophilids lacking the third antennal segments fail to respond to odors such as those emanating from normal foods. Sturtevant (1915) presented data indicating that the occurrence of a copulation within a container of limited volume would serve to stimulate the mating speed of the next pair of flies to be introduced into the space after removal of the first pair which had just completed the copulatory act. Ewing and Manning (1963) were not able to achieve confirmation of Sturtevant's findings. Observations both in the laboratory and in the field show that a number of the Hawaiian species do produce scents that are involved in courtships. These pheromones are produced by fluid secretions from the anal papilla and are used both to attract the females into the immediate vicinity of the male and during courtship activities. Some species utilize pheromones in both ways, others in only one of the two.

The air borne mechanical stimuli are difficult to separate from the air borne chemical stimuli, and it is probable that in some instances they may eventually be proven to be intertwined not only by the producing agent but also in the receptors of the individuals receiving the stimulus. Wing and leg vibrations are the most obvious sources of such mechanical stimuli. Observations and experiments by a number of investigators (starting with Sturtevant, 1915, and continuing through the studies of Bastock, 1956, Petit, 1958, 1959, and Ewing, 1964) have determined that the main stimulus produced by the male's wing vibration is not chemical but rather mechanical and is received via the air-current receptors located at the base of the female's antennae. *Drosophila* do not possess tympanal organs per se, and the mechanical stimuli such as air currents must therefore be perceived by sensory hairs. It is possible that leg vibrations such as those displayed by *D. picticornis* are different from the wing vibrations in that either: (1) the mechanoreceptors are located on the abdomen of the female, or (2) chemical stimuli (odors) rather than mechanostimuli are being transmitted by the leg vibrations.

A single isolated but sexually mature male, whether in an isolation vial in the laboratory or on a leaf surface, stem, or crevice in the field, does not engage in courting actions. In other words, his threshold is at such a level that the environment does not produce stimuli of the type or intensity that will cause him to react sexually. The known exceptions in *Drosophila* are the males of some of the Hawaiian species which, like many male birds, advertise their presence not by song as do birds, but by pheromones. In these instances an appropriate environmental site plus internal stimuli lowers the male's threshold to the level that he engages in display.

Excepting these, the typical sexually mature male when he is introduced into the presence of other individuals, either on the food site or in the observation cell in the laboratory, then receives stimuli which lower his threshold to the level that he begins courtship. Spieth (1966a) has shown that at least three types of stimuli are involved in this threshold change: (1) the configuration and size of the other individual; (2) the movements of the other individual, and (3) the tapping action. The relative importance of the three types of stimuli varies between species. Thus, a male of *D. auraria,* a light-dependent species, will not initiate courtship if the wings of the female have been completely removed (Grossfield, 1966) nor will he court a living but decapitated female unless he approaches her in such a manner that he can not visually determine that her head has been removed (Spieth, 1966a). It is obvious that tapping by the male and movement on the part of the female are not sufficient either individually or in combination to initiate courtship by *auraria*. Significantly, however, after an *auraria* male has initiated courtship with a normal female who refuses his advances, he will then turn to other females and begin courting them but will often omit the tapping action.

In comparison, *D. virilis* males will initiate courtship with nonmoving, decapitated females after they have tapped. With normal females, however, males that have had their fore tarsi amputated will also initiate courtship, thus indicating that movement and configuration plus perhaps other as yet unidentified stimuli are sufficient without the mediation of the stimuli received via tapping to lower the male's sexual threshold. Further, such mutilated males will court females of closely related species whom they would not court if their tarsi were unimpaired.

Males of *D. melanogaster* display a behavior pattern that is intermediate between that of *D. auraria* and *D. virilis*. Eventually *melanogaster* males will court decapitated females, but they are slow about initiating courtship and are obviously confused—sometimes even courting the anterior end of a decapitated female, a mistake they never make with normal females.

A male of *D. pseudoobscura,* whose courtship pattern is rather similar to that of *D. melanogaster,* is like *D. auraria* in that the male fails to court a decapitated, nonmoving female unless he approaches from the rear in such a manner that the female's lack of a head is not visually apparent to him. *D. pseudoobscura* is not light dependent and does inseminate normal females in total darkness. Under such conditions stimuli other than visually determined configuration of the female are sufficient to lower the male's threshold. The observations of Streisinger on etherized females (1948) and of Milani with models (1950a and 1950b) also confirm the fact that multiple stimuli are received by the male at the time of courtship initiation and that these have varying values with different species.

The female is passive until the male initiates courtship. She then responds in a binary fashion, either yes or no. Typically her sexual threshold at the beginning of a courtship appears to be high even though she is virginal and sexually mature. She first receives information from the orienting and tapping movements of the male. Infrequently she will give the acceptance response at this juncture, indicating that the male's action of orienting and tapping has sufficiently lowered her sexual threshold. Typically, however, the male must perform the major portion of his courting repertoire, producing at least once a variety of specific stimuli, before the female will give the acceptance response. A male may court a "reluctant" female for prolonged periods of time, or may repeatedly desist, back away, and then return to her. Such "reluctant" females may be unsuccessfully courted by several males for prolonged periods of time and then may unexpectedly give the acceptance response to one of them.

The nonreceptive female may display the full gamut of her repelling actions or only some of them, and, most important, she may display them in random sequence.

The courtship pattern of *Drosophila* is thus quite different from that of some other well-studied organisms in which the stimulus-response pattern forms an interlocking, sequential series or chain which is typically interrupted if one of the pair fails to respond to the preceding stimulus produced by the other member (e.g., the three-spined stickleback). In *Drosophila* only the initiation of courtship, where the male commences his display as a response to multiple stimuli received from the female, conforms to this interlocking pattern; from then on, the male may and often does display his full repertoire of actions regardless of the response of the female. The female may, if nonreceptive, engage in any one or more of her repelling actions. She may even "ignore" the male and show, so far as the observer can ascertain, no response.

Mating Behavior of Other Diptera

The courtship of most dipteran species, particularly cyclorrhaphous types, differs from that of the drosophilids in that the male first grasps ("assaults") the female and then attempts to achieve intromission. This involves the male first orienting upon the individual to which he has become alerted, be it male or female, and then typically manoeuvering to a position behind or at the side of the other individual, followed immediately by a lunge onto the body of the potential mate. Gruhl (1924), Thomas (1950), Cazier (1963), and others have described such behaviors. Thomas' observations upon a number of coprophagous flies are pertinent in that these sarcophagous species gather upon localized food masses (e.g., human feces). Mating usually occurs after the flies have fed and the male randomly assaults another individual, then vibrates his wings violently and attempts to hook his genitalia under the posterior tip of the abdomen of the other fly.

Comparable behavior is exhibited by the genera *Chymomyza* and *Scaptomyza* which are close relatives of *Drosophila*, both belonging to the family Drosophilidae (Wheeler, 1952; Hardy, 1965). The mating behavior of a number of species belonging to these two genera has been studied and found to conform basically to the assault pattern described for *Sarcophaga* and others. Both males and females are wary of any other nearby moving individuals and typically orient themselves so that they either face such individuals or escape from their immediate vicinity. Sexually active males of both genera seek to move to the rear of other flies and then, with a running lunge, attempt to mount. Having mounted, the male of some species vibrates one or both wings, but always elongates and curls the abdomen down and under the abdominal tip of the other individual and seeks to make genitalic contact, repeatedly drawing his genitalia across that of the female. A male or a nonreceptive female reacts vigorously to the mounted male by kicking, by twisting the tip of the abdomen away from the seeking movements of the courting male, by shaking the entire body, and by wing vibrations of varying amplitude. After a variable period of time, the courting male is either dislodged by the actions of the other fly or dismounts of his own accord. Males are persistent and vigorous suitors, going from individual to individual with the result that nonreceptive females, the great majority, are constantly being interrupted in the pursuit of their activities of feeding, preening, or resting.

There is no valid reason to assume that a simple appearing courtship of this type involves fewer stimuli than are involved in the pattern displayed by drosophilids. Hardy (1965) notes that the *Scaptomyza* species have "prominent, well developed and exposed male claspers" while the drosophilid male claspers "are usually hidden beneath the ventral lobes of the ninth tergum." He has also observed that in classifying scaptomyzids, it is necessary to mount and study the male genitalia in order correctly to identify the species, since these structures are not only complex but also uniquely constructed for each species. It is reasonable to assume that each female receives multiple and specific stimuli resulting from the complex and unique genitalia plus the other stimuli involved by the male's mounting activities.

The larvae of *Scaptomyza* are mostly leaf miners, although Wheeler (1952) notes that there are some exceptions. The adults are typically collected by sweeping, and observation indicates that only a few of them come to baits (Dobzhansky and Brncic, 1957). They appear not to be diurnal in their feeding habits, and, although gregarious or semigregarious, feeding and courting are not restricted to short periods of time as is true for drosophilids.

Significance of Drosophilid Mating Behavior

It is abundantly clear that *Drosophila* have evolved a mating behavior quite different from that of most Diptera and of a type that is more easily analyzed. Their behavior seems to be an adaptation to their short-time, typically diurnal feeding-courting-mating-ovipositing requirements. The drosophilid pattern allows the female to proceed with her important tasks of feeding and ovipositing, even in the inevitable presence of courting males, without the kinds of physical harassment that would pertain if the more primitive pattern of the male-lunging-onto-the-female pertained. There are a few species such as *D. auraria* and others in which the males do lunge onto the females, but these are the exceptions and in each case that has been investigated there is ample evidence that these are specialized and secondarily derived from the typical drosophilid pattern.

Once the pattern was established of using wings, legs, and proboscis to produce both the male's courting stimuli and the female's refusal signals, then the possibility of greater evolutionary flexibility was achieved. Without evolutionary remodelling of the structures themselves, as seems to have been necessary for the scaptomyzids and other flies, changes in tempo, amplitude of movement, specific use of a structure, and specific posture position could modify the total courting pattern. Such remodelling of the courtship pattern could be anticipated to occur as quantitative changes within a population. This is exactly what has happened under artificial selection, and Bastock (1956) showed that the yellow mutant of *D. melanogaster* reduced the frequency of certain movements without modifying their gross form. Others (see Manning, 1965) have achieved similar results involving both rates of performance and frequency of performance of specific courtship elements, as well as threshold changes. Further, as Manning (1965) notes, these are interrelated with other behavioral characteristics, and a shift in any particular aspect of behavior may and typically does affect the performance of other units. The behavioral fabric is all of one piece, and natural selection is constantly re-tailoring and modifying it to achieve a compromise optimum.

During courting periods, males of many animals, especially vertebrates, exhibit toward the other sex of their own species three functionally intertwined tendencies: attacking, fleeing from, and engaging in sexual actions. Tinbergen (1960) suggests that the differences in the courting displays of such organisms resides primarily in the relative intensities of these three components and that they not only vary between species but within a given species during the extended courtship periods that pertain in most instances.

Clearly the basic pattern of the Diptera does not conform to this tripartite pattern. As Thomas (1950) showed, the *Sarcophaga* males are indiscriminate in their selection and will physically mount a *Lucilia* female, a male *Sarcophaga,* or a female of any *Sarcophaga* species present. By trial and error, he seeks a receptive mate among individuals that possess the approximately proper configuration and size, toward whom he shows no tendency either to attack or flee. Most frequently he is thwarted in his attempt and the negative response he receives from the individual he has mounted results in his dismounting. The level of his sexual threshold then determines whether he shall desist from further sexual activity or seek another seemingly appropriate individual. Scaptomyzid males appear to utilize a type of mating behavior similar to that of the sarcophagids.

The pattern for drosophilid species differs basically from that just described in only one fundamental aspect, i.e., the male performs courting actions before he mounts the other individual. Like the males of *Sarcophaga* and *Scaptomyza,* the drosophilid male (excepting certain Hawaiian species, see subsequent discussion) shows no tendency to flee or fight. He can not *a priori* distinguish male from female, nor members of his own species from those of another species of similar configuration and size; he is usually thwarted in his courting attempts and he will break off a courtship after a time and then often turn to another nearby individual to recommence his courting actions.

Unfortunately, scant evidence is available for identifying the stimuli, or lack of stimuli, that cause a courting male to cease courting one female and then immediately seek out another one to court. Observation unequivocally indicates that the male receives repelling stimuli by the kicking action of the nonreceptive female's legs against his body and by the extrusion action which typically involves the female twisting the tip of her abdomen so that it is directed to the front of the male's face. Some females, however, neither kick nor extrude or decamp, and yet the males will cease courting. Individual males of *D. melanogaster* and other species which perform wing vibration during courtship, will occasionally have had their wings accidentally mutilated so that the total vane surface is much reduced in size. Such individuals typically are exceedingly persistent in their courting displays. Critical experiments have not been performed to show whether this persistence is due to internal effects in the male himself or whether the lack of normal vibration prevents reception by the male of unknown repelling stimuli produced by the female.

Lek Behavior

When drosophilids gather upon food masses and engage in feeding, ovipositing, and courting, they present an appearance that is in many ways similar to the lek behavior of various birds and mammals. They are actually, however, engaging in a type of gregarious behavior resulting from the individuals of all ages and of both sexes gathering together diurnally upon a food site. That courtship and mating also occur at this time is a secondary result.

True lek behavior involves a number of promiscuous males assembling in close proximity and engaging in strikingly hypertrophied displays which attract sexually mature and receptive females to communal breeding grounds. Since drosophilid males are promiscuous and since the pair bond exists only for the period of courtship and copulation, there are parallels between the pseudo-lek gregarious drosophilid behavior and true lek behavior. Clearly the drosophilid males compete with each other in their courtship of the females, and it might be considered that the complex and hypertrophied courting displays of the males can have evolved as a response to the selective pressure created by the pseudo-lek type of situation. Such a conclusion is not supported by the situation found in *Chymomyza* and *Sarcophaga* who, as shown earlier, also may collect on food masses and are promiscuous and have short bond pairing but display relatively simple appearing mating behaviors. It thus seems most plausible that the selective pressure that has caused the development of the drosophilid male courtship was the replacement of the assault type of courtship displayed by most cyclorrhaphous Diptera and the evolution of a pattern of specific courtship movements which are subsequently followed by mounting (see p. 165).

From the basic drosophilid pattern, the endemic Hawaiian *Drosophila* have evolved a true lek behavior, which has involved the spatial separation of courtship from feeding and oviposition (see p. 168) and intensified sexual selection since the males must attract the females to the lek or mating sites. As in other organisms, the result of sexual selection is sexual

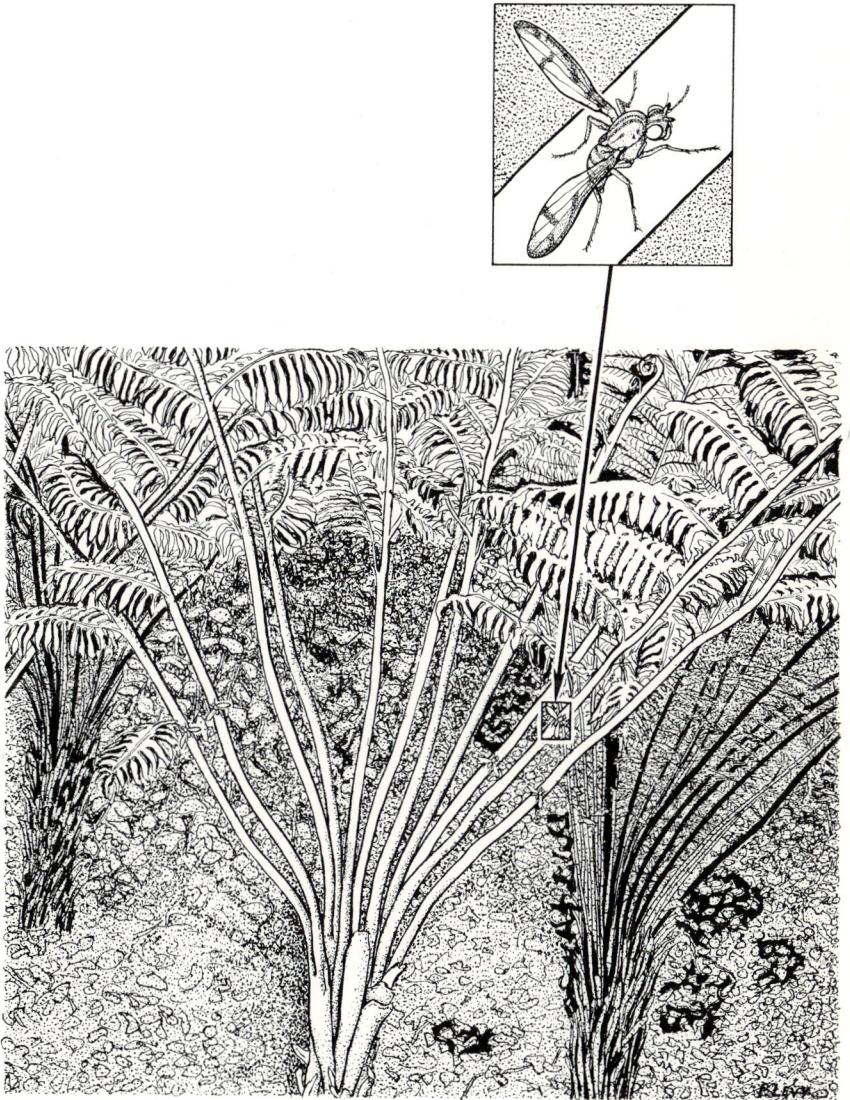

Fig. 1. Idiomyia planitibia Hardy males engaging in lek behavior on tree fern stem.

dimorphism and this has been developed to an extraordinary degree in the Hawaiian males.

As Sibley (1957) notes, "the genetic basis for such characters may, and probably does, involve but a few genes and these may control only relatively superficial characters" of male structure and display movements. Thus many Hawaiian species have females which can not be separated by expert taxonomists but these same species have males that differ clearly in their sexual dimorphic characters and especially in their courtship patterns. Furthermore, species that are quite distinct structurally and particularly in their courtship patterns, e.g., *D. crucigera* Grimshaw and *D. grimshawi* Oldenberg (Spieth, 1966c), under laboratory conditions produce viable, fertile hybrids.

It is impossible to determine with certainty the selection pressure that originally caused the spatial separation of courtship from the feeding sites, but the following pattern can be inferred. In the impoverished oceanic fauna, the Hawaiian drosophilids form a major element both in abundance of individuals and in protoplasmic volume. They dwell in the rain forests and congregate to feed and oviposit in exactly those areas and on those

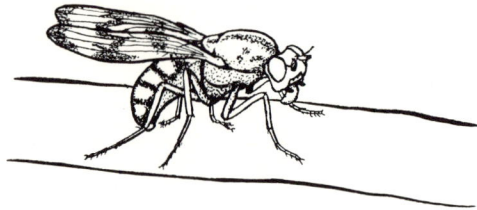

Fig. 2. Drosophila grimshawi Oldenberg *male exhibiting lek display. Note tip of abdomen being dragged against substrate.*

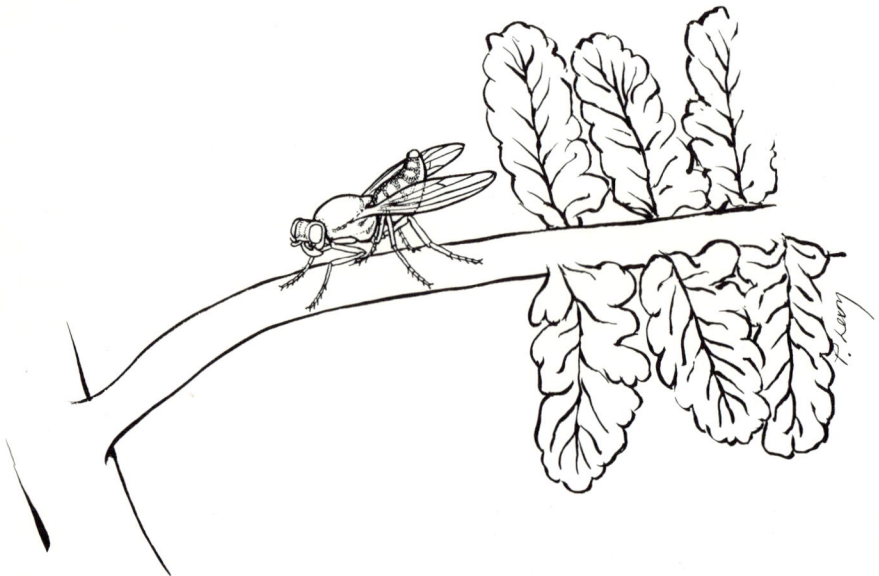

Fig. 3. Drosophila comatifemora Hardy *male exhibiting lek display. Note elevated tip of abdomen.*

plants that are frequented by many of the native birds, especially the insectivorous Elepaio (*Chasiempsis sandwichensis*) and the creepers (*Paroreomyza* spp.). Other honey creepers also catch insects and both the birds and the flies (Spieth, 1966b) are attracted to the lobeliad shrubs in numbers. All aspects of the Hawaiian drosophilid's behavior indicate extreme adaptation to the avoidance of predators, and the birds seem to be the only plausible candidates for this role. A parallelism to the unique behavior that the Hawaiian drosophilid fauna has evolved can be seen in the kittiwake (*Rissa tridactyla*) which displays a distinct type of nesting behavior by utilizing steep cliffs and is thus enabled to be out of reach of the predators that prey upon other gulls (Cullen, 1957).

When the ancestral drosophilid stock colonized the Hawaiian Islands, it found a habitat with many open niches into which it could radiate, and this, plus the lek type of mating behavior, has resulted in the enormously rich fauna of probably 400 plus species in an area of less than that of the state of New Jersey.

Sexual Isolation

In attempting to understand how sexual isolation has arisen within *Drosophila,* several basic considerations are to be noted.

1. Individuals of an interbreeding population of drosophilids perform similar courtship displays. Although few critical studies have been conducted, the evidence to date indicates that the variation occurring between the courting behavior of individuals of a given interbreeding population is no greater than the morphological variation that occurs between such individuals. In other words, the courtship pattern is species specific.

2. The more closely related two or more species are, as determined by morphology, genetics, or other criteria, the more similar are their courtship patterns. Thus, in the case of sibling species, which can be defined as "morphologically similar or identical natural populations that are reproductively isolated" (Mayr, 1963), the courtships are also essentially similar or identical. Any differences that can be detected between the courtship patterns of such populations are invariably of a quantitative character.

3. Closely related sympatric species, which rarely if ever hybridize in nature, will under the artificial conditions of the laboratory usually engage in courtship and often produce hybrids. In this they parallel the conditions found in many other organisms.

A variety of techniques has been employed by divers investigators in attempts to determine the amount of sexual isolation that exists between various populations of drosophilids. The typical procedure is to separate laboratory reared flies as to sex shortly after they have eclosed from the pupal stage, and isolate them until they have reached sexual maturity. Such males and females are then introduced to each other and allowed to remain together for a given period of time. The females are then dissected and determination is made as to what percent, if any, of the individuals have been inseminated. Four different combinations can be used for such studies:

1. No-Choice. Males of population A are placed with females of population B and males of B with females of A. With this procedure, an equal number of males and females is utilized, varying in number from pair matings to mass matings.

2. Male-Choice. Males of population A are placed with females of populations A and B, and reciprocally B males with females of A and B.

3. Female-Choice. Females of population A are placed with males of populations A and B, and reciprocally B females with males of A and B.

4. Multiple-Choice. Equal numbers of males and females of both populations are placed together.

Most investigators do not observe the flies during the time that the individuals are exposed to each other, and thus are unable to determine what part, if any, courtship behavior contributes to the final results. Observation of such experiments is time consuming but highly rewarding in gaining insight into the results obtained.

A more serious drawback with choice experiments is that the flies spend the period from shortly after eclosion to sexual maturity either in complete isolation from any other specimens or in the company only of specimens of their own sex and of their own population. In nature they would mature under quite different conditions; particularly significant is the fact that in most laboratory experiments they are deprived of any opportunity to become habituated to individuals of other species during the period from emergence to sexual maturity. Since flies of most species visit the feeding sites at least one or more times before they become mature, they are in nature not only exposed to mature, courting adults of their own population but also to those of other species (see also Manning, 1965).

Intraspecific Variations

Just as in geographically widespread species, individuals display quantitative morphological differences between parts of the population, so, likewise, comparable but not necessarily correlated differences can be detected in the mating behavior of some species. Such mating differences involve variations in the balance between the courtship threshold levels of the two sexes. In some areas the males of a given species may be aggressive suitors (low threshold) and the females lethargic or slow in responding to the male overtures (high threshold). In another part of the range, the opposite may pertain, i.e., lethargic males and highly receptive females. Patterson and Stone (1952) have shown that in wild-caught flies, the females of most species are inseminated, so that in any given population the courtship balance insures that the females do accept the males' overtures.

In an analysis of *Drosophila ananassae* stocks from various Pacific islands (Fig. 4; see also Spieth, 1966d), laboratory observations showed that males from Pago Pago, American Samoa, courted and mated with females from Rongerik Atoll, Marshall Islands, and from Majuro, Marshall Islands, as persistently and successfully as they did with their own females from Pago Pago. The Rongerik and Majuro females, however, displayed more accept-ance responses to the Pago Pago males than they did to their own males. The reciprocal crosses of Rongerik and Majuro males with Pago Pago

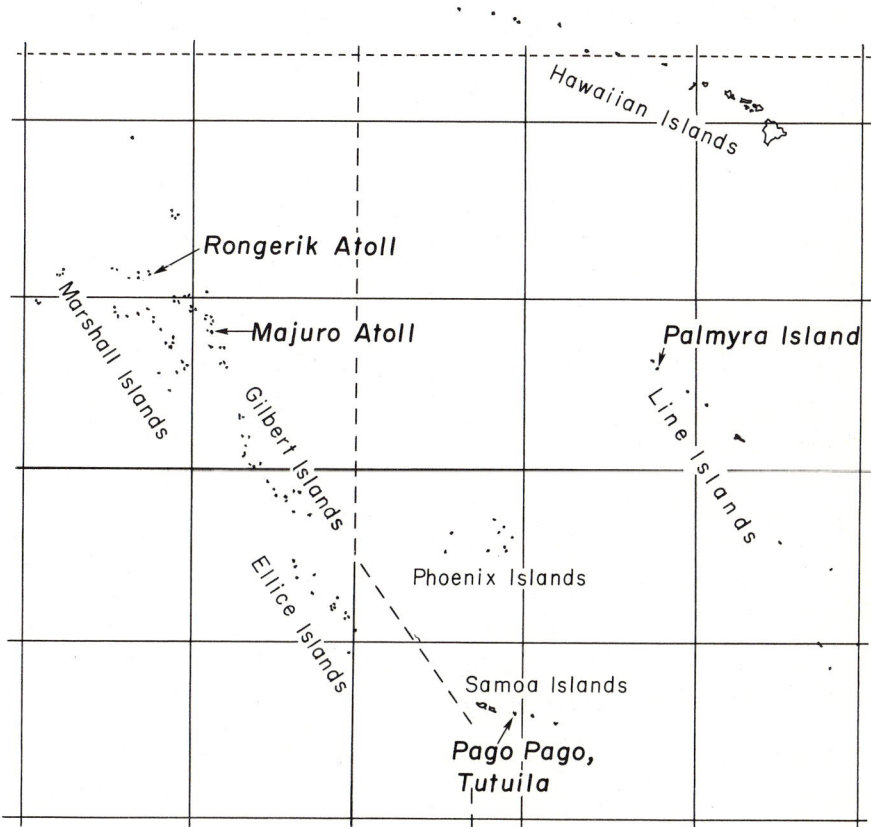

Fig. 4. *Central Pacific Islands, showing origin of* Drosophila ananassae *stocks discussed text.*

females resulted in no copulations during comparable observation periods. The Rongerik and Majuro males appeared lethargic (high threshold). Apparently the stimuli produced by the Pago Pago females was not sufficient to lower the threshold of the Rongerik and Majuro males to the level that they would court sufficiently to induce the females to engage in copulation. Diametrically opposite findings were derived from crosses involving Pago Pago individuals and those from Palmyra Island. The normally sexually aggressive Pago Pago males simply did not court the Palmyra females; in fact, to the observer they appeared to be repulsed at the time of tapping the females. The reciprocal cross of Pago Pago females and Palmyra males resulted not only in a high level of courtship by the Palmyran males but also ready acceptance on the part of the Pago Pago females. Futch (1966) has shown on the basis of a variety of evidences such as cytogenetics and morphology that the Palmyra, Majuro, and Rogerik populations must all be relatively closely related. These island populations are, of course, physically isolated from each other by essentially impassable barriers, but some continental species also show similar responses. Thus *D. americana americana,* a member of the *D. virilis* species group from continental North America, shows a similar pattern in that stocks collected from the eastern United States have sexually aggressive males and lethargic females, while the reverse is true for stocks from the western United States, except that unpredictably one western stock parallels its eastern congers in sexual behavior (Spieth, 1951). Significantly *americana americana,* like other members of the *virilis* group, are rigidly restricted ecologically to living along the edges of streams and lakes.

Various investigators have used the male-choice method rather than direct observation to check the sexual preferences of various geographical strains of species. The results were then analyzed by determining the isolation index of the various combinations. As an example, Dobzhansky and Streisinger (1944) utilized seven strains of *D. prosaltans.* All of the 47 combinations crossed and produced viable offspring, but 27 gave positive isolation indices and 15 yielded negative indices, thus indicating that the courtship balance varied from population to population within a single widespread species.

Geneticists have approached the problem of sexual isolation by comparing the mating ability of genotypically unique strains of flies with (1) that of "normal" flies or (2) another genetically "abnormal" strain. They have used sundry types of strains, e.g., single locus mutations, different chromosomal types, stocks that have been selected for morphological differences such as size, bristle number, or other comparable differences, or selected for behavioral differences, e.g., positive or negative geotaxis.

The pattern that emerges from these studies is that few, if any, genetic changes are neutral with respect to mating behavior. Mutations are mostly detrimental to the mating competence of the males, but their effects are less predictable in the females. Spiess and Langer (1964a) and Kaul and Parsons (1965) found that different chromosomal inversions of both *D. persimilis* and *D. pseudoobscura* affect the mating speed of the males. Del Solar (1966) studied strains of *D. pseudoobscura* that had been selected for positive or for negative geotaxis, and for positive or for negative phototaxis and found that all combinations of these strains display a significant preference for homogamic mating.

Few investigators have actually tried to determine by observation how such genetically unique strains differ in their actual mating behavior from normal flies. It is well known, however, that inbred lines of *Drosophila* and strains that have been kept in laboratory stocks for years show less sexual drive than do wild flies. Bastock (1956) determined that male yellow mutants of *D. melanogaster* are as persistent at courting as normal flies but that the mutants engage in less vibrating and licking. Spiess and Langer (1964b) ascertained that the Klamath inversion males of *D. persimilis* show low courting ability and the Klamath females are unresponsive sexually when compared with Whitney females.

But even in such instances as these and others where quantitative differences can be determined to exist in the courting behavior of either geographically separated populations such as *D. ananassae* flies or genetically unique flies, we can not exclude the possibility of the existence of chemical differences. That extremely subtle and as yet poorly understood factors are involved in courtship behavior is shown by the investigations of Petit (1958) on *D. melanogaster* and Ehrman (1966) on *D. pseudoobscura,* who found by multiple choice experiments that mating success depends upon the relative frequencies of the individuals of the strains used. The less frequent type, especially of the males, is invariably more successful in mating. This pertains to strains of mutants, to different chromosomal inversions, and to those of different geographical origin. Significantly Ehrman did not find this frequency effect when she utilized two sexually isolated strains of *D. paulistorum.* It would appear that female sexual preference, i.e., sexual selection, somehow is operating in such situations but just what the stimuli are that enables the rare individual in a population to be more successful is as yet unknown.

Interspecific Behavior

In wild populations under field conditions, males will orient upon and tap any other moving individual that approximates a drosophilid in size and configuration. Tapping transmits information to both indivduals, as shown by the fact that the males will turn away if a member of another species has been tapped and the female will respond with repelling actions to a foreign male. This type of discrimination must be based upon chemo-reception by both the males and the females. The receptors of the male's fore tarsi and the female's body surface must be stimulated by substances found on the surface of the cuticle of the other fly. Put another way, the "chemical configuration" of the cuticle of the female of the male's own species must have the unique competence to fit with the chemoreceptors of the male's fore tarsi, whereas that of other species of females does not. This type of sexual isolation is highly advantageous when compared to that displayed by other Diptera such as a scaptomyzid which indiscriminately mounts another fly before ascertaining whether it is the "right" individual or not.

Laboratory studies have given some insight into the effectiveness and nature of tapping as an isolation mechanism. When individuals of rela-tively closely related species which rarely or never hybridize in nature are isolated as to sex immediately after eclosion, aged until sexual maturity, and then utilized for no-choice experiments, the males will often "ride over" the inhibition of the tapping barrier and engage in courtship. Thus Spieth (1949) found that, with 30 crosses involving six species of the *Drosophila willistoni* group, 12 broke off at tapping but in the other crosses 18 of the males engaged in courtship in varying degrees with the foreign females. The majority of such courtships were less persistent, however, than normal and usually occurred only during the first part of the observation period. Sexually mature males when first introduced might court with almost normal intensity, but after one or more unsuccessful courtships would perceptibly decrease their courting aggressiveness and in a short time com-pletely cease. If, at that time, specimens of their own females were also introduced into the observation cell, the males would immediately begin to court them vigorously.

Clearly then there are other inhibitory stimuli that are involved beyond those received at tapping. If females of various ages are presented to mature males under no-choice conditions, then variations in male responses occur which show that the effectiveness of the tapping inhibition undergoes a definite ontogeny. Manning (1959b) showed that males of *D. melano-gaster* and *D. simulans* would court females of the other species if the females were less than two days old, but would break off with older ones at the tapping stage.

A comparable situation is found with the sibling species *D. pseudo-obscura* and *D. persimilis*. Under no-choice conditions, sexually mature males that have been isolated at the time of eclosion will readily court the females of the other species and the females will accept them. If, however, the flies are introduced together in a multiple choice situation immediately after eclosion, then by the time they reach sexual maturity the males turn away from the foreign females at the tapping stage (Spieth, 1958). In nature these flies are sympatric over most of the range of *persimilis* and yet, despite intensive collecting by Dobzhansky and his associates, a naturally occurring hybrid has been collected only once. Although the species differ somewhat ecologically, Carson (1951) found them both feeding and breeding in the same slime fluxes on the black oak. The basis for their isolation in the wild state must therefore be behavioral.

A study of the ontogeny of the sexual behavior shows that for a period of time after eclosion, the young males will not court and the young females are not attractive to the older males. When males of both species are 12 hr of age, they will occasionally court but are unable to achieve coition. At 24 to 28 hr of age, the males court older females and attempt mounting, but they can not achieve intromission until they are about 32 hr old. The females are not attractive to males until they are about 24 hr old and although males, especially older ones, will court them before they attain this age, the *D. pseudoobscura* females will not accept their own males until they are 32 to 36 hr old, and the *D. persimilis* females not until they are 44 to 48 hr old. Since, while undergoing sexual maturation, the individuals of both species assemble together on the feeding-mating sites, both the young males and the young females will normally engage in sexually "sampling" individuals of both species and thus become habituated to the stimuli of the foreign species. By the time they reach sexual maturity, males typically break off courtship with "foreign" females at the tapping stage, but even if the courtship should proceed past this point the female's sexual preferences will cause the courtship eventually to cease.

In an area where one of the species is relatively rare, it is probable that an occasional female might reach sexual maturity without having encountered males of both species, conceivably males of neither. She would then parallel the laboratory female that had been isolated from eclosion to sexual maturity before being introduced to a no-choice or male-choice type of experiment. Such a situation most probably accounts for the rare hybrid found in nature. It is to be noted that behavioral isolation between *D. pseudoobscura* and *D. persimilis* is also backed up by post mating mechanisms.

A further variation between the sexual development of *D. persimilis* and *D. pseudoobscura* is significant. A group of *pseudoobscura* females under laboratory situations when placed at the time of eclosion in a no-choice experiment with their own males will display 100 percent acceptance within 8 to 18 hr *after* they reach sexual maturity. In comparison, *D. persimilis* females will not achieve the same percentage until 36 to 48 hr *after* they have reached sexual maturity. Thus the *persimilis* females display a "wariness" or high threshold toward their own males. They must repeatedly sample the courtships of the males before their threshold is sufficiently lowered to result in acceptance.

The ontogeny of mating behavior has been carefully determined for few species, and the problem invites study by investigators. Incidental observations do indicate that various other species apparently have patterns comparable to those found in the four species discussed above.

The picture that emerges from interspecies studies is that the courtship of drosophilids consists of intertwined elements of both sexual isolation and sexual selection. Maynard Smith (1958) suggested that sexual selection was involved in drosophilid courtship and while present data are not adequate to analyze thoroughly and to separate these two elements, certain facts and conclusions can be deduced from the available data:

1. Sexual isolation seems dependent on chemosensory stimuli, perhaps abetted by concomitant but as yet unknown mechanical stimuli. Drosophilids have carried over, so to speak, the basic pattern of sexual isolation displayed by most cyclorrhaphous Diptera as exemplified by *Sarcophaga* males who mount other individuals indiscriminately and, if a wrong choice has been made, are eventually thwarted by improper stimuli received from the mounted individuals. An adaptive feature of drosophilid sexual behavior is that the male typically derives merely from tapping adequate information for making the binary choice of yes or no with respect to the subsequent time and energy consuming courtship display.

2. As promiscuous individuals, an adaptative advantage clearly exists for those individual males who are most effective in inducing females of their own species to accept their courting overtures and especially the young females who have just reached sexual maturity. Selection, therefore, has operated to produce a complex courtship display which directs multiple stimuli to the females.

3. The females, during their sexual ontogeny while on the feeding sites, repeatedly sample the tapping of many males both of their own species and of other related species. At the time they reach sexual maturity, they can be assumed to have become habituated to foreign males. From their own males, since their threshold is still high, they must receive relatively strong stimuli if they are to accept their overtures. Since, in nature, sexually mature males rapidly move from female to female, the females have adequate opportunity to sample courting overtures of a number of males. It is significant that as their maturation proceeds, the females' threshold decreases and under laboratory conditions older *inexperienced* virginal females will sometimes give the acceptance response as soon as the male taps and without waiting for his courting display. Female response to a male's overtures thus involves both sexual isolation and sexual selection. Since she can not prevent the male from attempting to court her, her

negative response is shown merely by refusal to mate and is identical in appearance regardless of whether sexual isolation or sexual selection is involved. The observer is unable to determine which is responsible for her thwarting action.

4. It can be anticipated that sexual isolation and sexual selection will evolve independently when a given population evolves into two or more separate, noninterbreeding populations. After sexual isolation has been achieved, changes in those elements of the behavior that pertain to sexual selection will later evolve. This is substantiated by the fact that many closely related but sexually isolated species of drosophilids do have quite similar courtship behaviors and, further, the pseudo lek behavior of most drosophilids does not apparently provide sufficient intensity of sexual selection to result in the evolution of great differences between closely related species. With the Hawaiian species that display true lek behavior which, as is known, does result in intensive selection, there do exist striking differences in the courting by the males of closely related species.

5. When the investigator studies intraspecific mating behavior and particularly when comparing the sexual competence of mutants to that of normal flies, he is dealing primarily with the effects of sexual selection and not sexual isolation. Thus typically the mutant males are less competent than are normal males which leads to one-sided preference on the part of the females. As Manning (1965) points out, it is easy to see how the disadvantaged males will be eliminated but difficult to conceive how true sexual isolation can be reached by such a mechanism.

Similar end results can also be anticipated from mixing of populations such as *D. ananassae* described above where differences in sexual balance between the aggressiveness of the males and receptivity of the females exist. If lethargic males are introduced into a population where the females display high resistance to the males' courtship, then the lethargic males would eventually be eliminated, whereas if aggressive males were introduced into an area where the females were highly receptive than a new sexual balance between the sexes would result. An example of how such modifications can occur is shown by the yellow mutant of *D. melanogaster*. Bastock (1956) found yellow males to be less competent than normal males, displaying less wing vibration and less licking during courtship, but significantly they were persistent suitors. A stock which had been homozygous for yellow for many generations possessed females that displayed high sexual receptivity. Bastock was able to show that the high receptivity of the female was not due to the direct effect of the yellow gene but rather that the reduced effectiveness of the yellow males had resulted in the sexual selection for females that were easily stimulated to sexual receptivity and thus the sexual balance had been restored within the laboratory population.

Origin of Sexual Isolation

The question can be raised, as it often has been (Muller, 1940; Dob-zhansky, 1951; Mayr, 1963) just how sexual isolation develops between two populations that are the descendants of a common ancestor. Two major points of view have been proposed. Some authors have suggested that it arises as a by-product of genetic divergence which occurs between geographically isolated populations; others believe that it results from the selection against hybrids whenever hybridization becomes disadvantageous. As del Solar (1966) points out, the two concepts are not mutually exclusive and confusion results if they are treated as either-or concepts.

In drosophilids closely related sibling species such as *melanogaster-simulans, pseudoobscura-persimilis,* and the races of *auraria* all display essentially similar courtship patterns. Careful study has shown, however, that quantitative differences do exist (Manning, 1959a, 1959b; Brown, 1966). Further, it has also been found that ecological and other types of behavioral differences exist between such pairs. The sum total of all these differences is such that it is difficult to conceive that they have arisen under sympatric conditions merely as a result of the disadvantages of hybridization. In sum, the evidence from drosophilids favors the geographical isolation by-product hypothesis.

If such closely related, but geographically isolated, populations subsequently become sympatric, then hybrids if they are produced and if they are maladapted will be selected against to the extent that the populations become reproductively isolated. Pre-mating mechanisms and especially behavioral mechanisms seem the most biologically economical methods, but the selection pressure that creates such isolation mechanisms ceases once isolation has been achieved for the conditions under which the populations normally live.

If samples of such populations are subjected to the "abnormal" conditions of the laboratory, the isolating mechanisms typically often break down. They can, however, then be intensified and built up by deliberate artificial selection. Koopman (1950) was able to do this by removing the hybrids for a number of generations and thus reduced the percentage of hybrids from 30 to 50 percent down to five percent. Just what aspect of the courtship was changed in this case is not known. Manning (1965) states that Crossley was able, by using stocks of *D. melanogaster* carrying the genes *ebony* and *vestigial* respectively, to reduce the number of hybrids under multiple choice conditions after 50 generations and that "there were significant changes to the sexual responses of the two stocks."

In nature, similar changes appear to occur when two populations are partially sympatric. Dobzhansky and Koller (1938) found that *Drosophila miranda* stocks derived from areas that were sympatric with *D. pseudo-obscura* and *D. persimilis* hybridized less readily than they did with the *persimilis* and *pseudoobscura* stocks that were allopatric to *miranda* in origin.

Similarly, Mettler and Nagle (1966) found a parallel situation with *D. arizonensis* and *D. mojavensis*. Race A of *mojavensis* is allopatric to *arizonensis* but Race B is sympatric. Hybrids are not found in nature, but under laboratory conditions Race B is much better isolated than Race A from *arizonensis*.

Conclusion

The behavior of a group of organisms, even if limited to one element such as mating behavior, has implications touching all aspects of the total biology of the group. I have tried to keep the thread of discussion related to the evolutionary implications that can be deduced from the known information for the large and diverse genus *Drosophila*. Incomplete as our knowledge is, certain major trends can be discerned. The kinds of questions that should now be asked are emerging and we can anticipate the types of investigations that are needed.

Many more detailed comparative studies are called for, not only of the sexual behavior of related species both in the field and in the laboratory, but also of the ontogeny of behavior. Then will we be able more effectively to determine how genetic changes affect behavior and particularly the physiology of behavior.

Behavioral studies have and will continue to contribute richly to the fuller understanding of evolution which is, indeed, the golden thread of biology. The genus *Drosophila* will continue to be a prime subject for such studies.

References

BASTOCK, M. 1956. A gene mutation that changes a behaviour pattern. Evolution, 10:421–439.

———, and A. MANNING. 1955. The courtship of *Drosophila melanogaster*. Behaviour, 8(2–3):85–111.

BRNCIC, D. 1957. The southernmost *Drosophila*. Amer. Nat., XCI(857):127–128.

———. 1966. Ecological and cytogenetic studies of *Drosophila flavopilosa*, a neotropical species living in *Cestrum* flowers. Evolution, 20(1):16–29.

BROWN, R. G. B. 1966. Courtship behaviour in the *Drosophila obscura* group. Part II. Comparative studies. Behaviour, 25:281–321.

CARSON, H. L. 1951. Breeding sites of *Drosophila pseudoobscura* and *Drosophila persimilis* in the transition zone of the Sierra Nevada. Evolution, 5(2):91–96.

CAZIER, M. A. 1963. The description and bionomics of a new species of *Apiocera*, with notes on other species (Diptera:Apioceridae). Wasmann J. Biol., 21:205–234.

CULLEN, E. 1957. Adaptations in the kittiwake to cliff-nesting. Ibis, 99:275–303.

DEL SOLAR, E. 1966. Sexual isolation caused by selection for positive and negative phototaxis and geotaxis in *Drosophila pseudoobscura*. Genetics, 56:484–487.

DETHIER, V. G. 1963. The Physiology of Insect Senses. London, Methuen.

DOBZHANSKY, TH. 1951. Genetics and the Origin of Species, 3d ed., New York, Columbia Univ. Press.

———, and D. BRNCIC. 1957. The southernmost Drosophilidae. Amer. Nat., 91:127–128.

———, and A. DA CUNHA. 1955. Differentiation of nutritional preferences in Brazilian species of *Drosophila*. Ecology, 36(1):34–39.

———, and P. C. KOLLER. 1938. An experimental study of sexual isolation in *Drosophila*. Biol. Zentralb., 52:589–607.

———, and G. STREISINGER. 1946. Experiments on sexual isolation in *Drosophila*. II. Geographic strains of *Drosophila prosaltans*. Proc. Nat. Acad. Sci. U.S.A., 30:340–345.

EHRMAN, L. 1966. Mating success and genotype frequency in *Drosophila*. Anim. Behav., 14:332–339.

EWING, A. W. 1964. The influence of wing area on the courtship behaviour. Anim. Behav., 12(2–3):316–320.

———, and A. MANNING. 1963. The effect of exogenous scent on the mating of *Drosophila melanogaster*. Anim. Behav., 11(4):596–598.

FLÜGGE, C. 1934. Geruchliche Raumorientierung von *Drosophila melanogaster*. Z. vergl. Physiol., 20:463–500.

FUTCH, D. 1966. A study of speciation in South Pacific populations of *Drosophila ananassae*. Univ. of Texas Publ., 6615:79–120.

GROSSFIELD, J. 1966. The influence of light on the mating behavior of *Drosophila*. Univ. of Texas Publ., 6615:147–176.

GRUHL, K. 1924. Paarungs Gewohnheit der Dipteren. Z. wissen. Zool., 122:205–280.

HARDY, D. E. 1965. Insects of Hawaii. Vol. 12, Diptera: Cyclorrhapha, II. Family Drosophilidae, Honolulu, Univ. of Hawaii Press.

HODGSON, E. S. 1965. The chemical senses and changing viewpoints in sensory physiology. *In* Carthy, J. D., and C. L. Duddington ed., Viewpoints in Biology, 4:83–124. London, Butterworths.

KAUL, D., and P. A. PARSONS. 1965. The genotypic control of mating speed and duration of copulation in *Drosophila pseudoobscura*. Heredity (London), 20(3):381–392.

KOOPMAN, K. F. 1950. Natural selection for reproductive isolation between *Drosophila pseudoobscura* and *Drosophila persimilis*. Evolution, 4:135–148.

MANNING, A. 1959a. The sexual isolation between *Drosophila melanogaster* and *Drosophila simulans*. Anim. Behav., 7(1–2):60–65.

———. 1959b. The sexual behavior of two sibling *Drosophila* species. Behaviour, 15(1–2):123–145.

———. 1965. *Drosophila* and the evolution of behaviour. *In* Carthy, J. D., and C. L. Duddington ed., Viewpoints in Biology, 4:125–169. London, Butterworths.

MAYNARD SMITH, J. 1958. Sexual selection. *In* Barnett, S. A. ed., A Century of Darwin, 231–244, London, Heinemann.

MAYR, E. 1963. Animal Species and Evolution, Cambridge, Mass., Harvard Univ. Press.

METTLER, L. E., and J. J. NAGLE. 1966. Corroboratory evidence for the concept of the sympatric origin of isolating mechanisms. *Drosophila* Information Service, 41:76.

MILANI, R. 1950a. Release of courtship display in *subobscura* males stimulated with dummies. *Drosphila* Information Service, 24:88.

————. 1950b. Sexual behavior of *D. subobscura, ambigua, bifasciata, tristis, obscuroides. Drosophila* Information Service, 24:88.

MULLER, H. J. 1940. Bearings of the *Drosophila* work on systematics. *In* Huxley, J. S. ed., The New Systematics, 185–268. Oxford, Oxford Univ. Press.

PATTERSON, J. T., and W. S. STONE. 1952. Evolution in the Genus *Drosophila.* New York, Macmillan.

PETIT, C. 1958. The genetic determination and psycho-physiology of the sexual competition among *Drosophila melanogaster.* Bull. Biol. France Belg., 92(3):249–329.

————. 1959. De la nature des stimulations responsables de la selection sexuelle chez *Drosophila melanogaster.* C. R. Acad. Sci. Paris, 248:3484–3485.

PHILIP, U., J. M. RENDEL, H. SPURWAY, and J. B. S. HALDANE. 1944. Genetics and karyology of *Drosophila subobscura.* Nature (London), 154:260–262.

SIBLEY, C. G. 1957. The evolutionary and taxonomic significance of sexual dimorphism and hybridization in birds. Condor, 59:166–191.

SPIESS, E. B., and B. LANGER. 1964a. Mating speed control by gene arrangements in *Drosophila pseudoobscura* homokaryo types. Proc. Nat. Acad. Sci., U.S.A., 51:1015–1019.

————. 1964b. Mating speed control by gene arrangement carriers in *Drosophila persimilis.* Evolution, 18:430–444.

SPIETH, H. T. 1949. Sexual behavior and isolation in *Drosophila.* II. The interspecific mating behavior of species of the *willistoni* group. Evolution, 3(2):67–81.

————. 1951. Mating behavior and sexual isolation in the *Drosophila virilis* species group. Behaviour, 3(2):105–145.

————. 1952. Mating behavior within the genus *Drosophila* (Diptera). Bull. Amer. Mus. Nat. Hist., 99(7):395–474.

————. 1958. Behavior and isolating mechanisms. *In* Roe, A., and G. G. Simpson, ed., Behavior and Evolution, 363–389, New Haven, Yale Univ. Press.

————. 1966a. Drosophilid mating behavior: the behavior of decapitated females. Anim. Behav., 14:226–235.

————. 1966b. Hawaiian honeycreeper, *Vestaria coccinea* (Forster), feeding on lobeliad flowers, *Clermontia arborescens* (Mann) Hillebr. Amer. Nat., 100(914): 470–473.

————. 1966c. Courtship behavior of endemic Hawaiian *Drosophila.* Univ. of Texas Publ., 6615:133–146.

————. 1966d. Drosophilid mating behavior: The behavior of *ananassae* and some *ananassae*-like flies of the Pacific. Univ. of Texas Publ., 6615:245–314.

STONE, W. S., W. C. GUEST, and F. D. WILSON. 1960. The evolutionary implications of the cytological polymorphism and phylogeny of the *virilis* group of *Drosophila.* Proc. Nat. Acad. Sci. U.S.A., 46:350–361.

STREISINGER, G. 1948. Experiments on sexual isolation in *Drosophila.* IX. Behavior of males with etherized females. Evolution, 2:187–88.

STURTEVANT, A. H. 1915. Experiments on sex recognition in *Drosophila.* Anim. Behav., 5(5):351–366.

————. 1942. The classification of the genus *Drosophila* with descriptions of nine new species. Univ. of Texas Publ., 4213:5–51.

THOMAS, H. T. 1950. Field notes on the mating habits of *Sarcophaga* Meigen (Diptera). Proc. Royal Entom. Soc. London, Series A25(pls.7–9):93–98.

THROCKMORTON, L. H. 1962. The problem of phylogeny in the genus *Drosphila.* Univ. of Texas Pub., 6205:207–343.

TINBERGEN, N. 1960. Behaviour, systematics, and natural selection. *In* Tax, S. ed,. Evolution After Darwin, Vol. I, The Evolution of Life, 595–613. Chicago, Univ. of Chicago Press.

WHEELER, M. 1952. The Drosophilidae of the Nearctic region, exclusive of the genus *Drosophila.* Univ. of Texas Publ., 5204:162–218.

5

Evolution of Photoreceptors

RICHARD M. EAKIN

Department of Zoology
University of California, Berkeley

Introduction 194
Ciliary Photoreceptors 196
Rhabdomeric Photoreceptors 201
Ciliary Line of Evolution 205
    Protista 205
    Coelenterata 207
    Ctenophora 209
    Echinodermata 209
    Chaetognatha 210
    Cephalochordata 214
    Vertebrata 216
Rhabdomeric Line of Evolution 216
    Platyhelminthes 217
    Rotifera 217
    Mollusca 221
    Annelida 222
    Onychophora 228
    Arthropoda 232
Other Taxa 233
Protostomes Versus Deuterostomes 234
References 235

Introduction

This essay will endeavor to demonstrate the relevance to photoreceptors of the Darwinian dictum: descent with modification. Considering the critical importance of radiant energy for almost all living organisms, it may be assumed that organelles adapted for activation by photons arose very early, perhaps not long after the evolution of the cell itself. These first photoreceptors, which made possible a new way of life, autotrophism, were probably elaborations of the cell membrane, similar perhaps to the vesicles or stacks of lamellae, rich in bacteriochlorophyll, found in the purple bacterium *Rhodospirillum* (Cohen-Bazire and Kunisawa, 1963; Gibbs, Sistrom, and Worden, 1965). From that point in time, thousands of millennia ago, to the present day most animal and plant photoreceptors have been membranous structures. This feature is the common evolutionary theme (the descent aspect of the dictum) upon which nature has played many variations (the modifications).

But why membranes? It is probable that the ordered disposition of molecules of photopigment in extensive monolayers which could be oriented toward the source of photons was of selective value. It is assumed that the most effective position of arrays of membranes bearing a photopigment, such as chlorophyll or rhodopsin, is normal to the direction of incident light (Wolken, 1963; Wald, Brown, and Gibbons, 1963). At least light-sensitive organelles, such as rod and cone disks in the vertebrate retina, are usually so arranged. Exceptions to the rule are known. For instance, the microvilli in the eyecup of a sea star are directed toward the opening of the ocellus (see Eakin, 1963). In actual function, however, the light probably strikes these receptors obliquely as the organism lifts the tip of an arm to expose the optic cushion upon which the ocelli are situated. Moreover, certain plants exhibit a tendency to orient their chloroplasts so that the broad surfaces of the plastids face the light of low or moderate intensity. If, however, the illumination is intense, the chloroplasts are rotated so that their narrow edges are presented to the light (see Granick, 1961).

There are other features which most, if not all, animal photoreceptors have in common. First, light sensitive organelles are closely associated with mitochondria. Presumably, the energy transfers which mitochondria implement are needed for the maintenance or function (probably both) of the membranous system. One of the primary features of maintenance is the resynthesis of the photopigment after exposure to light. An aspect of function is the transduction of the excitation, with amplification, into a nervous impulse or the inhibition of spontaneous activity. A second general characteristic of an animal photoreceptor is a neurite leading from the end of the cell opposite the photoreceptoral organelle. Again, the cell membrane is all important, being the carrier of the action potential to some other part of the organism, usually a neural center, or to a second order of neurones in the pathway. Finally, photoreceptors often exhibit prominent rough endoplasmic reticulum, Golgi centers, and deposits of glycogen suggesting a high synthetic activity of, as examples, photopigments, neurotransmitter substances, humors, lens materials, and structural proteins for membranes.

I turn now from the similarities among photoreceptors to their remarkable diversity and to the possibility that basic morphological patterns have evolutionary significance. I postulated earlier (Eakin, 1963, 1966a) that there are two basic kinds of receptors: (1) the ciliary type, in which the receptoral organelle is derived from the plasma membrane of a modified cilium or flagellum; (2) the rhabdomeric form, in which the organelle is an array of microvilli formed by multiple outfolding of the cell membrane independently of cilia. Before discussing the evolution of these two types let us first describe an example of each kind, and since the distinction between the two is embryological, it will be useful to sketch briefly the development of each type.

Ciliary Photoreceptors

As an example of the ciliary type I show an electron micrograph (Fig. 1) of a vertebrate photoreceptor, but not that of a rod or cone of the lateral eye which is usually selected for illustration. Instead, I present a receptor from the third or parietal eye of a reptile, the Western Fence Lizard, *Sceloporus occidentalis,* for the following reasons. It is less well known, a good micrograph of it has not been published heretofore, and its close similarity to a cone of the lateral eye will be demonstrated. Finally, a personal matter: it has been a great pleasure to study vertebrate third eyes (reptilian, amphibian, and cyclostome)—enigmatic organs that represent a legacy of times past when median eyes were probably more significant functionally than they are today. A few references to recent literature on the structure and function of third eyes are: Stebbins and Eakin (1958); Eakin and Stebbins (1959); Eakin (1960, 1961, 1964a, 1964b); Eakin and Westfall (1960); Steyn (1960); Heerd and Dodt (1961); Dodt and Heerd (1962); Miller and Wolbarsht (1962); Eakin, Quay, and Westfall (1963); Oksche and von Harnack (1963, 1964); Kelly (1965, 1967); Oksche (1965); Oksche and Vaupel-von Harnack (1965b); and Chugunov (1966).

Like the typical cone, a receptor of the parietal eye consists of an outer segment composed of a stack of disks or flattened sacs; a narrow connecting piece through which passes a set of microtubules, formerly called fibrils; and an inner segment containing the centriolar apparatus and a concentration of mitochondria. One can become confused on the identity of the disks of a cone or cone-like receptor because their flattened sacs, in most instances, communicate with extracellular space (see arrows, Fig. 1). The disks are the infoldings of the ciliary membrane, to be described shortly, and not the plates of cytoplasm between the invaginations. The receptoral processes of the vertebrate median eye project forward from the front

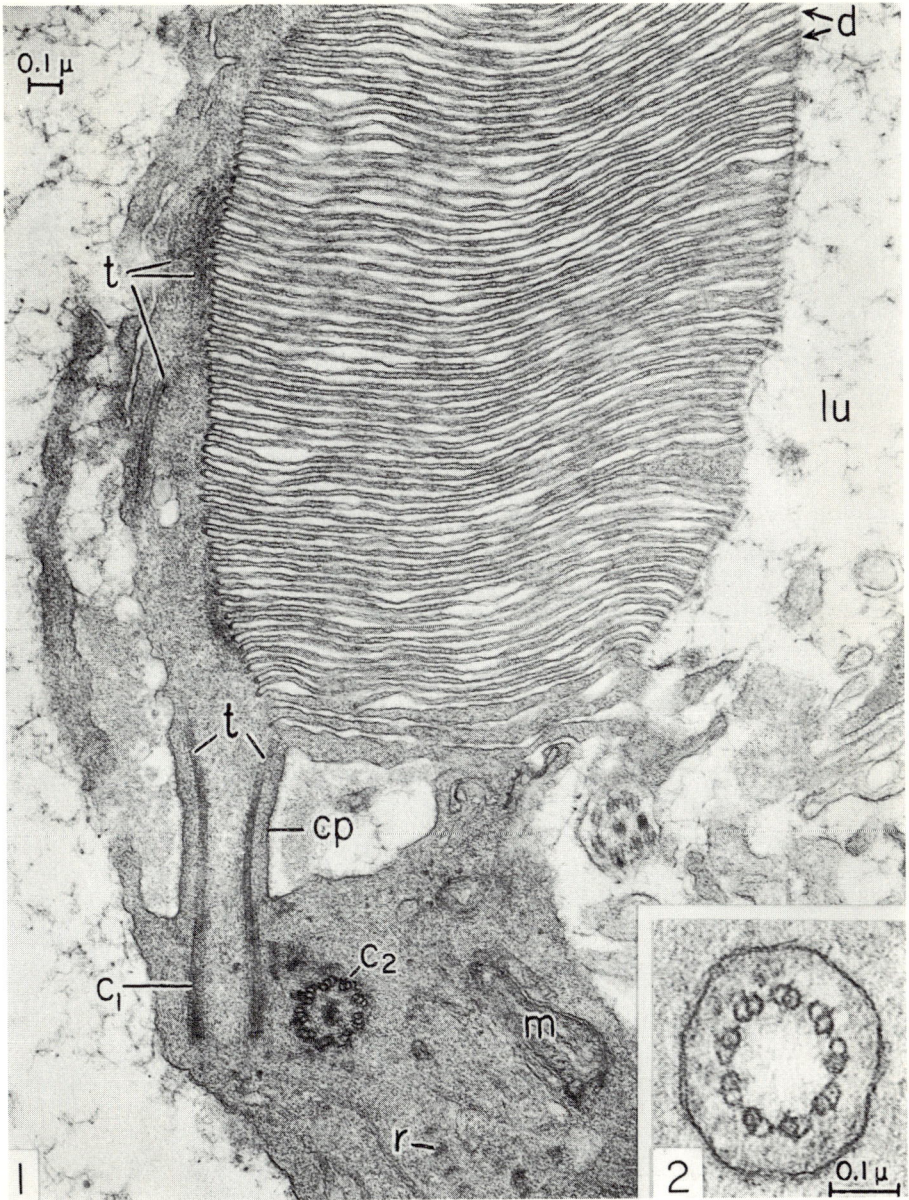

Fig. 1. Photoreceptoral process in parietal eye of Western Fence Lizard, Sceloporus occidentalis. Outer segment (above) consists of stack of disks (d), each open to lumen (lu) of eye. Inner segment (below) contains axial centriole or kinetosome (c_1), oblique centriole (c_2), mitochondria (m), striated rootlet (r). Connecting piece (cp) joins outer and inner segments. Microtubules (t) extend from kinetosome through connecting piece and alongside outer segment. X 45,000.

Fig. 2. Cross section of connecting piece of parietal eye photoreceptoral process showing ring of nine doublets of microtubules. Note absence of central tubules and arms on doublets normally found in motile cilia. X 95,000.

surface of the retina toward the lens rather than from the back of the retina as do rods and cones of a lateral eye. This difference in retinal polarity is, of course, the consequence of a fundamental dissimilarity in development of parietal and lateral eyes. The embryonic optic vesicle of the lateral eye invaginates to form an optic cup, the inner layer of which becomes the retina and the outer layer a pigmented epithelium. The median optic vesicle, on the other hand, remains spherical and its uninverted proximal wall differentiates into a retina (the homologue of the pigmented epithelium of the lateral eye) and its distal wall becomes the lens (the homologue of the retina of the lateral eye).

The microtubules, centrioles, and associated structures (e.g., rootlets) are the clues to the ciliary nature of the receptor. A cross section of the connecting piece (Fig. 2) shows a ring of nine doublets of microtubules as in a cilium or flagellum. Incidentally, these last two terms are used interchangeably in this chapter. The system of microtubules, however, is not precisely like that of a motile cilium (kinocilium). At least two significant differences should be noted. First, a pair of central microtubules is lacking in all but a few ciliary photoreceptors (see below). The pattern is thus 9 + 0 instead of 9 + 2, characteristic of motile cilia (see discussion in Barnes, 1961). Secondly, unlike kinocilia, a ciliary photoreceptor has no small arms projecting from each doublet. Gibbons (1965) has shown that these arms in motile cilia are composed of a protein, called dyenin, that is high in ATPase activity. The usual absence of the central tubules and arms on the doublets in ciliary photoreceptors suggests that in the course of evolution these modified cilia have lost their capability to move.

The centriolar apparatus of the ciliary photoreceptor also resembles that of a cilium. The distal or axial centriole is the kinetosome of the ciliary process, and from it extend the microtubules just described. The proximal or oblique centriole usually lies to one side and at right angles to the kinetosome. The typical centriolar pattern of nine triplets of microtubules may be seen in the cross-sectional view of the proximal centriole (Fig. 1). Often a striated rootlet may be observed extending from the lower end of the kinetosome deeper into the inner segment of the parietal eye receptoral cell. In the sensory cell figured, the rootlet is incompletely shown.

To learn how a cilium becomes modified into a vertebrate photoreceptor we must examine ultrastructurally sequential stages in the embryonic differentiation of a neuroepithelial (ependymal) cell. Using a third eye receptor again as an example, but this time that in an amphibian, I show three steps (Figs. 3 to 5) in the differentiation of a photoreceptor in the frontal organ (Stirnorgan) of the Pacific Treefrog, *Hyla regilla* (Eakin and Westfall, 1961). This organ is homologous with the reptilian parietal eye because each arises from a vesicular outpocketing at the anterior end of the

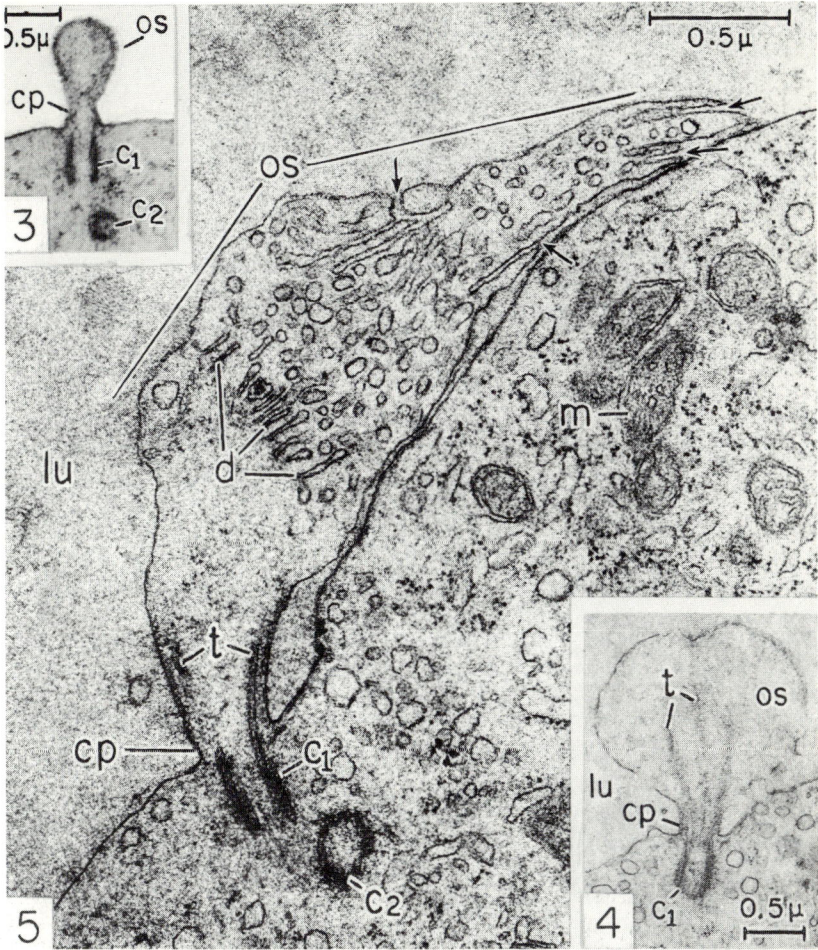

Fig. 3. Photoreceptoral process of frontal organ in first stage of formation in very early larva of Pacific Treefrog, Hyla regilla. c_1, axial centriole (kinetosome); c_2, oblique centriole; cp, future connecting piece; os, outer segment. X 16,000. (Figs. 3–5 from Eakin, 1963, after Eakin and Westfall, 1961)

Fig. 4. Photoreceptoral process of frontal organ, showing development of microtubules (t). c_1, axial centriole; cp, future connecting piece; lu, lumen of organ; os, outer segment. X 17,000.

Fig. 5. Photoreceptoral process of frontal organ at stage of formation of disks (d). c_1, axial centriole; c_2, oblique centriole; cp, connecting piece; lu, lumen of organ; m, mitochondrion; os, outer segment; t, microtubules. Arrows indicate points of infolding of plasma membrane to form disks. X 37,000.

pineal diverticulum which is in turn a dorsal evagination of the diencephalic region of the neural tube. The first step (Fig. 3) is the formation of a balloon-like evagination of the cell membrane and cortical cytoplasm immediately above the axial centriole (kinetosome). Figure 3 bears a striking resemblance to the first stage in the development of a cilium from the neural epithelium of a chick embryo (Sotelo and Trujillo-Cenóz, 1958). There seems to be little doubt that a kinetosome is an organizer or inducer of a ciliary process. The evidence for this statement is largely circumstantial: cilia, flagella, and ciliary receptors (mechano, chemo, and photo) are invariably associated with kinetosomes. Recently, Stubblefield and Brinkley (1966) have described and figured ciliary buds forming from vacuolar (Golgi) membranes overlying centrioles in hamster fibroblasts treated with colcemid. I have tried to test the relationship between centriole and cilium experimentally without success by using centrifugal force to keep centrioles away from the outer surfaces of presumptive ciliated epithelium. Figure 4 is a slightly later stage in the differentiation of the ciliary photoreceptor in the frontal organ of *Hyla*. Note that the evagination is larger, less cilium-like in shape, and that it now contains the complex of microtubules which have extended into the process (incipient outer segment) from the distal end of the kinetosome.

The origin of the double-membrane disks by multiple infolding of the ciliary membrane is shown in Figure 5. Several points of invagination are marked with arrows. The diverticula extend deep into the process in parallel array. At one time I considered the vesicles along the lines of invagination the consequence of undulations in the diverticula since serial sections of this specimen enabled us in many instances to demonstrate continuity of the vesicles along a given line. I now recognize an alternative or perhaps additional explanation, namely, fixation artifact, because similar lines of vesicles are frequently obtained by osmium treatment of cone disks (but not rod disks). Preservation with glutaraldehyde, however, does not break up cone disks into vesicles (Eakin, 1965). The diverticula (flattened sacs or double-membrane disks) later assume their definitive position at right angles to the longitudinal axis of the outer segment, and like cones of the lateral eye they retain their patency to the lumen of the eye vesicle. Further multiplication of disks occurs by repeated infolding of the ciliary membrane at the base of the outer segment. Meanwhile the microtubules become crowded to one side (to the reader's left in Fig. 5), the outgrowth is constricted basally to form the connecting piece (*cp*), and the distal end of the cell proper is elevated to form the inner segment of the receptoral process. The fully differentiated receptoral process of the amphibian frontal organ (see Fig. 1 of Kelly and Smith, 1964, p. 655) is essentially the same as that of the reptilian parietal eye shown here (Fig. 1).

Rhabdomeric Photoreceptors

The term rhabdomere was first used by Grenacher (1879) for a co-lumnar organelle on the medial border of each retinular cell in an arthropod ommatidium. In some instances (e.g., honeybee) these prismatic columns are fused together (closed type) to form a refractile rod which Grenacher called a rhabdom. In other arthropods (e.g., *Drosophila*) the rhabdomeres are unfused (open type), each projecting into a fluid-filled space in the center of the ommatidium. Later Grenacher (1886) observed comparable structures in cephalopods. Electron microscopy has now shown that the rhabdomeres of both arthropods and cephalopods are composed of arrays of microvilli from the surfaces of the retinal cells (see Moody's review, 1964). Finding that many invertebrate photoreceptors possess a phalanx of microvilli I extended the term rhabdomere to include any assembly of microvilli projecting from a surface of a cell presumed to be light-sensitive, irrespective of variation in size, shape, and arrangement of the villi (Eakin, 1962, 1963). Excluded from the term, of course, are the tubules or microvilli which project from a cilium (see below).

To illustrate the rhabdomeric type of photoreceptor, I have selected the sensory cell from the eye of a pulmonate snail, *Helix aspersa*. Figure 6 shows the distal end of a receptor cell bearing a crown of countless, long, slender microvilli, about 10μ in length and 0.1μ in diameter. The tips of the villi, especially the central ones, approach the under surface of the lens. Beneath the villi are aggregations of mitochondria and vesicles. Whereas the retinular cells of many arthropod and cephalopod eyes bear pigment granules, in gastropod eyes the sensory cells are usually unpigmented, and the pigment is carried by the supportive cells. Cilia or centrioles have not been observed as yet in the adult sensory cells of *Helix* (Eakin, 1963; Röhlich and Török, 1963; Schwalbach, Lickfeld, and Hahn, 1963; Eakin and Brandenburger, 1967). Cilia are present, however, in the embryonic eye of this snail. To determine the relationship, if any, between these cilia and the photoreceptoral microvilli, and to illustrate the development of a rhabdomere in general let us examine the development of the eye of *Helix aspersa*.

The optic vesicle of *Helix* arises, as in most invertebrates, by an invagina-tion of ectoderm that later closes to form a vesicle which becomes sepa-rated from the overlying epidermis (cornea). In some invertebrates, how-ever, the eye remains an open cup-shaped ocellus. Recent study of the fine structure of the newly formed optic vesicle of *Helix aspersa* shows one, sometimes two, short ciliary buds projecting from each ectodermal cell into the cavity of the organ (Eakin and Brandenburger, 1967). Simultaneously, the luminal surfaces of the cells send forth short, irregular projections, the

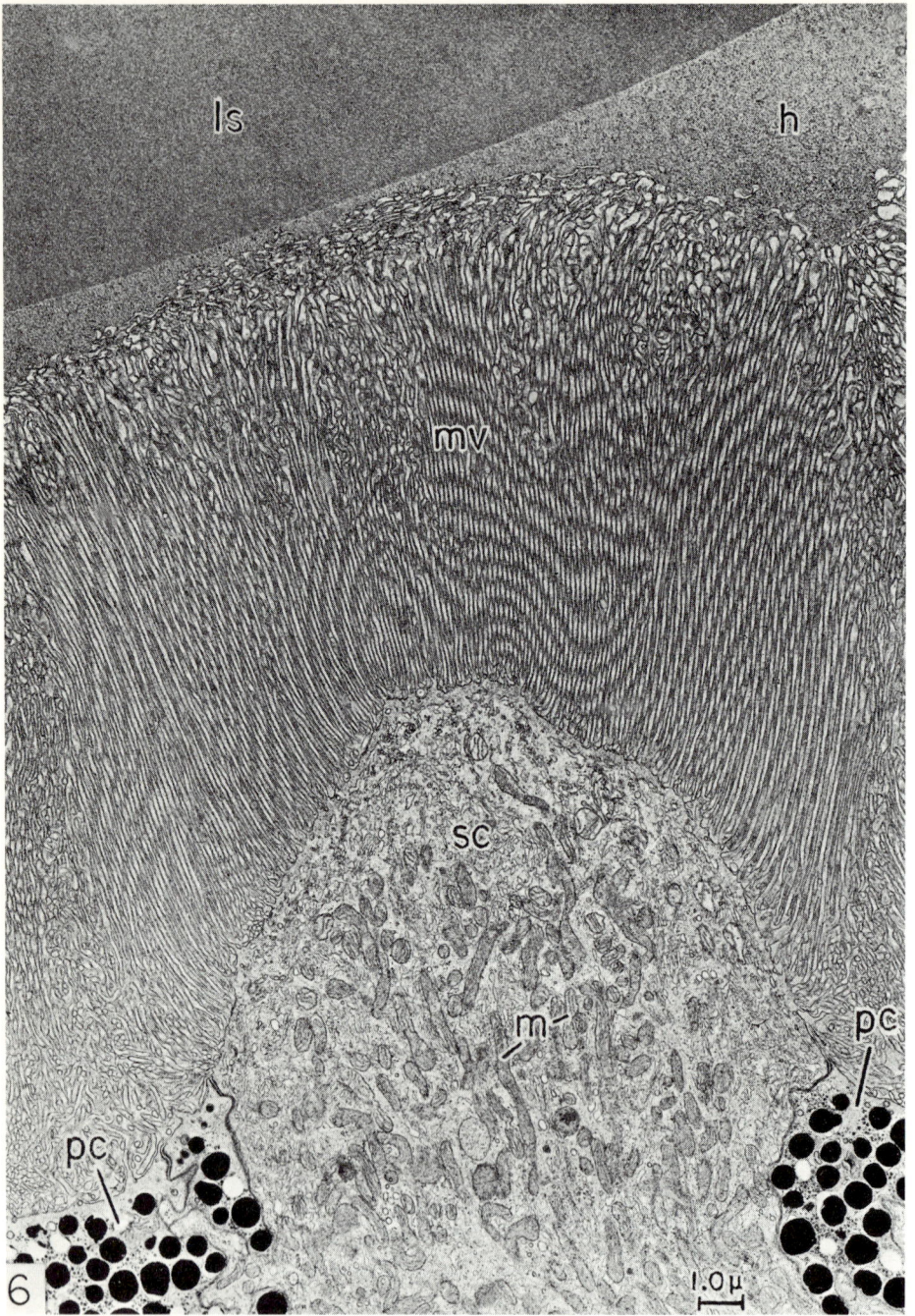

Fig. 6. (Legend on page 203)

forerunners of the rhabdomeric microvilli, which appear to bear no relationship to the ciliary buds (Fig. 7). The villi continue to increase in length and number, especially in those cells which become sensory, but the cilia remain rudimentary and are probably lost altogether in later stages of development. It is our conclusion that the cilia in the embryonic eyes of *Helix aspersa* are purely incidental. They are present because the eye is formed from ectoderm that was ciliated initially. Accordingly, the photoreceptors of this form, like those of arthropods and cephalopods, are regarded as rhabdomeric in type.

Because the above point of developmental relationship between cilia or centrioles and microvilli is critical to this chapter, it should be explored further. It might be argued that the distinction between a ciliary and a rhabdomeric photoreceptor is a superficial one because fundamentally all microvillar outgrowths of a cell may be initiated by centrioles. Although I cannot see any connection between the rudimentary cilia and the microvilli in the embryos of *Helix* (or, as we shall see below in the young of *Nereis* and of *Peripatus*), it is possible that some organizing influence spreads from a centriole to induce multiple evaginations of the cell surface. If this were true one would expect to find centrioles associated with microvilli without exception. As yet none has been noted in connection with the development of an arthropod rhabdomere. True, only a few embryological studies have been made to date. Neither Waddington and Perry (1960) noted centrioles in their paper on the formation of rhabdomeres in *Drosophila melanogaster* nor Eguchi, Naka, and Kuwabara (1962) in their investigation of the development of the rhabdom in the moth *Bombyx mori*. It is possible, of course, that centrioles were overlooked or seen but not mentioned. On the other hand, a goodly number of adult arthropod eyes have now been studied with the electron microscope. No evidence of centrioles and cilia—not even remnants of them—has been reported as yet. As we shall note shortly, centrioles and striated rootlets and even rudimentary cilia persist from embryonic stages into the adult in annelids and onychophorans. Finally, so far as I know, cilia or centrioles have not been shown to be associated with other villous structures such as brush borders, albeit centrioles are invariably found at or near the surfaces of most if not all differentiated, nondividing epithelial cells.

Fig. 6. Distal ends of one sensory cell (sc) and two pigmented cells (pc) in eye of pulmonate snail Helix aspersa. *Array of microvilli (mv) extends from dome-shaped tip of receptor cell toward under surface of lens (ls). Most villi cut obliquely, some (right and left) sectioned longitudinally. h, humor bathing lens and microvilli; m, mitochondria in large number as typically found near photoreceptive organelles. X 6,300.*

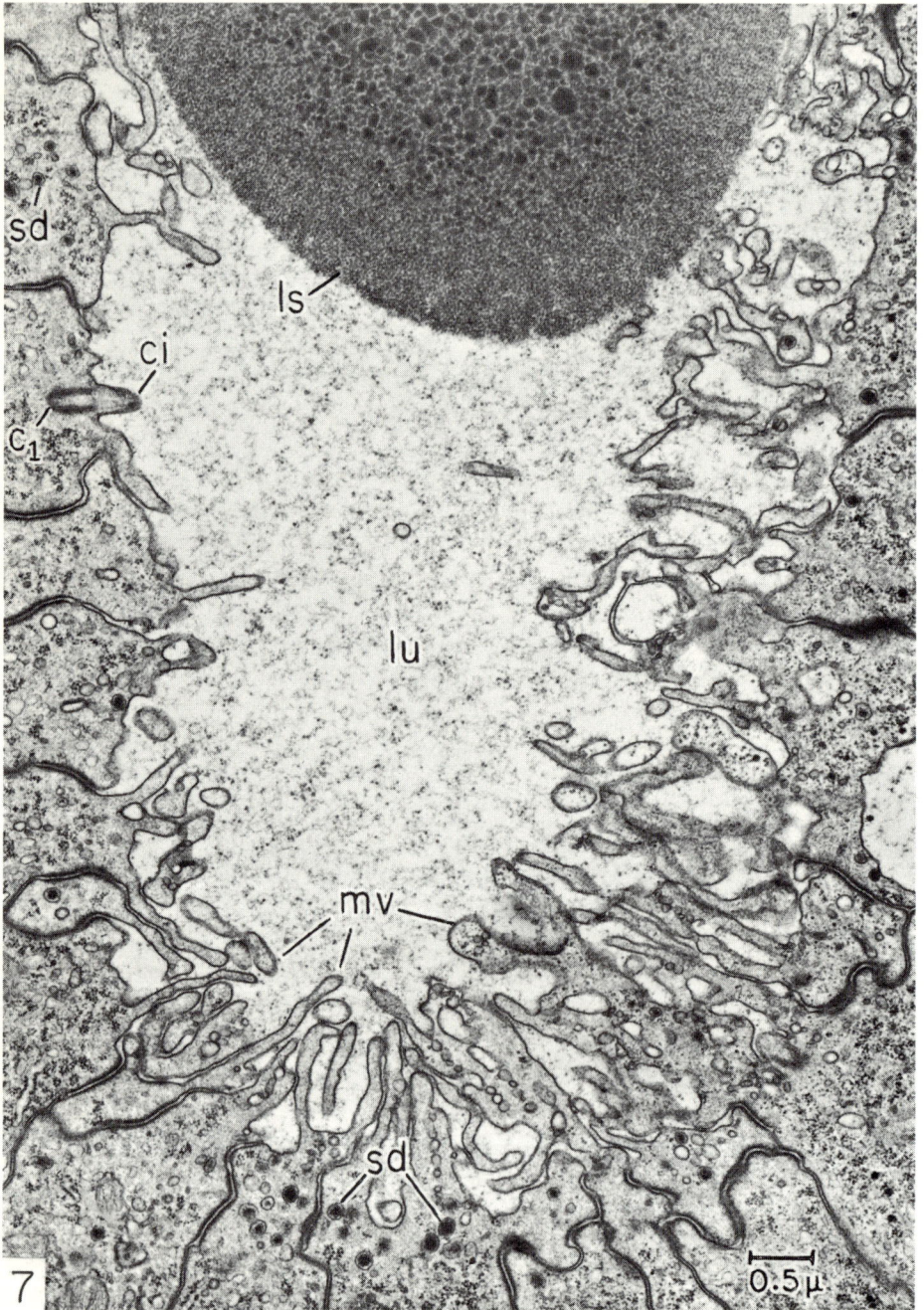

Fig. 7. (Legend on page 205)

Ciliary Line of Evolution

Having convinced the reader, I hope, that there are two fundamentally different types of photoreceptors, I shall now attempt to persuade him to accept tentatively an even more tenuous hypothesis, namely that ciliary and rhabdomeric photoreceptors have evolutionary significance. In so doing I am speculating like an election analyst trying to predict the outcome of a poll from early returns of a very few precincts.

Protista. I postulate that the elaborate and remarkably complex vertebrate photoreceptor described above is the climax of a ciliary line of evolution (see Fig. 8) which had its origin in the Protista. Perhaps, some ancient alga "discovered," by the evolution of appropriate DNA-dependent signals, the advantage of a close association between one of its chloroplasts and its locomotor organelle, a flagellum. Receptor and effector might then have worked together enabling the organism to orient and to move in relation to the source of light. The picture in *Chromulina psammobia* (Fauré-Fremiet and Rouiller, 1957; Rouiller and Fauré-Fremiet, 1958) would be a suitable model for this evolutionary stage, except for the fact that the flagellum associated with stigma and chloroplast in this form is the non-locomotory internal one.

Alternatively, or as a second step in evolution, perhaps a protist "learned" to incorporate a light-sensitive substance, a photopigment, within the locomotory flagellum itself. The organelle was now both receptor and effector. The green alga *Euglena* may offer a model for this hypothetical first ciliary photoreceptor. The chief flagellum of several euglenoids as well as that in other protists (e.g., spermatozoid of *Fucus,* Manton and Clarke, 1956) bears a swelling (paraflagellar body) at its base immediately adjacent to the eyespot or stigma. This body has long been considered light-sensitive (see Wager, 1900; Pitelka, 1963; Pringsheim, 1963; Clayton, 1964). The stigma merely serves as a shadow-casting organelle. The most convincing evidence for these conclusions come from the study of certain genetic mutants (see Clayton, 1964). Strains of euglenoids without the flagellar swelling but with a stigma are nonphototactic; those, on the other hand, with a paraflagellar body but without an eyespot are responsive to light. Gibbs (1960) and Leedale, Meeuse, and Pringsheim (1965)

Fig. 7. Luminal ends of cells composing early optic vesicle in embryo of Helix aspersa, showing beginning development of microvilli (mv). c_1, axial centriole (kinetosome); ci, rudimentary cilium; ls, lens formed by aggregation of particles in lumen (lu) of optic vesicle; sd, secretory droplets, believed to be lens and humor forming material, assembled in Golgi apparatus (not shown) and released into lumen of eye. X 17,000. (From Eakin and Brandenburger, 1967.)

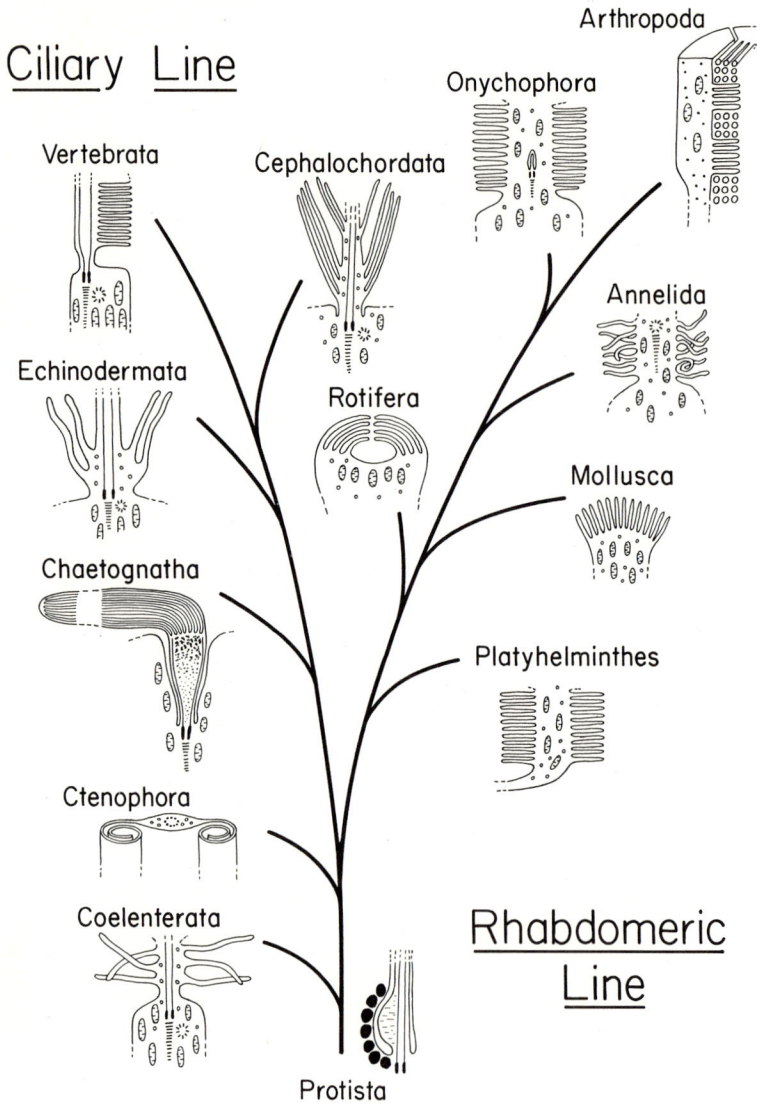

Fig. 8. Schematic representation of photoreceptoral organelles in selected examples of groups of animals along ciliary and rhabdomeric lines of evolution.

have conducted electron microscopic studies of both structures in *Euglena gracilis* and *E. spirogyra* respectively. Neither investigation gave a clear picture of the organization of the flagellar swelling, but both suggested the possibility of a finely lamellate substructure. It is hoped that new research providing higher resolution and information on the chemical nature of the paraflagellar body will be forthcoming in the near future.

Coelenterata. I speculate that, as metazoans evolved from protists, ciliated cells on the surface of the multicellular organism invaginated to form simple eyecups. Some of the epithelial cells differentiated into sensory cells with light-sensitive cilia projecting into the cavity of the ocellus and with neurites transmitting on or off excitations to neural centers which had evolved meanwhile. Other epithelial cells surrounding the sensory elements acquired accessory pigment and served to prevent light entering the eyecup except by way of its opening or primitive pupil. Moreover, the pigmented supportive cells permitted differential illumination of the sensory cells by casting shadows upon a multicellular retina. As the pattern of shade is more complex than that formed by a simple stigma upon a single paraflagellar body, an ocellus was presumably better adapted to determine the direction of light than was an eyespot.

Nature then began to "experiment," through selection of mutations, to increase the light-sensitivity of a cilium by enhancing its photon-gathering ability. The evagination of the covering of a cilium into tubular projections was one possibility of increasing the surface area of a photopigment-bearing membrane. The ocellus of a hydromedusan jellyfish serves as a model for this stage of evolution of eyes along the ciliary line (Fig. 8).

The first, and so far as I know the only, electron microscopical study of a coelenterate ocellus is that conducted by Dr. Jane Westfall and me several years ago on the hydromedusan *Polyorchis penicillatus* (Eakin and Westfall, 1962b). The eyecups of this beautiful little jellyfish are situated at the bases of the tentacles. They are presumed to be light-sensitive because the removal of the ocelli from other jellyfishes has a marked effect upon the responsiveness of the animals to light (Horstmann, 1934). Moreover, Yoshida (1963) and Yoshida and Ohtsuki (1966) have demonstrated a photolabile pigment in the ocelli of a hydromedusan.

An eyecup in *Polyorchis* measures about 180 μ in diameter and 90 μ in depth, and consists of two types of columnar cells: sensory and pigmented. From the distal end of each sensory element there extends into the cavity of the ocellus a long cilium bearing numerous villous branches about 0.2 μ in diameter (Fig. 9). These microvilli interdigitate with similar lateral extensions from the pigment-containing supportive cells. Many mitochondria and a typical centriolar apparatus including a long striated rootlet are found at the base of the cilium. A cross section of the ciliary shaft reveals the characteristic ring of nine doublets of microtubules (Fig. 10).

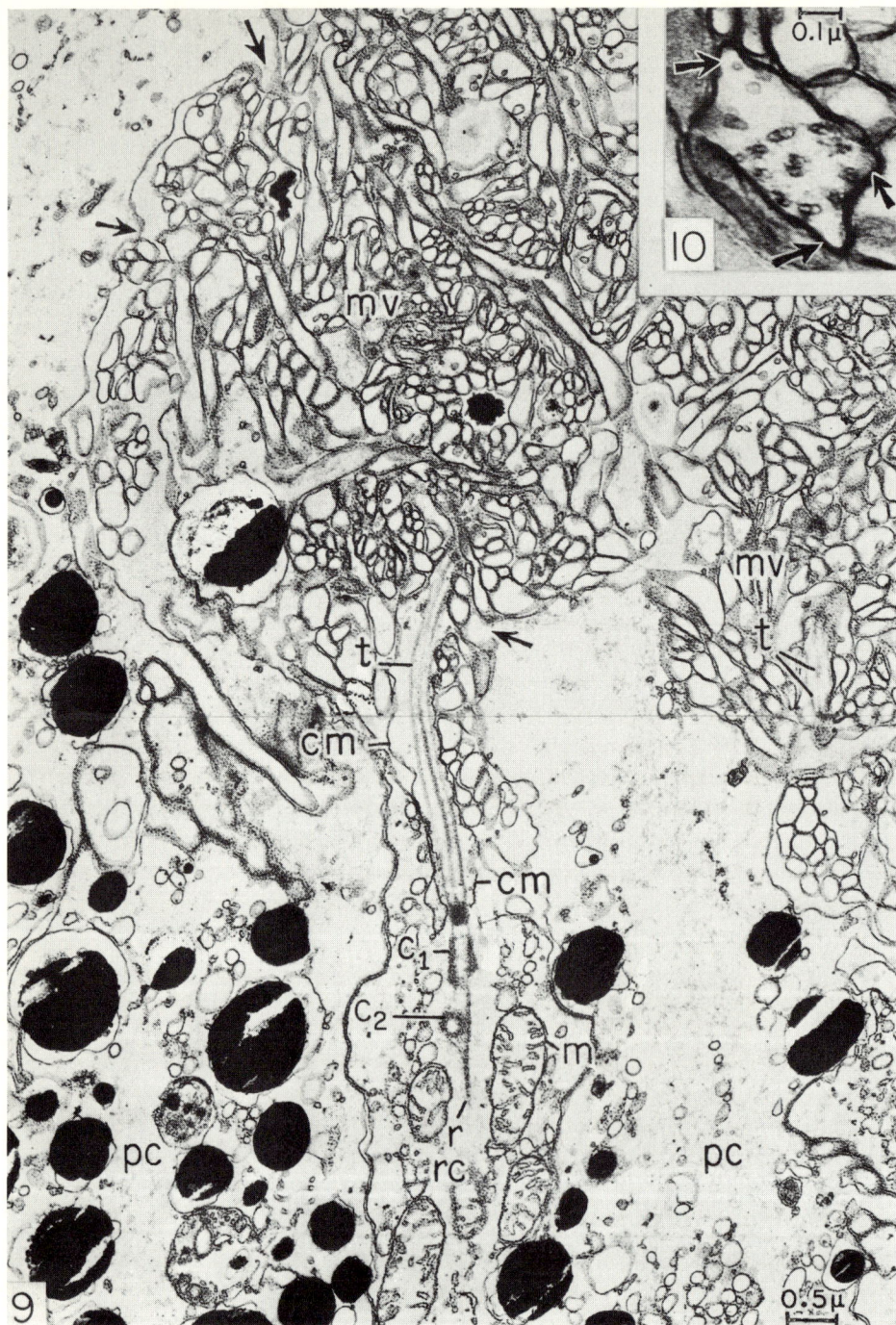

Figs. 9 and 10. (Legend on page 209)

And two central microtubules are present also! I regret that the resolution of our micrographs is not quite adequate to determine with certainty if arms are present on the doublets. They seem to be absent. Are the ciliary processes incapable of motility? Considering the interdigitation of microvilli from sensory and supportive cells there is little likelihood that much movement could occur in any event.

Ctenophora. Horridge (1964) has studied the fine structure of numerous laminated bodies in the floor of the apical organ of the comb-jelly *Pleurobrachia pileus.* Although there is no physiological or biochemical evidence that these structures are light-sensitive (and very little that ctenophores react to light), Horridge advances the suggestion that they are indeed photoreceptors because of their unique microanatomy. Each body is composed of whorls of double membranes which extend laterally from one or both sides of cilia (Fig. 8). These highly specialized organelles arise from an invagination in an epithelial cell bordering the cavity of the apical organ. Horridge suggests that granules found in the lamellate cilia might contain photopigment. Cross sections of the ciliary shafts reveal a $9 + 0$ pattern of microtubules and the arms on each doublet are said to be lacking. Other cilia with $9 + 2$ arrangement of microtubules and with arms are also found in the apical organ and, of course, elsewhere on the animal. These kinocilia do not possess the above mentioned granules. The kinetosomes of the lamellate cilia have neither striated rootlets nor oblique centrioles. Cilia of the comb-plates, on the other hand, exhibit prominent and branched rootlets, although they also lack accessory centrioles.

I agree with Horridge that these remarkable cilia with their sides extended and coiled might well be light-sensitive organelles, but I share also his opinion that "direct electrophysiological micromethods would be necessary to prove the case."

Echinodermata. Most sea stars bear an optic cushion or eyespot on the oral surface of the tip of each arm. A cushion contains many pigmented (red) ocelli, about the size of those in *Polyorchis.* There is good evidence that the eyecups are light-sensitive from studies of the behavior of sea stars or of isolated arms in relation to light (see Hyman, 1955). Hartline, Wagner, and MacNichol (1952) recorded light-induced potentials from the eyespot of an undesignated species of *Asterias.* Further observations on

---

Fig. 9. Longitudinal section of distal ends of one receptoral cell (rc) and two pigmented cells (pc) in ocellus of hydromedusan Polyorchis penicillatus. c_1, axial centriole (kinetosome); c_2, oblique centriole; cm, membrane of ciliary process; m, mitochondrion; mv, microvilli from ciliary process variously cut because of irregular arrangement; r, striated rootlet; t, microtubules. Arrows indicate processes of pigment cells. X 14,000.

Fig. 10. Cross section of shaft of ciliary process showing nine plus two pattern of microtubules, vesicles and bases of several microvilli (see arrows). X 62,000. (Figs. 9 and 10 from Eakin and Westfall, 1962b.)

the reactions of arms, with and without ocelli, to light of varying wave length may be found in a recent paper by Yoshida and Ohtsuki (1966) on the function of eyecups in *Asterias amurensis.*

The ocelli of several species of sea stars have been examined with an electron microscope (Eakin and Westfall, 1962b; Vaupel-von Harnack, 1963; Eakin, 1963). The lumen of the eyecup is filled with ciliary shafts and long microvilli (Fig. 11). In one species of sea star, *Henricia levius-cula,* Dr. Jane Westfall and I found the microvilli emerging from the bases of the ciliary shafts *distal* to the kinetosomes (Fig. 8). I classified the photoreceptoral apparatus of this sea star as ciliary in type because of the abundance and prominence of the cilia and their basal association with microvilli. A similar picture was found in *Leptasterias pusilla* and in *Patiria miniata.* Vaupel-von Harnack (1963) studied the ocelli in *Asterias rubens* and reported many cilia and microvilli. None of the villi, however, were observed to be connected to the bases of cilia. Unfortunately, most of her micrographs are too low in magnification for easy interpretation. We both agree that microvilli in sea star ocelli do not extend from a ciliary shaft along its length, as in the hydromedusan eyecup. Some of the cilia in the ocelli of sea stars studied by us exhibited a 9 +2 pattern (Fig. 11), others a 9 + 0 configuration (Eakin, 1963). Vaupel-von Harnack (1963) states in her discussion that it was difficult to determine whether central microtubules are present in the ocellar cilia of *Asterias.* Some micrographs were said to show them.

Studies on other echinoderms are greatly needed to elucidate further the relationship of cilia and microvilli in light-sensitive structures in this phylum. We have endeavored without success to identify photoreceptors by electron microscopy in the following: bipinnaria of the sea star *Pisaster ochraceous;* echinoplutei of the sea urchins *Strongylocentrotus purpuratus* and *S. franciscanus;* body wall of a sea cucumber, *Leptosynapta clarki;* and tegmen from a feather star, *Florametra serratissima.* Kawaguti and Kami-shima (1964) made an electron microscopic study of the blue patches in the integument of the sea urchin *Diadema setosum.* These spots, thought at one time to be eyes and responsible for the animal's shadow reaction, are iridophores possessing lamellae which contain ellipsoidal bodies regularly arranged in a double row along the length of each lamella.

Chaetognatha. Up to this point on the ciliary line the areal increase in membranes presumed to bear photopigments has been achieved by outfold-ing of the cilium. In the small phylum Chaetognatha (arrowworms), however, the light-sensitive organelle is a tubule formed, apparently, by infolding of the ciliary membrane. I say apparently because as yet the development of the organelle has not been studied, and all that we know about the fine structure of the eye of an adult arrowworm is based upon one study (Eakin and Westfall, 1964b) on one species, *Sagitta scrippsae.*

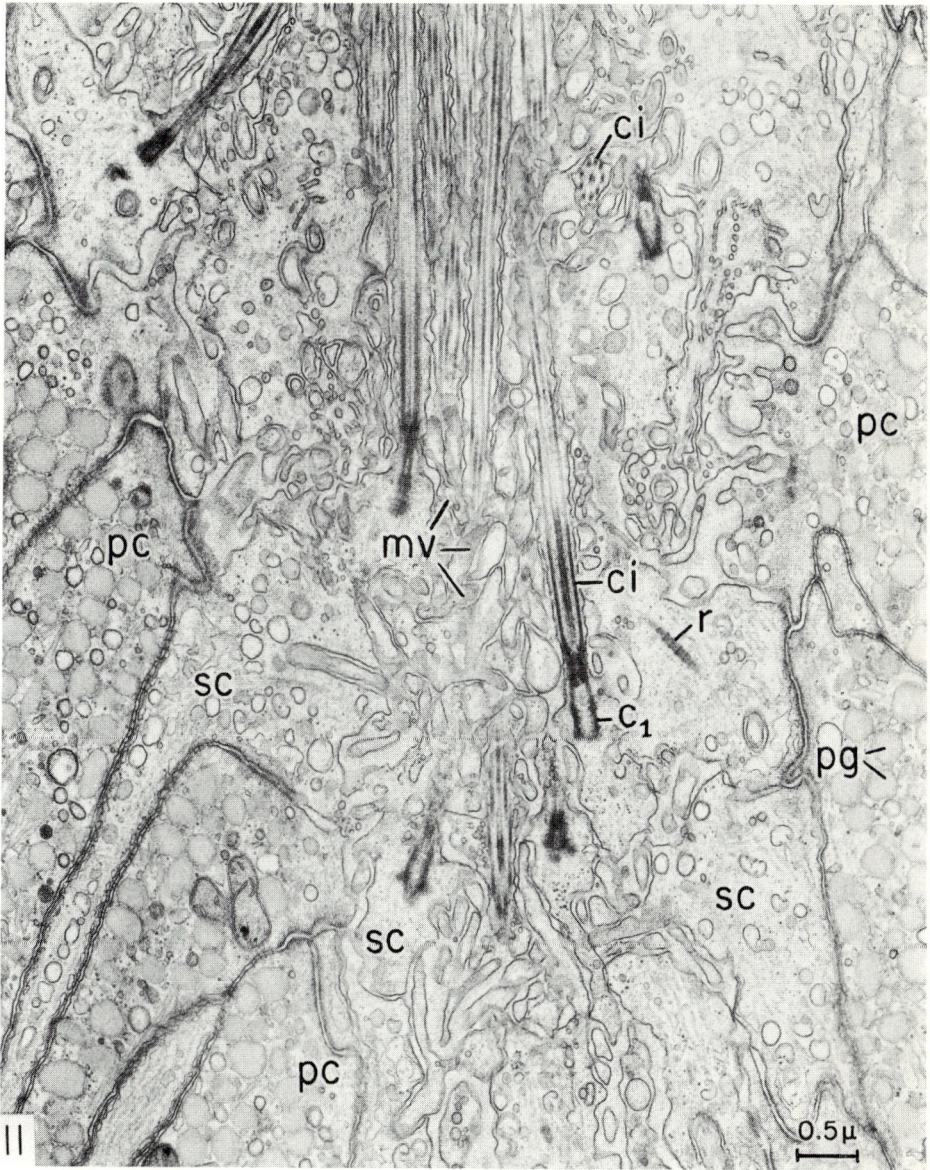

Fig. 11. Base of an ocellus of sea star Henricia leviuscula *showing distal ends of several sensory cells* (sc) *and pigmented cells* (pc). c_1, *axial centriole (kinetosome)*; ci, *cilia, one in cross-sectional view*; mv, *microvilli*; pg, *pigment granules*; r, *striated rootlet.* X 17,000.

Arrowworms, which live in the subsurface waters of the sea where the light is weak and diffuse, have a pair of eyes on the dorsum of the head. Each eye contains several so-called ocelli (Hesse, 1902; Burfield, 1927). We found that in *Sagitta scrippsae* the ocelli are merely clusters of receptoral cells situated between arms of the single pigment-containing cell which lies at the center of the eye. Each sensory cell is a long, columnar element bearing a ciliary process at its inner end (i.e., adjacent to the pigment cell) and a neurite distally. The process is most unusual, consisting of two distinctly different segments: an apical region composed of longitudinally oriented tubules only 500 Å in diameter but as long as 20 μ, and a cone-shaped basal region of densely packed granules (Fig. 8, 12, 13). The tubular section abuts the broad end of the cone; the narrow end of the cone joins the cell proper immediately above a kinetosome. Nine doublets of microtubules extend from the kinetosome into and along the sides of the conical part of the ciliary process. A striated rootlet may be present; accessory centrioles were not observed.

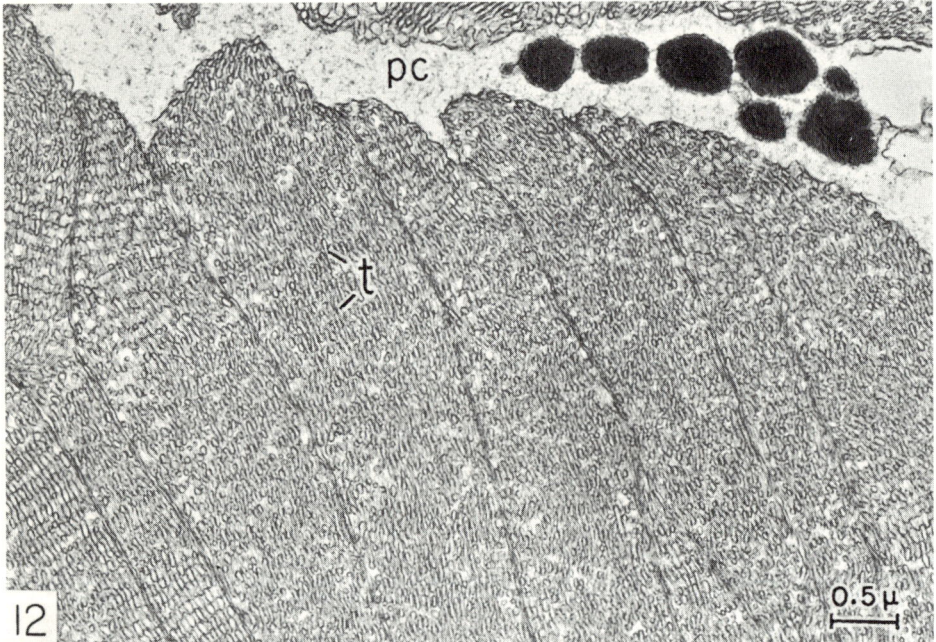

Fig. 12. Tips of tubular segments (see Fig. 8) of several receptoral processes bordering an arm of the pigment cell (pc) in eye of Sagitta scrippsae. Rows of receptoral tubules (t) sectioned obliquely. X 18,000.

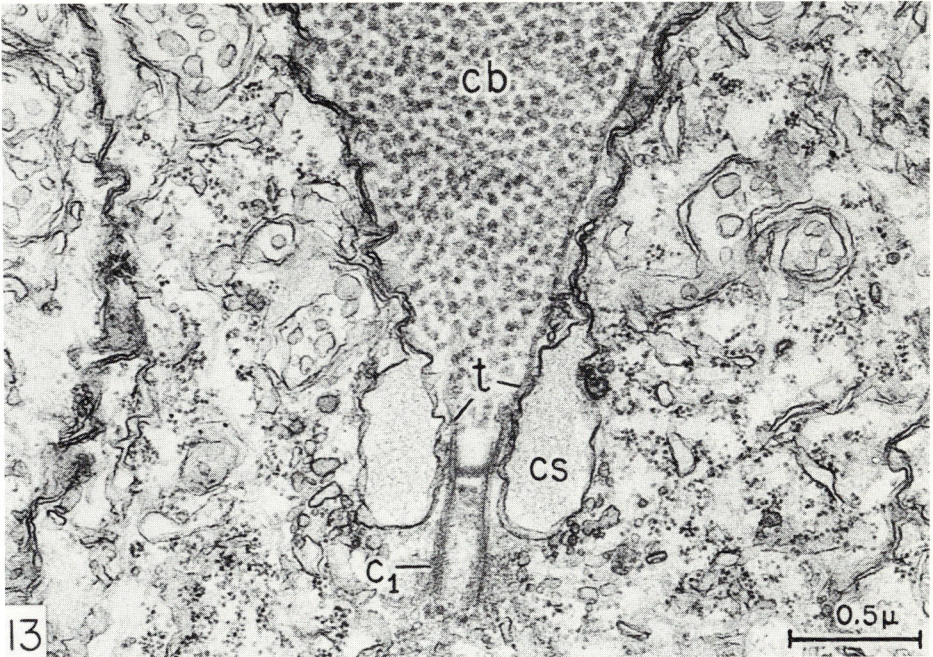

Fig. 13. The base of one receptoral process, at higher magnification, showing basal end of conical body (cb) along which run nine doublets of microtubules; parts of two may be seen (t). c_1, axial centriole (kinetosome); cs, circumciliary space. X 36,000. (From Eakin and Westfall, 1964b)

I presume that the tubules are the light-sensitive organelles. In this connection Burfield (1927) observed that the processes imparted a faint pink color in a living arrowworm (*Sagitta bipunctata*). But what could be the function of the conical body at the base of the ciliary process? We have offered several suggestions (Eakin and Westfall, 1964b), the most tempting being the possibility that the body might have wave guide effects or that it might act as a resonator in the manner of a laser.

The finding of a ciliary type photoreceptor in a chaetognath is, I think, useful to the student of phylogeny. The evolutionary relationships of the arrowworms have been a subject of speculation for a long time. They have been regarded as relatives of no less than eight other invertebrate taxa (Hyman, 1959). Although Libbie Hyman considers the arrowworms more like aschelminths in morphology, she classifies them in the deuterostomia because of certain embryological characteristics (see subsequent discussion). At long last we have at least one adult feature, namely ciliary photoreceptors, which the chaetognaths share with other deuterostomes.

Cephalochordata. Several electron microscope studies of photoreceptors in the evolutionarily significant protochordate, amphioxus, have been made but, frankly, as yet no really good electron micrographs have been published, at least nothing that I would care to reproduce here. It is hoped that someone will rework this material with improved techniques. We have, nevertheless, good evidence that in the roof of the cerebral vesicle of amphioxus there are ependymal cells with long ciliary processes projecting into the lumen. These were first seen by Satir (as reported by Miller, 1960) and then by Eakin and Westfall (1962a) and Olsson (1962b). The shaft of each process bears along its length narrow branches formed apparently by the outfolding of the ciliary membrane (Fig. 8). In some respects these lamellae from the sides of cilia resemble those in the apical organ of the ctenophore discussed above. The best demonstration of the branched cilia in amphioxus may be seen in Figures 21–13 to 21–15 in Eakin (1963), as found in *Branchiostoma californiense*. The shafts possess a 9 + 0 arrangement of microtubules, the kinetosome is continuous with a striated rootlet, and accessory centrioles are present. Whether these dorsal ependymal cells with unique ciliary processes are actually light-sensitive is yet to be proved.

The eyecups of Hesse which occur in segmentally arranged clusters along the ventral wall of the spinal cord are undoubtedly photoreceptors, considering their ocellus-like nature and the experimental studies of investigators such as Parker (1908). Electron microscopy by Eakin and Westfall (1962a), Eakin (1963), and Nakao (1964) has shown that the sensory cells have an array of microvilli projecting from the cell surface adjacent to the cup-shaped pigment cell. Cilia have not been observed arising from the receptor cells. Although Nakao noted flagella in the space between sensory and pigmented cells, he did not know their origin. In the instance of the eyecups of Hesse we have a rhabdomeric type photoreceptor—an exception to the rule. And there are large cells (called Joseph cells) in the dorsal wall of the cerebral vesicle which also have an array of microvilli very similar to those of a Hesse cell in the spinal cord (Eakin and Westfall, 1962a; Eakin, 1963; Nakao, 1964). If a Joseph cell is light-sensitive (it seems likely considering the similarity of this cell to the sensory element of the Hesse ocellus) we have an additional example of a rhabdomere in amphioxus.

There has been speculation regarding the origin of vertebrate eyes, both median and lateral, from the brain of some amphioxus-like ancestor. The ependymal cells with branched cilia might have been the evolutionary precursor of the vertebrate median eye. And at one time I suggested the possibility that certain ependymal cells in the infundibular organ (a region on the floor of the cerebral vesicle of amphioxus) might be the antecedent of the vertebrate lateral eye (Eakin and Westfall, 1962a). This hypothesis was based upon the following: earlier notions (e.g., those of Franz, 1923) regarding the sensory function of the infundibular organ; our findings that infundibular cilia in *Branchiostoma californiense* possess many small vesicles and also basal branches with strings of vesicles (Eakin and West-fall, 1962a); and the knowledge that the paired vertebrate lateral eyes develop from a single median anlage in the gastrula and neurula (see Man-gold, 1931). Several investigators, notably Olsson (Olsson and Wing-strand, 1954; Olsson, 1962a) have advanced with cogency the proposition that the infundibular organ of amphioxus is purely secretory. I am reluc-tant to abandon my hypothesis, however, until it has been proven that the infundibular organ of amphioxus is not light-sensitive.

Parenthetically, it should be pointed out that the idea of ciliary vesicles containing a photopigment was first suggested by Grimstone, Horne, Pantin, and Robson (1958). They observed clusters of small vesicles within cilia on epithelial cells of the gastrodermis in the sea anemone *Metridium senile*. They postulated that these vesicles might be photo-receptoral in view of the demonstration that the musculo-epithelial system of *Metridium* is light-sensitive. North and Pantin (1958) recorded action potentials from the gastrodermis of this animal upon illumination. We have observed ciliary vesicles in photoreceptors of a hydromedusan and of several sea stars as well as in the ciliary processes from dorsal ependymal cells in the cerebral vesicle of amphioxus (see Fig. 8). And Horridge (1964) noted similar vesicles in the laminated cilia of the apical organ of a ctenophore.

The only other protochordate photoreceptor examined with the electron microscope is the larval ocellus of the ascidian *Ciona intestinalis* by Dilly (1961, 1964). The tunicate tadpole, which is negatively phototactic shortly before metamorphosis, has an eyespot at the posterior end of the cerebral vesicle made up of a one-cell pigment cup and four to nine retinal cells. Each sensory cell possesses a process composed of two parts: a tubular section containing 50 to 100 filaments and an outer segment consisting of a stack of double membrane disks, resembling the disks of vertebrate photoreceptors according to Dilly. The latter segment, presumed to carry the photopigment, is situated inside the concavity of the pigment cup and immediately behind the lens. No centriolar structures were described or figured, but Dilly (per-sonal communication) has observed centrioles in association with the pro-cesses. I wish that I knew more about the relationship between centriole and filaments in this case. Dilly (1964) homologizes the tubular sector of the process with the connecting piece of the vertebrate rod or cone, recogniz-ing, of course, that a tangle of 50 to 100 filaments is grossly unlike the 9 + 0 pattern of microtubules in other ciliary photoreceptors. Finally, to make things interesting, Dilly (personal communication) informs me that he has also found in the cerebral vesicle of tadpoles of *Ciona* structures that look like Hesse cells, that is, possessing microvilli.

Vertebrata. The zenith and end of the ciliary line, until evolution makes further advances, is the vertebrate rod and cone which endow respectively the owl with its great retinal sensitivity and the eagle with its remarkable visual acuity. Or to cite a more anthropocentric example, I remind the reader that he is using modified cilia—his foveal cones—as he scans these lines. Indeed, I do not know of any well differentiated vertebrate photoreceptor which is not of the ciliary type. This includes the receptors of lateral eyes, median eyes, and other diencephalic structures thought to be light-sensitive, such as the epiphysis of reptiles (Steyn, 1960; Oksche and Kirschstein, 1966a), amphibia (Kelly, 1962; Oksche and Vaupel-von Harnack, 1963), fishes (Oksche and Kirschstein, 1966b; Rüdeberg, 1966), and birds (Oksche and Vaupel-von Harnack, 1965a, 1966; Collin, 1966). The literature on the fine structure of vertebrate photoreceptors has become so extensive that a review of it will not be undertaken here. Some general works which may be consulted are: Sjöstrand, 1959; Smelser, 1961; Cohen, 1963; Missotten, 1964; Moody, 1964.

In the sweeping generalization just made regarding the ciliary nature of all vertebrate photoreceptors, I am not overlooking the fact that there are various tissues without this kind of receptoral apparatus from which electrical responses can be recorded when they are illuminated. For example, flashes of light evoke signals from the pigmented epithelium of the retina (Brown, 1965). The myeloid bodies may be the photoreceptoral organelles (Yamada, 1961). Melanophores have long been known to be light-sensitive, the melanin presumably acting as a photopigment. Recently, Becker and Cone (1966) have obtained electrical responses from the skin of several vertebrates upon illumination with intense flashes of light. These authors believe that melanin when present augments a response but since recordings can be made from the skins of albino animals they postulate that simple heating of a cellular structure may generate the response.

Rhabdomeric Line of Evolution

A safe journey has been made, I hope, along the ciliary line from protist to vertebrate despite some rough terrain in the echinoderm province and some rhabdomeric sidepaths and uncertain routes through the land of the cephalochordates. I now invite the reader to accompany me on another exciting but more perilous adventure along the rhabdomeric lineage. I warn him that ciliary photoreceptors are lurking on the way. Our destination is the citadel of rhabdomeres, the arthropods, where microvilli are beautifully ordered and complexly arranged, in some instances permitting the detection of polarized light, and where cilia have not yet been seen. But where do we begin? Frankly, I do not know. My best guess at the moment is somewhere above the radiates, the Cnidaria and Ctenophora. There is a vague suggestion, however, that a rhabdomere or something foreshadowing it, may have been "tried out" by some protist.

Greuet (1965) has described and figured the fine structure of a remarkable ocellus in a highly specialized dinoflagellate, *Erythropsis pavillardi*. Behind a crystalline lens lies a choroid body consisting of several strata. The next to last layer is an array of long (about 2 μ) lamellae which project away from the lens toward a final layer of carotenoid globules. These lamellae may be analogous to the microvilli of a rhabdomere.

Platyhelminthes. The first unmistakable rhabdomere is that of turbellarian worms, the ocelli of which contain the distal ends of sensory cells bearing a myriad of microvilli (Figs. 8, 14). Although there is no doubt that the ocelli are light-sensitive (Taliaferro, 1920), only anatomical evidence so far indicates that the microvilli are the responsible organelles. The fine structure of platyhelminth eyes has been reported by several workers (Wolken, 1958a; Press, 1959; Röhlich and Török, 1961, 1962; Wetzel, 1961; Eakin, 1963; MacRae, 1964, 1966). In some instances (e.g. *Dugesia*) the villi arise like a crown from the distal end of the sensory cell (Fig. 15), a picture similar to that presented earlier (Fig. 6) for the rhabdomere of the snail *Helix*. In other species of turbellarians, however, the microvilli occur not only at the tip but also along the sides of the receptoral process (see Fig. 8, based upon the findings in *Dendrocoelum lacteum* by Röhlich and Török, 1961). Neither cilia nor centrioles were observed associated with the microvilli of flatworm ocelli. A similar picture has been described for ocelli in miracidia (Kümmel, 1960; Isseroff, 1964), and in cercariae of trematodes (Pond and Cable, 1966).

Some students of evolution (e.g., Hadži, 1963) regard the platyhelminths as ancestral to the coelenterates, and hence not properly shown phylogenetically in Figure 8. I refer the reader to Libbie Hyman's critical and pithy comments on this interpretation in her review of recent phylogenetic speculation (Hyman, 1959).

Rotifera. The only study of the fine structure of an eye in an aschelminth, is our report (Eakin and Westfall, 1965a) on the cerebral ocellus of a rotifer, *Asplanchna brightwelli,* situated on the posterior ventral border of the brain. The eye consists of two cells (Fig. 16): a cup-shaped element possessing flat plates containing a red pigment, and a sensory cell, the receptoral villi of which lie in the concavity of the cup. The villi are arched lamellae overlying one another like the petals of an unopened flower (Fig. 8). As the ocellus is directed dorsally, light must pass through brain and body of sensory cell to reach the microvilli. A search for cilia and centrioles was conducted by careful examination of serial sections of several eyes. None was found, although cilia abound elsewhere in rotifers.

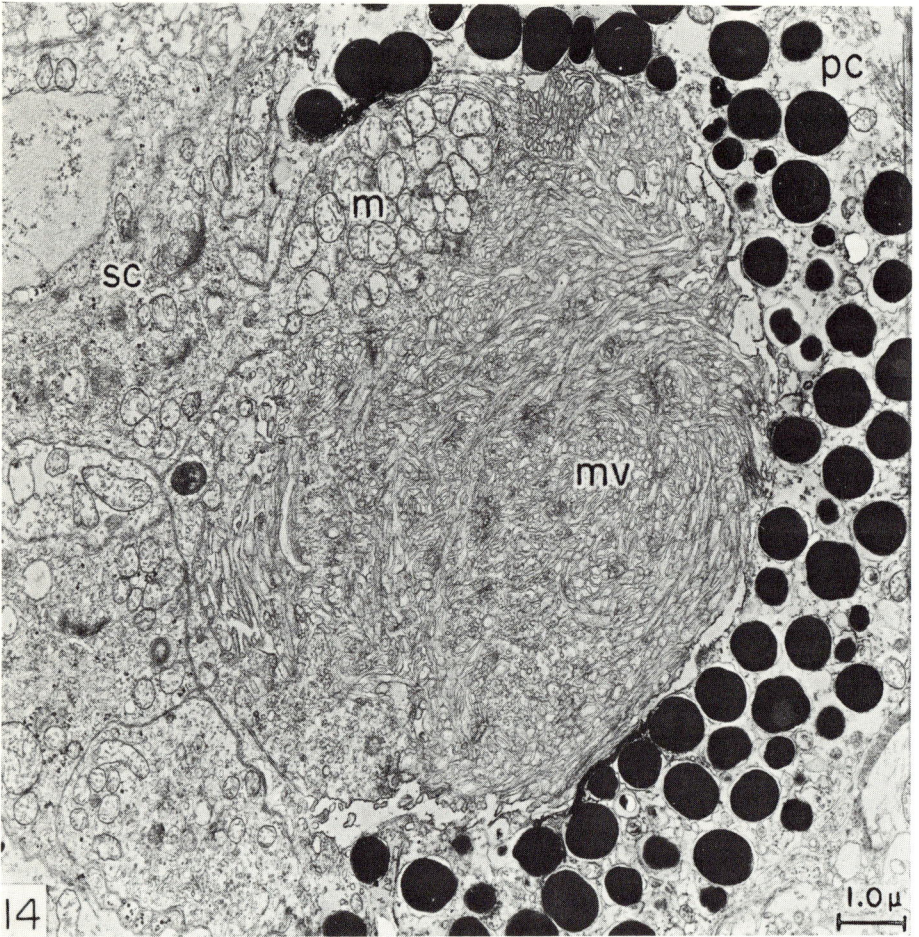

Fig. 14. Ocellus of marine flatworm Notoplana acticola, showing cup-shaped pig-mented cell (pc) enclosing the photoreceptoral microvilli (mv) of sensory cell (sc). Note abundance of mitochondria (m) at bases of microvilli. X 9,000. (From MacRae, 1966.)

The assignment of rotifers to the rhabdomeric line of evolution of photo-receptors is not inconsistent with other conclusions on their evolutionary affinities. Hyman (1951) states that "Rotifers are related to the annelid-mollusk stem," and that "the Rotifera show a greater resemblance to the Turbellaria than to any other invertebrate group and may be considered as relating the Aschelminthes to the Platyhelminthes" (pp. 150–151). Ocelli of other aschelminths need study by electron microscopy.

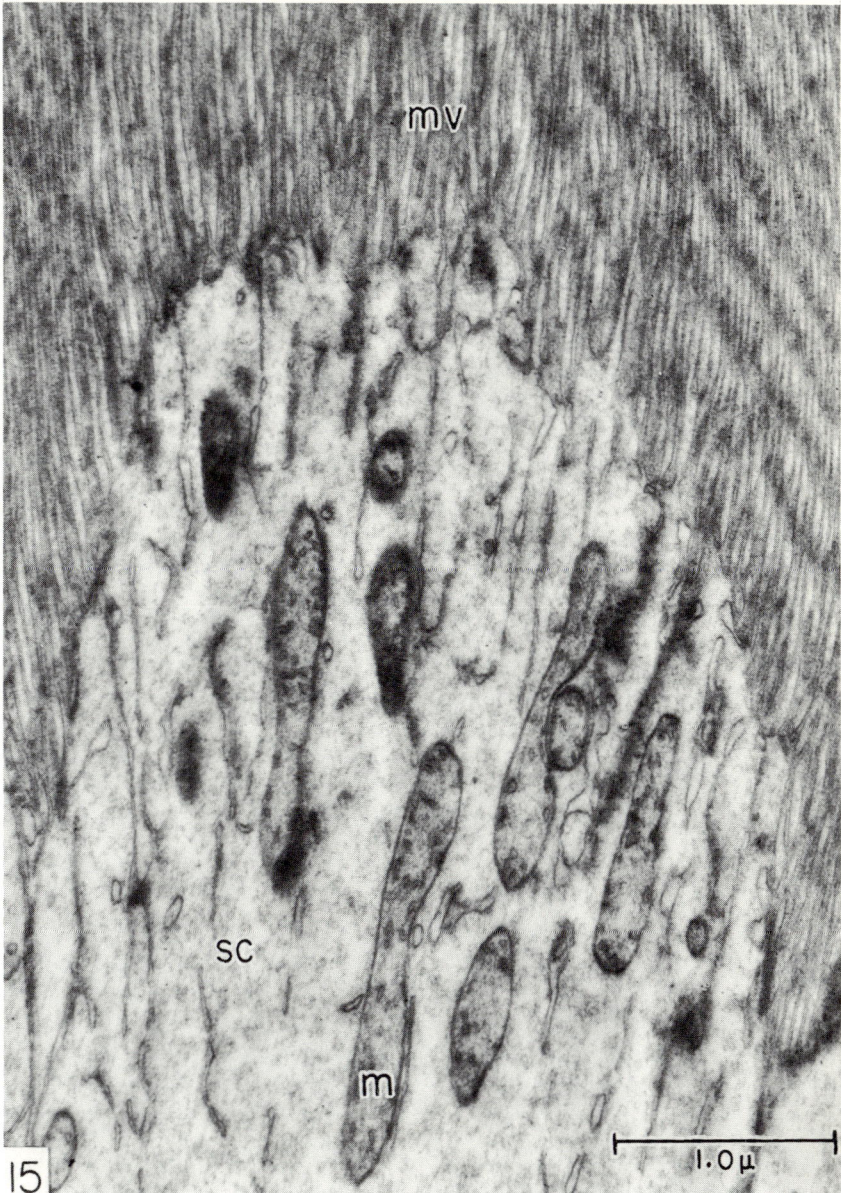

Fig. 15. Distal part of sensory cell (sc) from eyecup of planarian (Dugesia tigrina) showing bases of receptoral microvilli (mv) cut lengthwise (left side) and obliquely (right side). m, mitochondrion. The continuity between the cytoplasm of sensory cell and that of microvilli may be observed in several places. X 29,300. (From Wetzel, 1961.)

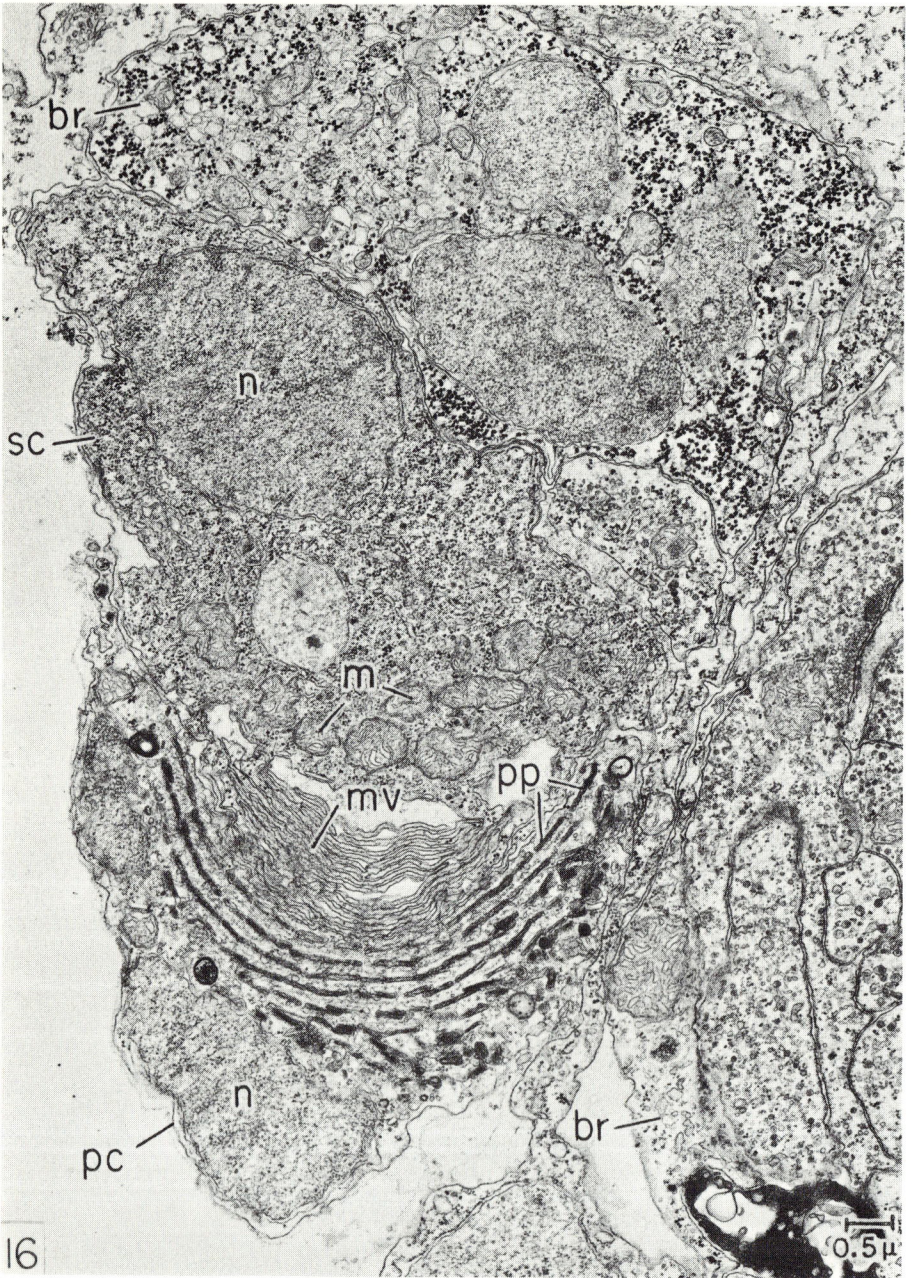

Fig. 16. (Legend on page 221)

Mollusca. As noted earlier, cephalopod eyes show rhabdomeres that are very similar to those in arthropods (Fig. 8), namely, well ordered arrays of microvilli projecting laterally from the sides of the sensory cells. These organelles have been examined with the electron microscope in several kinds of squid and octopus (Wolken, 1958b; Moody and Robertson, 1960; Zonana, 1961; Nishioka, Hagadorn, and Bern, 1962; Yamamoto et al., 1965). Cilia or centrioles have not been reported.

The ultrastructure of the eyes of various gastropods has been studied. The rhabdomere in the pulmonate snail *Helix* was described above (see Figs. 6 to 8). Our research on the development of the eye of *H. aspersa* disclosed the presence of rudimentary cilia in the embryonic optic vesicle, unrelated, so far as could be determined by observation, to the differentiation of microvilli (Eakin and Brandenburger, 1967). In another pulmonate gastropod, *Onchidium verruculatum,* however, Yanase and Sakamoto (1965) found both ciliary and rhabdomeric type photoreceptors. The dorsal ocellus exhibits laminated bodies, reminiscent of those described by Horridge (1964) in a ctenophore, consisting of whorled membranous extensions from the sides of cilia. The ciliary shafts have a ring of nine doublets of microtubules which terminate basally in typical kinetosomes. Yanase (personal communication) obtained electrophysiological evidence of sensitivity of the dorsal eye of *Onchidium* to light. I am further informed by him that the stalked eye of this mollusk, on the other hand, possesses rhabdomeres.

Finally, three other investigations on gastropod eyes should be mentioned. Eakin, Westfall, and Dennis (1967) have just completed a study of an interesting eye in a nudibranch, *Hermissenda crassicornis.* This organ has only five receptor cells, each bearing a rhabdomere at one end, adjacent to the lens, and a neurite at the other end of the cell. A few rudimentary cilia were observed in both sensory and pigmented cells. Clark (1963) reported the presence of cilia in the adult eye of a live-bearing snail, *Viviparus maleatus,* but the relationship of these organelles to the microvilli of the rhabdomere has not yet been analyzed. And Tonosaki (1966) has described the ultrastructure of the eyecups of an abalone, *Haliotis.* The sensory cells bear clusters of short villous projections along the sides of an apical extension from each cell. No cilia were seen.

Among the lamellibranchs there is one eye which has been examined ultrastructurally, that of the scallop *Pecten.* Miller (1958) first pointed

Fig. 16. Cerebral eye of rotifer Asplanchna brightwelli, showing cup-shaped pigmented cell (pc) containing flat pigmented plates (pp) and enclosing leaf-like microvilli (mv) of sensory cell (sc). br, brain in which eye is partly embedded; m, mitochondria abundant above microvilli; n, nuclei. Concavity of ocellus directed dorsally. X 13,000. (From Eakin and Westfall, 1965a.)

out that the whorled membranous processes of the distal sense cells of *Pecten irradians* appeared to be derivatives of cilia because the appendages possessed basal bodies and in cross-sectional view exhibited a ring of nine dense loci. Barber, Evans, and Land (1967) have recently obtained more convincing evidence of the correctness of this interpretation of the distal receptors in their study of the eye of *Pecten maximus*. On the other hand, the proximal sense cell in the retina of *Pecten* appears to be fundamentally rhabdomeric in type (Barber, Evans, and Land, 1967). It is worthy of note that the distal sense cells give "off responses" and perhaps serve to detect shadows whereas the proximal receptors show "on responses" and may monitor the intensity of light striking the retina of the eye (Hartline, 1938; Land, 1965, 1966).

In conclusion, it seems safe to say on the basis of present information that in the Mollusca the predominant photoreceptoral apparatus is the rhabdomere, but that in some instances, as in *Onchidium* and *Pecten,* both ciliary and rhabdomeric types exist—a situation analogous to amphioxus on the ciliary line of evolution.

Annelida. A similar conclusion seems in the making for the phylum Annelida. Eakin (1962, 1963) first reported rhabdomeres consisting of highly irregular and twisted microvilli extending from the sides of sensory processes in the eyes of a polychaete worm, *Nereis vexillosa* (Fig. 8, 17). But near the tip of the process was a good centriole from which a prominent striated rootlet extended to the base of the process. Suspecting that cilia might be found in embryonic or juvenile eyes we examined developmental stages in the eyes of *Neanthes succinea* and *Nereis limnicola* (Eakin and Westfall, 1964a). Indeed, cilia were observed, but they appeared to have no relationship to the microvilli which were formed *below* the kinetosomes. Subsequently Fischer (1963) and Fischer and Brökelmann (1966) found essentially the same picture in adults of *Platynereis dumerilii,* except that rudimentary cilia were observed. In a restudy of *Nereis vexillosa* we also found a few rudimentary cilia at or near the tips of the receptoral processes.

In the brain of *Nereis pelagica,* on the other hand, Dhainaut-Courtois (1965) finds an organ, thought to be photoreceptive, which contains cilia with villous branches. Each ciliary shaft has nine doublets of microtubules which are said to be distributed to the branches. Unfortunately, the electron micrographs do not fully demonstrate the author's findings, in my opinion. I do not doubt the correctness of her interpretation (see her schematic representation, Fig. 1, p. 1085), yet I hope that a restudy with higher magnification of the microvilli and the system of microtubules will be forthcoming.

Hermans and Cloney (1966) have recently published an informative and stimulating study of prostomial ocelli in the polychaete *Armandia brevis.*

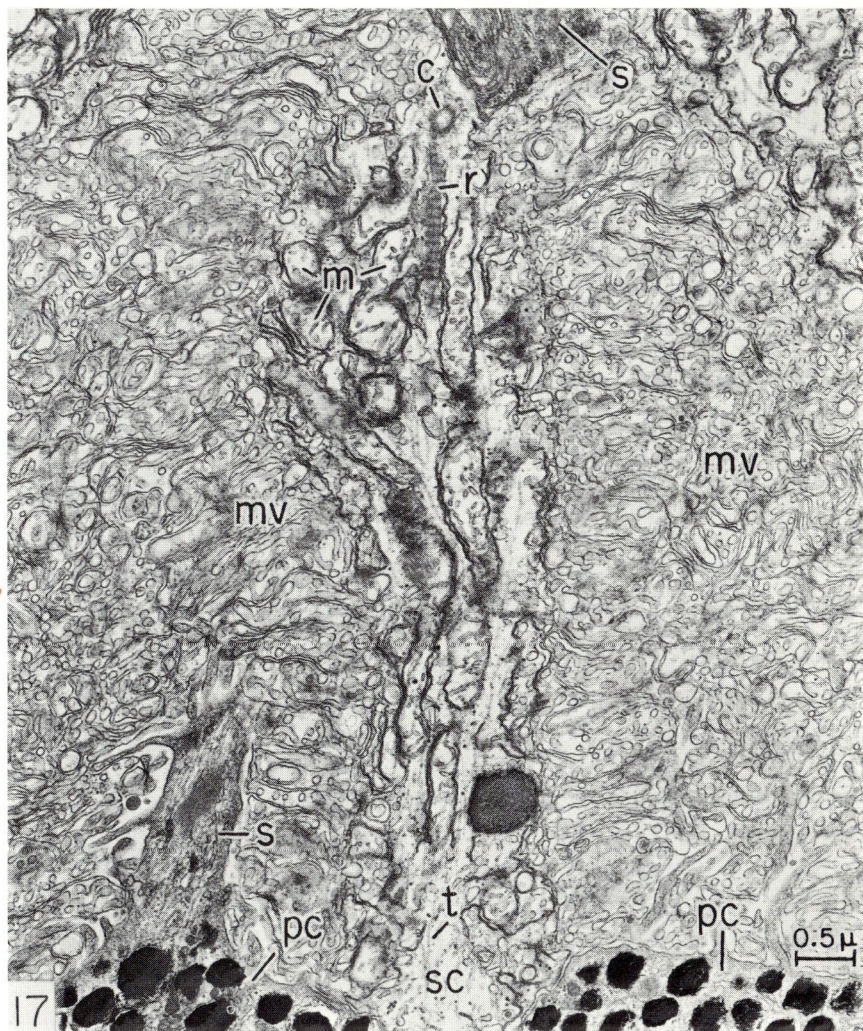

Fig. 17. Distal end of sensory cell (sc), in eye of polychaete annelid Nereis vexillosa, and its long process bearing many highly irregular microvilli (mv) on all sides. c, centriole; m, mitochondria; pc, pigmented cells flanking sensory cell; r, striated rootlet; s, septa interconnecting pigmented cells and lens; t, microtubules. X 16,000. (From Eakin, 1963.)

These eyecups, three in number and each consisting of one pigmented and one sensory cell, are embedded in the brain of the worm. They are remarkably like the eyecups of Hesse in amphioxus. These workers, after noting at least seven points of close similarity, draw the judicious conclusion that the resemblance is due to evolutionary convergence. Moreover, they sound the warning that invoking homology as the cause of similarity at the fine structural level is risky business—a warning which I might well heed. Incidentally, Hermans and Cloney could have pointed out the similarity between "photoreceptor-like cells partially enclosed by unpigmented epitheloid cells" which they describe in the brain of *Armandia* to Joseph cells in the brain of amphioxus (Eakin, 1963). Hermans (personal communications and Hermans and Cloney, 1966) makes a very good case for the homology between the so-called converse eyes of nereid worms, in which receptoral villi are directed toward the light, and the inverse eyes of opheliids, the group of polychaete annelids to which *Armandia* belongs, in which the receptors face toward the pigmented cup and away from the light. And, lest I forget, Hermans and Cloney did not observe cilia and centrioles in the eyes of *Armandia* although a special search for them was not made. *Armandia* also possesses segmentally arranged ocelli along the sides of the body. I eagerly await Hermans' study of these organs.

We come now to an unmistakable ciliary photoreceptor in a sabellid worm, *Branchiomma vesiculosum* (Lawrence and Krasne, 1965; Krasne and Lawrence, 1966). The eyes of this animal are borne at the tips of the filaments of the branchial crown. As shown in Figures 18 and 19 (from Krasne and Lawrence, 1966), each ocellus consists of two cells: a superficial lens element and a large elongate receptor. The latter is specialized into a distal segment filled with closely packed, long, hexagonal mitochondria and a basal segment containing about 450 disk-shaped membranous sacs. Mitochondria and sacs are oriented perpendicularly to the long axis of the cell. The lamellar sacs are actually cilia, extended and flattened, which project into an extracellular space. A typical kinetosome, situated at the base of each sac, sends forth into the sac a bundle of nine doublets of microtubules. Central tubules are lacking, and the peripheral ones become disorganized as they spread into the flattened sac. The authors discuss the limited evidence that the eyes are light-sensitive structures, that the ciliary structures are the receptoral organelles, and that the mitochondria might endow the eye with dichroic properties which would permit it to function as an analyzer of polarized light. A very interesting eye indeed! A similar but less well documented picture is described by Kernéis (1966) in the closely related *Dasychone bombyx,* which should be regarded as *Branchiomma bombyx,* according to Hartman (1959).

There is yet one final annelid eye which has been studied ultrastructurally, the eye of leeches (Hansen, 1962; Jung, 1963; Röhlich and Török,

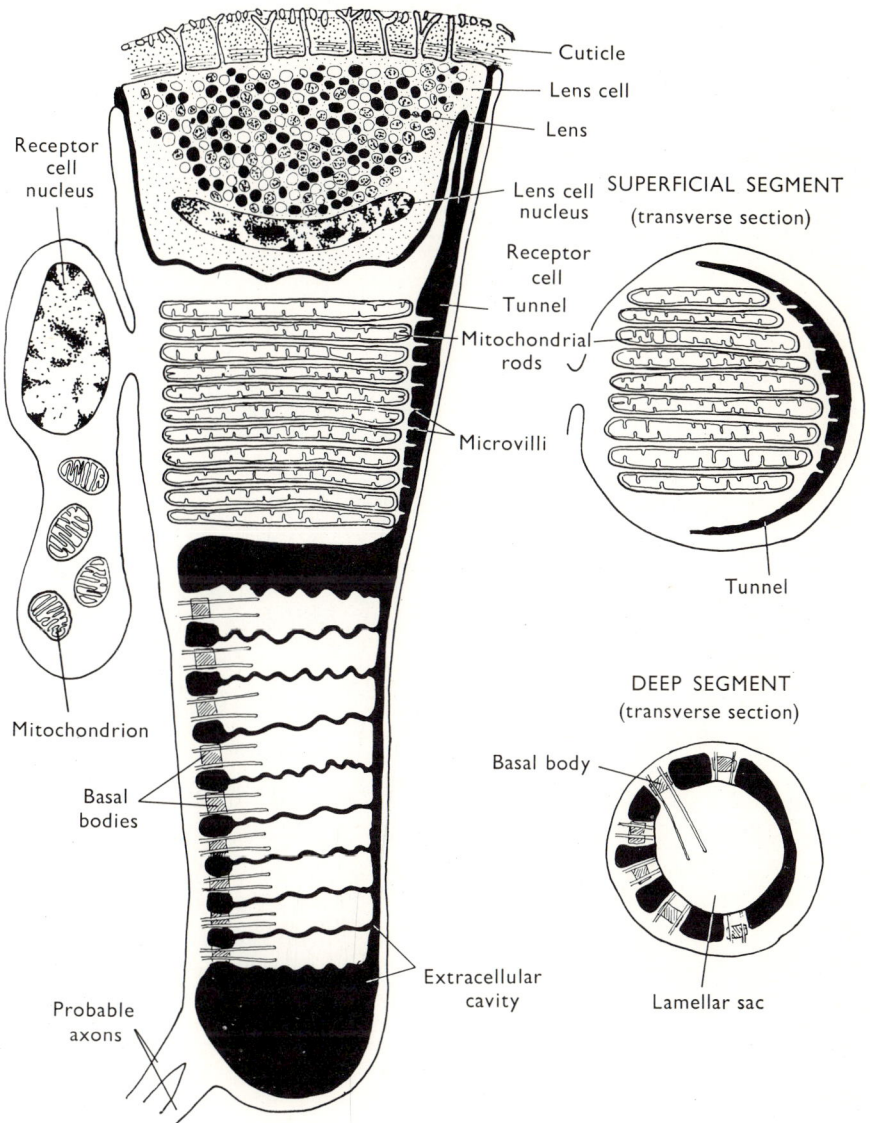

Fig. 18. Diagram of receptor and lens cells in eyespot of *sabellid* annelid Branchiomma vesiculosum. (*From Krasne and Lawrence, 1966.*)

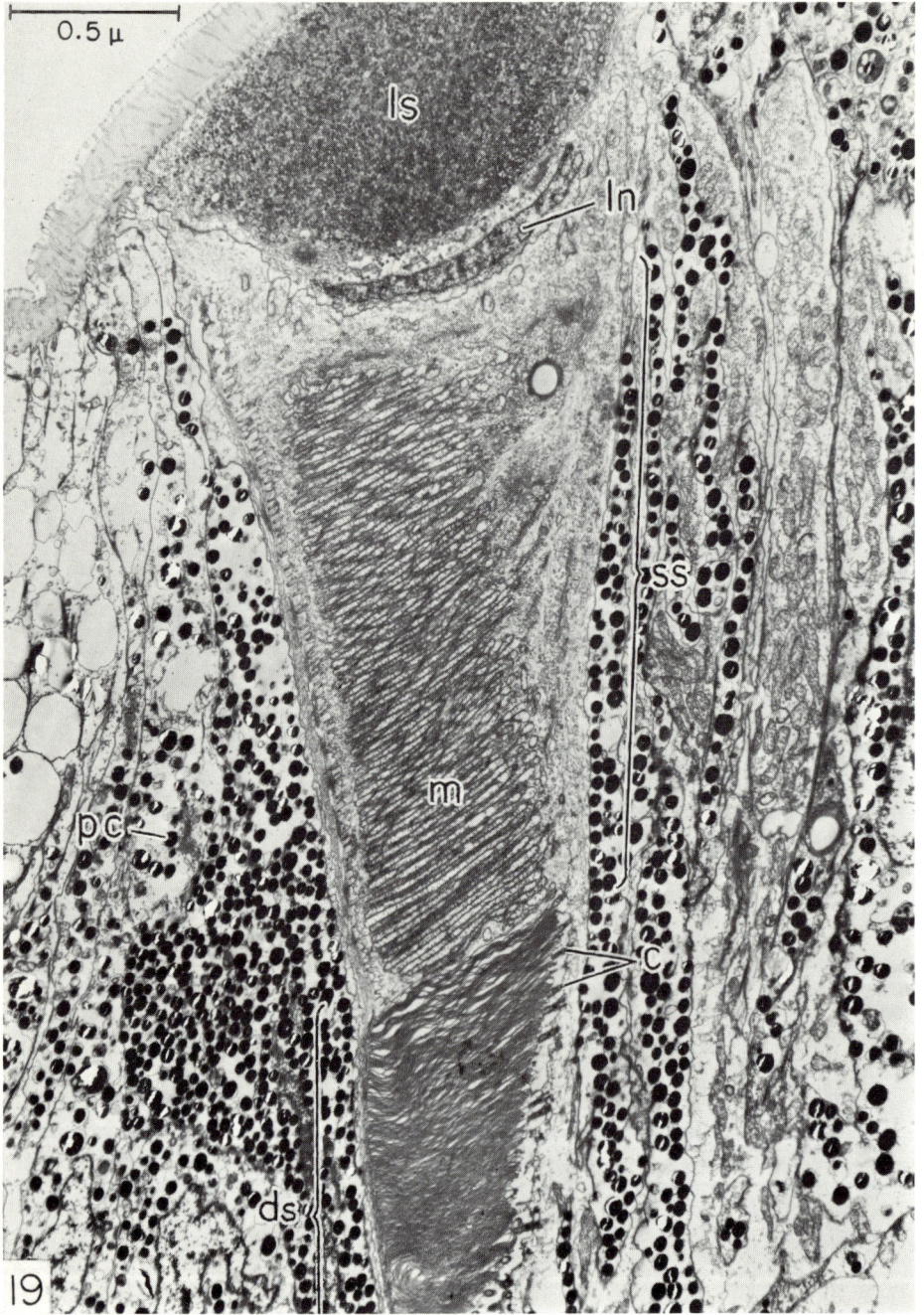

Fig. 19. (Legend on page 227)

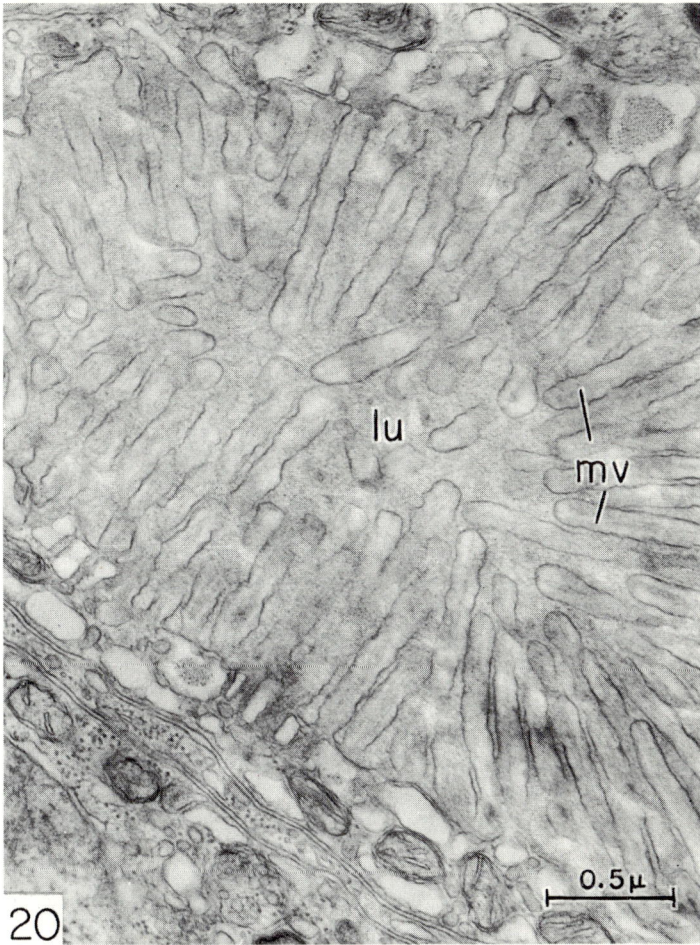

Fig. 20. Photoreceptoral apparatus of sensory cell in eye of leech Helobdella stagnalis, showing many microvilli (mv) projecting into an intracellular cavity (lu). X 35,000. (From Clark, 1967.)

Fig. 19. Longitudinal section through eyespot of B. vesiculosum showing lens (ls), superficial segment (ss) of receptor cell containing mitochondria (m), and deep segment (ds) of receptor cell consisting of stack of ciliary sacs. c, centrioles (kinetosomes) of ciliary receptors; ln, lens cell nucleus; pc, pigmented cell. X 4,500. (From Krasne and Lawrence, 1966.)

1964; Yanase, Fujimoto, and Nishimura, 1964; Clark, 1963, 1967). The photoreceptoral apparatus in this eye is a rhabdomere of microvilli which project into an intracellular (!) cavity (Fig. 20). In the studies just cited only Clark has observed any structures related to cilia, in this instance centrioles near the bases of the microvilli (Fig. 21). Clark (1967) poses the same question that I did upon finding centrioles near the microvilli of the nereid eye: were cilia once present in the embryo? I look forward to the results of his present embryological study bearing not only on this point but upon the origin of the intracellular cavity which encompasses the microvilli.

Onychophora. The small phylum Onychophora consisting of but a few genera, the most famous being *Peripatus,* is of much interest to evolutionists because it possesses characteristics in common with both annelids and arthropods. The presence of cilia, nephridia, a thin flexible cuticle, thick muscular body wall, and unjointed appendages are annelid features, whereas an open circulatory system, heart with ostia, and a pair of append-ages modified for feeding are found in the Arthropoda. Is the eye of an onychophore annelidan or arthropodan?

Peripatus bears a prominent eye at the base of each antenna. The only study to date on its fine structure (Eakin and Westfall, 1965b) reveals long sensory processes extending from the distal face of the retina to the under surface of the lens. Each process is clothed with countless micro-villi (about 800 Å diameter and 2 to 4μ in length) that are straight and thickly set like the bristles of a testtube brush (Figs. 8, 23). At the base of each process we found a rudimentary cilium, sometimes two, enclosed in an extracellular space (Figs. 8, 22, 23). The same picture was observed in three different species, one from New Zealand (*Perpatonder novaezea-landiae*) and two from Panama (*Epiperipatus braziliensis* and *Macro-peripatus geayi*). A study of the development of the eye in one form (*Macroperipatus geayi*) confirmed (Eakin, 1966b) our earlier hypothesis that the embryonic optic vesicle is composed initially of ciliated ectodermal cells, from the distal ends of which a rhabdomere of microvilli differentiate with no apparent relationship to the cilia (Fig. 24). Subsequently, the cilia become recessed or overgrown by the extended distal end of the cell in the formation of the long sensory process. Some lumen about each cilium is likewise engulfed by the sensory cell. This explains the presence of circumciliary spaces about the rudimentary cilia at the bases of the adult sensory processes.

The eye of Peripatus resembles that of an annelid, such as *Nereis,* in several respects. It is cup-shaped. The retina is composed of two kinds of cells intermingled: supportive (pigmented) and sensory. The receptoral processes bear microvilli from all sides, unlike a retinula cell in an arthro-pod which carries microvilli along limited regions of the cell surface. And

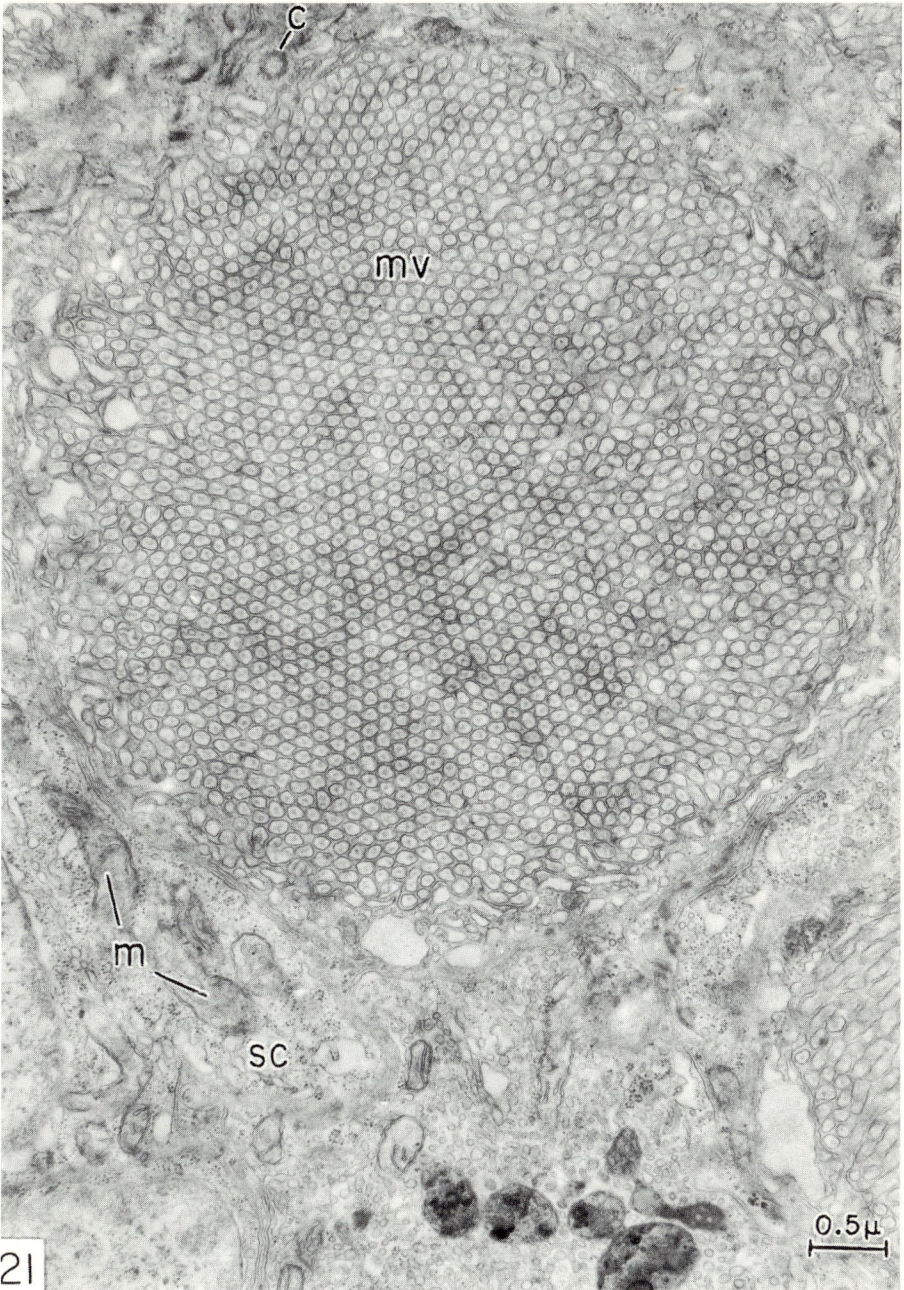

Fig. 21. Cross-sectional view of microvilli (mv) of photoreceptoral cell (sc) of leech
H. stagnalis. Note centriole (c) near microvilli; m, mitochondria. X 21,000. (From
Clark, 1967.)

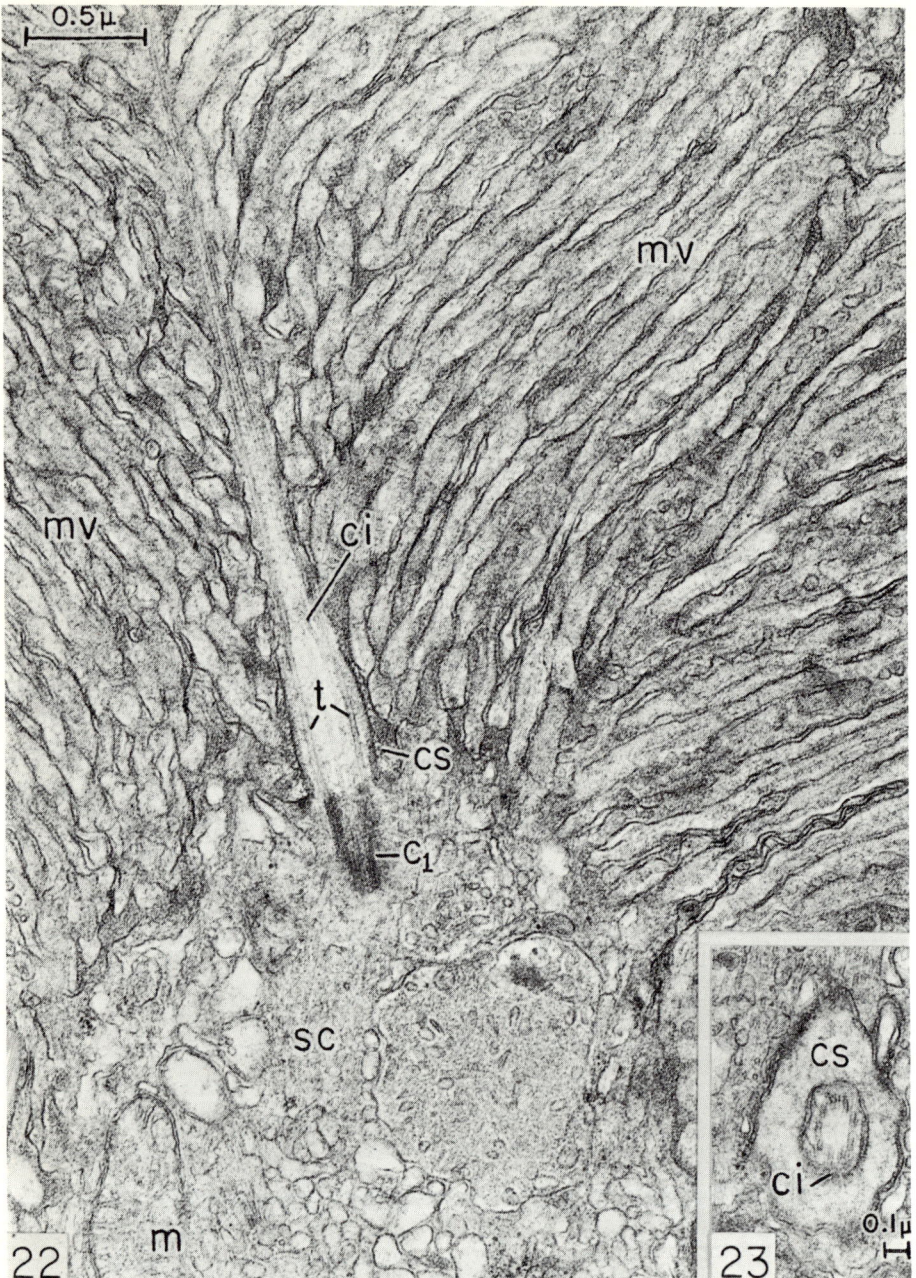

Fig. 22. Distal end of sensory cell (sc) in eye of adult onychophoran *Macroperipatus geayi* bearing many microvilli (mv) and a rudimentary cilium (ci). Note circumciliary space (cs) separates cilium from villi. c_1, axial centriole (kinetosome); m, mitochondrion; t, microtubules. X 33,000.

Fig. 23. Cross section of rudimentary cilium (ci) of sensory cell. cs, circumciliary space. X 35,000.

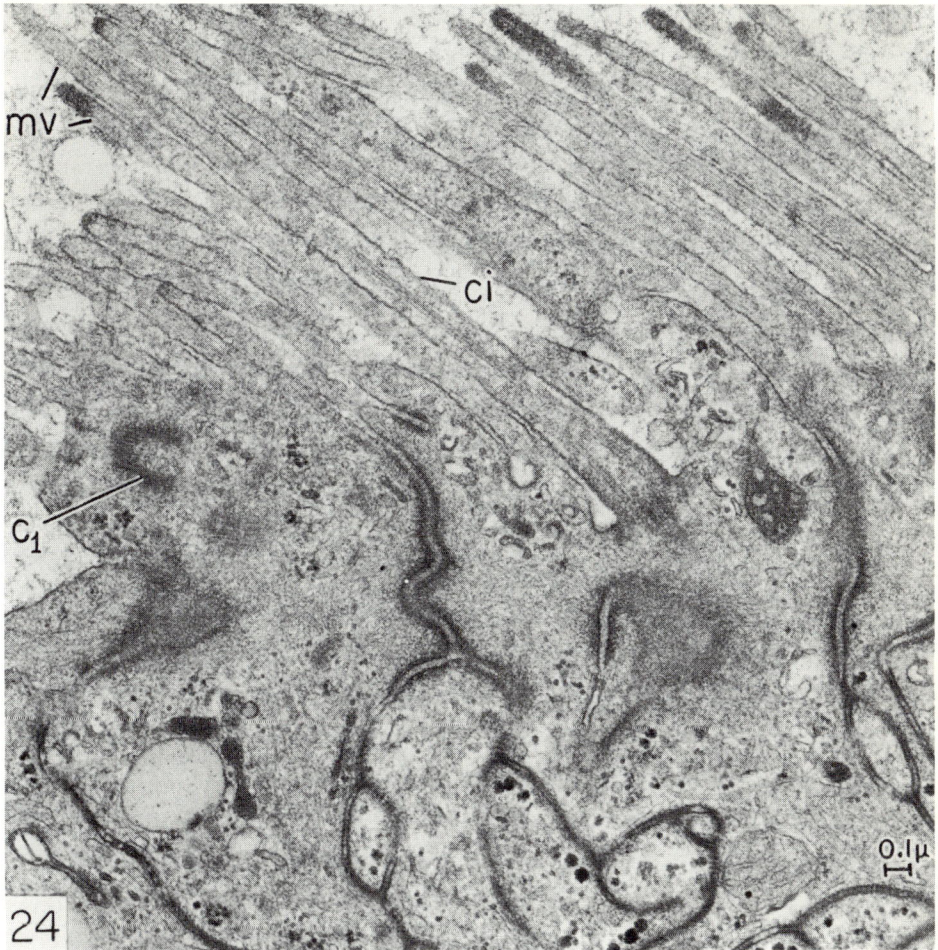

Fig. 24. Distal ends of three sensory cells in embryonic eye of Macroperipatus geayi bearing microvilli (mv) among which occur cilia, one of which is shown (ci). c_1, axial centriole (kinetosome) of another cilium not shown. X 33,000. (From Eakin, 1966b)

the sensory cells possess rudimentary cilia, kinetosomes, and striated root-lets, although the rootlet is poorly differentiated. Ciliary structures have not been observed in arthropod eyes (see below).

On the other hand, there are some features of the eye of Peripatus which are arthropod-like. Most importantly, the microvilli are straight and or-derly arranged, whereas in most annelids, with the exception of leeches, the microvilli appear to be highly irregular and unordered, so much so that good views of them are difficult to obtain. The lens of Peripatus is secreted as are lenses of arthropods, whereas those of annelids are composed of processes of the supportive cells (Eakin and Westfall, 1965b). Some authors (e.g., Hermans and Cloney, 1966) prefer not to use the term lens for the material filling the cavity of the annelid eye in agreement with Hesse (1899) who called it *Füllmasse* in his classic studies of annelid eyes by light microscopy.

Robson (1964) has recently studied the fine structure of the skin of an onychophoran, *Peripatopsis moseleyi*. She remarks in conclusion: "Peripatus emerges as a real arthropod" (p. 297). She finds the cuticle of *Peripatopsis* structurally and chemically similar to those of arthropods, disregarding some specialized adaptive features of the onychophoran skin. I cannot agree, however, on the basis of our studies in which we find similarities to both annelids and arthropods, with the former being more numerous and perhaps more significant than the latter. I conclude that on the basis of the eye of peripatus the phylum status of Onychophora is fully justified.

Arthropoda. At long last we come to journey's end and to the climax of the rhabdomeric trail, the arthropods, where cilia are as yet unknown in photoreceptors, although they have been found in recent years in chemo- and mechanoreceptors (e.g., Slifer and Sekhon, 1963; Thurm, 1966). More electron microscopy has been conducted on the eyes of arthropods than upon photoreceptors of any other group of animals, except the vertebrates.

Crustacean eyes consist of rhabdomeres of straight microvilli (see Fig. 8) projecting from the sides of retinal cells as shown in copepods by Vaissière (1961) and Fahrenbach (1964), in a shrimp by Mayrat (1962), in a crayfish by Eguchi (1965), in a lobster by Rutherford and Horridge (1965), in the stomatopod *Squilla* by Sertorio (personal communication), and in a cladoceran by Wolken and Gallik (1965). Simpler arrays of microvilli have been described in ganglia in barnacles (Fahrenbach, 1965) which Gwilliam (1963) demonstrated neurophysiologically to be light-sensitive. And piles of lamellae, thought to be photoreceptive, have been found in the ventral nerve cord of crayfishes (Hama, 1961; Uchizono, 1962). The interpretation of Hama has been rejected by Hermann and Stark (1963).

Insect eyes have been objects of study by electron microscopy for more than ten years. In all so far examined the photoreceptoral apparatus is a rhabdomere of straight microvilli very similar to those in crustacean eyes. A partial list of the insects investigated is as follows: housefly, the first to be studied (Fernández-Morán, 1956; Khalaf, 1958; Pedler and Goodland, 1965); pomace flies or the Drosophilidae (Danneel and Zeutzschel, 1957; Wolken, Capenos, and Turano, 1957; Wolken, Mellon, and Contis, 1957; Yasuzumi and Deguchi, 1958; Waddington and Perry, 1960); fleshflies

(Goldsmith and Philpott, 1957; Trujillo-Cenóz, 1965); dragonfly (Gold-smith and Philpott, 1957; Ruck and Edwards, 1964); honeybee (Gold-smith, 1962); bristletails (Brandenburg, 1960; Pipa, Nishioka, and Bern, 1964); cockroach (Wolken and Gupta, 1961); cranefly (Sotavalta, Tuurala, and Oura, 1962); and silkworm moth (Eguchi, 1962; Eguchi, Naka, and Kuwabara, 1962). And Fernández-Morán (1958, 1961) has studied a variety of insects.

The basic rhabdomeric pattern of a phalanx of straight microvilli so typical of crustacean and insect eyes, continues in the Arachnida and Xiphosura. Several studies on the eyes of spiders have been conducted (Miller, 1957; Baccetti and Bedini, 1964; Trujillo-Cenóz, 1966; Melamed and Trujillo-Cenóz, 1966), one on scorpion eyes (Bedini, 1965), and one on the eye of the xiphosuran *Limulus* (Miller, 1957).

So, throughout the phylum Arthropoda, rhabdomeres without any trace of ciliary structures are the invariably characteristic sensory apparatus, whether the photoreceptors lie in eyes or within nervous tissue, whether the eye is a simple, unicorneal ocellus or compounded of many ommatidia, whether the rhabdom is closed (rhabdomeres fused) or open (rhabdomeres separated by an ommatidial cavity), or whether the eye is superpositional (refractile elements not in contact with rhabdomeres) or appositional (dioptric structures adjacent to rhabdomeres).

Other Taxa

As stated earlier in this chapter, very few photoreceptors have been studied with the electron microscope, even though the literature here cited seems considerable. With the exception of the vertebrates and the arthropods the number of types examined is exceedingly limited. Many more samples from the major phyla need to be made to ascertain the characteristic type of photo-receptoral organelle in each taxa and the range of variation. And there are a number of smaller phyla along both lines which have not yet been explored at the ultrastructural level for light-sensitive structures. In some instances physiological and behavioral studies indicate that the members of these groups are responsive to light or to shadows, e.g., hemichordates, brachio-pods, ectoprocts, sipunculids, and others (Hyman, 1959). The photosen-sitivity of one of these, the bryozoan *Bugula avicularia,* has been studied by Schneider and his associates (Kaissling and Schneider, 1962; Kaissling, 1963; Schneider and Kaissling, 1964). A preliminary electron microscopical study by R. A. Steinbrecht in Schneider's laboratory revealed no mem-branous or villous structure in the reactive cells of this bryozoan (see dis-cussion at end of Eakin, 1966a). Light sensitivity in this instance may be like that in certain vertebrate tissues (see page 216).

Protostomes versus Deuterostomes

The ciliary and rhabdomeric lines which we have traced are in agreement with the classic concept of two lineages within the coelom-bearing animals: the Protostomia (mollusks, annelids, arthropods) and the Deuterostomia (echinoderms, chordates). The distinguishing features are predominantly embryological. One of them is embodied in the names of these divisions of the eucoelomates. The protostomes (first mouth) have one embryonic opening into the archenteron, the blastopore, that by subdivision forms both mouth and anus. The deuterostomes (second mouth), on the other hand, have two embryonic apertures: the blastopore which becomes the anus, and the stomodeum, formed by an anterior invagination of ectoderm, which establishes the mouth. A second differentiating feature is the mode of formation of coelom: schizocoelous (splitting of mesoderm) in the annelid line, enterocoelous (outpocketing of enteron) in the echinoderm line. The pattern of cleavage is often characteristic: spiral in the former, radial in the latter. A fourth character frequently cited is the fundamental nature of development. The protostomes tend to be determinate and mosaic with regenerative power absent or limited, whereas the deuterostomes are typically indeterminate and regulative, and they often exhibit remarkable ability to regenerate. Finally, two biochemical features may be mentioned: 1) chitin is usually present in protostomes but absent from most deuterostomes, and 2) arginine phosphate is more abundant than creatine phosphate in the protostomatous phyla, whereas the converse is characteristic of the deuterostomatous groups.

But these traditional ways of distinguishing the two major divisions of the eucoelomates are not always reliable. There are exceptions. For example, the blastopore does not invariably transform into the anus. The blastopore may be completely closed and a new anal aperture formed by an invagination of the ectoderm near the obliterated blastopore. Again, not all deuterostomes are enterocoelous. Vertebrates form the body cavity by separation of the mesoderm into somatic and splanchnic layers. It is only because amphioxus, a model of the theoretical ancestral vertebrate, exhibits enterocoelous formation of the coelom that we have a basis for regarding the chordates enterocoelous. The chaetognaths are actually pseudocoelous as adults (i.e., the gut is not clothed with a mesodermal epithelium). The coelom of the embryo, however, is enterocoelous, although the method of its formation differs significantly from that in other deuterostomes. Still again, not all protostomes show spiral cleavage (arthropods, for instance), and some deuterostomes (ascidians, for example) show mosaic development.

In conclusion, characters of evolutionary significance are not infallible guides, even such a conservative one as the mechanics of coelom forma-tion. Nature is extraordinarily versatile, so much so that her diversity (the modifications) often obscure the threads of continuity (the descent). So, the kind of photoreceptor—ciliary or rhabdomeric—is not an indubitable clue to phylogenetic relationships, but I believe that the hypothesis ad-vanced in this essay is useful in our thinking and speculation about the evolution of animals.

Acknowledgments

I am greatly indebted to many persons for assistance in my studies on photorecep-tors and the preparation of this chapter. I single out a few: Dr. Jane A. Westfall and Mrs. Jean L. Brandenburger, research associates; Mrs. Emily E. Reid, artist; and Professors Ralph I. Smith and Timothy H. Goldsmith who gave the manuscript a critical reading. I thank also several authors and publishers for permission to reproduce illustrations from their publications, cited in the explanation of figures. Lastly, I am grateful to the United States Public Health Service for a grant-in-aid of research and to the University of California which nurtured me and supported me for more than thirty years.

References

BACCETTI B., and C. BEDINI. 1964. Research on the structure and physiology of the eyes of a lycosid spider. I. Microscopic and ultramicroscopic structure. Arch. Ital. Biol., 102:97–122.

BARBER, V. C., E. EVANS, and M. LAND. 1967. The fine structure of the eye of the mollusc, *Pecten maximus*. Z. Zellforsch. Z. Zellforsch, 76:295–312.

———, and M. LAND. 1967. Eye of the cockle, *Cardium edule:* Anatomical and physiological investigations. Experientia, 23:677–678.

BARNES, B. G. 1961. Ciliated secretory cells in the pars distalis of the mouse hypophysis. J. Ultrastruct. Res., 5:453–467.

BECKER, H. E., and R. A. CONE. 1966. Light-stimulated electrical responses from skin. Science, 154:1051–1053.

BEDINI, C. 1965. Ultrastruttura degli occhi di uno scorpione *Euscorpius carpathicus* (L.). Annali dell' Istituto Superiore di Sanità, 1:107–110.

BELL, A. 1966. The fine structure of the eye of the scallop *Pecten irradians*. Biol. Bull., 131:385.

BRANDENBURG, J. 1960. Die Feinstruktur des Seitenauges von *Lepisma saccharina* L. Zool. Beitr., N. F., 5:291–300.

BROWN, K. T. 1965. An early potential evoked by light from the pigment epithelium-choroid complex of the eye of the toad. Nature (London), 207:1249–1253.

BURFIELD, S. T. 1927. Sagitta. Proc. Tran. Liverpool Biol. Soc., 41, Memoir 28, 1–104.

CHUGUNOV, IU. D. 1966. СТРОЕНИЕ И ВОЗМОЖНЫЕ ФУНКЦИИ ТЕМЕННЫХ ОРГАНОВ ПОЗВОНОЧНЫХ ЖИВОТНЫХ. (Structure and possible functions of the parietal organs of vertebrate animals). Progress in Contemporary Biol., 62: 264–273.

CLARK, A. W. 1963. Fine structure of two invertebrate photoreceptor cells. J. Cell Biol., 19:14A.

———. 1967. The fine structure of the eye of the leech, *Helobdella stagnalis*. J. Cell Sci., 2:341–348.

CLAYTON, R. K. 1964. Phototaxis in microorganisms. *In* Giese, A. C., Photophysiology, 2:51–77, New York, Academic Press.

COHEN, A. I. 1963. Vertebrate retinal cells and their organization. Biol. Rev., 38:427–459.

COHEN-BAZIRE, G., and R. KUNISAWA. 1963. The fine structure of *Rhodospirillum rubrum*. J. Cell Biol., 16:401–419.

COLLIN, J.-P. 1966. Étude préliminaire des photorécepteurs rudimentaires de l'épiphyse de *Pica pica* L. pendant la vie embryonnaire et postembryonnaire. C. R. Acad. Sci. Paris, 263:660–663.

DANNEEL, R., and B. ZEUTZSCHEL. 1957. Über den Feinbau der Retinula bei *Drosophila melanogaster*. Z. Naturforsch. [B], 12b:580–583.

DHAINAUT-COURTOIS, N. 1965. Sur la présence d'un organe photorécepteur dans le cerveau de *Nereis pelagica* L. (Annélide polychète). C. R. Acad. Sci. Paris, 261:1085–1088.

DILLY, N. 1961. Electron microscope observations of the receptors in the sensory vesicle of the ascidian tadpole. Nature (London), 191:786–787.

————. 1964. Studies on the receptors in the cerebral vesicle of the ascidian tadpole. 2. The ocellus. Quart. J. Micr. Sci., 105:13–20.

DODT, E., and E. HEERD. 1962. Mode of action of pineal nerve fibers in frogs. J. Neurophysiol., 25:405–429.

EAKIN, R. M. 1960. Number of photoreceptors and melanocytes in the third eye of the lizard, *Sceloporus occidentalis*. Anat. Rec., 138:345.

————. 1961. Photoreceptors in the amphibian frontal organ. Proc. Nat. Acad. Sci. USA, 47:1084–1088.

————. 1962. Lines of evolution of photoreceptors. J. Gen. Physiol., 46:357A–367A.

————. 1963. Lines of evolution of photoreceptors. *In* Mazia, D., and A. Tyler General Physiology of Cell Specialization, 393–425, New York, McGraw-Hill.

————. 1964a. The effect of vitamin A deficiency on photoreceptors in the lizard *Sceloporus occidentalis*. Vision Res., 4:17–22.

————. 1964b. Development of the third eye in the lizard *Sceloporus occidentalis*. Revue Suisse Zool., 71:267–285.

————. 1965. Differentiation of rods and cones in total darkness. J. Cell Biol., 25:162–165.

————. 1966a. Evolution of Photoreceptors. Cold Spring Harbor Symposia on Quant. Biol., 30:363–370.

————. 1966b. Differentiation in the embryonic eye of peripatus (Onychophora). 6th Int. Cong. Electron Micr. Kyoto, 2:507–508.

————, and J. L. BRANDENBURGER. 1967. Differentiation in the eye of a pulmonate snail *Helix aspersa*. J. Ultrastruct. Res., 18:391–421.

————, and R. C. STEBBINS. 1959. Parietal eye nerve in the fence lizard. Science, 130:1573–1574.

————, and J. A. WESTFALL. 1960. Further observations on the fine structure of the parietal eye of lizards. J. Biophys. Biochem. Cytol., 8:483–499.

————, and J. A. WESTFALL. 1961. The development of photoreceptors in the stirnorgan of the treefrog, *Hyla regilla*. Embryologia, 6:84–98.

————, and J. A. WESTFALL. 1962a. Fine structure of photoreceptors in amphioxus. J. Ultrastruct. Res., 6:531–539.

————, and J. A. WESTFALL. 1962b. Fine structure of photoreceptors in the hydromedusan, Polyorchis penicillatus. Proc. Nat. Acad. Sci. USA, 48:826–833.

————, and J. A. WESTFALL. 1964a. Further observations on the fine structure of some invertebrate eyes. Z. Zellforsch., 62:310–332.

————, and J. A. WESTFALL. 1964b. Fine structure of the eye of a chaetognath. J. Cell Biol., 21:115–132.

————, and J. A. WESTFALL. 1965a. Ultrastructure of the eye of the rotifer *Asplanchna brightwelli*. J. Ultrastruct. Res., 12:46–62.

————, and J. A. WESTFALL. 1965b. Fine structure of the eye of peripatus (Onychophora). Z. Zellforsch., 68:278–300.

————, W. B. QUAY, and J. A. WESTFALL. 1963. Cytological and cytochemical studies on the frontal and pineal organs of the treefrog, *Hyla regilla.* Z. Zellforsch., 59:663–683.

————, and J. A. WESTFALL, and M. J. DENNIS. 1967. Fine structure of the eye of a nudibranch mollusk *Hermissenda crassicornis.* J. Cell Sci., 2:349–358.

EGUCHI, E. 1962. The fine structure of the eccentric retinula cell in the insect compound eye (*Bombyx mori*). J. Ultrastruct. Res., 7:328–338.

————. 1965. Rhabdom structure and receptor potentials in single crayfish retinular cells. J. Cell Comp. Physiol., 66:411–430.

————, K. Naka, and M. Kuwabara. 1962. The development of the rhabdom and the appearance of the electrical response in the insect eye. J. Gen. Physiol., 46:143–157.

————, and T. H. WATERMAN. 1965. Fine structure patterns in crustacean rhabdoms. Int. Symp. Functional Organization of the Compound Eye, 105–124.

————, and T. H. WATERMAN. 1967. Changes in retinal fine structure induced in the crab *Libinia* by light and dark adaptations. Z. Zellforsch., 79:209–299.

FAHRENBACH, W. H. 1964. The fine structure of a nauplius eye. Z. Zellforsch., 62:182–197.

————. 1965. The micromorphology of some simple photoreceptors. Z. Zellforsch., 66:233–254.

FAURÉ-FREMIET, E., and CH. ROUILLER. 1957. Le flagelle interne d'une Chrysomonadale: *Chromulina psammobia.* C. R. Acad. Sci. [D] (Paris), 244:2655–2657.

FERNÁNDEZ-MORÁN, H. 1956. Fine structure of the insect retinula as revealed by electron microscopy. Nature (London), 177: pt. 2, 742–743.

————. 1958. Fine structure of the light receptors in the compound eyes of insects. Exp. Cell Res., Suppl., 5:586–644.

————. 1961. The fine structure of vertebrate and invertebrate photoreceptors as revealed by low-temperature electron microscopy. *In* Smelser, G. K., The Structure of the Eye, 521–556, New York, Academic Press.

FISCHER, A. 1963. Über den Bau und die hell-dunkel-Adaptation der Augen des Polychäten *Platynereis dumerilii.* Z. Zellforsch., 61:338–353.

————, and J. BRÖKELMANN. 1966. Das Auge von *Platynereis dumerilii* (Polychaeta) sein Feinbau im ontogenetischen und adaptiven Wandel. Z. Zellforsch., 71:217–244.

————, and J. BRÖKELMANN. 1965. Morphology and structural changes of the eye of the *Platynereis dumerilii* (Polychaeta). Structure of the eye, II. Symposium, ed. Johannes W. Rohen, 171–174. Stuttgart, F. K. Schattauer.

FRANZ, V. 1923. Haut, Sinnesorgane und Nervensystem der Akranier. Jen. Z. Naturwiss. Med. Grundlagenforsch, 59:401–526.

GIBBONS, I. R. 1965. Chemical dissection of cilia. Arch. Biol. (Liege), 76:317–352.

GIBBS, S. P. 1960. The fine structure of *Euglena gracilis* with special reference to the chloroplasts and pyrenoids. J. Ultrastruct. Res., 4:127–148.

————, W. R. SISTROM, and P. B. WORDEN. 1965. The photosynthetic apparatus of *Rhodospirillum molischianum.* J. Cell Biol., 26:395–412.

GOLDSMITH, T. H. 1962. Fine structure of the retinulae in the compound eye of the honey-bee. J. Cell Biol., 14:489–494.

————, and D. E. PHILPOTT. 1957. The microstructure of the compound eyes of insects. J. Biophys. Biochem. Cytol., 3:429–440.

GRANICK, S. 1961. The chloroplasts: Inheritance, structure and function. *In* Brachet, J., and A. E. Mirsky, The Cell, 2:489–602, New York, Academic Press.

GRENACHER, H. 1879. Untersuchungen über das Sehorgan der Arthropoden, Insbesondere der Spinnen, Insecten und Crustacean, Göttingen, Vandenhoek and Ruprecht.

————. 1886. Abhandlungen zur vergleichenden Anatomie des Auges. I. Die Retina der Cephalopoden. Abh. Naturforsch. ges. Halle, 16:207–256.

GREUET, C. 1965. Structure fine de l'ocelle d'*Erythropsis pavillardi* Hertwig, Péridinien Warnowiidae Lindemann. C. R. Acad. Sci. Paris, 261:1904–1907.

GRIBAKIN, F. G. 1967. Ultrastructural organization of photoreceptive cells of the complex eye in the honey-bee, *Apis mellifera*. Zh. Evol. Biokh. Fiziol., 3:66–73.

GRIMSTONE, A. V., R. W. HORNE, C. F. A. PANTIN, and E. A. ROBSON. 1958. The fine structure of the mesenteries of the sea-anemone *Metridium senile*. Quart. J. Micr. Sci., 99:523–540.

GWILLIAM, G. F. 1963. The mechanism of the shadow reflex in Cirripedia. I. Electrical activity in the supraesophageal ganglion and ocellar nerve. Biol. Bull., 125:470–485.

HADŽI, J. 1963. The Evolution of the Metazoa, Oxford, Pergamon Press.

HAMA, K. 1961. A photoreceptor-like structure in the ventral nerve cord of the crayfish, *Cambarus virilus*. Anat. Rec., 140:329–336.

HANSEN, K. 1962. Elektronenmikroskopische Untersuchung der Hirudineen-Augen. Zool. Beitr. N.F., 7:83–128.

HARTLINE, H. K. 1938. The discharge of impulses in the optic nerve of Pecten in response to illumination of the eye. J. Cell. Physiol., 11:465–478.

———, H. G. WAGNER, and E. F. MACNICHOL, Jr. 1952. The peripheral origin of nervous activity in the visual system. Cold Spring Harbor Symposia on Quant. Biol., 17:125–141.

HARTMAN, O. 1959. Catalogue of the polychaetous annelids of the world. Part II. Allan Hancock Foundation Publ., No. 23:355–628.

HEERD, E., and E. DODT. 1961. Wellenlängen-Diskriminatoren im Pinealorgan von *Rana temporaria*. Pflueger Arch. Ges. Physiol., 274:33–34.

HERMANN, H., and L. STARK. 1963. Prerequisites for a photoreceptor structure in the crayfish tail ganglion. Anat. Rec., 147:209–217.

HERMANS, C. O., and R. A. CLONEY. 1966. Fine structure of the prostomial eyes of *Armandia brevis*. (Polychaeta: Opheliidae). Z. Zellforsch., 72:583–596.

HESSE, R. 1899. Untersuchungen über die Organe der Lichtempfindung bei niederen Thieren. V. Die Augen der polychäten Anneliden. Z. Wiss. Zool., 65:446–516.

———. 1902. Untersuchungen über die Organe der Lichtempfindung bei niederen Thieren. VIII. Weitere Thatsachen. Allgemeines. Z. Wiss. Zool., 72:565–656.

HORRIDGE, G. A. 1964. Presumed photoreceptive cilia in a ctenophore. Quart. J. Micr. Sci., 105:311–317.

HORSTMANN, E. 1934. Untersuchungen zur Physiologie der Schwimmbewegungen der Scyphomedusen. Pflueger. Arch. Ges. Physiol., 234:406–420.

HYMAN, L. H. 1951. The Invertebrates: Acanthocephala, Aschelminthes, and Entoprocta, III. New York, McGraw-Hill.

———. 1955. The Invertebrates: Echinodermata, IV, New York, McGraw-Hill.

———. 1959. The Invertebrates: Smaller Coelomate Groups, V, New York, Mc-Graw-Hill.

ISSEROFF, H. 1964. Fine structure of the eyespot in the miracidium of *Philophthalmus megalurus* (Cort, 1914). J. Parasit., 50:549–554.

JUNG, D. 1963. Bau und Feinstrucktur der Augen auf dem vorderen und hinteren Saugnapf des Fischegels *Piscicola geometra* L. Zool. Beitr. N.F., 9:121–172.

KAISSLING, K. E. 1963. Die phototropische Reaktion der Zoide von *Bugula avicularia* L. Z. Vergl. Physiol., 46:541–594.

———, and D. SCHNEIDER. 1962. Aktionsspektrum und Intensitätsabhängigkeit des phototrophischen Wachstums von *Bugula avicularia* (Bryozoa) Verhandl. Dtsch. Zool. Ges. (Wien):286–296.

KAWAGUTI, S., and Y. KAMISHIMA. 1964. Electron microscopic structure of iridophores of an echinoid, *Diadema setosum*. Biol. J. Okayama Univ., 10:13–22.

KELLY, D. E. 1962. Pineal organs: photoreception, secretion, and development. Amer. Sci., 50:597–625.

———. 1965. Ultrastructure and development of amphibian pineal organs. *In* Ariëns Kappers, J., and J. P. Schadé, Progress in Brain Research, 10:270–287, Amsterdam, Elsevier Publishing Co.

————. 1967. The circumventricular organs. *In* Haymaker, W., and R. Adams, Histology and General Pathology of the Human Nervous System, Chapter 26, Springfield, Illinois, Charles C Thomas. In press.

————, and S. W. SMITH. 1964. Fine structure of the pineal organs of the adult frog, *Rana pipiens*. J. Cell Biol., 22:653–674.

KERNÉIS, A. 1966. Photorécepteurs du panache de *Dasychone bombyx* (Dalyell), Annélides Polychètes. Morphologie et ultrastructure. C.R. Acad. Sci. Paris, 263:653–656.

KHALAF, K. T. 1958. Electronmicroscopy of the compound eyes of the fly (*Calliphora vicina* (= erythrocephala) R./−D.). Mikroskopie, 13:206–210.

KISHIDA, Y. 1967a. Electron microscope studies on the planarian eye. I. Fine structure of the normal eye. Science Reports of Kanazawa University, 12:75–109.

————. 1967b. Electron microscope studies on the planarian eye. II. Fine structures of the regenerating eye. Science Reports of Kanazawa University, 12:111–142.

KRASNE, F. B., and P. A. LAWRENCE. 1966. Structure of the photoreceptors in the compound eyespots of *Branchiomma vesiculosum*. J. Cell Sci., 1:239–248.

KÜMMEL, G. 1960. Die Feinstruktur des Pigmentbecherocells bei Miracidien von *Fasciola hepatica* L. Zool. Beitr. N. F., 5:345–354.

LAND, M. F. 1965. Image formation by a concave reflector in the eye of the scallop, *Pecten maximus*. J. Physiol., 179:138–153.

————. 1966. Activity in the optic nerve of *Pecten maximus* in response to changes in light intensity, and to pattern and movement in the optical environment. J. Exp. Biol., 45:83–99.

LAWRENCE, P. A., and F. B. KRASNE. 1965. Annelid ciliary photoreceptors. Science, 148:956–966.

LEEDALE, G. F., B. J. D. MEEUSE, and E. G. PRINGSHEIM. 1965. Structure and physiology of *Euglena spirogyra*. I and II, Arch. Mikrobiol., 50:68–102.

MACRAE, E. K. 1964. Observations on the fine structure of photoreceptor cells in the planarian *Dugesia tigrina*. J. Ultrastruct. Res., 10:334–349.

————. 1966. The fine structure of photoreceptors in a marine flatworm. Z. Zellforsch., 75:469–484.

MANGOLD, O. 1931. Das Determinationsproblem. Das Wirbeltierauge in der Entwicklung und Regeneration. Ergebn. Biol., 7:193–403.

MANTON, I., and B. CLARKE. 1956. Observations with the electron microscope on the internal structure of the spermatozoid of Fucus. J. Exp. Bot., 7:416–432.

MAYRAT, A. 1962. Premiers résultats d'une étude au microscope électronique des yeux des crustacés. C. R. Acad. Sci. Paris, 255:766–768.

MELAMED, J., and O. TRUJILLO-CENÓZ. 1966. The fine structure of the visual system of *Lycosa* (Araneae: Lycosidae). Part I. Retina and optic nerve. Z. Zellforsch., 74:12–31.

MILLER, W. H. 1957. Morphology of the ommatidia of the compound eye of Limulus. J. Biophys. Biochem. Cytol., 3:421–428.

————. 1958. Derivatives of cilia in the distal sense cells of the retina of *Pecten*. J. Biophys. Biochem. Cytol., 4:227–228.

————. 1960. Visual photoreceptor structures. *In* Brachet, J., and A. E. Mirsky, The Cell, 4:325–364, New York, Academic Press.

————, and M. L. WOLBARSHT. 1962. Neural activity in the parietal eye of a lizard. Science, 135:316–317.

————. 1965. The anatomy of the neuropile in the compound eye of *Limulus*. Structure of the eye, II. Symposium, ed. Johannes W. Rohen, 159–170. Stuttgart, F. K. Schattauer.

MISSOTTEN, L. 1964. L'ultrastructure des tissus oculaires. Bull. Soc. Belg. Ophthal., 136:1–200.

MOODY, M. F. 1964. Photoreceptor organelles in animals. Biol. Rev., 39:43–86.

————, and J. D. ROBERTSON. 1960. The fine structure of some retinal photoreceptors. J. Biophys. Biochem. Cytol., 7:87–92.

NAKAO, T. 1964. On the fine structure of the amphioxus photoreceptor. Tohoku J. Exp. Med., 82:349–369.

NISHIOKA, R. S., I. R. HAGADORN, and H. A. BERN. 1962. Ultrastructure of the epistellar body of the octopus. Z. Zellforsch., 57:406–421.

NORTH, W. J., and C. F. A. PANTIN. 1958. Sensitivity to light in the sea-anemone *Metridium senile* (L.): adaptation and action spectra. Proc. Roy. Soc. London, Ser. B., 148:385–396.

OKSCHE, A. 1965. Survey of the development and comparative morphology of the pineal organ. *In* Ariëns Kappers, J., and J. P. Schadé, Progress in Brain Research, 10, Structure and Function of the Epiphysis Cerebri, 3–29, Amsterdam, Elsevier Publishing Co.

————, and M. VON HARNACK. 1963. Elektronenmikroskopische Untersuchungen am Stirnorgan von Anuran. (Zur Frage der Lichtrezeptoren), Z. Zellforsch., 59:239–288.

————, and M. VON HARNACK. 1964. Die Elektronenmikroskopische Feinstruktur des Stirnorgans (Epiphysenendblase) der Anuren. *In* Bargmann, W., and J. P. Schadé, Progress in Brain Research, 5, Lectures on the Diencephalon, 209–222, Amsterdam, Elsevier Publishing Co.

————, and H. KIRSCHSTEIN. 1966a. Zur Frage der Sinneszellen im Pinealorgan der Reptilien. Naturwissenschaften, 53:46.

————, and H. KIRSCHSTEIN. 1966b. Elektronenmikroskopische Feinstruktur der Sinneszellen im Pinealorgan von *Phoxinus laevis*. L. (Pisces, Teleostei, Cyprinidae) (Mit vergleichenden Bemerkungen). Naturwissenschaften, 53:591.

————, and M. VAUPEL-VON HARNACK. 1963. Elektronenmikroskopische Untersuchungen an der Epiphysis cerebri von *Rana esculenta* L. Z. Zellforsch., 59:582–614.

————, and M. VAUPEL-VON HARNACK. 1965a. Über rudimentäre Sinneszellstrukturen im Pinealorgan des Hühnchens. Naturwissenschaften, 52:662–663.

————, and M. VAUPEL-VON HARNACK. 1965b. Vergleichende Elektronenmikroskopische Studien am Pinealorgan. *In* Ariëns Kappers, J., and J. P. Schadé, Progress in Brain Research, 10, Structure and Function of the Epiphysis Cerebri, 237–258, Amsterdam, Elsevier Publishing Co.

————, and M. VAUPEL-VON HARNACK. 1966. Elektronenmikroskopische Untersuchungen zur Frage der Sinneszellen im Pinealorgan der Vögel. Z. Zellforsch., 69:41–60.

OLSSON, R. 1962a. The infundibular cells of *Amphioxus* and the question of fiber-forming secretions. Arkiv. Zool. (Stockh.), 15:347–355.

————. 1962b. Lancettfiskens hjärnblåsa och ryggradsdjurshjärnans ursprungliga byggnad. Svensk Naturvetenskap, 346–357.

————, and K. G. WINGSTRAND. 1954. Reissner's fibre and the infundibular organ in Amphioxus—results obtained with Gomori's chrome alum haematoxylin. Univ. Bergen Årbok (Publ. Biol. Stat.), Nr. 14:1–15.

PARKER, G. H. 1908. The sensory reactions of Amphioxus. Proc. Amer. Acad. Arts Sci., 43:415–455.

PEDLER, C., and H. GOODLAND. 1965. The compound eye and first optic ganglion of the fly. A light and electron microscopic study. J. Roy. Micr. Soc., 84:161–179.

PIPA, R. L., R. S. NISHIOKA, and H. A. BERN. 1964. Thysanuran median frontal organ: Its structural resemblance to photoreceptors. Science, 145:829–831.

PITELKA, D. R. 1963. Electron-Microscopic Structure of Protozoa. New York, Pergamon Press.

POND, G. G., and R. M. CABLE. 1966. Fine structure of photoreceptors in three types of ocellate cercariae. J. Parasit., 52:483–493.

PRESS, N. 1959. Electron microscope study of the distal portion of a planarian retinular cell. Biol. Bull., 117:511–517.

PRINGSHEIM, E. G. 1963. Farblose Algen. Stuttgart, Gustav Fischer.

ROBSON, E. A. 1964. The cuticle of *Peripatopsis moseleyi*. Quart. J. Micr. Sci., 105:281–299.

RÖHLICH, P., and I. TÖRÖ. 1965. Fine structure of the compound eye of Daphnia in normal and dark-, and strongly light-adapted state. Structure of the eye, II. Symposium, ed. Johannes W. Rohen, 175–186.

————, and L. J. TÖRÖK. 1961. Elektronenmikroskopische Untersuchungen des Auges von Planarien. Z. Zellforsch., 54:362–381.

————, and L. J. TÖRÖK. 1962. The effect of light and darkness on the fine structure of the retinal clubs of *Dendrocoelum lacteum*. Quart. J. Micr. Sci., 103:543–548.

————, and L. J. TÖRÖK. 1963. Die Feinstruktur des Auges der Weinbergschnecke (*Helix pomatia* L.) Z. Zellforsch., 60:348–368.

————, and L. J. TÖRÖK. 1964. Elektronenmikroskopische Beobachtungen an den Sehzellen des Blutegels, *Hirudo medicinalis* L. Z. Zellforsch., 63:618–635.

ROUILLER, CH., and E. FAURÉ-FREMIET. 1958. Structure fine d'un flagellé chryso-monadien: *Chromulina psammobia*. Exp. Cell Res., 14:47–67.

RUCK, P., and G. A. EDWARDS. 1964. The structure of the insect dorsal ocellus. I. General organization of the ocellus in dragonflies. J. Morph., 115:1–26.

RÜDEBERG, C. 1966. Electron microscopical observations on the pineal organ of the teleosts *Mugil auratus* (Risso) and *Uranoscopus scaber* (Linné), Pubbl. Staz. Zool. Napoli, 35:47–60.

RUTHERFORD, D. J., and G. A. HORRIDGE. 1965. The rhabdom of the lobster eye. Quart. J. Micr. Sci., 106:119–130.

SCHNEIDER, D., and K. E. KAISSLING. 1964. Wachstum und Phototropismus bei Moostieren, Naturwissenschaften, 51:127–134.

SCHWALBACH, G., K. G. LICKFELD, and M. HAHN. 1963. Der mikromorphologische Aufbau des Linsenauges der Weinbergschnecke (*Helix pomatia* L.). Protoplasma, 56:242–273.

SJÖSTRAND, F. S. 1959. The ultrastructure of the retinal receptors of the vertebrate eye. Ergebn. Biol., 21:128–160.

SLIFER, E. H., and S. S. SEKHON. 1963. Sense organs on the antennal flagellum of the small milkweed bug, *Lygaeus kalmii* Stal (Hemiptera, Lygaeidae). J. Morph., 112:165–193.

SMELSER, G. K. 1961. The Structure of the Eye, New York, Academic Press.

SMITH, T. G. JR., and J. E. BROWN. 1966. A photoelectric potential in invertebrate cells. Nature (London), 212:1217–1219.

SOTAVALTA, O., O. TUURALA, and A. OURA. 1962. On the structure and photo-mechanical reactions of the compound eyes of craneflies (Tipulidae, Limmobiidae). Ann. Acad. Sci. Fenn., A, IV, 62:5–13.

SOTELO, J. R., and O. TRUJILLO-CENÓZ. 1958. Electron microscope study of the development of ciliary components of the neural epithelium of the chick embryo. Z. Zellforsch., 49:1–12.

STEBBINS, R. C., and R. M. EAKIN. 1958. The role of the "third eye" in reptilian behavior. Amer. Mus. Novitates, No. 1870:1–40.

STEYN, W. 1960. Electron microscopic observations on the epiphysial sensory cells in lizards and the pineal sensory cell problem. Z. Zellforsch., 51:735–747.

STUBBLEFIELD, E., and B. R. BRINKLEY. 1966. Cilia formation in chinese hamster fibroblasts in vitro as a response to colcemid treatment. J. Cell Biol., 30:645–652.

TALIAFERRO, W. H. 1920. Reactions to light in Planaria maculata with special reference to the function and structure of the eyes. J. Exp. Zool., 31:59–116.

THURM, U. 1966. An insect mechanoreceptor. Part I. Fine structure and adequate stimulus. Cold Spring Harbor Symposia on Quant. Biol., 30:75–82.

TONOSAKI, A. 1966. On the differentiation of plasma membrane in Haliotis visual cells. 6th Int. Cong. Electron Micr. (Kyoto), 2:509–510.

————. 1967. The fine structure of the retina in *Haliotis discus*. Z. Zellforsch., 79:469–481.

TRUJILLO-CENÓZ, O. 1965. Some aspects of the structural organization of the inter-mediate retina of dipterans. J. Ultrastruct. Res., 13:1–33.

————. 1966. Some aspects of the structural organization of the arthropod eye. Cold Spring Harbor Symposia on Quant. Biol., 30:371–382.

UCHIZONO, K. 1962. The structure of possible photoreceptive elements in the sixth abdominal ganglion of the crayfish. J. Cell Biol., 15:151–154.

VAISSIÈRE, R. 1961. Morphologie et histologie comparées des yeux des crustacés copépodes. Arch. Zool. Exp. et Gén., 100:1–125.

VAUPEL-VON HARNACK, M. 1963. Über den Feinbau des Nervensystems des Seesternes (*Asterias rubens* L.) III. Mitteilung. Die Struktur der Augenpolster. Z. Zellforsch., 60:432–451.

WADDINGTON, C. H., and M. M. PERRY. 1960. The ultra-structure of the developing eye of *Drosophila*. Proc. Roy. Soc. [Biol.], 153:155–178.

WAGER, H. 1900, On the eye-spot and flagellum in *Euglena viridis*. J. Linnean Soc., Zool., 27:463–481.

WALD, G., P. K. BROWN, and I. R. GIBBONS. 1963. The problem of visual excitation. J. Opt. Soc. Amer., 53:20–35.

WETZEL, B. K. 1961. Sodium permanganate fixation for electron microscopy. J. Biophys. Biochem. Cytol., 9:711–716.

WHITE, R., and C. D. SUNDEEN. 1967. The effect of light and light deprivation upon the ultrastructure of the larval mosquito eye: I. Polyribosomes and endoplasmic reticulum. J. Exp. Zool., 164:460–478.

WOLKEN, J. J. 1958a. Studies of photoreceptor structures. Ann. N.Y. Acad. Sci., 74:164–181.

———. 1958b. Retinal structure. Mollusc cephalopods: *Octopus, Sepia*. J. Biophys. Biochem. Cytol., 4:835–838.

———. 1963. Structure and molecular organization of retinal photoreceptors. J. Opt. Soc. Amer., 53:1–19.

———, and G. J. GALLIK. 1965. The compound eye of a crustacean, *Leptodora kindtii*. J. Cell Biol., 26:968–973.

———, and P. D. GUPTA. 1961. Photoreceptor structures. The retinal cells of the cockroach eye. IV. *Periplaneta americana* and *Blaberus giganteus*. J. Biophys. Biochem. Cytol., 9:720–724.

———, J. CAPENOS, and A. TURANO. 1957. Photoreceptor structures. III. Drosophila melanogaster. J. Biophys. Biochem. Cytol., 3:441–448.

———, A. D. MELLON, and G. CONTIS. 1957. Photoreceptor structures. II. Drosophila melanogaster. J. Exp. Zool., 134:383–410.

YAMADA, E. 1961. The fine structure of the pigment epithelium in the turtle eye. *In* Smelser, G. K. The Structure of the Eye, 73–84, New York, Academic Press.

YAMAMOTO, T., K. TASAKI, Y. SUGAWARA, and A. TONOSAKI. 1965. Fine structure of the octopus retina. J. Cell Biol., 25:345–359.

YANASE, T., K. FUGIMOTO, and T. NISHIMURA. 1964. The fine structure of the dorsal ocellus of the leech, *Hirudo medicinalis*. Memoirs Osaka Gakugei Univ., B, No. 13:117–119.

———, and S. SAKAMOTO. 1965. Fine structure of the visual cells of the dorsal eye in molluscan, *Onchidium verruculatum*. Zool. Mag., 74:238–242.

YASUZUMI, G., and N. DEGUCHI. 1958. Submicroscopic structure of the compound eye as revealed by electron microscopy. J. Ultrastruct. Res., 1:259–270.

YOSHIDA, M. 1963. A photolabile pigment from the ocelli of *Spirocodon* an anthomedusa. Photochem. Photobiol., 2:39–47.

———, and H. OHTSUKI. 1966. Compound ocellus of a starfish: Its function. Science, 153:197–198.

ZONANA, H. V. 1961. Fine structure of the squid retina. Bull. Hopkins Hosp., 109:185–205.

6

Evolution and Domestication
of the Dog

J. P. SCOTT

Department of Psychology
Bowling Green State University
Bowling Green, Ohio

Introduction 244
Taxonomy and Distribution of the Genus *Canis* 244
Origin of the Dog 245
Further Evidence for the Wolf as Ancestor to the Dog 246
    Comparative Anatomy 246
    Chromosomes 247
    Evidence from Behavior 247
    Admixture of Wild Genes 249
Time of Domestication 249
The Place of Domestication 250
    Archeological Evidence 251
The Process of Domestication 252
    The Process of Primary Socialization in the Dog 253
    Socialization in the Wolf 255
    Similar Ecological Niches of Man and the Wolf 256
    Mutually Understandable Behavior Patterns 257
    Polymorphism in the Wolf 257
    Probable Method of Domestication 258
Evolution of the Wolf 259
Evolution of the Dog 260
Evolution of Behavior in Dogs and Wolves 264
    Social Selection in the Dog and Wolf 265
Theoretical Considerations 268
    Variation and Selection 268

Group Survival and Individual Survival 269
The Origin of Variation in the Dog 270
Synergistic Relationship between Genetic Variation and Complexity
of Social Organization 271
Genetics and the Evolution of Human Social Organization 272
References 274

Introduction

We ordinarily think of the dog as an animal whose biological history has been determined by artificial selection enforced by human masters, whereas evolution is usually considered a process which takes place without human direction. However, for the greater part of the history of this species, which now can be authentically timed as beginning at least 10,000 to 12,000 years ago, dogs were probably not subject to the conscious selection practised by dog breeders within the last century. Even within this latter period the direction of selection has changed many times. Furthermore, most of the main varieties of dogs were not produced by scientific breeding, but had their origin in the remote historical past as local varieties. It seems most likely that most of these local strains were produced largely by accident and were only later recognized as having valuable special characteristics, after which the variety would be spread over a larger area by travellers and traders.

The dog is, therefore, not in any large degree a conscious product of human ingenuity. Rather, it has evolved under the influence of countless thousands of interactions with human masters. We can therefore think of the dog as a species which, on domestication, entered a new habitat and underwent a process of adaptive radiation similar to that of a wild species entering a vacant ecological niche. It subsequently underwent further modification and diversification as it became divided into small local populations and selection pressure became relaxed in certain directions and increased in others under the influence of the human social environment.

Taxonomy and Distribution of the Genus Canis

Linnaeus placed all breeds of domestic dogs in one species, *Canis familiaris,* and modern scientific opinion supports this view. He placed the wolf, *Canis lupus,* in the same genus. Although wolves have in historical times been distributed all over Eurasia and much of North America, they, too, are considered one species (Lawrence, 1967). Other members of the genus are the coyotes of North America (*C. latrans*), and the jackals of the eastern hemisphere.

The black-backed jackal, *C. mesomelas,* is an entirely African species.

The golden jackal, *C. aureus,* lives today in North Africa and Southern Asia, and was once found in Southeastern Europe. Jackals are sufficiently different from other members of the genus that they are sometimes placed in a separate genus, *Thos.* Jackals and coyotes occupy somewhat similar ecological niches on different continents, being plains or desert dwellers living on small game and carrion.

When European explorers came to Australia they found a wild canid, the dingo, which is sometimes temporarily domesticated by the aborigines. It appeared to be a feral domestic dog. Chiefly because of its long isolation from other domestic dogs, it is placed in a separate species, *C. dingo.*

Wolves, coyotes and jackals have a long separate history, being found as bones or fossils as far back as Pliocene times (Matthew, 1930). Other members of the family Canidae are still more remotely related to domestic dogs. The so-called "wild dogs," such as the dhole of India (*Cuon*) and the Cape hunting dog of Africa (*Lycaon*), trace back to a common ancestor with wolves in the Oligocene epoch. The family includes such diverse forms as the foxes (*Vulpes*) and the peculiar "raccoon dog" (*Nyctereutes*) of Europe, as well as various tropical and desert families. In addition to being different from the genus *Canis* in form, other genera vary widely in chromosome numbers (Matthey, 1954).

Origin of the Dog

One of the principal problems of the evolutionary history of the dog is its origin, and three principal theories have been advanced. One of these, popular in the early part of this century (Allen, 1920), is that the dog was domesticated from a wild species which later became extinct. Since no trace of such a wild dog species has ever been found, this hypothesis is no longer taken seriously. Darwin (1859) thought that the dog must have been derived from at least two species, the wolf and the golden jackal, in order to account for the great variation between breeds and individuals. Darwin was working from an assumption of blending inheritance, however, and was ignorant of the degree of variation which is possible through mutation and Mendelian segregation. Early naturalists thought that dogs of the North American Indians might have been domesticated coyotes, on the basis of superficial resemblances in size and color, but anatomical studies on the bones of the Amerind dogs show them to be clearly identical with the European dog breeds. Lorenz (1955) recently revived the theory of the dual origin of the dog, but has since changed his opinion on the basis of behavioral evidence.

This leaves only one tenable theory for the origin of the dog, that it was domesticated from a local variety of small wolf. This immediately raises the question of whether this took place only once or whether dogs were domesticated on several different occasions. As will be seen later, most of the evidence is in favor of a single domestication, although there may have been some subsequent admixture of genes from wild wolf populations on later occasions.

Further Evidence for the Wolf as Ancestor to the Dog

Comparative Anatomy

All members of the genus *Canis* have very similar body proportions, being large-chested, slim-waisted, and long-legged animals. Therefore the body skeleton is of little use in determining the species to which a specimen belongs, and in any case, these bones are frequently not well preserved. The part of the skeleton which is most often preserved is the skull, and especially the teeth, which are the hardest bones of all. Therefore, a great deal of work has been done on tooth characteristics, and one of the outstanding characteristics of the large northern wolves is the size of the teeth in relation to the skull. Some breeds of dogs, such as the St. Bernards, have skulls as wide as those of wolves, and the Irish wolfhound has an even longer skull. Their teeth, however, are considerably smaller (Wagner, 1930; Scott and Fuller, 1965).

The situation is different with respect to the smaller Indian wolf, whose tooth sizes may overlap with that of dogs (Lawrence, 1967). One of the basic characteristics of dogs is, therefore, a set of relatively small teeth. In addition to the genetic heritage of the original ancestor, there must have been, in many cases, a tendency for dog owners to select those animals having smaller teeth and a less fearsome appearance.

It is also reported that wolves bred in captivity tend to show changes in their skull shape, noticeably toward shortening the jaw. Since the teeth are not similarly reduced in size, this results in overlapping of molar teeth. Presumably some of the same effect should be observed in domestic dogs which are normally reared in captivity, but whose teeth are always smaller than those of wolves. These results with alteration of wolf jaws in captivity were obtained from old data gathered before the modern science of nutrition was developed, and it would be interesting to see how wolves reared on modern dog food would compare with them and, indeed, whether modern nutrition has produced differences in the skeletal growth pattern of dogs themselves.

The dog breeds themselves vary considerably in the straightness or curvature of the upper jaw and the crowding of the teeth. For example, in a group of beagles, the sum of the widths of the molar teeth averaged 60 mm whereas the average longest distance from front to back of the jaw was only 57 mm. On the other hand, a group of Shetland sheep dogs tended to have straight jaws with wide spaces between the teeth, with the result that the sum of the tooth measurements added up to only 56 mm compared to 62 mm jaw length (Scott and Fuller, 1965).

One of the difficulties with this kind of evidence is that typological methods do not give clear-cut results, as in almost any measurement there is overlap between populations. One possible answer is that of discriminate-function analysis, in which many measurements are considered at the same time, and are based on entire populations. Using such methods Lawrence and Bossert (1967) have concluded that the red wolf is not a distinct species from *Canis lupus,* but these methods have not yet been applied to the problem of distinguishing the different dog breed populations from populations of wild canids.

Chromosomes

Typological methods are still being used with comparative studies of chromosomes, most authors collecting one or two specimens from a population and drawing conclusions accordingly. Thus Benirschke and Low (1965) reported results from two coyotes, a male and a female, and concluded that the chromosome number was the same as that of the dog (39 pairs), and that the total karyotypes were indistinguishable from those of dogs.

The modern methods of preparing chromosomes have been relatively little used on dogs, but all reports confirm the number established earlier by less adequate techniques, namely 39 pairs (Reiter et al., 1963; Borgaonakar et al., 1967). An early report by Ahmed (1941) indicated that breed differences, again based on small numbers of animals, existed in chromosome shape. Further work should be done with modern techniques on larger populations, as it would appear quite likely that chromosome anomalies might be associated with some of the gross abnormalities of physique found in certain breeds.

Fertile hybrids have been reported between dogs and all three of the closely related species, wolves, coyotes, and jackals. This last report is inconsistent with Matthey's (1954) finding that the yellow jackal has only 37 pairs of chromosomes. On the basis of more recent evidence, Matthey is inclined to think that he was mistaken. It now appears that all members of the genus *Canis* have a diploid chromosome number $=78$, but this conclusion will obviously be more firm when the chromosome complements of these species have been studied on a population basis.

Evidence from Behavior

Intensive comparative studies of the behavioral patterns of dogs and wolves reveal close resemblances in all patterns observed in both species, and very few patterns that have not been found in both (Scott and Fuller,

1965). The behavioral evidence thus indicates that dogs and wolves are closely related. Similar detailed studies have not yet been done with the coyote and jackal, but such evidence as exists indicates differences from both dogs and wolves. For example, one of the outstanding characteristics of wolves and dogs is their highly social nature. Wolves typically run in packs, which may contain as many as 20 or 25 individuals in the case of packs habitually hunting large herd animals such as moose, and the "lone wolf" is an extreme rarity. Both coyotes and jackals, on the other hand, seldom run in packs, the typical social group being a mated pair with a litter which breaks up when the animals become mature. Jackals are sometimes seen in large numbers around garbage dumps, but these animals do not belong to the same social group.

The evidence is even more clear with respect to patterns of vocalization. Both wolves and dogs show similar patterns of barking and howling. The bark of a wolf is primarily an alarm signal which is readily elicited when a strange animal comes into the territory around the den, whereas the clear howl is used either in chorus with a group of animals or is given by individual animals who become separated on a hunt and howl back and forth.

From hunter's reports, Darwin thought that wolves did not bark, and if they did, learned it from domestic dogs. However, all observers of wolves agree that they bark, no matter what their proximity to or isolation from domestic dogs. The tradition probably got started because hunters relatively rarely came upon the small territory around a wolf den, but frequently heard wolves howling on the hunt.

Coyotes show a typical pattern of vocalization in which barks, howls, yips, and other vocalizations are mingled together in varying pitches and loudness, with a ventriloqual effect. The result is that two animals will sound to the uninitiated listener like a pack of forty. Nothing like this is ever heard in dogs or wolves.

Jackals also make more elaborate noises than dogs or wolves. The African black-backed jackal is reported to make a sound like "ke-ke-ke-kek" when cornered, and in the mating season the female makes a sound described as "a hearty laugh." Jackals howl on occasion, but also make peculiar noises when hunting (Van der Merwe, 1953).

If there is any possibility that jackals have been crossed with dogs, the most likely case would be that of the African basenji. These animals make vocalizations which are distinctly different from other dog breeds. They are supposed by breeders' standards to be barkless, but occasionally do bark, both those in Africa and the improved breed in Europe and this country. Basenjis howl like other dogs and also produce a crowing noise which appears to be a modified and extended bark. Whether or not there is any possibility that basenjis may have been produced by a jackal cross will await further detailed studies and comparison with jackals.

Admixture of Wild Genes

One indication of the common ancestry of our domestic breeds is the universal occurrence in all dogs of a sickle-shaped or curly tail, contrasted with the drooping tail carriage of wild canids. We can hypothesize an early mutation which was preserved because it was useful in distinguishing wild from domestic animals, and that this has been maintained with some variation in all domestic dogs. This does not, however, eliminate the possibility that genes from wild populations may have been introduced from time to time. Degerbøl (1927) points out that the northern dog breeds such as the Siberian Laika, the Scandinavian gray deerhound, and the Esquimo dogs all have relatively large teeth and that this may have resulted from crossing with the northern wolves. Northern travellers frequently bring back stories of such hybrids. It is also possible that the large teeth of northern dogs resulted from accidental selection of animals whose teeth were more competent to deal with large bones. Whether or not there has been the possibility of crossing with jackals to produce the southern variety of dogs is another problem, and one for which we may never find any final answer. In any case, admixture of genes from wild species is likely to have taken place only on the fringes of the species and is unlikely to have produced any major shift in gene frequencies in the total gene pool.

Time of Domestication

Until relatively recently the origin of domestic animals has been only a minor problem for archeologists, who were primarily concerned with human prehistory and origins. Animal bones were often collected in connection with human remains, but were frequently neglected or carelessly classified. Furthermore, the early workers concerned with this problem were dominated by typological techniques so that each new dog skeleton was considered a separate subspecies of *Canis familiaris*. Dahr (1937) was one of the first workers to attack the problem from a population standpoint, and he pointed out that the Stone Age dogs of Europe, far from being specialized breeds, could all be included in one population showing considerably less variation than modern dogs.

For many years the earliest known remains of the Stone Age dogs were those described by Degerbøl (1927) from the Danish kitchen middens. By methods then in use it was estimated that these bones were deposited between 8,000 and 10,000 BC. Since then the more accurate carbon dating method has been developed. With this process these same remains have been dated at 6810 ± 70 BC. Thus the Denmark dogs are probably not older than 9,000 BP (Degerbøl, 1961).

Canine remains were also found at the earlier site of Star Carr, in Yorkshire, England. Degerbøl (1961) has concluded that a skull fragment found at this hunting camp is that of a true dog. Carbon-dating methods place this settlement at 7538 ± 350 BC, so that this dog may have lived as recently as 7200 BC, or as early as 7900 BC, or approximately 10,000 years ago. A contemporary specimen from Turkey has been dated at approximately 7000 BC (Lawrence, 1967), and until recently the timing of these finds was strong evidence that the first domestication of the dog occurred in Europe or Asia.

However, the oldest known dog remains that have been definitely identified come from a specimen in a cave in the Beaverhead mountains in Idaho (Lawrence, 1967). Carbon-14 dates place this specimen at the latest 8300 BC and at the earliest 9500 BC. Similar carbon dating of the earliest human remains in North and South America indicates that a simple hunting culture became established between 14,000 and 9000 BC (Johnson, 1967). Some of these early settlers probably brought the dog with them, which would mean that domestication must have taken place at an earlier date, perhaps around 10,000 BC (Lawrence, 1967). Dogs have thus been domesticated for at least 10,000 years and possibly as long as 12,000.

These conclusions are based on the best evidence available at the moment. While it seems unlikely that the above estimates of timing will be upset by future discoveries, it should be remembered that most parts of the world have been explored by archeologists in only a superficial way, and particularly so with respect to animal remains. Future discoveries should not only pinpoint the time of domestication more closely, but also give more accurate indications of place.

The Place of Domestication

The possible areas in which dogs could have been domesticated are limited by the distribution of the wild species which could have been the possible ancestors. The wolf has an almost world wide distribution in the northern hemisphere, but has never been found south of the Equator. This means that South America and much of Central America can be eliminated, as well as Africa, Australia, and South East Asia, as the known geographical ranges of wolves do not enter into these areas. North America and parts of Eurasia are therefore the only possible centers of origin.

Dogs, on the other hand, were found in historical times associated with man on every continent except Australia, where the wild dingo was found.

According to all anatomical evidence, this species is descended from a dog which escaped from domestication and was able to prosper on the new continent with no competition from other wild species except the native marsupial predators. The distribution of dogs therefore gives no clue as to their point of origin.

Archeological Evidence

New finds are continually being reported from all over the world, and since 1958 it has been possible to date these relatively accurately with the Carbon-14 method. There are many areas of the world which have still not been thoroughly explored, particularly in China and other parts of the Far East, and it is therefore possible that conclusions drawn from the present evidence may be radically changed in the future. As recently as 1948, Haag estimated that the earliest dog remains in the Western Hemisphere were as recent as 1500 BC, but this date has now been pushed back to at least 8300 BC, slightly older than the earliest known dog remains from Europe and closely resembling them. However, it is unlikely that these North American animals represent the original ancestors, on two counts. One is that, as far as is known, the early human migrations were all out of Asia into the empty continent of North America, and not vice versa. Second, and more convincingly, the early American dogs were all small or medium sized animals, quite unlike the large North American wolves. Although there was a smaller variety of wolf in the southeastern part of the United States, formerly classified as a separate species (*Canis niger*), Lawrence (1967) reports that there is no overlap in tooth size between these forms and domestic dogs.

Otherwise, the oldest remains of true dogs are found in western Europe, in England and Denmark respectively. Again, it is not likely that these dogs were domesticated locally, as the northern wolves of Europe are large in size and have very large teeth compared to those of dogs.

In the Near East and Southern Asia, there are two subspecies of small wolves, the Indian wolf *Canis lupus pallipes,* and the Arabian wolf *Canis lupus arabs.* Both of these varieties show close resemblances to dogs, and Lawrence suspects that *arabs* may actually be a hybrid between domestic dogs and the Indian wolf. This area in the Near and Middle East therefore seems to be the most likely center for the original domestication of the dog.

The dog remains that have been located in the Near and Middle East are all associated with agricultural communities (Reed, 1959). Currently, the earliest known remains from this general area of Asia come from a Turkish site which has been dated around 7000 BC. Another set of remains were found in Jericho, dated at approximately 6500 BC, and in the contemporary farming village of Jarmo several canine clay figurines with curly tails indicate that the artists were acquainted with dogs. However, judging from the association of dogs with the prehistoric hunters of Star Carr, the dog was probably domesticated by hunters and food gatherers before the beginning of the agricultural revolution.

This does not eliminate the Far East as a possible center of origin. According to Lawrence (1967), there was a now extinct subspecies of wolf in China which closely resembled the Indian wolf and which could have been a possible ancestor, and there are also a few reports of very early dog remains in Japan which have not yet been thoroughly studied. However, on the basis of present evidence it is most likely that the dog was first domesticated somewhere in the vicinity of the Near East or Central Europe and spread out in all directions from this point.

The Process of Domestication

All domestic mammals and birds are highly social animals, with the possible exception of cats, which have a tendency to become attached to places rather than people and so arrive at domestication by a somewhat different road. A correspondingly high degree of sociality is found in the wild species from which they were derived. Among birds, mallard ducks, Indian jungle fowl, wild geese, and wild turkeys are all animals that constantly live in social groups. The wild ancestors of some of the herd animals, such as horses and cattle, are now extinct, but all information that we have indicates that they lived in herds like their domestic descendents. The wild species of pigs, sheep, and goats are highly social, and wolves are no exception to the rule. Even the European rabbit, the presumed ancestor of our domestic forms, is a group-living animal and much more social than the North American forms which have never been domesticated (Hale, 1962).

One characteristic of these highly social mammals and birds is a short period in early life in which social relationships can be readily formed, normally with the members of the same species, but also with other species with whom they come into contact. Most of the domestic animals readily become attached to human beings as well as their own species during this critical period of primary socialization.

This phenomenon was first noticed scientifically in young chicks and its general importance in birds was first appreciated by Lorenz (1937), who gave it the name of imprinting. A similar phenomenon takes place in the herd animals like sheep and goats, where the mother forms a more specific attachment to her young than vice versa (Hersher et al., 1963), and the process has been extensively studied in the dog.

The Process of Primary Socialization in the Dog

Reproduction and the development of behavior in the dog and wolf is obviously adapted to the life of a hunting animal. The period of pregnancy is short, some 63 days, and implantation is delayed so that the fetus does not begin rapid growth until pregnancy is halfway along. The pups are born in a small and immature state, weighing perhaps half a pound each at birth in a medium sized dog. The size of the litter tends to be relatively small, averaging 4 or 5, although litters may be much larger in some of the big domestic breeds. The result is that the female is handicapped in hunting activity for only a relatively short period toward the end of pregnancy. Although not particularly useful to a household pet, this has obvious adaptive value for a wolf or a hunting dog.

When the pups are born the mother takes constant care of them during the first few days, when nursing is being established, and then begins to leave them for longer and longer periods. In the wild species the mother during this period can obtain food readily from meat cached around the den or brought back by other members of the pack. When the pups are approximately three weeks of age the mother supplements the milk with regurgitated food from her own stomach, and the pups are usually weaned between 7 and 10 weeks of age. In wolves, adults of both sexes feed the pups in this way, whether or not they are the actual parents. While the pups at this age are by no means mature themselves, the mother is left free to leave them for long periods.

Behavioral development in dogs and wolves has evolved in two directions. During the first two weeks of life (the neonatal period), behavior patterns are adapted for neonatal nutrition, or nursing, and for an existence in which all care and protection is provided by the mother. Even urination and defecation are induced in a reflex fashion by the licking of the mother, an adaptation which has the effect of keeping the den clean. Otherwise, their social behavior is limited to distress vocalization, or care-soliciting behavior, and a slow crawl, throwing the head from side to side, which is a primitive form of investigative behavior.

The pups are both blind and deaf and thus have relatively little sensory contact with the outside world. Puppies in this period are slow learners and require many more repetitions of experience to form fixed habits than do older dogs.

The neonatal patterns of behavior are so different from those of an adult dog that one would have great difficulty in recognizing puppies as members of the same species from their behavior alone. A similar phenomenon is, of course, well known in those insects whose larvae and adults live in different habitats, such as the caterpillar which metamorphoses into an adult butterfly.

In dogs, the first overt change in behavior takes place on the average at about two weeks, when the eyes open and the pups first begin to crawl backward as well as forward. Within the next five or six days, comprising the transition period in development, the puppy changes quickly but not entirely from the neonatal to the adult forms of behavior. In sensory capacities, both the ears and eyes open and become at least partially functional. In motor development, the puppies stand and run, begin to cut their teeth and, associated with this, begin to mouth and chew objects as well as to suck them. By 19 or 20 days of age they become capable of being conditioned at a rate similar to that of an adult. Most significantly, they begin to show many adult patterns of social behavior in immature form, such as playful fighting, running in groups, and even immature sexual behavior.

At this time the pups also become capable of rapidly forming social attachments. The first indication that this has taken place is the response of distress vocalization when isolated in the home pen, indicating that the puppies notice the absence of familiar individuals (Scott and Bronson, 1964). At the same time they begin to react with distress to being placed in strange surroundings, indicating that an attachment has been formed to particular places.

From approximately 3 until 12 weeks of age is the critical or sensitive period for rapid establishment of social relationships, with a peak in this phenomenon between 6 and 8 weeks of age. Various isolation experiments indicate that the period is brought to a close by the development of a fear response to strange individuals, beginning about seven weeks of age. Up until this time puppies ordinarily give only momentary fear responses to strange individuals, and their chief emotional response is distress vocalization caused by the absence of the familiar.

If a puppy is taken from the litter at the beginning of the period of socialization and raised entirely by human beings from this point onward, he will develop all social relations with human beings and will later show very few responses to other dogs other than those of fear and antagonism. If, on the other hand, he is reared exclusively with dogs for the entire period of socialization, i.e., as long as 14 weeks of age, he will develop social attachments only with dogs and will respond to human beings in a fearful fashion, forming attachments only with great difficulty. Finally, if a puppy is left with the litter and removed to human companionship between 6 and 8 weeks of age, he has had the opportunity to form relationships with dogs and still capable of doing this easily with people. The general result is an animal which is considered a normal dog, being a part of both dog and human society.

Socialization in the Wolf

Numerous examples of persons who have reared wolf cubs and made pets of them are present in the scientific literature, from the day of Buffon (1804) down to the present (Fentress, 1967). Unless the cubs are obtained from a zoo it is impossible to determine their exact age, but as far as it is known, the course of development runs the same time sequence as that in the dog. If the wolf cub is obtained about the time the eyes open it can be raised like a puppy and grows up looking and acting very much like a large dog. With few exceptions the end result is an animal that becomes very much attached to people and behaves in a fashion indistinguishable from that of a normally reared pet dog, with the exception that wolves appear to be more reactive to slight noises and sudden movements. The owners are able to take the animals on the streets without attracting any attention, so dog-like is their appearance and behavior.

Wolf cubs appear to develop strong fear reactions sooner than puppies, which usually discourages people from adopting them at later ages than the period when the eyes are just open. However, Woolpy and Ginsburg (1967) have been able to socialize wild-caught adult wolves by patient and careful contact over a long period of time. The essentials of the method are confinement and passive contact. The wolf is penned in a solitary cage, and the experimenter enters and remains quietly for a long period each day. At first the wolf shows every sign of extreme fear, urinating, defecating, salivating, and attempting to climb the walls to escape. He gradually becomes more quiet and after a period of several weeks may finally make a positive approach to the experimenter. If the latter reacts appropriately at this time, neither frightening the wolf nor acting fearful himself, the wolf will establish a tolerant and friendly relationship which is extended to other human beings as well. This relationship, although amicable and tolerant, is different from that which a wolf cub develops with a human being, especially in that the wolf does not occupy a subordinate position. The relationship is also different from that which it enjoys with other wolves.

These results indicate that the process of forming an emotional attachment can take place at any age, but that it requires a much longer period than in infancy, because of the length of time required to overcome the interfering fear responses. The ability of the wolf to adapt to people also has some significance in connection with possible contacts between wild wolves and prehistoric human hunters and food gatherers.

Similar Ecological Niches of Man and the Wolf

In Eurasia and North America wolves occupy the ecological niche of hunters of the large herd animals, and the size of the packs seems to be partially correlated with the size of the prey, very large packs developing when the prey is moose, and smaller ones for mountain sheep, deer, or caribou. However, wolves do not confine their diet to these prey animals and at different seasons of the year will eat mammals as small as mice and a certain amount of vegetable material such as berries. They will also eat carrion. They thus show considerable adaptibility in their food habits, although perhaps not as great as that of man.

However, as hunters of the large herd mammals, wolves occupy the same ecological niche as early man, and, indeed, must have been active competitors in prehistoric Eurasia and North America. We have, therefore, the theoretically interesting situation of two different species each occupying the niche of a dominant predator. We have no evidence of what actually happened when the two species came in contact except from historical records of hunters in North America a century or so ago. There are frequent records of wolves hanging around hunting camps and eating the remains of animals killed by the hunters. There are also many records of wolves following hunters, who sometimes became very frightened. Investigating all of these cases, Young and Goldman (1944) were able to find only one authentic case of a North American wolf attacking a human being. Like most carnivores, wolves appear to be highly traditional concerning their prey species, although they adapt very well to domestic live stock which are similar to the wild prey animals.

The situation may be somewhat different in Eurasian wolves, where there are not only numerous legends of man-killing wolves, but also, according to Pulliainen (1967, private communication), authentic cases of wolf attacks on human beings in contemporary Finland and Russia. Further information is obviously needed, but it looks as if the North American and Eurasian wolves may be either genetically or culturally different from each other.

The point here is that an early hunting society in Eurasia would easily come into contact with wolves, and that in some circumstances wolves are capable of forming a tolerant relationship with people if this is permitted, especially if there is sufficient food for both.

It is obvious that stone age hunters were not able to exterminate wolves in North America. Indeed, even with modern weapons and poisoning techniques, modern men have only been able to push wolves back toward the Arctic, assuming the position of sole dominant predator only in areas where domestic stock are reared. In the long period before the domestication of herd animals, when the two species were engaged in similar occupations in the same area, the usefulness of a domestic wolf in hunting would become immediately apparent.

Mutually Understandable Behavior Patterns

Besides being highly sociable and occupying similar ecological niches under certain conditions, human beings and both dogs and wolves have other points in common, especially the existence of behavior patterns which are mutually recognizable. With respect to agonistic behavior, it is easy for a human being to recognize the intent of a growling and snarling dog, and it is easy for a dog to understand a shouting and threatening human being. Living in groups as they do, both species have evolved the capacity to develop dominance-subordination relationships, with the result that individuals can learn to live in close association without destructive violence.

Even more striking are the resemblances in allelomimetic behavior and the motivation connected with it. Both people and dogs are strongly motivated toward companionship and doing what the other individual does, with some degree of mutual imitation (Scott, 1967b), and both species are capable of making group attacks on an individual, whether of the same or a different species. Certain other kinds of behavior have no counterparts, such as the sexual tie in dogs and reaction of the male dog to the urine of a female in estrus. However, the behavior patterns of dogs and wolves are sufficiently similar to those of human beings that it can be said that wolves and men were strongly preadapted for life in a combined social group. To put this in other words, the wolf is preadapted for domestication.

Polymorphism in the Wolf

As a dominant predator the wolf is protected from certain kinds of selection pressure, thus permitting the survival of individuals with a considerable variation from the mean. As a highly social species, wolves should be subject to selection favoring variation useful in cooperative enterprises, as a greater degree of variation permits a greater degree of division of labor. For an example, a wolf pack might benefit both by the presence of individuals that were highly timid and reacted to danger quickly and effectively, and also by the presence of other more stolid individuals who did not run away but stayed to investigate the perhaps nonexistent danger.

Although there is still little more than anecdotal evidence to support the existence of such behavioral polymorphism, the existence of anatomical polymorphism is much better verified. Murie (1944) was easily able to distinguish the individuals in a wolf pack by their different appearance in form and color. In addition, local populations of wolves vary considerably in form and size, with a general tendency for the northern or Arctic wolves to be larger than the southern variety (Jolicoeur, 1959).

Assuming that behavioral as well as anatomical polymorphism existed in the wild wolf populations of 12,000 years ago, there must have been a considerable variety of individuals and subpopulations from which the first domestic dogs could have been selected. If one attempt did not succeed, another might, and it is very likely that the first successfully domesticated animals were smaller, less aggressive, and less fearful than the average wild wolf.

Polymorphism also carries the implication that the gene pool of the species is adapted to permit viable variants. One can hypothesize a condition in which almost any variation from the standard gene complement would be distinctly nonviable, and other situations in which the genetic complement was buffered against the effects of variation.

Probable Method of Domestication

The archeological evidence indicates that the wolf was first domesticated by Stone Age hunters sometime between 8000 and 10,000 BC. At this time the Ice Age was coming to an end, the supply of game was plentiful, and the human population quite small. These Stone Age hunters therefore must have lived under prosperous conditions similar to those of the North American plains Indians during the early part of the 19th century, with the exception that the Eurasian hunters did not have the horse.

Scavenging wolves would have come around the hunting camps, looking for offal and attempting to steal stored supplies of meat. The hunters may, on occasion, have even hunted wolves and dug the young cubs out of their dens. Some of these may have been brought home alive and escaped the soup pot perhaps by attracting the attention of a woman who had lost her baby and was suffering discomfort from persistent lactation. Such a wolf cub could be very easily reared on the breast by a human mother for a few weeks, after which it could subsist on scraps and bits of cooked food. In a time of ample meat supplies there would have been plenty to go around. The adopted cub would have become rapidly attached to human beings, as wolf cubs do today, if taken at the right time, and it would have been friendly and playful with the children. By the time it was three months old it would have been largely self-sufficient, living on scraps of food and becoming a member of the human group. And unless human behavior has changed markedly, the foster mother would have become strongly attached to it.

If the original puppy was a female it could have mated with a wild male and brought up her own puppies within the camp, or perhaps a mate was provided by bringing in other puppies in subsequent years. Or the adoption and taming of wolf cubs could have become a standard part of the magical and religious practices of the tribe. Once such a tame wolf reached maturity the practical advantages of possessing one would be immediately apparent. It would act as if the hunting camp were its den and its boundaries a territory. It would respond at night to both strange animals and human beings entering the area, and give the alarm by barking. Furthermore, if taken on a hunt it would join in group attacks on a game animal without any particular training and also aid in locating small game, whether birds or mammals.

In short, the original domesticated wolf must have immediately been appreciated as an extremely useful social invention, and once a breeding population had been established it must have spread rapidly from tribe to tribe. In fact, the archeological evidence indicates that the dog spread throughout the world wherever people were living within a thousand years or so.

Considering the ease with which wolves can be tamed, it is, of course, possible that domestication took place on many occasions. Against this is the evidence of the common characteristic of all dogs, the curly tail, which may have been a very early mutation by which domestic animals could be distinguished from wild ones. There is also the evidence that all dogs have relatively small teeth and are basically smaller animals. Even the giant breeds, such as St. Bernards and Great Danes, do not have the bodily proportions of Arctic wolves but are obviously overgrown dogs, clumsy as the result of their large size. Until better evidence is available, we can adopt the hypothesis of a single domestication, with possible admixture of genes from wild populations of wolves from time to time.

Evolution of the Wolf

The major controlling factors in evolutionary change are mutation pressure, selection pressure, isolation, and inbreeding. Wright (1931, 1965) has pointed out that the ideal conditions for rapid evolution, based on the assumption of polygenic effects and differential effects of genes in different combinations, are provided by a species which is divided into a number of small local subpopulations that are nearly but not completely isolated and small enough so that a moderate degree of inbreeding takes place. Selection pressure should also be moderate, otherwise each population will become stabilized around the particular gene combination which is most advantageous. As local populations become extinct, evolutionary changes will inevitably occur, guided by the genetic make-up of the surviving populations. The division of a species into two or more subspecies is of course dependent on complete isolation being achieved in some way.

The organization of wild populations of wolves is in many ways closely comparable to this ideal situation. As dominant predators and adaptable animals able to live in a variety of habitats, wolves are ordinarily subject to only a moderate selection pressure. While the mating system of wolves is known only by inference, it appears that new groups are probably formed by litters leaving an older pack in a body and hence having a considerable opportunity for inbreeding. In any case, the social group itself is always small and semi-isolated from others by distance and territorial boundaries.

On the other hand, there is no evidence that local populations of wolves frequently became extinct except as the result of intensive extermination programs by their human competitors. Wolves are large and mobile animals and are restricted by few natural barriers except oceans and deserts. If one area becomes unlivable, they can easily migrate into another. This same mobility prevents any complete isolation between populations, and taxonomists consider that all wolves belong to the same species, although the connection between North American and Eurasian wolves at the Bering Strait must be a tenuous one. Rather than becoming divided into separate species, wolves have evolved along the lines of numerous local populations, each intergrading with the next. The overall result has been considerable stability for the species, and fossil wolves are not reported to be greatly different from modern ones. The only area of the world where wolves have come close to separating into two species is South Asia, where the Indian wolves are separated from the northern ones by the great barrier of the Himalayas and adjacent deserts and are sufficiently different to be considered a subspecies. It is this same population, or one closely similar to it, which is the most probable ancestor of the dog and therefore formed the raw material for the further evolution by the dog in the new habitat provided by human culture.

Evolution of the Dog

Domestication of the dog by early hunting and agricultural populations did not drastically change the organization of populations from that seen in the wolf. The dogs in a hunting tribe or agricultural village comprise a small local population with a considerable amount of inbreeding and some opportunity for cross-breeding with groups from adjacent tribes. Yet there has been a much greater differentiation of local populations than in the parent species.

The obvious explanation is deliberate selection for unusual variation, such as is practised by many modern dog breeders. This is at best only a partial explanation, and the origin of diversity in dogs can be viewed in more general terms as the result of penetration of a species into a vacant habitat, in this case the human cultural environment. Human cultures themselves evolve and differentiate, and from the viewpoint of the dog represent a great variety of habitats. Each culture represents a different habitat, exercising a different sort of selection pressure, and as the original populations of dogs entered such habitats the small numbers in each new group gave a great opportunity for accidental selection of unusual variations as well as for deliberate selection pressure exerted by the human masters.

Dogs can reproduce at one year of age, and it is theoretically possible to pass through a generation once every year. Estimating the average dog generation as a conservative two years, dogs have had perhaps 5,000 generations in which to accumulate mutations and to differentiate into sub-population. While this has been going on, human cultures have also changed and differentiated, and the Stone Age hunting camp has been replaced first by agricultural villages and latterly by the towns and cities of our modern industrial society.

Historical records of what has actually happened to dogs in this time are incomplete and not always reliable. The best historians are usually uninterested in dogs, and even the crudest written records extend over little more than half of the period of domestication. The first extensive treatise on dogs was written not quite 400 years ago (Caius, 1576). Nevertheless, some general conclusions can be drawn, and we can also make some enlightened guesses on the basis of some of the more reliable information.

In the first place, the Stone Age dogs appear to have been medium-sized animals, similar in form and probably serving as all-purpose hunting and guard dogs. Their center of origin was probably somewhere in Central Europe or the Middle East. In Mesopotamia today there are two kinds of dogs: the salukis, which are used for gazelle hunting, and a large, stockily-built, long-haired dog which is used to guard flocks of sheep and goats (Hatt, 1959). Early carvings indicate that these same two kinds of animals were present in the pre-Christian civilizations of that area, and that the large animals were then used for war dogs.

The salukis appear to have been the ancestors of the modern greyhounds and were first brought back to western Europe by the returning Crusaders. The Afghan hounds of nearby Afghanistan and the borzois of Russia were probably also derived from this original Mesopotamian stock.

Judging from the resemblances between African basenjis and Australian dingos, and the occurrence of similar dog breeds in southeastern Asia and the East Indies (Werth, 1944), we can postulate the early origin of a short-haired dog adapted to tropical living and that it was taken into Africa and across southern Asia, eventually reaching Australia.

There is also a northern group of dogs with some characteristics in common; these include the sled dogs of the Eskimos in both Eurasia and North America. Whether these have a common origin or not, they go back to a relatively recent date, according to the timing of skeletons associated with those of prehistoric Eskimos.

A third group of early dogs reached the Americas, and the earliest remains so far discovered indicate that they were similar to the undifferentiated Stone Age dogs of Europe (Lawrence, 1967). At the time the white explorers reached the Americas there were dogs living all over the two continents, from Hudson's Bay to Tierra del Fuego. These rapidly disappeared under the pressure from European immigrants, and the introduction of the European breeds. Studying their remains, Allen (1920) described three groups, chiefly distinguishable on the basis of size. These included the large Eskimo dogs, medium-sized dogs in at least eight varieties, and small-sized dogs of at least five distinguishable populations. The Eskimo dogs or Huskies and Mexican hairless are the sole surviving breeds, and even their ancestry is in some doubt.

Beyond these scanty historical indications and what we can infer from prehistoric remains, there is little authentic information about ancient dogs. The modern breed associations are of quite recent origin, and accurate written information began to be collected less than 100 years ago. The Kennel Club of England was founded in 1873, and the American Kennel Club in 1884. It was only at this time that breeders began to limit cross-breeding, to set standards for the selection of parents, and to keep complete pedigrees. Dog breeding as a modern pastime has been practiced chiefly in Great Britain and the United States, and to a lesser extent in western European countries. The classification of breeds reflects the cultures of the country. In Great Britain the emphasis is on sport, or hunting, and the Kennel Club of England recognizes two major classes of breeds, the sporting dogs, including hounds, gun dogs, and terriers, and the nonsporting breeds which include working dogs, toys, and others. The French list recognizes hunting dogs but also watch dogs, running dogs, 17 kinds of shepherd dogs, and 24 "ladies" dogs, including mostly toy and lap dogs. The German list emphasizes working dogs and watch dogs of a relatively ferocious sort, whereas the Swedish one recognizes nine different Spitz breeds.

The American Kennel Club lists over a hundred breeds, originating from all over the world, and classifies them into six groups, chiefly on the basis of function rather than ancestry. In some cases a group includes breeds that are historically known to be genetically related, but the same group may also include animals with similar uses coming from opposite ends of the earth. These groups are worth looking at, however, as an example of the numerous ways in which dogs are now used.

In this classification, the *sporting breeds* chiefly include the bird dogs, or gun dogs, of which we have many historical records in England. The modern setters and pointers were developed from the medieval spaniels. By tradition, the spaniels were named because they originally came from Spain, and there are many modern breeds bearing the same name. Pointers are known to have some admixture of hound ancestry, and it is possible that this has occurred in other breeds as well.

The *hound breeds* are traditionally divided into sight hounds and scent hounds. The former includes the greyhounds and their various relatives, and probably comprises a natural group with a common ancestor similar to the gazelle hounds or Salukis of the Middle East. Among scent hounds, there is a group of English breeds, including the fox hounds and beagles, made up of similar dogs which probably have considerable common ancestry. The fox hounds and coon hounds of the United States have also been chiefly derived from these breeds. However, hound breeds from other parts of the world have probably been developed independently, as we have no records of transportation of these animals from one country to another. The Norwegian elkhound, for example, is most similar to the Spitz and other northern breeds.

The *working breeds* include three main kinds of dogs from all over the world. One of these includes shepherd and farm dogs, such as the Scottish collie, Old English sheepdog and German shepherd. There is probably no common ancestry except in animals coming from neighboring areas. Another group of working dogs is made up of the guard dogs, of miscellaneous origin and including various giant breeds such as the mastiff, Great Dane, and St. Bernard. The third group includes the sled dogs from various Arctic regions. As indicated above there is some evidence that these last animals have remote common ancestry in an early variety of dogs adapted for life in a far northern climate.

Terriers, as indicated by their names (Scottish terrier, Welsh terrier, Irish terrier, etc.), are largely dogs from the British Isles. From medieval times hunting has been a popular sport in these countries, but only the nobility were allowed to hunt deer and use hounds. The common people had to content themselves with hunting ignoble game, or vermin, and for this purpose the smaller terriers were developed. As a group, terriers are the most aggressive of all the dog breeds, and some of them were used in the once popular sport of dog fighting.

The *toy breeds* have miscellaneous origins from widely different geographic regions and have in common chiefly their small size. They are true dwarfs rather than a separate species, as they have disproportionately large brains and genital organs. The puppies are also disproportionately large compared to the adults.

The American Kennel Club's final group is the *nonsporting breeds,* which include a variety of animals now used entirely as companions and show dogs. Some of these formerly had special uses. Dalmatians were coach dogs, and bulldogs were once used in bull-baiting.

As well as recognized breeds, there are a large number of varieties of dogs in the United States which have either never been under the control of professional breeders or which have never been entered in the American Kennel Club. In almost every part of the world there are local native varieties of dogs, often distinctly different from the commonly recognized breeds, although many of these are disappearing under pressure of European breeds brought in by explorers and immigrants. Most of the native varieties of dogs in South Africa were destroyed during an epidemic of rabies and have been replaced by either European breeds or dogs of mixed ancestry. Almost all of the Amerindian breeds have long since disappeared, and the aboriginal dogs of Malaysia are rapidly declining in numbers.

Most primitive cultures support only one kind of dog, or at most two, so that the early differentiation of dog populations had a largely geographic basis. At the present time, a large urban culture can support and keep separate dozens of breeds (the American Kennel Club recognizes over 100), maintained as separate populations by restrictions in a way which is reminiscent of caste systems in human societies.

Such breed populations are often quite large, containing 40,000 individuals or more, but many are quite small. Selection for breed standards is based on typological ideas, and changes from time to time according to the current ideas of the breeders, with the result that great changes of form have taken place since accurate records have been kept. The popularity of different breeds waxes and wanes according to human cultural changes. The genetic situation is not a stable one, and we can predict that dogs will continue to change in the future in both appearance and behavior.

Evolution of Behavior in Dogs and Wolves

One of the most important factors affecting evolution in a highly social animal is that of the nature of the social environment itself. Darwin chiefly thought of this problem in terms of sexual selection, referring to competition between members of the same sex for the opportunity to reproduce, and the tendency for an animal to prefer one kind of mate over another. These phenomena are examples of a broader phenomenon which may be called *social selection,* and which may be defined as any sort of selection exerted by the social environment and including the selective survival of individuals of any age, irrespective of sex. Social selection is an obviously important factor in the wolf (Scott, 1967a) and even more important in the dog, which belongs to two societies, canine and human.

Social Selection in the Dog and Wolf

We have already pointed out that the evolution of behavior in dogs and wolves has proceeded in two different directions in early and late development, with puppies in the neonatal period being selected for behavior adapted to survival under conditions of highly protective maternal care, and later behavior evolving toward adaptation to the adult social situation involving greater independence of the individual, but also considerable social cooperation and caregiving behavior. In between there is a brief transition period accompanied by a metamorphosis of behavior almost as spectacular as that which accompanies metamorphosis from the tadpole to the frog.

These general evolutionary tendencies have not been seriously modified by social selection from the human social environment, probably because similar, though not as clear-cut, tendencies are found in human development. In human infants the process of primary socialization takes place before most of the transition processes, and these are not concentrated in one brief period as they are in the dog (Scott, 1963). However, there is in early human development a neonatal period very similar to that of the dog, whose behavioral development has been modified least of all in the neonatal period. There is likewise little difference in the transition period between the behavior of wolves and dogs. For the most part, the early care of puppies is still left to their canine mothers, and it is only after the transition period is over (at approximately three weeks) and the puppy begins to take supplemental food, that human care begins to have an important effect.

In the period of socialization there has evolved in the wolf the capacity to form rapid emotional attachments to familiar objects and individuals, an attachment that is strengthened over a period of several weeks. This process is so fundamental for the survival of a highly social animal born in an immature state that there is no reported variation in wolves, and even in dogs it takes place with great uniformity with respect to time.

The nature of this process is such that it permits the formation of attachments to different species, even though in the normal course of development of a wolf no contact with animals other than the parent species would be possible during the appropriate period. As indicated above, the existence of such a process is a major factor determining the fitness of any animal species for domestication.

Certain modifications of the primary socialization process seem to have occurred in dogs as a result of human social selection. One is a less rapid development of the fear response to strangers which tends to bring the period of socialization to an end by preventing close and prolonged contact with strange animals. More complete studies of wolf development will determine how far development in dogs has changed from that in the ancestral species.

Most dogs are valued for their capacity to make rapid and close attachment to human beings, but there is some variation between breeds. In the strain of African basenjis studied by us there was both a tendency for fear responses to appear more readily in the early part of the period of socialization, and a tendency for less close emotional attachments to develop when basenjis were given the same amount of human handling as other breeds (Scott and Fuller, 1965).

Once the puppies have been partially weaned from the breast there is a tendency in human societies to replace the parental care of the dogs with human care, and in a puppy taken from a litter and adopted as a pet, parental care is taken over completely. As a result, there has been a relaxation of selection for animals that give the puppies later parental care. While dogs still bring back food to their home areas and attempt to cache it, there is seldom any effort to hook this up with feeding younger animals, and the feeding of vomited food to pups by adult males is almost never seen.

Nevertheless, when the adult behavior patterns of dogs and wolves are compared there is no evidence of any fundamental change in behavioral organization, since almost every behavior pattern that is observed in wolves can also be seen in dogs, at least in a low frequency. Rather than completely suppressing some behavior patterns, or creating new ones, domestication has had the effect of exaggerating or diminishing the frequency of occurrence of behavior patterns in different populations of dogs, with a resulting enormous increase in variation. The other result is that dogs are almost never given the opportunity of developing among themselves the high degree of social organization seen in wolf packs.

In addition to care-giving and parental behavior, dogs exhibit extended variation in three other major behavioral systems. For example, dogs and wolves have similar patterns of agonistic behavior, but in beagles and certain other hound breeds the occurrence of actual fighting is reduced almost to zero, probably as a result of selection for individuals that will tolerate each other in pack hunting. Two strange adult males can usually be placed together with little result other than a certain amount of barking and growling. At the opposite extreme, the terrier breeds have been selected for insensitivity to pain and ease of arousal for attack, so that they become highly intolerant of each other. When wirehaired fox terrier puppies were raised with their mothers throughout the period of socialization we found that they would not tolerate each other in groups larger than three, even as early as seven weeks of age (Fuller, 1953). Thus beagles are much more peaceful than wolves and terriers are much more aggressive, although the actual behavior patterns of agonistic behavior remain relatively unchanged.

In sexual behavior, dogs still show the same patterns of courtship and mating as wolves, given the opportunity to express them. Both the time of sexual maturity and the seasonal cycle, however, have been altered. Female dogs of various strains and breeds will come into the first estrus as early as five or six months and ordinarily not later than 15 months, whereas wolves show the first estrus not earlier than the end of the second year, and sometimes not until the third year. The normal annual cycle of wolf sexual behavior comes in the spring, with the pups being born in late April or May, and similar cycles take place in both the coyote and jackal. The cycle in the dog has been modified in two directions. In basenjis and dingos the estrus period has been shifted to the autumn of the year rather than the spring, but an annual cycle has been maintained, influenced by declining day length (Fuller, 1956). In all other known breeds of dogs the estrus periods of females are not related to the season of the year and occur at roughly six month intervals. This again is an argument for common ancestry of all dogs, with an early split between the southern varieties and the rest of the species.

Perhaps the greatest modification has occurred in investigative behavior, which is closely related to the hunting and predation which are the major activities of a carnivorous mammal. Wolves hunt birds, when they are available, and mammals in all sizes from mice to moose. Various dog breeds have been selected for specialized hunting activities, the scent hounds and bird dogs for finding game, the sight hounds for pursuit, and the terriers for attack. Shepherd dogs have been selected for their ability to learn to herd large mammals, a practise sometimes seen in wolves in the course of hunting. All these activities are specializations of behavior patterns seen in wolves. While most dogs have more general capacities than is commonly supposed (many shepherd dogs will readily learn to hunt deer), the general result of selection has been to exaggerate the frequency of certain patterns of investigative behavior in some breeds and to diminish it in others.

Thus the general effect of human selection upon domestic dogs has been to increase enormously the amount of variation seen in the wild species. This human selection, whether conscious or not, is equivalent to the phenomenon of social selection seen in any animal society, and we can now consider its theoretical basis.

Theoretical Considerations

Variation and Selection

As Wright (1931) has pointed out, the theoretical effect of strong selection pressure of a consistent sort is to limit variation within a population and consequently to inhibit evolutionary change. However, in any natural situation there are always a large number of selection pressures operating upon a population, and these are not always consistent from generation to generation.

From the viewpoint of genetics, the important basic phenomenon is the differential survival of genes, which are subject to selection pressures from several sources. The first of these is the genic environment, since the action of a gene is dependent on the nature of the gene complex of the individual which is in turn determined by the nature of the gene pool of the species. Any gene whose action is incompatible with that of others will be selected against. A second source of selection pressures is the prenatal environment. In mammals, homeostatic processes keep this environment relatively stable under most conditions, and one would expect a consequent restriction in the range of variation of processes going on in prenatal development. The postnatal social environment likewise tends to be stable because of maternal care and the protection provided by social groups, especially in the more highly social mammals.

The biotic environment, from which a large number of selection pressures originate in the form of competing species and available food supplies, is only relatively stable. The "balance of nature" is a shifting and unsteady balance of forces rather than a steady state. Likewise, climatic and other factors in the physical environment vary widely from year to year as well as from season to season.

Because of the complexity of these selection pressures, no gene or gene complex is likely to contribute equally to survival under all the conditions which are experienced by an individual or population. Opposing selection pressures should lead to the preservation of variation. Further, because many selection pressures vary widely from time to time, there is a tendency in any species population to preserve those genes and gene combinations which are buffered against the effects of selection, the phenomenon which Lerner (1958) has called genetic homeostasis. Thus, while selection has the theoretical effect of limiting variation under certain simple situations, it can also have the effect of preserving and maintaining variation under more complex conditions.

While the survival of the gene is the ultimate effect determined by selection, its pressures are exerted directly against gene carriers at different levels of organization: the individual at all times in development, the social group to which it belongs, the subspecies population made up of social groups in a certain geographic area, and the entire species population including all individuals. This brings up the possibility that certain characteristics may have different and even conflicting survival values for the units of organization at different levels. While the number of theoretical conflicts is quite large, the general principles involved are illustrated by situations in which behavior that contributes to the survival of a social group may be dangerous to the individual involved, and vice versa.

Group Survival and Individual Survival

This problem is usually posed in the form of the evolution of altruistic behavior, in which an animal sacrifices his own safety for the benefit of his social group. The actual problem is much more general and complex, but we can present it in broad outline by adding three other simplified theoretical possibilities. The first of these is the evolution of behavior which promotes both group and individual survival. An example of this is allelomimetic behavior in dogs and wolves, where cooperative group action increases success in hunting, allows successful defense of a den (in wolves), and alerts the whole group to sources of danger. Every individual in the group benefits by his own behavior and contributes to the survival of others. This is perhaps the commonest, and certainly the strongest, situation which would lead to evolution of social behavior and organization, and selection would obviously favor it under almost any circumstances.

A second theoretical situation is one in which behavior promotes individual survival and group death. An example of this is the occasional cannibalism of newborn infants seen in dogs, and there should be strong selection pressure against the survival of genes that contribute to this kind of behavior. The obverse of this situation is one in which behavior results in individual death but group survival ("altruistic" behavior), and an example is seen in the defense of the den by wolves against predators such as bears. As Hamilton (1964) has pointed out, this situation should theoretically result in some sort of balance between the two pressures, and what we actually see in wolves is a vigorous but at the same time somewhat cautious defense of the den. The same sort of cautious behavior is seen in predation, where dangerous animals such as moose are approached carefully and only attacked if sufficiently weak to be killed without danger to the individual.

Finally, there is the theoretical possibility of behavior that promotes the death of both the individual and the group. This is the result of violent fighting, and there are almost no instances of unrestrained destructive fighting in wild animal societies (Lorenz, 1964; Scott, 1962). In the terrier breeds of dogs which have been selected for their ability to fight in this manner, these animals have to be protected from each other in order to survive (Scott and Fuller, 1965).

The results of this analysis are clear. Where selection pressures at two levels of organization coincide (and this would have to be extended to all levels of organization to establish a clear-cut case), the effect is to produce strong selection pressure and to limit variation. Where there is a conflict between two pressures, the net effect is to set up a balanced situation. This could be achieved in a number of ways, one being intermediate behavior pattern based on a fixed gene combination, and another being a balanced gene frequency in the population. In the latter case, the conflict in selection pressures should have the effect of preserving variation.

The Origin of Variation in the Dog

The above considerations apply to a wild social species such as the wolf from which the dog was derived. The dog has become, however, a different animal from the ancestral species in one outstanding way—through an enormous increase in variation, not only in behavior but also in physical form and appearance. To take a dramatic example, a full grown Chihuahua may weigh as little as 4 pounds, while a St. Bernard may weigh as much as 120. This degree of variation is not exceeded by any other domestic species, including man himself.

This phenomenon is related to several theoretical considerations. In a highly social animal such as the dog, important environmental selection pressures are exercised by three groups of factors, those belonging to the physical, biotic, and social environments. As pointed out above, the social environment (including the prenatal environment) in the highly social species of wild mammals tends to be more stable than the other two over periods of many generations, and hence to exert constant rather than varying and fluctuating selection pressures. A lamb always grows up in the environment of a sheep flock, generation after generation, and wolf cubs normally grow up among members of a wolf pack. The result is that social behavior tends to become conservative in an evolutionary sense, being stable over long periods and less variable at any given time.

This rule is violated in the dog because the species has become involved with the phenomenon of human cultural evolution. Instead of remaining constant generation after generation, the human social environment varies from one generation to the next and from place to place. These changes produce shifting selection pressures on the dog and in part account for the great genetic diversity of form and behavior in this species.

Social selection indirectly increases variation in another way. The social group most likely to survive is one whose members contribute most effectively to social organization. One of the obvious effects of social organization is to increase the chances of survival of individuals by lessening the selective pressure from physical and biotic factors. The result of this relaxation of selection pressure should be an increase in variation in those adaptive characteristics related to the physical and biotic environments. This relaxation can be maintained only as long as effective social organization is maintained, and the latter is itself dependent upon social selection pressure. Thus there is a reciprocal or compensating relationship between social selection pressure and selection pressure arising from other parts of the environment.

This result in the dog is complicated by the fact that the major factor exerting social selection pressure is human social organization, but the results of the relaxation of other pressures are essentially the same. Breeds such as the Chihuahua with its tiny size and paper-thin skull obviously could not survive without the protection of human society, and individuals with physical defects from any breed often live to a ripe old age.

There is still another way in which social selection contributes to the origin of variation. Social organization implies division of labor, and once a useful diversity has arisen within a social group, there must be strong selection pressure in favor of maintaining this diversity.

Synergistic Relationship between Genetic Variation and Complexity of Social Organization

The existence of different varieties of animals within a social group makes greater division of labor and diversification of social behavior possible, and thus contributes to social organization. In turn, increasing complexity of organization provides useful niches or social roles for a still greater variety of individuals. This suggests that a synergistic relationship exists between complexity of social organisation and genetic variation. Once such a relationship is established it should set in motion a process of continuous change, limited only by whatever other factors limit the development of social organization.

In the case of dog and man, it is probable that most of the increased variation in dogs has come from changes in human social organization, with relatively little effect of genetic changes in dogs upon the organization of human behavior. In the early prehistory of man, genetic variation in the dog may have made an important contribution to human social organization, by making the domestication of herd animals possible through the use of shepherd dogs. Today, variation in dogs is chiefly a response to human social change. As soon as a new social niche for dogs appears, it is easy to find an animal suitable for the purpose, and it is easy for our prosperous human society to support a variety of breeds having no special use.

There is another qualification to this principle, arising from conflicting tendencies in social selection pressure. In a relatively stable animal society such as that of the wolf there must be strong selection pressure in favor of behavior that maintains social organization, with a consequent tendency to decrease variation in the area of social behavior. Only those social variants whose behavior is neutral or immediately useful would be permitted to survive. A large or rapid increase of genetically determined variation in social behavior is therefore dependent on the relaxation of social selection pressures. It is only when social organization becomes relatively independent of genetics, as it has in human cultural evolution, that a strong synergistic relationship between genetic variation and complexity of social organization becomes possible.

Genetics and the Evolution of Human Social Organization

The evolution of form and behavior in the domestic dog cannot be understood except with reference to the human societies of which the dog has become a part. One of the outstanding characteristics of human societies is verbal communication that vastly increases the possibilities of cultural transmission of information from one generation to the next. This in turn makes cultural evolution a major phenomenon in human societies. While this process has many similarities to biological evolution, it has become almost entirely independent of it, except with respect to the synergistic relationship between genetic variation and complexity of social organization described above.

Much of our progress toward understanding the process of biological evolution has come about because of the discovery of the mechanism by which genetic information is transmitted, namely, the genes and chromosomes. Likewise, the nature of cultural evolution can only be appreciated by taking into account the nature of its transmittal mechanism, that of learning. Unlike the chromosomal system, which transmits the same quantity of genic material from generation to generation, the learning mechanism produces a cumulative effect, and the theoretical extent of accumulation of information is almost unlimited, depending only on physical capacities for the storing of written records. Since little is lost, this could lead to an increasingly stable situation except for a built-in factor of instability, the fact that each individual in every generation must learn everything anew and must start from a different point in time. Because learning is an organizational process, this inevitably produces a reorganization of information in each new individual. And, unlike genetic information, which is transmitted only at the initiation of development, cultural information can be transmitted and accumulated by an individual throughout his lifetime.

The result is a rate of change which vastly outstrips that of genetic change. Major cultural changes often take place within the lifetime of an individual, whereas biological changes produced by selection usually take several generations to accomplish. The result is that social selection pressures generated by cultural conditions are likely to fluctuate and change direction from generation to generation, or even within the same generation. The dog, with its short generation span, can respond biologically to these shifts in selection pressures, but the longer-lived human being cannot. This in itself is one explanation of the fact that the human species is less variable than that of the dog.

In addition, there are no human populations that have been subjected to deliberate genetic manipulations and control as have the different breeds of dogs. The so-called human "races" are not genetically equivalent to dog breeds, nor are any other human populations. If there is any genetic correspondence between dog and human, it is that the dog breeds may be taken as models of the kind of genetic variation *between individuals* that exists within human populations.

As in the dog, the outstanding characteristic of human populations is the immense amount of variation in form and behavior compared to other animal species. Excluding deliberate selection, much the same factors account for it: adaptive radiation in prehistoric man, the relaxation of selection pressures from the physical and biotic environments as the result of effective social organization, and, finally, the synergistic relationship between the development of complex social organization and the survival of genetically variant individuals who can make a contribution to social organization by making increased division of labor possible.

I am suggesting that there is not only a positive relationship between social organization and genetic variation, but that this relationship is in part a reciprocal one. Once this process is set in motion, there is no reason why it should stop, barring a complete breakdown of social organization and its replacement by a simpler form. From this viewpoint, mankind's genetic future should be one of increasing individual diversity and greater complexity of organization, but involving no fundamental shifts in averages. Genes will be added that will extend variation in all directions, but very few will be lost from the gene pool. Social selection resulting from cultural change occurs so rapidly and in such a fluctuating fashion that there is little opportunity for making great changes in the genetic constitution of the species. In any case, the immense size of modern human populations in itself assures genetic stability.

As for dogs, they should continue to respond to the shifts and changes of human cultural evolution, but in a more extreme fashion than we do ourselves. As we look at them we shall see an exaggerated reflection of our own genetic condition, but true only to the extent that we extrapolate correctly. We can do this on the broad general dimension of variation, but not in its details, for dogs are still dogs, with genetic constitutions most similar to other members of the family Canidae, even after 12,000 years as a part of human society.

References

AHMED, I. A. 1941. Cytological analysis of chromosome behaviour in three breeds of dogs. Proc. Roy. Soc. Edinburgh [B] 61:107–118.

ALLEN, G. M. 1920. Dogs of the American aborigines. Bull. Museum Comp. Zool. Harvard Coll. 63:431–517.

BENIRSCHKE, K., and R. J. LOW. 1965. Chromosome complement of the coyote. Mammalian Chromosomes Newsletter, No. 15, Feb., 1965, 102.

BORGAONKAR, D. S., O. S. ELLIOT, M. WONG, and J. P. SCOTT. 1967. Chromosome study of four breeds of dogs. J. Hered. In press.

BUFFON, G. L. L. 1804. Histoire naturelle, Paris, de l' Imprimerie Royale.

CAIUS, J. 1576. Of Englishe dogges. Copied and reprinted in modern type by A. Bradley. London, 1880.

DAHR, E. 1937. Studien über Hunde aus primitiven Steinzeitkulturen in Nordeuropa, Lunds Universitets Arsskrift, 32(4):1–63.

DARWIN, C. 1859. The Origin of Species. Reprinted. New York, Modern Library.

DEGERBØL, M. 1927. Über prähistorische dänische Hunde. Vidensk, Meddel. Dansk. Naturhist. For. København, 84:17–72.

――――. 1961. On a find of a preboreal domestic dog (*Canis familaris* L.) from Star Carr, Yorkshire, with remarks on other Mesolithic dogs. Prehistoric Soc. for 1961, New Ser., 27:33–55.

FENTRESS, J. C. 1967. Observations on the behavioral development of a hand-reared timber wolf. Amer. Zool. 7:339–351.

FULLER, J. L. 1953. Cross-sectional and longitudinal studies of adjustive behavior in dogs. Ann. NY Acad. Sci. 56:214–224.

――――. 1956. Photoperiodic control of estrus in the basenji. J. Hered., 47:179–180.

HAAG, W. G. 1948. An osteometric analysis of some aboriginal dogs. Rep. in Anthropol. Univ. of Kentucky. 7:107–264.

HALE, E. B. 1962. Domestication and the evolution of behavior. *In* HAFEZ, E. S. E. ed., The Behaviour of Domestic Animals, London, Ballière, Tindall & Cox.

HAMILTON, W. D. 1964. Genetical evolution of social behavior. J. Theor. Biol. 7:1–52.

HATT, R. T. 1959. The mammals of Iraq. Misc. Publ. Museum Zool. Univ. of Michigan. 106:1–113.

HERSHER, L., J. B. RICHMOND, and A. U. MOORE. 1963. Maternal behavior in sheep and goats. *In* RHEINGOLD, H. ed., Maternal Behavior in Mammals, New York, John Wiley & Sons.

JOHNSON, F. 1967. Radiocarbon dating and archeology in North America. Science, 155:165–169.

JOLICOEUR, P. 1959. Multivariate geographical variation in the wolf, *Canis lupus* L. Evolution, 13:283–299.

LAWRENCE, B. 1967. Early domestic dogs. Z. Säugethierkunde. In press.

――――, and W. H. BOSSERT. 1967. Multiple character analysis of *Canis lupus, latrans,* and *familiaris,* with a discussion of the relationships of *Canis niger.* Amer. Zool., 7:223–232.

LERNER, I. M. 1958. The genetic basis of selection. New York, John Wiley & Sons.

LORENZ, K. 1937. Der Kumpan in der Umwelt des Vogels. J. Ornithol., 83:137–213, 289–413.

LORENZ, K. 1955. Man Meets Dog. Boston, Houghton-Mifflin.

———. 1964. Ritualized Fighting. *In* CARTHY, J. D., and F. J. EBLING, ed., The Natural History of Aggression. New York, Academic Press.

MATHEW, W. D. 1930. The phylogeny of dogs. J. Mammal., 11:117–138.

MATTHEY, R. 1954. Chromosomes et systématique des Canidés. Mammalia, 18:225–230.

MURIE, A. 1944. The wolves of Mt. McKinley. *In* Fauna of the National Parks of the U.S., Fauna Series no. 5. Washington, U.S. Dept. Interior, U.S. Govt. Printing Office.

REED, C. A. 1959. Animal domestication in the prehistoric Near East. Science, 130:1629–1639.

REITER, M. B., V. H. GILMORE, and T. C. JONES. 1963. Karyotype of the dog (*Canis familiaris*). Mammalian Chromosomes Newsletter, 12:170.

SCOTT, J. P. 1962. Hostility and aggression in animals. *In* BLISS, E. L. ed., Roots of Behavior, New York, Harper & Row.

———. 1963. The process of primary socialization in canine and human infants. Monograph. Soc. Res. Child Develop. 28(1):1–47.

———. 1967a. Evolution of social behavior in dogs and wolves. Amer. Zool. In press.

———. 1967b. The development of social motivation. Nebraska Symposium on Motivation. Lincoln, Univ. of Nebraska Press. In press.

———, and F. H. BRONSON. 1964. Experimental exploration of the et-epimeletic or care-soliciting behavioral system. *In* LEIDERMAN, P. H., and D. SHAPIRO, ed., Psychobiological Approaches to Social Behavior, Stanford Univ. Press.

———, and J. L. FULLER. 1965. Genetics and the Social Behavior of the Dog. Chicago, Univ. of Chicago Press.

VAN DER MERWE, N. J. 1953. The jackal. Fauna and Flora, Transvaal Prov. Admin. Publication No. 4.

WAGNER, K. 1930. Rezente Hunderassen: eine osteologische Untersuchung, Oslo Videnskaps-Akademi, 3(9):1–157.

WERTH, E. 1944. Die primitiven Hunde und die Abstammungsfrage des Haushundes, Z. Tierzüchtung Züchtungsbiologie, 56:213–60.

WOOLPY, J. H., and B. E. GINSBURG. 1967. Wolf socialization: a study of temperament in a wild social species. Amer. Zool. 7:357–363.

WRIGHT, S. 1931. Evolution in Mendelian populations. Genetics, 16:97–159.

———. 1965. Factor interaction and linkage in evolution. Proc. Roy. Soc. B, 162, 80–104.

YOUNG, S. P., and E. A. GOLDMAN. 1944. The Wolves of North America. Washington, Amer. Wildlife Institute.

7

A Sampler of Human Behavioral Genetics

IRVING I. GOTTESMAN

Department of Psychology and Behavioral Genetics
University of Minnesota
Minneapolis

Introduction 276
The Nature of Psychological Traits 278
    The Trait Approach 281
    Pathways between Genes and Traits 283
    Psychological Tests as Signs and Samples of Behavioral Traits 285
The Twin Method 287
    Strategic Implications of Twin Results 288
    Twin Data Statistics 289
Intellectual Abilities 292
    Mental Retardation 294
    Intelligence and Fitness 296
    A Polygenic Model for the Transmission of Intelligence 300
Schizophrenia 302
    Family and Twin Studies 303
    Schizophrenia and Natural Selection 307
    The Transmission of Schizophrenia 309
Summary 313
References 314

Introduction

A chapter devoted to a consideration of genetic and evolutionary aspects of some *behavioral* characters in man in a book such as this may come as a surprise to many evolutionary biologists. There is, however, a ferment in the behavioral sciences. Any temptation to invoke Samuel Johnson's simile of a woman's preaching and a dog's walking on its hind legs, followed by his comment—"It is not done well; but you are surprised to find it done at all."—might be suppressed by awareness of the array of energies devoted to the demonstration and elucidation of heritable variations in animal and human behaviors. Organisms subjected to behavior genetic analyses range from *Schizophora* (Hirsch and Erlenmeyer-Kimling, 1962; Dobzhansky and Spassky, 1967) to schizophrenics (Gottesman and Shields, 1966a,b,c; Huxley et al., 1964). Evolutionary thinking is slowly but surely returning to a deservedly seminal position in the formulations of some behavioral scientists (e.g., Erlenmeyer-Kimling and Paradowski, 1966; Freedman, 1967; Ginsburg and McLaughlin, 1966; Gottesman, 1965, 1967) after the near extinction that followed its emotional abandonment in the wake of injuries from social Darwinism and insults from the Third Reich (cf. Dunn, 1962; Haller, 1963). Simpson has said that the modern theory of evolution ". . . reinstates behavior not merely as something to which evolution has happened but as something that is itself one of the essential determinants of evolution" (Simpson, 1958, p. 9). Psychology as a whole, however, is not alone in its reluctance to embrace evolution. "There are vast areas of modern biology, for instance biochemistry and the study of behavior, in which the application of evolutionary principles is still in the most elementary stage." (Mayr, 1963, p. 9)

Publications are the artifacts of the cultural evolution in the behavioral sciences that bear witness to the ferment referred to above. Conferences jointly sponsored by the American Psychological Association and the Society for the Study of Evolution led to a wealth of new ideas (Roe and Simpson, 1958). Fuller and Thompson (1960) produced the first textbook in behavior genetics, and it included a masterful review of over 800 pertinent references from disparate sources covering almost a century of research. The *Annual Review of Psychology* now provides occasional coverage of new developments in behavioral genetics (Kallmann and Baroff, 1955; Fuller, 1960; McClearn and Meredith, 1966) and the authors always have more to review than can be comfortably managed. Other interdisciplinary conferences have resulted in an expansion of the primary literature (e.g., Vandenberg, 1965, 1967a,b; Hirsch, 1967; Tax, 1966; Kallmann, 1962; Glass, 1967).

Advances in human cytogenetics and in knowledge about specific metabolic defects associated with mental retardation impinge upon the field of human behavioral genetics in ways that are beyond the scope of this review (Anderson, 1964; Gottesman, 1963a; Lejeune, 1964; Penrose, 1963). Progress in a psychopharmacology attuned to the utility of genetics will assume a growing role in human behavioral genetics (e.g., Benaim, 1960; Kalow, 1962; Porter, 1966), but this area too must be slighted in a "sampler" that tries to shy away from superficiality. Finally the psycho-logical-mindedness of some leaders in genetics (David and Snyder, 1962; Dobzhansky, 1962; Stern, 1966) guarantees the viability of the obviously fertile F_1, behavioral genetics.

The goal of this chapter will be limited to introducing the reader from outside the specialty of human behavioral genetics but with an interest in evolutionary biology to the data, methods, enigmas, and conclusions from selected studies within selected areas of the former field. After a neces-sarily curtailed discussion of the nature of psychological characters or traits, samples of research on intelligence and mental illness will be re-viewed. The content areas will be preceded by an introduction to the twin method since it constitutes one of the most popular and efficient ways of screening traits that are continuously distributed in human populations for significant genetic variation.

The Nature of Psychological Traits

Psychology lacks an adequate taxonomy of behavior. As a character for genetic analysis, behavior presents the investigator with several unique impediments that distinguish it from the simple traits that permitted the formalization of genetics as a science. Definitions of behavior *qua* behavior are usually avoided except in introductory textbooks on psychology because something pertinent might be excluded. Thompson (1966) focused on three properties of behavior that contribute to the difficulties of human behavioral genetics. First, behavior traits that are important to society are *continuous* rather than discrete. This leads immediately to the use of models from quantitative genetics (Falconer, 1960) but ones which have not been expressly adapted to human genetics owing to the infeasibility of the usual kinds of breeding experiments. Continuity of behavior also leads to problems of sampling and of scaling. Not all available human samples will have a particular trait to the degree that measurement becomes possible. Except for identical twins, no two humans have the same genotype. An individual with an IQ of 120 is not twice as intelligent as one with an IQ of 60 and the notion of zero intelligence has no meaning. Enhanced perform-ance of a skill in one part of the range of variation is often not comparable to the same increment in another part of the range. For example, lowering the running time in the hundred yard dash from 15 sec to 14 sec is not note-worthy, but lowering it by one sec when the original time was 10 sec would result in a new world record. Many human traits can probably be usefully construed within the framework of threshold characters (e.g., Grüneberg, 1952; Fuller, Easler, and Smith, 1950) but the concept is a foreign one to most psychologists.

The second vexing property of behavior is its *complexity*. There are no natural basic units of behavior; each trait can be fragmented into a very large number of bits at various levels of analysis, from biochemical and biophysical to choreographic. Such intricacies of the gene to behavior pathway make the distinction between genotype and phenotype especially crucial to all formulations in this area of endeavor (Fuller, 1957; Thompson, 1957). Genetic heterogeneity may also be anticipated for patent phenotypic homogeneity as has been demonstrated for syndromes of mental retardation. Behavioral phenocopies are just as likely to be found, if not more so, as are morphological phenocopies in species other than man. As shown below, the concept of *trait* does not solve the problem of finding *the* unit of behavior, but it at least provides an interim solution to the stalemate.

Thirdly, Thompson focused on *fluidity* of behavior. By this he meant to call attention to the observation that measurements of a behavior vary over time. For some psychological traits a large part of this variation over time is simply due to imperfections in the measuring instrument. Age-related or developmental changes in behavior are so important as to probably require the eventual specialization of developmental behavioral genetics; a trait may show differing degrees of genetic determination depending on whether the population sampled consists of infants, children, adults, or octogenerians. Many traits show fluidity because they are *very* sensitive to the environment and will change as a function of learning and/or the conditions of testing. A great deal of new information may be expected by contrasting static with dynamic measures of some traits. Resting state measures are not likely to give the kinds of data that bear on questions of adaptive value. Procedures analogous to the determination of blood sugar under conditions of fasting and glucose loading will have to be developed for behavioral traits whose values under stressful conditions are likely to be important for a genetic analysis. Other temporally-dependent sources of variation are the systematic fluctuations in traits that are autochthonous or circadian. It is obvious, for example, that measurements of such traits as emotionality in females would have to take into account not only the specific point in the menstrual cycle but also the individual differences in response to the physiological changes. What has been termed the fluidity of behavior may also be construed as *plasticity*. It seems logical at this point to infer that plasticity was a general property of behavior that has evolved for many but not all traits because of the adaptive advantages conferred in an environment that did itself fluctuate daily, seasonally, and randomly.

The difficulties of specifying natural units of behavior are further highlighted by our current ignorance of the neuronal mechanism of either learning or memory. Bullock discussed the evolution of neurophysiological mechanisms related to behavior and concluded that, "The concept of emergent properties, believed to inhere in the properties of the lower levels but not readily predicted with an incomplete knowledge, is an old and familiar one. It offers only a refuge, but it does offer that." (Bullock, 1958, p. 166) Our current ignorance is understandable. Bullock noted that even at the level of the higher invertebrates there were such phenomena as the graded synaptic potential. He went on to say, "Indeed, so much can be done by means of this graded and nonlinear local phenomenon prior to the initiation of any postsynaptic impulse that we can no more think of the typical synapse in integrative systems as being a digital device exclusively, . . . but rather as being a complex analogue device which finally converts into a digital output." (Bullock, 1958, p. 172) The output of many neurons may even be a graded event. Bullock's observations prompted Simpson to comment, "The potentialities of a device that is both digital and analog, that has millions of units, and that can be interconnected in more millions of ways stagger the imagination—mine, at least!" (Simpson, 1958, p. 513)

From both a genetic and evolutionary point of view, an adequate taxonomy of behavior, i.e., one that allows for explanation, must make provision for its function and physiology. As noted some time ago by Cattell, description of behavior must precede measurement; however, to stop there would impede progress towards a more powerful science of behavior. Learning theory, one of the major stocks in trade of psychology, is bankrupt when it comes to meeting the demand for laws of behavior that are other than descriptive.

Nihilism aside for the moment, there *are* distinguishable categories of behavior (cf. Roe and Simpson, 1958). This obvious fact has led to a kind of metataxonomy wherein the categorization of possible categories has been set out. Categories suggested by Pribram (1958) and Pittendrigh (1958) are subsumed under the more comprehensive scheme of Nissen (1958). He proposed three classes of categories: (1) *Functional,* which indicate the usefulness and purpose of the behavior, e.g., mating and food-getting; (2) *Descriptive,* which depict the mechanisms or processes for accomplishing behaviors in the first category, e.g., locomotion, perception, and manipulation; (3) *Explanatory,* by which is meant a concern for phylogenetic precursors, physical substrates, or an appropriate model or analog. Examples of explanatory categories would include hypothalamically mediated behaviors, androgen influenced behaviors, and instincts. It is apparent that behavior is multidimensional but this need not lead to an impasse. Ginsburg (1958) has suggested that genetic lesions, such as phenylketonuria (PKU), be used to "naturally dissect" the nervous system and the strategy has been endorsed by Hirsch (1964). This is not the same as creating a natural unit of behavior, but can sometimes be a powerful tool analogous to stereotaxic brain stimulation. The human behavioral geneticist finds himself oscillating between staying close to the gene end of the gene-to-behavior pathway and the more familiar ground of the behavior end. Both ends have their own advantages.

The Trait Approach

In order to even launch a science of behavioral genetics, we must insist that traits have a real existence in people and reject the elusive suggestion from some quarters of psychology that traits exist only in the observations made by some person other than the subject. The former view is termed biophysical, the latter, biosocial. There are three implicit steps by which a trait acquires a name (Guilford, 1959). We observe that people differ in what they do and in how they do it, infer a quality common to their actions drawing on the richness of our everyday language, and attach an adverb: Tom behaves aggressively. Then the quality is applied to Tom instead of just his behavior by means of an adjective: Tom is aggressive. After confirmation that Tom behaves that way more or less consistently under more or less similar situations, we abstract the property as a thing and give it a noun form: Tom has the trait of aggression. Allport (1931, 1937, 1966), a major figure in psychology, believed that traits as a generic term had withstood the pessimistic attacks of positivism. "Traits are cortical, subcortical, or postural *dispositions* having the capacity to gate or guide specific phasic reactions. It is only the phasic aspect that is visible; the tonic is carried somehow in the still mysterious realm of neurodynamic structure" (Allport, 1966. p. 3). It is reassuring to note the consistency with an earlier definition when a trait was defined as ". . . a generalized and focalized neuropsychic system (peculiar to the individual), with the capacity to render many stimuli functionally equivalent, and to initiate and guide consistent (equivalent) forms of adaptive and expressive behavior" (Allport, 1937, p. 295).

Since traits must be inferred, we must expect difficulties in addition to those mentioned above in regard to behavior per se as a character for genetic analysis. When Allport and Odbert (1936) found almost 18,000 terms in the English language that described human behavior (e.g., genial, obstinate, secretive, humble), proponents of a trait approach to behavior were confronted by an embarrassing and unwieldy abundance. Since only ten scalable traits each having ten specifiable levels can lead to the identification of 10^{10} unique individuals, and since most of the 18,000 terms tend to correlate with many others, some economy of description was called for in order to initiate a viable taxonomy of behavior. Factor analysis had been devised as early as 1904 for the express purpose of reducing a correlation matrix from many variables into the minimum number necessary to account for the observed phenotypic interrelationships (Guilford, 1954; Anastasi, 1958). Cluster analysis (Tryon, 1958) is another method that leads to parsimonious description of behavior and it has the advantage of being closer to what common-sense tells us about the importance of certain blocks of behavior under which other behaviors are subsumed. A consensus based upon the past thirty years of research in the domains of abilities and personality has led to a preference among psychologists for an intermediate number of traits and these have been organized into a hierarchy (Thurstone, 1938; Cattell, 1950; Vernon, 1961; Eysenck, 1947).

Guilford (1966) has developed a structure of intellect model that theoretically has 120 distinct abilities discoverable by factor analysis. So far 80 of the abilities have actually been specified by Guilford, including 24 kinds of comprehension (e.g., of printed words, of spoken words, kinesthetic-spatial orientation) and 15 memory factors. A less differentiated view of the structure of intelligence was preferred by Thurstone and his followers; it involves six separate abilities—vocabulary size, verbal fluency, numerical, spatial, reasoning, and memory. Vandenberg (1966, 1967a) concluded from his multivariate analyses of twins' scores on the six Primary Mental Abilities that there was evidence for four genetically independent abilities. The most widely used intelligence tests for both children and adults consist of a mixture of separate abilities and yield a very useful practical estimate of *general intelligence* (cf. Maher, 1963). A dozen or more personality dimensions have been isolated by the use of factor analysis (Guilford, 1959; Cattell, 1965) and these do not include considerations of motivational aspects of personality.

As implied above, the exact number of traits any investigator wishes to adopt as sufficient for the description of human behavior cannot be legislated and will depend upon his purpose. A reasonable solution is to have available a schema for organizing traits, whether derived by factor analysis or not, into a hierarchy of varying levels of differentiation, and then to choose a level consonant with one's purpose. Figure 1 illustrates one such hierarchy for the broad traits of general social adjustment (alpha) and general emotional adjustment (beta) as envisaged by Guilford (1959). The diagram is based on the correlations that have been obtained for various kinds of personality measurements. The habitual response level consists of traits inferred from observing consistent behavior in a limited range of situations. Examples would include preferences for attending social gatherings (*a*), starting conversations with strangers (*c*), and being able to hold one's own in an argument (*j*). The network of lines above the habitual response level shows that such responses can contribute to more than one primary trait. The latter are inferred from the intercorrelations among habitual responses, for example, *a, b, c, d, f,* and *g* "go together" to form *S*. *S* is named sociability because it seems to be the element that the habitual responses have in common. Thus primary traits exist in the eye of the trained beholder and their validity as independently organized entities subject to biological or genetic analyses must rest on other grounds. The intercorrelations among primary traits lead to the positing of the more general traits at the type level. Thus *S, A* (ascendance), and *I* (confidence)

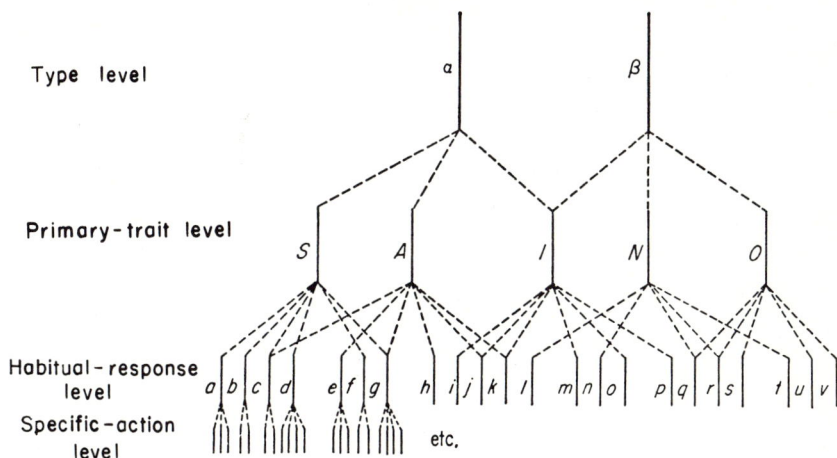

Fig. 1. Diagram of a hierarchical organization of traits with varying levels of specificity (after Guilford, 1959, p. 100).

intercorrelate positively and yield the type level broad trait of alpha or general social adjustment. At the lowest level, that of specific actions, we do not have traits but rather trait indicators.

Pathways between Genes and Traits

Fuller and Thompson (1960) have provided a model that elucidates the possible relationships between genes and behavioral traits. Unlike the model proposed by Royce (1957), they do not believe in a congruence between blocks of genes and the primary traits of Figure 1.

> Single genes characteristically affect many forms of behavior; single psychological traits have variance ascribable to many genes. Between the genetic and behavioral levels of description is a network of physiological processes. We propose that behavior is insensitive to a substantial portion of the metabolic shunting which occurs in this network, for a neuron may not be concerned with the source of its energy provided it gets enough. In such a situation any one of a number of alternative genetically controlled pathways may be equivalent from the viewpoint of behavior integration (Fuller and Thompson, 1960, pp. 343–344).

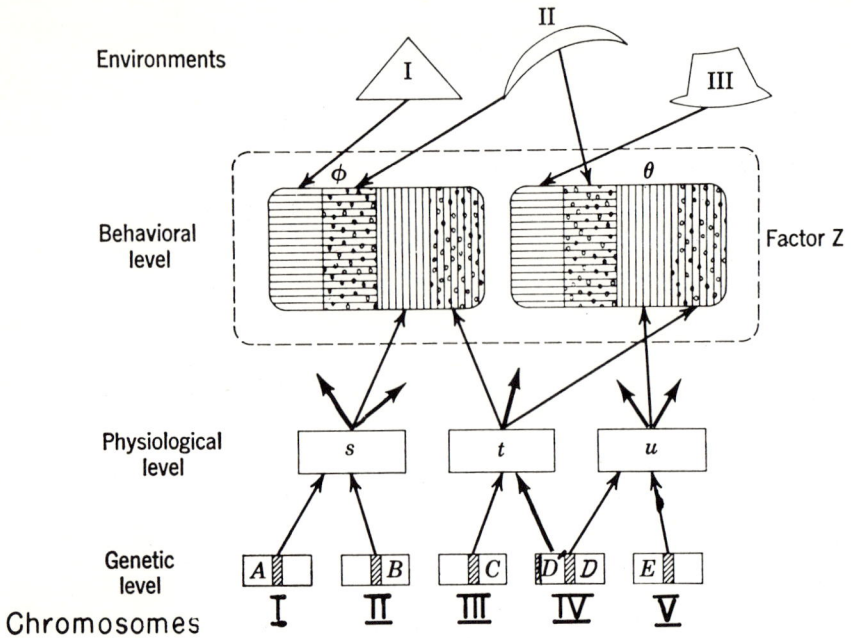

Fig. 2. Gene and chromosomal communality and specific and common variances of genetic and environmental origins illustrated schematically (after Fuller and Thompson, 1960, p. 343).

An understanding of the gene to trait pathway as well as the genetic meaning of factors or clusters is aided by reference to Fig. 2 adapted from Fuller and Thompson (1960, p. 343).

It can be seen that there is no precise correspondence between the genes shown in the bottom of Fig. 2 and the traits ϕ and θ at the behavioral level; thus the model is noncongruent. Enzymes, hormones, and neurons mediate the path through the physiological level. More complications may be expected for this model when mechanisms that evolved corresponding to regulator and operator genes at lower levels of life become known for more complex organisms. Any genetic meaning for primary traits derived from factor analysis resides in the causes of the correlation coefficients that were processed. Fuller and Thompson point out that correlations between behavorial traits may arise from one or more of the following kinds of communalities: genetic, chromosomal, gametic, and environmental. Genetic communality is shown in Fig. 2 and is simply pleiotropism; i.e., part of the correlation between the two traits is a result of the contribution of physiological character t to both ϕ and θ and t is under the control of gene C. Chromosomal communality is illustrated by that correlation between the two traits arising from the linkage between genes D' and D in chromosome IV; this kind of covariation is only important in other than large, random-mating populations. Gametic communality is not shown but originates from assortative mating for initially independent but heritable traits. The correlation between the two traits that stems from environmental communality is illustrated in Fig. 2 by the arrows from event II to ϕ and θ. Since ϕ and θ do correlate with each other for the various reasons mentioned, they will show up as a general trait, Z, when subjected to the procedures of factor analysis. Although the factor permits an economical summary of trait covariation, it can be seen from the figure that no single element in the genotype or the environment accounts for the covariation. The short arrows in Fig. 2 from the physiological level traits but not connected to behavioral traits are intended to convey pleiotropic effects on still other traits. Horizontal lines and vertical lines in the traits signify environmental and genetic variances respectively. Dotted areas signify covariances between the two behavioral traits.

Psychological Tests as Signs and Samples of Behavioral Traits

When the results of behavior genetic analyses are presented, they are usually in terms of test scores. Thus, discussions about the genetic variation in traits such as general intelligence, reasoning, aggression, social introversion, or schizophrenic tendencies derive not from direct ratings by skilled observers, but from questionnaires or inventories constructed by psychologists to measure the trait. When a child is said to be very intelligent, we mean, for instance, that a six-year-old is able to answer questions that are ordinarily only mastered by the age of seven or eight. The meaning of more "introverted" or more "dominant" is less obvious and has led to the development of a large body of literature on test development. Good summaries are provided by Cronbach (1960), Anastasi (1961), Loevinger (1957), and Cronbach and Meehl (1955).

Psychologists distinguish between tests as *samples* of a trait and as *signs* of a trait. If we were interested in measuring typing ability or spelling ability, tests could be constructed simply by permitting the subject to type for two min or to spell a series of words of known difficulty; we would be using performance on an obvious sample of the ability in question to predict more broadly to typing or spelling. If we were interested in measuring introversive tendencies, however, we could accumulate a series of self-descriptive statements that had been answered more often in a certain direction by individuals who had been rated by close observers as socially introverted than by individuals in general. Each such item would constitute a sign of introversion when answered in the "scored" direction by a previously untested subject. The inference that the new person has a certain rank on the dimension of introversion is made on a strictly actuarial basis and there need be no necessary rational connection between a particular item response and the criterion. Such empirically derived *construct* measures serve their original purpose of prediction quite well but do little, as they stand, to contribute to an explanation of behavior in terms of the underlying biophysical trait.

The process whereby such constructed measures of traits having ambiguous criteria for their existence accrue meaning is termed *construct validity* (Cronbach and Meehl, 1955).

1. "A construct is defined implicitly by a network of associations or propositions in which it occurs. Constructs employed at different stages of research vary in definiteness.

2. Construct validation is possible only when some of the statements in the network lead to predicted relations among observables.

3. The network defining the construct, and the derivation leading to the predicted observation, must be reasonably explicit so that validating evidence may be properly interpreted.

4. Many types of evidence are relevant to construct validity, including content validity, inter-item correlations, inter-test correlations, test-(criterion) correlations, studies of stability over time, and stability under experimental intervention. . . ." (Cronbach and Meehl, 1955, pp. 299–300)

While this type of validity cannot be expressed numerically, upper and lower bounds for the proportion of test variance which can be attributed to the construct may be garnered from its nomological network. The garnering will involve some subjectivity. Cronbach and Meehl go on to say, "Constructs may vary in nature from those very close to 'pure description' . . . to highly theoretical constructs involving hypothesized entities and processes, or making identifications with constructs of other sciences" (Cronbach and Meehl, 1955, p. 300). The whole process is not unlike that for developing and confirming theories in other scientific areas.

The provision for *hypothesized* entities gives a discipline struggling with problems of taxonomy the kind of flexibility it needs to encompass newly observed combinations of traits that may be casually related to genetic or chromosomal errors. The problem of admitting a new behavioral trait is not unlike that of admitting a new species (Dobzhansky, 1955b; Mayr, 1963). Too strict an adherence to operationism may impede the progress of a science as dependent on intuition and "open concepts" as psychology is at this stage of its formalization. Without delving further into the philosophy of science that underlies the discussion in this section, the thought may be advanced that unoppressive philosophers of science have provided psychology with a healthy "fisik" that might even be health-promoting for evolutionary biology (Feigl and Brodbeck, 1953; Pap, 1958).

The Twin Method

Of the few available methods for human behavioral genetics, the twin method is the most fruitful and economic. In 1875 Sir Francis Galton, a cousin of Darwin, called attention to the uniqueness of twins and suggested that studies of them might lead to an appraisal of the hereditary contribution to intelligence. Galton did not know that some same-sex (SS) pairs of twins were dizygotic (DZ) and others, monozygotic (MZ), nor was he aware of Mendelian genetics. Essen-Möller (1963) thought that Saint Augustine deserved some kind of priority for using the different fortunes of the DZ twins, Jacob and Esau, to refute the claims of astrology.

Since MZ twins have identical genotypes, any dissimilarity between pair members must be due to the action of the environment, either prenatal or postnatal. DZ twins, either SS or opposite-sex (OS), on average will have half their genes from a common source but also have certain environmental similarities in common such as birth rank and age of mother, thus providing a means for environmental control not otherwise possible. When both MZ and DZ SS twins from sizeable and representative samples are contrasted on measures of a trait, a means is available for the evaluation of the effect of different environments on the same genotype or the expression of different genotypes under the same environment. The determination of the heritability of a trait will be discussed below. It should be noted that some pairs of DZ twins will have more than half their genes identical by descent and others have much less gene overlap. Stafford (1965) has calculated that the chances are 1 in 100 that a pair of DZ twins will have 31/46 or more chromosomes in common, and 99 in 100 that they will have 15/46 or more.

The twin method depends on the accurate separation of SS DZ pairs from MZ pairs. Zygosity diagnosis by means of extensive blood-grouping has been perfected over the past decade (Smith and Penrose, 1955; Race and Sanger, 1962) and allows 100 percent accuracy in the diagnosis of DZ pairs who differ on any antisera reactions and about 95 percent accuracy for the remaining SS pairs of both types. Blood grouping has also permitted an evaluation of other, earlier means of determining zygosity, such as fingerprint analyses or physical resemblances (Gottesman, 1963a; Slater, 1963). Gottesman found that judgments of zygosity from photographs agreed with serological findings 78 percent of the time with a similar figure for fingerprint analyses. Experienced observers of healthy twins, examining both simultaneously, will seldom make errors.

The frequency of twin births varies according to the genotype and age of the mother. The expected proportion of MZ twins at birth can be calculated by Weinberg's differential method; the frequency of SS DZ pairs should be the same as the easily observed frequency of OS pairs so that subtracting twice the OS number from the total pairs leaves the expected number of MZ pairs. It is a unique feature of the twin method that the derived proportions of twin types in the general population permit a check on the representativeness of a sample of twins selected for the occurrence of one or more traits. Conservativeness is called for in the rejection of data from any particular sample not fitting expectations because there is not sufficient knowledge about the number of pairs surviving unbroken at a given age according to sex, twin type, and social class in different countries. In the U.S.A., as of 1958, there were 9.9 twin births per thousand deliveries from white women; 38.4 percent of the twin pairs were MZ. Corresponding figures for Negro Americans were 14.0 and 27.8 percent.

Strategic Implications of Twin Results

Basically the comparison of MZ and DZ intra-pair resemblance provides evidence for possible genetic factors, whereas the histories of pairs that differ give evidence as to environmental factors. The possibilities inherent in twin research are immense and they go far beyond mere nose-counting and the calculation of concordance rates. Some of the possible deductions may be mentioned; the list is far from exhaustive (cf. Allen, 1965).

a. If genetic differences are of no importance, and family environment is the major explanation for the tendency of qualitative traits or trait scores to cluster in families, there should be no essential difference between the incidence of a trait in the MZ and DZ co-twins of index twins. Such is the case with infectious illnesses, like measles and with strongly environmentally influenced behaviors like delinquency (Shields and Slater, 1961).

b. If the genes are important in the manifestation of a trait, MZ co-twins should be similar more often or more similar than DZ co-twins. The establishment of such a difference does not by itself prove the importance of variations in genotype; but the latter remains the most likely explanation, unless it can be shown either that MZ twins as such are predisposed to show a particular trait or that the environments of MZ twins are systematically more alike than those of DZ twins in features which can be *shown* to be causally significant for the trait.

c. The comparison of MZ and DZ intra-pair resemblance could also lead to the refinement of trait taxonomy by revealing that subtypes of psychosis, or neurosis, or some ability or temperament factors may be either more under genetic influence than others or free of genetic influence (e.g., Blewett, 1954; Gottesman, 1962; Inouye, 1961).

d. Given evidence of gross genetic predisposition, the comparison of illnesses of twins who are both affected could lead to the discovery of which aspects of the illness, such as onset, symptoms, course, response to treatment, and outcome, may be most genetically influenced, and which aspects most susceptible to environmental influence. A parallel analysis may be made of the stages in the acquisition of a skill. In appropriate circumstances the co-twin control method of Gesell and Thompson (1929) can be used. For instance, treating only one of a pair of similarly affected schizophrenic twins by a given method (Benaim, 1960) or both with multiple methods.

e. The variability of psychiatric abnormality in the MZ co-twins of typical cases could lead to the identification of schizophrenic or affective "equivalents" or of an underlying "schizotypic" deviation (Meehl, 1962). Such a development could lead to an improved classification of mental disorders, as well as to the detection of "carriers" (Gottesman and Shields, 1966c).

f. The comparison of MZ twins who differ with respect to a trait either totally or quantitatively, allows deductions to be drawn regarding differences in the twins' previous life experiences that might relate to differences in their phenotype. Evidence from twins on these lines may identify interpersonal and environmental factors important in trait variation (Pollin et al., 1966; Rosenthal, 1963).

g. Under certain conditions, MZ twins reared apart from early in life provide important clues to both environmental and genetic sources of trait stability (Shields, 1962; Juel-Nielsen, 1965).

Although the goal of many classical human genetic studies has been to establish the mode of inheritance of a condition, the twin method does not claim to do this. It is as a screening device that results in the more eager pursuit of biological causes of variation for some traits and the abandonment of such hypotheses for others that the twin method serves an important function in human behavioral genetics.

Twin Data Statistics

Behavioral geneticists who use the twin method have elected different ways of expressing their findings. No single method can adequately summarize the data from twins so that multiple means are preferred (Allen, 1965). When a trait is qualitative or can arbitrarily but usefully be dichotomized, the simplest way of communicating the findings is in terms of the percentage of MZ and then DZ pairs that are concordant for the presence of the trait. Even this simple statistic, however, can vary from report to report depending on whether the author counted pairs or cases (i.e., some pairs would appear twice) and whether he chose to apply some type of age correction (Allen, Harvald, and Shields, 1967; Gottesman and Shields, 1966b). Estimates of the *amount* of genetic variance in a trait that has been treated dichotomously have been made using a formula for heritability (H) recommended by Neel and Schull (1954). Such values of H bear little resemblance to the usual concepts of heritability and are best abandoned. A procedure developed by Falconer (1965) for the estimate of the heritability of threshold characters does more justice to concordance data and yields approximate but useful estimates of the total amount of additive variance in such characters (Gottesman and Shields, 1966c).

Inasmuch as few behavioral traits are truly dichotomous, another approach is needed for the processing of twin-derived measures of continuous behaviors that will tell us, first, *whether* heredity is playing a significant part. The answer may be inferred from the comparison of the intra-class correlation coefficients (R) for the scores of the MZ and DZ pairs (Haggard, 1958). Once the between-pairs variance (BPV) and the within-pairs variance (WPV) have been calculated, R may be found from the following formula:

$$R = \frac{BPV - WPV}{BPV + WPV} \qquad (1)$$

The statistical significance of R is the same as that of the variance ratio (F) formed by BPV/WPV. The difference between the two R's, R_{MZ} and R_{DZ} is readily tested (Gottesman, 1963a). Another method often used for the same task consists simply of demonstrating a significant value of F for the comparison of the two within-pairs variances:

$$F = \frac{WPV_{DZ}}{WPV_{MZ}} \qquad (2)$$

Secondly, how great is the part played by heredity? The quantification of the genetic variation in a trait for a particular sample, or the relative effects of genetic and environmental factors on the differences observed within twin pairs raised in the same family, may be estimated with caution from the twin method. Holzinger (1929) proposed a measure of heritability that may be estimated from either MZ or DZ R's or from their WPV's. The symbol H will be used to denote twin-based estimates of heritability in contrast to h^2, the symbol used in animal genetics (Falconer, 1960). H gives the proportion of trait variance produced by genotypic differences within families only (unlike h^2) and underestimates the overall effects of heredity by a factor of two; furthermore H does not distinguish between additive and nonadditive genetic variance and so is closer to heritability in the "broad" sense (Lush, 1945). Only when Falconer's (1965) method for dealing with threshold characters is applied to twin data does the twin method yield estimates of h^2 rather than H.

Two formulas were proposed by Holzinger for calculating H and they are statistically equivalent under certain conditions (cf., Clark, 1956):

$$H = \frac{R_{MZ} - R_{DZ}}{1 - R_{DZ}} \qquad (3)$$

$$H = \frac{WPV_{DZ} - WPV_{MZ}}{WPV_{DZ}} \qquad (4)$$

Falconer (1960) suggested that the difference between MZ and DZ R's could by themselves be taken as an estimate of half the heritability if there were no nonadditive genetic variance. Since this assumption is not safe, such differences may only be taken as an index to half the broad sense heritability. Table 1 (Vandenberg, 1966) shows values of H for various combinations of R. It can be seen that unless trait measure reliability is high enough to permit a correlation of at least .50 between MZ pairs, H cannot attain large values.

Table 1. Values of h^2 for All Combinations of Values of Fraternal and Identical Twin Intraclass Correlations [a] (from Vandenberg, 1966)

R_{DZ}	.10	.20	.30	.40	.50	.60	.70	.80	.90
.90	—	—	—	—	—	—	—	—	.00
.80	—	—	—	—	—	—	—	.00	.50
.70	—	—	—	—	—	—	.00	.33	.67
.60	—	—	—	—	—	.00	.25	.50	.75
.50	—	—	—	—	.00	.20	.40	.60	.80
.40	—	—	—	.00	.17	.33	.50	.67	.83
.30	—	—	.00	.14	.29	.43	.57	.71	.86
.20	—	.00	.13	.25	.38	.50	.63	.75	.88
.10	.00	.11	.22	.33	.44	.56	.67	.78	.89
R_{MZ}	.10	.20	.30	.40	.50	.60	.70	.80	.90

[a] Dashes represent negative values of h^2.

If H has been calculated from Formula (4), there is a relationship between H and F from Formula (2) that permits converting one to the other if only one has been reported:

$$H = 1 - \frac{1}{F} \text{ and } F = \frac{1}{1 - H} \qquad (5)$$

The very simplicity of computation of heritability estimates from twin data invites misinterpretation. Some twin researchers have suggested a ban on reporting values of *H,* but this makes no more sense than not reporting probability levels or not using matrix algebra because some people misinterpret the findings. So long as *H* values are accompanied by caveats, and bolstered by *R*'s and *F*'s, they can be quite useful in comparing one twin study with another and one trait with another within the same study (cf. Kempthorne and Osborne, 1961; Harris, 1965). Until proven otherwise, a value of *H* for any trait is sample-specific, but its replication in other studies greatly strengthens genetic interpretations (Gottesman, 1965; Vandenberg, 1966). The simultaneous treatment of more than one variable at a time from twin samples, multivariate analyses, add a new and useful dimension to our understanding of behavioral characters; the few applications of such techniques may be expected to increase with the availability of computers (e.g., Cattell, 1965; Gottesman, 1963a; Vandenberg, 1965; Loehlin and Vandenberg, 1966).

Intellectual Abilities

Evidence for the crucial importance of genetic factors to the observed variation in normal and mentally retarded individuals' abilities has accrued from studies of pedigrees, familial correlations, foster children, and twins (Fuller and Thompson, 1960; Gottesman, 1963b; Vandenberg, 1967a). The results consistently show that, *with a more or less uniform environment,* the variation observed among individuals in general intelligence is largely accounted for by genetic factors. Galton's application of the pedigree method to study the inheritance of "genius" in 1883 was historically important but incorrect for a quantitative trait. He went on, with Karl Pearson, to develop correlation techniques that permitted an exact statement of the degree of resemblance between relatives. Since they found the same degree of resemblance between siblings, .52, for teachers' ratings of intellectual capacity as were found for various anthropometric characters, they concluded that mental traits were no less genetically determined and were inherited in the same mode (Pearson, 1904). Although both tests and sampling methods have improved since the days of Galton, the results of familial correlations have varied little. Conrad and Jones (1940) observed an *r* of .49 between siblings and between parents and offspring; the *r* between mothers and fathers was .52 showing that people do not mate randomly with respect to general intelligence.

If a homogeneous environment erased individual differences in intelligence, this would argue against the importance of genetic variation. Lawrence (1931) found that among children reared in an orphanage, there was as much variation as among children reared in their own homes. In some sound studies of foster children, the resemblance between such children and their foster parents is measured and compared with matched samples of parents and their biological offspring. Leahy (1935) found r's of about .20 for foster mothers and fathers with their children and the usual ones of .50 in the control group. For unrelated children reared together as sibs, Burt (1958) found an r of .27 for tested intelligence and one of .54 for general scholastic attainments. Comparable r's for true sibs reared together were .54 and .81.

As discussed above, a phenotypic correlation by itself cannot tell much about the relative roles of genetic and environmental factors since so many things go into its makeup. But the pattern of all correlation studies fits quite well with the idea that the greater the gene overlap, the greater the similarity in intelligence. A compilation of 99 such r's (Erlenmeyer-Kimling and Jarvik, 1963) from different tests and samples ranging from unrelated persons reared apart to MZ twins reared apart showed median phenotypic correlations for IQ that were the same as the known genetic correlations for different degrees of relationship. The data from twins is of special significance since environmental variables are better controlled and since the twin method generates values of H. The findings of Newman, Freeman, and Holzinger (1937) are typical. For their sample of 50 pairs each of MZ and SS DZ adolescent twins they found R's of .91 and .64 respectively. These data yielded a heritability estimate of .69 for the proportion of within family variation in IQ associated with genetic factors. These same authors located 19 pairs of MZ twins reared apart and the within pair differences in Binet IQ points ranged from 1 to 24. In terms of correlation for these special pairs, the value was .67. In a remarkably large sample of 40 MZ pairs reared apart, Shields (1962) found a correlation of .77 on a nonverbal test of intelligence. The average intra-pair difference, as computed by Shields and the author, was 10 points for the identicals reared apart compared to 9 points in a control sample of identicals reared together. At least 25 percent of the sample of reared apart MZ pairs had within pair IQ differences on one of two tests administered which exceeded 16 points. Thus it is obvious that individuals with exactly the same genotype can differ widely on the phenotypic trait called intelligence when their environments differ in relevant ways. The synthesis of such findings pose no problem to the behavior geneticist who possesses the schema of the norm of reaction (Dobzhansky, 1955a) or the reaction range (Gottesman, 1963b). A lack of clarity has been perpetuated in discussions about the origins of individual differences by the failure to specify routinely the environmental conditions when describing a phenotype. And conversely, the attribution of an effect to an environmental manipulation can be very misleading unless the genotype is specified. The work of Cooper and Zubek (1958) with rats and of Freedman (1958) with inbred dogs provide excellent examples of the interplay of heredity and environment in producing behavioral variation or constancy.

The discussion above has been limited to considerations of general intelligence, but recent attempts to use the twin method take cognizance of the work of the factor analysts in partitioning general intelligence into its "natural" components. No work on Guilford's 120 factors has been done from a genetic point of view, but Vandenberg (1967a) has reviewed twin studies using the six "primary mental abilities" (PMA) and other specific abilities. Seldom did researchers analyze the two sexes separately; it is possible that a specific ability may be highly heritable for one sex and not the other. Blewett (1954), using the PMA, tested the hypothesis that the heritability of the total score would be greater than any of the components (in keeping with the views of British factor analysts) but could not support it with his twin data. Vandenberg's work with the Wechsler and Differential Aptitude Test does support the hypothesis. The opposite hypothesis was tested by Nichols (1965) with different tests and received some support. The four studies using the PMA are in good agreement in showing the Verbal, Space, and Word Fluency factors to be highly heritable with values for H around .60. Only Blewett found Reasoning to have a significant and high heritability which is disconcerting in view of its high correlation, .84, with general intelligence (Bischof, 1954). The general idea that traits found by factor analysis may be biologically more basic than those chosen emipirically has not received sound support in the opinion of the author. Fuller and Thompson (1960) suggested that direct behavioral measures may be the only reliable indicators of certain kinds of inherited organic characters.

Mental Retardation

Genetic aspects of mental retardation have been greatly clarified in the past decade by advances in cytogenetics and biochemistry (Anderson, 1964; Lejeune, 1964). The older idea that mental retardation was a homogeneous condition manifested by low IQ was already weakened by Fraser Roberts' (1952) data showing that the siblings of probands with IQs less than 45 had their intelligence scores normally distributed with a mean of 100, some 20 points *higher* than the mean of the siblings of the more mildly retarded (cf. Penrose, 1963). He also showed that the number of school age children in a total sample with IQs less than 45 exceeded the number expected on the basis of normal curve statistics by a factor of 18. Taken together these findings suggest two distinct classes of causes for the severely and mildly retarded. For the former, such discrete factors are implicated as brain injury and diseases, chromosomal anomalies, and single dominant or recessive genes. The mildly retarded group, making up as much as 75 percent of all retarded, can be encompassed etiologically within the framework of polygenic inheritance of a graded character, i.e., the mildly retarded are simply those individuals at the tail end of the normal distribution of intelligence and differ quantitatively rather than qualitatively from the remainder of the population.

Twin data on mental retardation were, for the most part, collected before the taxonomic revolution that led to the discovery of discrete subtypes such as Down's Syndrome, phenylketonuria, and some four dozen other syndromes. In a subset of twins with retardation uncomplicated by such symptoms as epilepsy, delinquency, or psychosis (Rosanoff, Handy, and Plesset, 1937) an MZ concordance rate of 73 percent was found compared to 29 percent in SS DZ pairs; the concordant MZ pairs were within five IQ points of each other. Juda in 1939 as reported by Shields and Slater (1961) excluded twin probands with known exogenous causes for their retardation and found an MZ concordance rate of 100 percent, with 58 percent for the DZ pairs. Allen and Kallmann (1962) report a paradoxical finding in their series of high-grade, undifferentiated retarded twins. They found 41/41 MZ pairs and 35/39 SS DZ pairs concordant. Such results taken at face value would rule out genetic factors for so-called familial mental retardation, surely an unlikely conclusion in the light of other data.

In their monumental family study of mental retardation, Reed and Reed (1965) provided more indirect information regarding the genetic basis of retardation. Their data permit the inference that five million of the six million retarded in the U.S.A. are the offspring of either a retarded parent or of a normal parent who had a *retarded sibling*. A more detailed look at the empiric risk figures for their sample further implicates polygenes as the major source of high grade (IQ 50 to 70) retardation. If genetic factors were of little importance it would be irrelevant as to whether the sibling of a normal parent were retarded when it came to looking at the children of such normal parents. However, 7.3 percent of normally intelligent persons with retarded siblings produced one or more retarded children themselves while only 1.3 percent of normals without retarded siblings met this fate. It is unlikely that some qualities of the environment account for this almost sixfold increase in offspring retardation since aunts and uncles have little influence on their nieces and nephews compared to parents. It is a fact of gene segregation though, that 25 percent of the genes of avuncular relatives will be identical because they are from a common source. Thus it seems reasonable to conclude that polygenes for lowered intelligence were transmitted with a higher frequency by phenotypically normal parents whose retarded siblings provided a truer picture of the stored variability in the parent's genotypes. In the same study, the Reeds found that the entire distribution of IQs for children with retarded avunculars was shifted toward the low end compared to children with normally intelligent aunts and uncles. The mean IQ of the latter was 108 (N = 5988), of the former, 102 (N = 1047); 6.4 percent of the normal group's children had IQs under 80 compared with 18.2 percent of the children with retarded avunculars.

In those few unions between retardates and normals with retarded siblings, 23.8 percent of the offspring were retarded while only 0.53 percent of the offspring of normal by normal matings, with all siblings normal, had such an outcome. The empiric risk figures for normally intelligent or retarded individuals who have already produced at least one retarded child are shown in Table 2. These data complete this section and set the stage for a brief discussion of the relationship between intelligence and fitness. Since differential reproduction is the essence of natural selection, any trait determined in part by the genotype and associated with fitness commands our attention.

Table 2. Empiric Risk Figures for Retardation Subsequent to One Retardate
(after Reed and Reed, 1965)

Mating Type	% Retarded after 1 Retardate
Retardate × Retardate	42.1
Retardate × Normal	19.9
Normal × Normal with Retarded Avunculars	12.9
Normal × Normal	5.7

Intelligence and Fitness

Mayr (1963) summed up the role of behavioral traits in evolution: "The point that is important for us is that new habits and behavior always start in a concrete local population. If the new behavior adds to fitness, it will be favored by selection and so will be all genes that contribute to its efficiency" (Mayr, 1963, p. 605). Unfortunately, as Mayr also pointed out, selection favors genotypes that are *merely* successful at reproduction and which may not add to the overall survival value of the species.

Differential fertility was recorded in Europe for urban vs. rural and rich vs. poor strata of society as early as the 1600's. With the advent of intelligence testing and the construction of valid instruments, surveys relating a child's IQ to the number of his siblings became feasible. It was not until the second Scottish survey of 1947 that such research was done on a truly large scale (see Maxwell, 1954 for references). The results from testing almost every eleven-year-old child in all of Scotland confirmed earlier findings. The more brothers and sisters a child had, the lower was his IQ.

The correlation between family size and IQ in various studies clusters around a value of $-.30$. From these kinds of evidence, many scientists predicted a gradual decline in the intellectual level of the population of from two to four IQ points per generation. If true, it meant that the total forces of selection were favoring lower intelligence.

From a comparison of the 1932 Scottish survey with the one done in 1947, it was apparent that not only had the mean IQ not declined, it had undergone a small improvement (for the group test only, not on the Stanford-Binet). Similar findings were reported by Cattell (1951) for English children. A paradox existed and efforts towards its solution finally paid off in the work of Higgins, Reed, and Reed (1962).

Direct studies of the relationship between IQ and fertility had been impossible because early IQ tests were designed for children, and no tested children had been followed to the completion of their reproductive lives. Other commentators had objected to the conclusions about the decline in IQ because a survey of children excludes the unmarried and infertile adults from the data. Other objections were reviewed by Anastasi (1959) but they turn out not to be crucial to the solution of the paradox. Penrose (1948) and Willoughby and Goodrie (1927) entertained models which may have anticipated the data provided by Higgins and his co-workers.

The Minnesota geneticists in their study referred to above (Reed and Reed, 1965) covering 6 generations of 289 families (82,217 persons) also directed their attention to fertility and IQ. They reasoned that if the average intelligence of those who failed to reproduce in each generation was appreciably lower than those who did reproduce, the negative r between family size and IQ could not be valid. Among their total population they had IQ values recorded for 1,016 families in which both parents and at least one child had been tested thus avoiding reliance on educational attainment as a *sign* of ability (cf. Bajema, 1966). The parents had been tested when *they* were schoolchildren. In addition, the investigators had IQ test data for 884 married siblings of the parents as well as for 66 unmarried and childless siblings of the parental generation.

It should be noted that most of the individuals in this large subsample were unrelated to the original retarded subjects except by marriage. The IQ distribution for the 2,032 parents was essentially normal with the mothers' mean equal to 103, and the fathers', 101. The relationship between the IQs and size of family for the 2,039 children of the parental sample was quite in line with the earlier Scottish and English surveys. Up to a sibship of five, no marked difference exists in the mean IQs of the children. For the entire sample of children, the r between family size and IQ was $-.30$.

A direct test of the relationship between IQ of parents and children was then made. The usual correlation of about $+.5$ was obtained. The larger families with the lower IQ children were being produced by the lower IQ parents. The mentally retarded parents, as defined by IQ 70 and below, had an average of 3.81 children. The latter average was by far the highest of any of the parent groups. Does this mean that the dire predictions about the decline of intelligence are indeed coming true? Not quite. At this point the geneticists proceeded to use their unique data on the other siblings in the parental generation.

First the spouses who married into the sibships under study were removed from the analysis. To the remaining parents were added *their* married brothers and sisters to form a sample of 1,900 married siblings. Persons under IQ 55 averaged 3.64 children; those from 56 to 70 averaged 2.84; and, parents with IQs above 130 averaged 2.96. The key to the mystery must then be associated with the 66 unmarried siblings in the parental generation. When they are added to the 1,900 married siblings the distribution of fertility as a function of intelligence changes markedly. Table 3 shows that when *all* the siblings are followed-up, the lowest IQ

Table 3. Intelligence of all Siblings and Reproductive Rate
(from Reed and Reed, 1965)

I.Q. of Siblings	Number of Siblings	Average Number of Children
55 and below	29	1.38 ± 0.54
56 – 70	74	2.46 ± 0.31
71 – 85	208	2.39 ± 0.13
86 – 100	583	2.16 ± 0.06
101 – 115	778	2.26 ± 0.05
116 – 130	269	2.45 ± 0.09
131 and above	25	2.96 ± 0.34
Totals	1966	2.27

range produced the fewest children and the highest, the most. The average number of children ranged from 1.38 to 2.96. Previously obtained negative correlations of −.30 disappear when the single siblings are included.

It would thus appear that the net direction of selective forces for intelligence is probably in a favorable direction for the species. Some 43 percent of an unselected sample of retardates never reproduced (N = 1,450). Another means by which this comes about is for the persons with low IQ to remain unmarried. The Reeds (1965) found that only 38 percent of their total sibling group with IQs 55 and below married. Between IQs 56 and 70 the proportion married jumped to 86 percent, still below the remainder of the sample which ranged from 97 percent to 100 percent married. The latter figure of 100 percent married was for the brightest group with IQs

131 and above. The mean IQ for the unmarried siblings was 80, while it was 100 for the total sample of married siblings.

Other recent work supports the suggestion that the direction of selection for intelligence is not dysgenic. Carter (1962) reports a study by Quensel in Sweden of the fertility and marriage of a large sample of IQ tested recruits born in 1924. Although none of the men had completed their reproduction at age 29 (when the data were collected), trends were already evident. The dullest group had the highest fertility within marriage, but the lowest proportion married, 57 percent. In the study by Bajema (1963) in a Michigan city, the completed fertility of all native white *Ss* born in 1916 and 1917 who had reached the beginning of the seventh year of schooling was examined. The adults had been tested at an average age of 11.6 years. Although the Bajema sample was smaller than that of Higgins and his co-workers, the results are quite close. For IQ range greater than 130 the average number of children was 3.00, dropping to 2.05 for the range 71 to 85 (the 3 *Ss* under IQ 70 did not reproduce). *Relative fitness* is defined as the ratio of population growth rate per individual of a specific IQ group to the same rate for the optimum phenotype (IQ 120 and above) in the Bajema study. Thus the relative fitness value for the IQ 120 and above group was 1.0; for the IQ range 69 to 79 it was only .58. An intriguing bimodal distribution for IQ and fertility was found by both the Reeds and Bajema. It should serve as a challenge to other researchers.

Man's current level of intellectual ability must have been the outcome of a lengthy process of directional selection. Questions about the rate of evolution of higher intelligence, given the kinds of data on selection reviewed in this section, have rarely been faced. Reed (1965) has advanced some speculations on this matter after assuming, for purposes of illustration, that our ancestors of 35,000 generations ago had an average IQ of at least 30.

> If we assume that the average intelligence evolved from an IQ of 30, equated to the present average of 100, there has been an increase of about 70 IQ points in about 35,000 generations. This is 0.002 of an IQ point for the average rate of change *per generation*. Thus, while our imaginary rate of two thousandths of an IQ point per generation is not a practical figure, it does serve notice that, even during the greatest spurts of man's evolution, the largest change in intelligence for any one generation must have been modest indeed. . . . It is hard to imagine any reversals of intelligence which would correspond to the drops in population size due to disastrous epidemics. Changes in intelligence in each generation should be much more conservative than fluctuations in population size (Reed, 1965, pp. 320–321).

While this brief review of some recent findings on the direction of selection for intelligence may quiet the fears that society is headed for a chaos where the dull would inherit the Earth, it should not lull us into complacency about the quality of the species.

A Polygenic Model for the Transmission of Intelligence

Following in the wake of deserved enthusiasm over the rediscovery of Mendelian genetics and its immediate explanatory power, biologically-minded psychologists tried to explain many normal and abnormal behavioral traits by major genes with or without modifiers. Such naivete was not, however, confined to psychologists (cf. Carlson, 1966). It seems that the "classical Mendelian gene" *Zeitgeist* overdetermined the thinking of prototypic behavioral geneticists, leaving us an inertial heritage in need of exposure to contemporary views of the complexities of the gene to behavior trait pathway. The time is overdue for embracing the facts of pleiotropy and polygeny; i.e., a gene product may affect many traits, and a trait is affected by many gene products. Now that examples are available (e.g., Lerner, 1958; Falconer, 1960) we may expect more progress along these lines.

While specific polygenic models may be premature for explaining the transmission of intelligence, their heuristic value may compensate for this fault if they are not too unreasonable. Burt and Howard (1956) presented and defended a general polygenic model involving n pairs of genes. After allowing for assortative mating and incomplete dominance, their observed familial correlations agreed quite well with those predicted from Fisher's (1918) formulas. Two other attempts have been made to construct polygenic models, but they tried to explain too much. Hurst (1932), combining data from IQ scores of families he studied with ability ratings of royal families from as far back as the eleventh century, proposed that a pair of major genes, *Nn*, and five pairs of minor modifiers (*Aa*, . . . , *Ee*) could account for his observations. *N* was supposed to be dominant and led to middle of the range ability regardless of the modifier complement. For *nn* genotypes, however, each dominant allele in the modifier system added 20 IQ points so that the model covered the IQ range 0 to 200 in 21 phenotypic classes. Hurst's model had numerous shortcomings (Conrad and Jones, 1940; Fuller and Thompson, 1960) but it could account for children who resembled neither parent.

Pickford (1949) proposed a simple polygenic model involving 10 pairs of genes with equal and additive effects. As evidence he cited the fact that expansion of the binomial $(.5 + .5)^{20}$ resulted in a very good fit to the distribution of IQ's obtained in Terman and Merrill's *standardization* population for the 1937 revision of the Binet intelligence test. The range covered by a 10 pair model extends from plus to minus 4.5 standard deviations. Fewer pairs cover a smaller range and hence were rejected for the favored model. Fitting curves to a specifically nonrandom population seems like a strategic error when the goal is formulating a general theory. Terman and Merrill (1937) used a sample of children representative of the census distribution for fathers' occupational grades. Children in institutions were not tested. The lowest two grades were grossly under represented, comprising 16 percent of the total rather than the 30.8 percent in the census. One effect of this bias is to reduce the number of less intelligent children; the marked relationship between the IQs of children and their fathers' occupation is a well-documented fact (McNemar, 1942; Anastasi, 1958) with about a 20 point difference between the means of children from the two extremes. The point of this criticism is that a polygenic model need only cover the patently continuous range of observed IQs, 50 to 150 (about plus and minus three standard deviations) to encompass virtually the entire population.

A five pair polygenic model, assuming random mating and no dominance, provides a parsimonious starting point in theorizing about the mode of transmission for intelligence. Let the genotype for each parent be heterozygous at each of five loci and be represented by $A^1 A^2 B^1 B^2 C^1 C^2 D^1 D^2 E^1 E^2$. To such genotypes assign a phenotypic index of 100. For each additional allele with superscript 2, the phenotype is enhanced by 10 IQ points, while those with superscripts of 1 have a neutral effect. Relative frequencies of the offspring genotypes from the matings of such parents are given by the expansion of the binomial $(.5 + .5)^{10}$. Table 4 shows that such a model yields 11 phenotypic classes, the relative frequency of each, and the phenotypic IQ associated with each genotype. The total number of genotypes possible with a five locus model is 3^5 or 243. Plotting the frequencies from Table 4 would not lead to a smooth bell-shaped curve. The necessary provision for environmental sources of phenotypic variability would lead to a close fit with empirically obtained data.

Table 4. Theoretical Predictions of IQ from a Five-Pair Polygenic Model

Number of enhancing genes	Relative frequency of genotypes	Phenotypic IQ
0	1	50
1	10	60
2	45	70
3	120	80
4	210	90
5	252	100
6	210	110
7	120	120
8	45	130
9	10	140
10	1	150

The major shortcomings of the simplified model would have to be overcome in the next approximation. Provision would have to be made for the high degree of assortative mating for intelligence, for the dominance of normality implied by the regression of offspring IQs toward the mean, for the possible selection against excessive homozygosity, and for heredity *x* environment correlations (Cattell, 1965; Loehlin, 1965). As Stern (1960) has noted, similar models can be constructed where the posited factors are environmental events rather than genetic ones. The weight of the evidence considered in this section, however, given that the environment is more or less uniform, favors a polygenic model. The oft heard criticism from purists that polygenic proposals for any trait are merely confessions of our ignorance that lead nowhere seems unduly pessimistic and nonconstructive. Many loci in a polygenic system may not be heterotic thus simplifying future analyses (cf. Lewontin and Hubby, 1966). Even when many genes affect a trait, a manageable few may be making major contributions to the phenotype (Mayr, 1963). A soundly based science of quantitative genetics now exists with examples of progress in the analyses of polygenic characters (e.g., Falconer, 1960). Finally, it is difficult to forecast the long range effects of the recent rapid advances in molecular biology, including information about genetic fine structure, upon the problems of human genetics.

Schizophrenia

Among the mental disorders so serious as to result in hospitalization, schizophrenia is the most common with some 50 percent of all mental hospital beds in the U.S.A. occupied by persons carrying this diagnosis. Most estimates of the life-time morbid risk for developing schizophrenia range from one to three percent, with the disorder showing up between the ages of 15 and 55 in most instances. If diabetes mellitus may be termed a geneticist's nightmare (Neel, Fajans, Conn, and Davidson, 1965), schizophrenia deserves the appellation in psychiatric genetics (Gottesman and Shields, 1966b). Opinion is still sharply divided on both sides of the Atlantic as to whether genetic factors are even important in the etiology of this disorder (Bleuler, 1965; Lidz et al., 1958). Family and twin studies starting with Rüdin (1916) and Luxenburger (1928) provide a large body of data that bears on the issue of the relative importance of hereditary and environmental factors in the etiology of schizophrenia. More recently work has begun on following the children of schizophrenic mothers who were fostered out to normal mothers (Heston, 1966) or who have not yet developed overt signs of the disorder (Mednick and Schulsinger, 1965). Other strategic populations such as the children of two schizophrenics are studied whenever possible (Kallmann et al., 1964; Elsässer, 1952; Rosenthal, 1966). Hoped-for evidence from biochemical studies of blood and urine fractions has not yet proved itself useful (Kety, 1965; Himwich, Kety, and Smythies, 1966) to genetic theories about etiology, nor can the changes in symptom severity brought about by phenothiazine medication be used to argue directly against environmental causes.

So long as we limit ourselves to a simple Mendelian framework with dominant and recessive gene effects and construe schizophrenia as a homogeneous disease entity, we have difficulty in constructing a genetic model for the transmission of the disorder. It has been known all along that the classical ratios were not found among the relatives of probands (e.g., Gregory, 1960). Arguing by analogy with advances in research on mental deficiency, thus leading to genetic and environmental heterogeneity, would allow room in a taxonomy of schizophrenia for some Mendelian forms. The latter would leave most schizophrenias unexplained, paralleling the situation in mental retardation, and so far no Mendelian forms have been discovered. If there were some reason to perpetuate a strictly mono-lithic view about the genotype underlying schizophrenia, it could be done by invoking the concept of incomplete manifestation. An example of such a theory will be presented below. To invoke continued mutation as the cause of a large proportion of cases, as for example with neurofibromatosis (Crowe, Schull, and Neel, 1956), puts a strain on current beliefs about the mutation rate for human loci of about 10^{-5}. The mutation rate required, depending on the mode of inheritance and calculation, ranges from 7.5×10^{-5} (recessivity) to about 8.6×10^{-3} (incomplete dominance or recessivity) according to Kishimoto (1957). Evidence for the completely opposite view that social factors are sufficient to account for the etiology of schizophrenia stem from such observations as those of Hollingshead and Redlich (1958) that the prevalence of the disorder is eight times higher in the lowest social class than in the highest. Genetic interpretations of the same data involving gene migration and assortative mating have also been advanced, however (Gottesman, 1965; Moran, 1965).

Family and Twin Studies. It is very easy to criticize the work that has been done thus far in psychiatric genetics, but most often the criticisms are not informed ones. A kind of double standard has emerged in some quar-ters of the behavioral sciences wherein the genetic argument must be proved beyond the shadow of a doubt before it will be entertained as a useful working hypothesis. Meanwhile thinking continues in terms of unproved psychological causes. Certainly problems of sampling and diagnosis are paramount in family and twin studies of schizophrenia, but they do not nullify the fairly consistent pattern that has emerged with respect to the frequency of the disorder in relatives of probands (Gottesman and Shields, 1966b; Zerbin-Rüdin, 1967). Rosenthal (1963) observed that very few colleagues, if pressed, would not admit that they gave weight to both genetic and environmental factors in the etiology of schizophrenia. It may not be generally appreciated that positing a *specific etiology* for the disorder that is genetic means only that the gene or genes are necessary, not that they are sufficient, for schizophrenia to occur (Meehl, 1962).

The average frequency of schizophrenia in various degrees of relatives of schizophrenic probands is presented in Table 5. The averages were calculated by Shields and Slater (1967) from the comprehensive survey of virtually all systematic studies conducted since 1916 in Europe and the United States reported by Zerbin-Rüdin (1967). Studies were combined that may have used different criteria for the diagnosis of schizophrenia in the proband but all probands were at least hospitalized for serious mental illness. Such a criterion will limit the kinds of cases entering a sample to usually clear instances of schizophrenia; at the same time such a criterion excludes mild or compensated cases or those tolerated by their families thereby weakening the data base for a general genetic model. Somewhat

Table 5. Frequency of Schizophrenia in the General Population and in the Relatives of Schizophrenics: Pooled Data [a]

Number of Investigations	Number of Countries	Relationship to Schizophrenic	Number of Relatives investigated (corrected)	Expectation of Schizophrenia in %
19	6	Unrelated (gen. pop.)	330752	0.86
14	8	Parents	6622	5.07
12	7	Sibs	8484.5	8.53
6	2	Children	1226.5	12.31
4	4	Uncles and aunts	3376	2.01
5	4	Nephews and nieces	2315	2.24
4	4	First cousins	2438.5	2.91
		Twin Pairs	*Both twins affected*	
6	5	DZ OS	24/430	5.6%
9	7	DZ SS	71/593	12.0%
10	8	MZ	252/437	57.7%

[a] Data on general population and relatives other than twins based on Zerbin-Rüdin (1967); correction for age usually by shorter Weinberg method, risk period 15–39; only cases of definite schizophrenia counted. Data on twins based on Gottesman and Shields (1966b) with omission of Kallmann's childhood schizophrenics and substitution of later data in current Norwegian and Danish studies; no age-correction, but some doubtful schizophrenias counted (after Shields and Slater, 1967).

different and usually broader diagnostic criteria are used for the relatives of probands since in many instances they will not have been hospitalized but will have been subjected to close psychiatric scrutiny solely because they are relatives. Since control cases (they are the people hospitalized) were not subjected to equally thorough screening, the population general incidences (0.35 percent to 2.85 percent in 19 studies) will be underestimates and thus exaggerate the contrasts in frequency between probands and their relatives compared to the population incidence. Relatives of a control group matched on such characteristics as social class have not been reported.

Further information comes from the few studies where the offspring of two schizophrenics have been examined for the disorder. A figure of about 35 percent (Rosenthal, 1966) affected can be taken as the expected morbid risk although a figure as high as 68 percent has been reported (Kallmann, 1938). The data in Table 5 taken together with the latter make it clear why no simple Mendelian hypothesis does justice to the observed frequencies. Evidence against the crucial importance of environmental factors in the *etiology* of schizophrenia comes from the fact that not more of the children of two schizophrenics are affected. It is difficult to imagine a worse child-rearing psychological climate, yet some 60 percent of such children remain unaffected, partly because they do not inherit the necessary genes. In a study by Heston (1966), remarkable for the thoroughness of follow-up, the children of schizophrenic mothers reared without contact with them were examined at a mean age of 36. The experimental "children" were compared with a matched group of children of nonschizophrenic mothers who had had comparable experience in institutions or foster homes. He found that 5 of the 47 experimentals had developed a schizophrenia while none of the controls had. The proportion of affected children (morbid risk of 16.6 percent) is not different from that reported in Table 5 for the children of schizophrenics who have not been systematically removed from their parents. Again, genetic arguments seem most plausible in accounting for the data, but the nature of such arguments must require some form other than the traditional ones.

The data summarized from studies of twins does not reveal the complexities involved. An intensive analysis of such data is provided elsewhere (Gottesman and Shields, 1966b). Suffice it to say that concordances for identical twins range from 0 to 69 percent in the ten studies averaged. The range would have been greater had we not removed the effects of age correction and reported all data in terms of simple pairwise concordance. Heterogeneity among the results of the ten systematic studies was largely accounted for when provision was made for the possibility of symptomatic schizophrenias being phenocopies and for differences in the composition of the twin samples. First, females tended to be more often concordant than males. Such a finding could be an artifact of sampling since only the Rosanoff et al. (1934) and Slater (1953) female pairs showed markedly higher concordance rates. These two investigations were based on resident hospital populations with unrepresentative sex distributions for their twin pairs. Additional explanations for the sex difference that could be proposed included other selection biases, such as differential mortality, the exposure of males to greater environmental variability thereby decreasing the probability of equivalent stresses, and psychological factors facilitating identification in females.

A second source of heterogeneity was related to the severity of schizophrenia in the proband. Internal analyses of the data of five of the samples showed a direct relationship between severity and concordance. Correlated with the dimension of severity are the dimensions of sampling from a standing, resident population versus sampling consecutively, and sampling from long-stay versus short-stay hospitals. Although concordances tended to be lower when sampling was done consecutively, the type of hospital was much more crucial to the magnitude of MZ concordance. Kallmann's (1946) adult twins (69 percent concordant) were mostly from consecutive admissions to long-stay hospitals, and Slater's series in long-stay hospitals analyzed by consecutive vs. resident status showed no difference in concordance rates for schizophrenia (64 vs. 65 percent). Twins from consecutive admissions to a short-stay hospital (Gottesman and Shields, 1966a) had a significantly lower MZ rate (42 percent).

A third source of heterogeneity remains unaccounted for by the analysis above; the four Scandinavian studies, as they stand, have the lowest MZ concordance rates for diagnosed and hospitalized schizophrenia. The differences are not accounted for by the method of ascertainment since each of the four used different methods. For the Scandinavian studies reported in detail, the co-twins tended to be characterized by milder psychiatric problems with schizophrenic-like features, or they presented eccentricities of character termed "borderline" or "schizoid." The possibility arises that there may be real reasons, genetic and/or environmental, why "concordance" in Scandinavian twin studies does not show up at the level of a clear-cut psychosis.

There is, however, no reason to expect the degree of genetic determination for a trait to be invariant throughout the world, even in studies using the same design. Heritability can vary according to the frequency of the genes involved and their penetrance (Huizinga and Heiden, 1957). It can vary according to the frequency and distribution in a population of the environmental influences *relevant in* schizophrenia. Since the latter are currently indeterminate, comparisons among studies done in different cultures must necessarily be somewhat elastic. The possibility remains that the continuing evolution of populations (Dobzhansky, 1962; Mayr, 1963) has led to a genotype in subarctic or low population density areas more adapted by natural selection to harboring the genetic component of schizophrenia (Böök, 1953; Gottesman, 1965).

The dimension of severity seems most important in accounting for heterogeneity of findings. In the sample of Gottesman and Shields they analyzed the relationship between severity and concordance. Using criteria for severity such as length of hospitalization and outcome on follow-up, they found that concordance was strikingly lower when the disorder in the index twin was mild. For example, outcome was determined as *good* if the index case had managed to stay out of the hospital for at least six months and could hold a paying job; other outcomes were called *poor*. There were 12 cases with good outcomes and 16 with poor among the 28 *identical* twin index cases. Only two of 12 co-twins in the good outcome group had been diagnosed as schizophrenic, compared with 12 of 16 in the poor outcome group. The concordance rates for the two subsamples of *identical* twins was 17 and 75 percent. The rate for the poor group was very similar to that found in the earlier studies which had sampled mainly from admissions to state-type hospitals. Kallmann (1946) had broken down his MZ pairs into those with deteriorating courses and other courses and found concordances of 100 and 26 percent respectively. Inouye (1961) found a rate of 86 percent in MZ pairs with the index case called "relapsing schizophrenia" and one of only 39 percent when the MZ index was "mild chronic" or "transient" schizophrenia.

Such data as these on the relationship between severity and concordance can be explained by a polygenic theory. We could infer that a proband with a good outcome had had few of the genes in the system, and we would expect his co-twin to have a much lower probability of encountering a stress sufficient to exceed the manifestation threshold than the co-twin of a severely affected schizophrenic. Most instances of schizophrenia, from the most mild to the most severe, could then be regarded as biologically related.

Is it still possible that the large difference between MZ and DZ rates, that is between individuals with the same sex, age, parents, social class, and more or less the same encounters with life, is largely accounted for in nonbiological terms? A plausible environmental explanation is that such factors as mutual identification, weak ego formation, or confusion of identity increase the risk of MZ twins in particular for schizophrenia. The simple reply is that neither MZ or DZ twins are more often schizophrenic than the general population. Furthermore, studies of the personality and intelligence of identical twins reared apart (Juel-Nielsen, 1965; Shields, 1962) show that similarities between those reared together (Gottesman, 1963a) is not simply due to their being raised by the same mother or to mutual identification (cf. Scarr, 1964).

Schizophrenia and Natural Selection. Differential reproduction of both female and male schizophrenics is well documented, showing them to be at a marked reproductive disadvantage (Erlenmeyer-Kimling et al., 1966; Lewis, 1958). Kallmann's (1938) group of severe (hebephrenic and catatonic) schizophrenics had a marriage rate 55 percent that of the general population; their birth rate per marriage was 42 percent that of census-based fertility. Interesting details of the nature of the reproductive disadvantage in more recent years and changes associated with improvements in chemotherapy and hospital policies are provided in Erlenmeyer-Kimling et al. (1966). Over the period 1934 to 1956 in New York State, the marriage rate of male schizophrenics rose from 26.4 to 40.8 percent, for females, from 54.2 to 64.3 percent. The number of children per fertile marriage or for all marriages did not differ appreciably from the general population figures, suggesting that those schizophrenics who marry are more like normals than they are like single schizophrenics in a number of ways. The number of children per person is the more important information from a selection point of view as is the relative fitness of schizophrenics. The number of children per schizophrenic male went from 0.5 to 0.6 over the period 1934 to 1956; for the females, 0.7 to 1.3. Relative fitness is more difficult to evaluate since the general census data do not provide a control group matched on variables known to be strongly related to fertility such as social class, urban vs. rural residence, religion, and age at marriage. Since the prevalence of schizophrenia is highly correlated with social class and since social class is highly correlated with fertility (Mitra, 1966), the schizophrenics may be at a greater reproductive disadvantage than may be inferred from comparisons with the combined data from the census. Data from Norwegian schizophrenic females admitted to hospital from 1946 to 1955 showed them to have a reproduction rate (fertility and marriage rates multiplied) 25.2 percent of that expected from census data (Odegaard, 1959). The exact meaning to be attached to combined data on the fertility of schizophrenics is difficult for other reasons too. Essen-Möller (1959) in his 1935 Bavarian sample of schizophrenics found that recovered patients did not differ in celibacy from the population at large nor did they have demonstrable reduction in fertility. In a London study by Macsorley (1964) of fertility in mental patients matched with controls for age and social class, male nonparanoid schizophrenics did not differ from their controls in percentage of childless marriages (both 18.7 percent) while this category of married female patients had more childless marriages (28.5 vs. 15.8 percent). Married male and female *paranoid* schizophrenics had comparatively more childlessness, 23.5 and 47 percent respectively, the females exceeding the matched controls by a factor of three. Macsorley's samples by subtype were small so that the replication of her findings will add important information about the problems surrounding the topic of fertility in psychotics.

Given the strong negative selection against schizophrenia from reduced fertility and the additional selection from early mortality in the children of schizophrenics (Kallmann, 1938), a genetic explanation for the maintenance of the gene or genes seems difficult to support at first sight. Erlenmeyer-Kimling and Paradowski (1966) review most of the possible hypotheses that might account for the continuation of a genetic disorder under strong negative selection. They point out that the conflicting interpretations about the mode of transmission as well as failure of replication in the biochemical sphere may stem from genotypic heterogeneity. The suggestion by Huxley, Mayr, Osmond, and Hoffer (1964) that schizophrenia may be another polymorphism is an intriguing one but lacks direct support. It should be noted that labeling the disorder as a morphism prejudices the case against considering schizophrenia within the context of quantitative genetics since morphisms are discontinuous. With one possible exception, reproductive superiority for "heterozygotes" analogous to the situation in sickle-cell anemia has not been demonstrated. The exception is the report by Erlenmeyer-Kimling and Paradowski (1966) that the unaffected sisters of schizophrenic females in their 1956 sample surpassed the reproduction per woman in the general population. Again it might be that controlling for class and religion will attenuate this difference.

As noted above, mutation alone could not maintain the gene or genes although granting heterogeneity and a very large number of loci, it might permit some fraction of cases to be accounted for in this manner. Huxley et al. (1964) suggested that the schizophrenia gene confers a selective advantage via immunity to pain or infection thus permitting differential survival and eventual reproduction. Moran (1965) has estimated that the advantage would have to amount to 5 to 10 percent. As we have seen this advantage appears to be counterfactual. One of the possibilities not given sufficient attention up to this time involves positing a large proportion of the cases of schizophrenia as being polygenically determined (Gottesman and Shields, 1966b, c). Schizophrenia would then be treated as a threshold character (Grüneberg, 1952) whose phenotypic appearance would depend both on the number of genes present and on the amount of relevant environmental stress. This model would then predict the continual appearance of segregants in the offspring of normal parents, termed phenodeviants by Lerner (1958), increased rates of schizophrenia in high risk families, and a slow response to negative selection. Schizophrenics could be thought of as part of the genetic load (Dobzhansky, 1964), the price paid for conserving genetic diversity. After briefly mentioning other models that have been proposed for the mode of transmission of schizophrenia, the polygenic model will be explored in more detail and evidence will be furnished that tends to support it being put forward as another working hypothesis.

The Transmission of Schizophrenia. The high frequency of schizophrenia in the general population, whether it be one percent or as much as three percent, leads to classifying it as a *common* disorder compared to other genetic entities. The genetics of common disorders are quite complex and have been left in a kind of limbo for the most part. Penrose (1953) and Edwards (1960) have outlined some of the issues involved. Potentially useful guidelines for schizophrenia have been provided by considerations of club-foot and pyloric stenosis (Falconer, 1965; Carter, 1965) and especially diabetes mellitus (Neel, Fajans, Conn, and Davidson, 1965). As with diabetes and formerly with mental retardation, almost every possible theory of genetic transmission has been suggested for schizophrenia. Each theory has successfully accounted for some of the empirical data as they have been gathered and stand in the literature.

Theories of simple dominance are embarrassed by the low frequency of schizophrenia in parents and in the children of affected probands. Recessive theories (i.e., Garrone, 1962; Kallmann, 1946) do not fit with the observed higher frequencies in children than in siblings of probands and the incidence in parents of probands is higher than expected. Rüdin advanced the first genetic theory, one involving two loci and recessive genes, in 1916 and variants of the idea are still extant. Burch (1964) subscribed to a two locus model with the proviso that the potential was triggered by two somatic mutations. In 1958 Slater, impressed by Böök's (1953) theory and conclusions, proposed a model that combined the features of dominance and recessive theories. On this scheme a single partially dominant gene led to schizophrenia in all homozygotes and a proportion of heterozygotes. The specific model advanced by Slater required fixing the frequency of schizophrenia in the general population at .008 and then generating expected frequencies in relatives as a function of various values of the gene frequency and the manifestation rate in heterozygotes. Slater decided to try to fit a model to the following observed frequencies: siblings, 14.2 percent, children of probands, 16.4 percent, children of dual mating probands, 39.2 percent, and children of normal first cousins, 1.1 percent. After allowing for some error of measurement, he concluded that the best fit was obtained with a gene frequency of about 0.015 and a manifestation rate in the heterozygote of 26 percent. From the relationship between gene frequency and trait frequency he could then calculate the proportion of all schizophrenics who were homozygotes: $h = p^2/0.008$, so that 2.81 percent were homozygotes. From these calculations it follows that some 97 percent of all schizophrenics are heterozygotes for the partially dominant gene. The theory is an elegant one and it goes a long way in accounting for a circumscribed set of observations. It cannot be taken as proved and Slater said only that the hypothesis merited further investigation. It would appear that Huxley et al. (1964) and Moran (1965) are less critical of the hypothesis than Slater himself.

Polygenic theories, with or without a specific major gene, can also account for many of the empirical observations about the frequency of schizophrenia in relatives of probands. The lack of popularity of such theories stems in part from the lack of statistical methods suitable for human threshold characters with an underlying continuous distribution of polygenes with "microphenic" effects (Morton, 1967). Recent advances in the explication of appropriate methods (Falconer, 1965) should lend substance to such theorizing which has justifiably been termed exasperatingly loose and vague (Rosenthal, 1963). The theory can contribute to explaining the distribution of clinical subtypes and the variability seen in pairs of affected relatives (Slater, 1953). So-called process and reactive schizophrenics would be construed as extremes on a continuum of severity, the latter determined by the number of genes. The irrational and schizoid relatives of probands would be viewed as being near but below the threshold number of genes. It would also be feasible to infer some adaptive advantages to individuals carrying some of the polygenes but not so many as to be labeled schizoid (e.g., Gottesman, 1965, 1967). "Spontaneous" remissions and symptom improvement under medication would then follow from raises in the threshold consequent on changes in psychosomatic state. Finally, with a polygenic theory we would expect the MZ co-twins of probands to be more abnormal than their DZ counterparts without necessarily being overtly concordant.

Although schizophrenia is treated like an all-or-none character, usually as a function of hospitalization and formal diagnosis, clinical contact with "pre-schizophrenics" (e.g., Rado, 1962) shows clearly the artificiality of the dichotomy. Many, but not all, such persons deviate along a dimension of schizoidness (Planansky, 1966) or show evidence of a specific "characterological defect" (Essen-Möller, 1941). The situation with schizophrenia makes the model and methods proposed by Falconer (1965) especially apropos. His techniques bring data on the incidences of disease in the relatives of probands into the fold of quantitative genetics. So-called threshold characters are treated as if they were determined by an underlying graduation of some attribute causing the disease. The latter attribute has been termed the *liability* by Falconer and is intended to convey both an innate tendency to develop the disease and the environmental milieu to which a person is exposed. The point on the scale of liability above which all persons are affected overtly is called the threshold, and it is the heritability of the liability to schizophrenia that is of concern.

Essentially Falconer's method converts incidences into regression coefficients; the latter then lead to an estimate of the heritability of liability (h^2). Standard errors of h^2 can also be calculated, thus permitting comparisons among estimates derived from different degrees of relatives. In the last analysis heritability, for any particular degree of genetic relationship, becomes a joint function of the incidence of a disease in the correct reference population and the incidence in a class of proband relatives. Falconer applied the method to relatives other than twins for renal stone disease, congenital pyloric stenosis, club foot, and peptic ulcer. He obtained h^2 values ranging from 37 per cent ± 6 percent for ulcer to 79 percent ± 5 percent for pyloric stenosis.

When Gottesman and Shields (1966c) adapted the Falconer method to MZ and DZ co-twins of schizophrenics as well as to other relatives, they used population incidences for schizophrenia of first 1 percent and then 2 percent to examine the effect on h^2 estimates. They found that the heritability of the liability to schizophrenia, however defined in representative studies, was quite substantial for both incidence figures. Furthermore, the estimates were remarkably consistent when estimated independently from MZ twins, DZ twins, siblings, or aunt and uncles. Heritability of the liability to schizophrenia from their own sample of MZ and DZ twins was 85 percent ± 3 percent and 72 percent ± 23 percent respectively. Estimated from aunts and uncles it was 61 percent ± 9 percent. Some estimates based on severely ill twins ran over 100 percent. The sharing of trait-relevant environmental factors is a source of error that inflates the regression computed from siblings. Falconer (personal communication) doubts that his method can be applied directly to MZ twins without obtaining inflated estimates of h^2. Noting a number of parallels between diabetes mellitus and schizophrenia, Gottesman and Shields went on to obtain h^2 estimates from the twins and other relatives of diabetics and found results very similar to those they had obtained for schizophrenia.

A polygenic theory for the transmission of schizophrenia is advanced not as a replacement for other viable theories, but rather as an idea deserving of equal consideration based on its merits. Data less ambiguous than those extant need to be collected before the various models mentioned above can be evaluated for the accuracy of their predictions. Morton's (1967) variations on Falconer's theme that sometimes permit the distinction between major gene and polygene models for such disorders as muscular dystrophy have yet to be applied to a phenomenon as complex as schizophrenia. We do not know the "true" frequency of schizophrenia in its protean forms in the population nor in the relatives of such an ideal sample of probands. Whenever possible in the future, clinically heterogeneous (e.g., classified by symptoms, course, outcome, or social class) subjects should be handled separately in data analyses as well as grouped so as to cast light on possible genetic heterogeneity. At best, we can hope to attain a more accurate view of the nature of schizophrenia by successive approximations. Although it has become commonplace to advocate the principle of multiple working hypotheses, the principle appears to grate against the instinct of "hypothesis-formation territoriality," and thus is in need of reendorsement in the biobehavioral sciences.

Summary

An attempt was made in this chapter to introduce the reader familiar with evolutionary biology to the problems and data of an emerging field known as human behavioral genetics. Emphasis was placed on the unique problems associated with the use of behavioral traits as a class of phenotypic characters because of their continuity, complexity, and fluidity. The trait approach was endorsed as a fruitful one for a behavioral taxonomy but preferably within the context of some hierarchical scheme with the level of complexity left as a choice to the investigator. The pathways between the genes and behavioral traits were conceptualized as were the sources of phenotypic trait correlations. Mention was made of the use of psychological tests as signs of a behavior and the process of construct validation wherein the tests accrued meaning. Since the twin method is so often used as a source of data in human behavioral genetics, it was described as were its strategic implications and summarizing statistics. Two content areas were selected for the sheer magnitude of effort that had been expended along the lines of interest to students of evolution and behavioral genetics. The structure of intelligence was discussed and then data on mental retardation and the relationship between fitness and intelligence presented. A heuristic, polygenic model for the transmission of intelligence was outlined. Finally, the complex trait of schizophrenia was presented from a genetic and evolutionary point of view with emphasis on the results of twin and family studies, the maintenance of the behavior in spite of strong negative selection, and the theories proposed for its genetic transmission. Again, a polygenic theory was suggested as worthy of serious consideration and some data not in conflict with such a model were demonstrated.

J. B. S. Haldane, pondering the problems surrounding data that might be gathered to gain an understanding of natural selection in man, made an observation that is just as appropriate for questions about the behavioral genetics of man—"We do not know. We shall know if enough people want to know" (Haldane, 1961, p. 37).

Acknowledgment

The author is indebted to S. C. Reed and D. T. Lykken for their helpful comments on the manuscript. I must acknowledge an intellectual debt to J. Shields, my colleague at the Maudsley Hospital, London. I have drawn freely on our correspondence, deliberations, and published papers for much of the material in the sections on twins and genetic aspects of schizophrenia. Preparation of this chapter was aided in part by a research grant from the National Institute of Mental Health (MH 13117).

References

ALLEN, G. 1965. Twin research: problems and prospects. *In* Steinberg, A. G., and A. G. Bearn, eds., Progress in Medical Genetics, vol. IV, 242–269, New York, Grune and Stratton.

———, and F. J. KALLMANN. 1962. Etiology of mental subnormality in twins. *In* F. J. Kallmann ed., Expanding Goals of Genetics in Psychiatry, 174–211, New York, Grune and Stratton.

———, B. HARVALD, and J. SHIELDS. 1967. Measures of twin concordance. Acta Genet. Basel. In press.

ALLPORT, G. W. 1931. What is a trait of personality? J. Abnorm. Psychol., 25:368–372.

———. 1937. Personality: a Psychological Interpretation. New York, Holt.

———. 1966. Traits revisited. Amer. Psychol., 21:1–10.

———, and H. S. ODBERT. 1936. Trait-names: a psycholexical study. Psychol. Monogr., 47, 1 (Whole No. 211).

ANASTASI, ANNE. 1958. Differential Psychology, 3rd ed., New York, Macmillan.

———. 1959. Differentiating effects of intelligence and social status. Eugen. Quart., 6:84–91.

———. 1961. Psychological Testing, 2nd ed., New York, Macmillan.

ANDERSON, V. E. 1964. Genetics in mental retardation. *In* Stevens, H. A., and R. Heber, eds., Mental Retardation, 348–394, Chicago: Univ. Chicago Press.

BAJEMA, C. J. 1963. Estimation of the direction and intensity of natural selection in relation to human intelligence by means of the intrinsic rate of natural increase. Eugen. Quart., 10:175–187.

———. 1966. Relation of fertility to educational attainment in a Kalamazoo public school population: a follow-up study. Eugen. Quart., 13:306–315.

BENAIM, S. 1960. The specificity of Reserpine in the treatment of schizophrenia in identical twins. J. Neurol. Neurosurg. Psychiat., 23:170–175.

BISCHOF, L. J. 1954. Intelligence: Statistical Concepts of its Nature. New York, Doubleday.

BLEULER, E. 1911. Dementia Praecox oder Gruppe der Schizophrenien, Leipzig, Deuticke. (1950. Dementia Praecox or the Group of Schizophrenias, New York, International Universities Press.)

BLEWETT, D. B. 1954. An experimental study of the inheritance of intelligence. J. Mental Science, 100:922–933.

BÖÖK, J. A. 1953. Schizophrenia as a gene mutation. Acta Genet. (Basel), 4:133–139.

BULLOCK, T. H. 1958. Evolution of neurophysiological mechanisms. *In* Roe, Anne, and G. Simpson eds., Behavior and Evolution, 165–177, New Haven, Yale Univ. Press.

BURCH, P. R. J. 1964. Schizophrenia: some new aetiological considerations. Brit. J. Psychiat., 110:818–824.

BURT, C. 1958. The inheritance of mental ability. Amer. Psychol., 13:1–15.

———, and M. HOWARD. 1956. The multifactorial theory of inheritance and its application to intelligence. Brit. J. Stat. Psychol., 9:95–131.

CARLSON, E. A. 1966. The Gene: a Critical History. Philadelphia, Saunders.

CARTER, C. O. 1962. Human Heredity. Baltimore, Penguin.

———. 1965. The inheritance of common congenital malformations. *In* Steinberg, A. G., and A. G. Bearn eds., Progress in Medical Genetics, Vol. IV, 59–84, New York, Grune and Stratton.

CATTELL, R. B. 1950. Personality. New York, McGraw-Hill.

———. 1951. The fate of national intelligence: test of a thirteen-year prediction. Eugen. Rev., 42:136–148.

———. 1965. Methodological and conceptual advances in evaluating hereditary and environmental influences and their interaction. *In* S. G. Vandenberg, ed., Methods and Goals in Human Behavior Genetics, 95–139, New York, Academic Press.

CLARK, P. J. 1956. The heritability of certain anthropometric characters ascertained from twins. Amer. J. Hum. Genet., 8:49–54.

CONRAD, H. S., and H. E. JONES. 1940. A second study of family resemblances in intelligence: environmental and genetic implications of parent-child and sibling correlations in the total sample. Yearbook Nat. Soc. Stud. Educ., 39, Part II, 97–141.

COOPER, R. M., and J. P. ZUBEK. 1958. Effects of enriched and restricted early environments on the learning ability of bright and dull rats. Canad. J. Psychol., 12:159–164.

CRONBACH, L. J. 1960. Essentials of Psychological Testing, New York, Harper.

———, and P. E. MEEHL. 1955. Construct validity in psychological tests. Psychol. Bull., 52:281–302.

CROWE, F. W., W. J. SCHULL, and J. V. NEEL. 1956. Multiple Neurofibromatosis. Springfield, C. C. Thomas.

DAVID, P. R., and L. H. SNYDER. 1962. Some interrelations between psychology and genetics. *In* S. Koch, ed., Psychology: A Study of a Science, Vol. 4, 1–50, New York, McGraw-Hill.

DOBZHANSKY, T. 1955a. Evolution, Genetics, and Man, New York, John Wiley & Sons.

———. 1955b. A review of some fundamental concepts and problems of population genetics. Cold Spring Harbor Symposium on Quantitative Biology, 20:1–15.

———. 1962. Mankind Evolving. New Haven, Yale Univ. Press.

———. 1964. Genetic diversity and fitness. *In* Proceedings XI International Congress of Genetics, 541–552, London, Pergamon.

———, and B. SPASSKY. 1967. Effects of selection and migration on geotactic and phototactic behavior of Drosophila. I. Proc. Roy. Soc. [Biol.], In press.

DUNN, L. C. 1962. Cross currents in the history of human genetics. Amer. J. Hum. Genet., 14:1–13.

EDWARDS, J. H. 1960. The simulation of Mendelism. Acta Genet. (Basel), 10:63–70.

ELSÄSSER, G. 1952. Die Nachkommen Geisteskranker Elternpaare. Stuttgart, Thieme.

ERLENMEYER-KIMLING L., and L. F. JARVIK. 1963. Genetics and intelliegnce: a review. Science, 142:1477–1479.

———, and W. PARADOWSKI. 1966. Selection and schizophrenia. Amer. Naturalist, 100:651–665.

———, J. D. Rainer, and F. J. KALLMANN. 1966. Current reproductive trends in schizophrenia. *In* Hoch, P. H., and J. Zubin, eds., Psychopathology of Schizophrenia. New York, Grune and Stratton.

ESSEN-MÖLLER, E. 1941. Psychiatrische Untersuchungen an einer Serie von Zwillingen. Acta Psychiat. Scand., Suppl. 23.

———. 1959. Mating and fertility patterns in families with schizophrenia. Eugen. Quart., 6:142–147.

———. 1963. Twin research and psychiatry. Acta Psychiat. Scand., 39:64–77.

EYSENCK, H. J. 1947. Dimensions of Personality. London, Paul Kegan.

FALCONER, D. S. 1960. An Introduction to Quantitative Genetics. New York, Ronald.

———. 1965. The inheritance of liability to certain diseases, estimated from the incidence among relatives. Ann. Hum. Genet., 29:51–76.

FEIGL, H., and MAY BRODBECK. 1953. Readings in the Philosophy of Science. New York, Appleton-Century-Crofts.

FISHER, R. A. 1918. The correlation between relatives on the supposition of Mendelian inheritance. Trans. Roy. Soc. Edin., 52:399–433.

FREEDMAN, D. G. 1958. Constitutional and environmental interactions in rearing of four breeds of dogs. Science, 127:585–586.

———. 1967. Personality development in infancy: a biological approach. *In* Brackbill, Yvonne ed., Infancy and Childhood, New York, Free Press. In press.

FULLER, J. L. 1957. The genetic base: pathways between genes and behavioral characteristics. *In* The Nature and Transmission of the Genetic and Cultural Characteristics of Human Populations, 101–111. New York, Milbank Foundation.

———. 1960. Behavior genetics. Ann. Rev. Psychol., 11:41–63.

———, C. EASLER, and M. E. SMITH. 1950. Inheritance of audiogenic seizure susceptibility in the mouse. Genetics, 35:622–632.

———, and W. R. THOMPSON. 1960. Behavior Genetics. New York, Wiley.

GARRONE, G. 1962. Étude statisque et génétique de la schizophrénie à Genève de 1901 à 1950. J. Génét. Hum., 11:89–219.

GESELL, A., and H. THOMPSON. 1929. Learning and growth in identical infant twins, an experimental study by the method of co-twin control. Genet. Psychol. Monogr., 6:1–124.

GINSBURG, B. E. 1958. Genetics as a tool in the study of behavior. Perspect. Biol. Med., 1:397–424.

———, and W. S. LAUGHLIN. 1966. The multiple bases of human adaptability and achievement: a species point of view. Eugen. Quart., 13, 240–257.

GLASS, D. C., ed. 1967. Biology and Behavior: Genetics, New York, Russell Sage College Press. In Press.

GOTTESMAN, I. I. 1962. Differential inheritance of the psychoneuroses. Eugen. Quart., 9:223–227.

———. 1963a. Heritability of personality: a demonstration. Psychol. Monogr., 77, No. 9 (Whole No. 572).

———. 1963b. Genetic aspects of intelligent behavior. *In* Ellis, N. ed., Handbook of Mental Deficiency: Psychological Theory and Research, 253–296. New York, McGraw-Hill.

———. 1965. Personality and natural selection. *In* Vandenberg, S. ed., Methods and Goals in Human Behavior Genetics, 63–80, New York, Academic Press.

———. 1966. Genetic variance in adaptive personality traits. J. Child Psychol. Psychiat, 7:199–208.

———, and J. SHIELDS. 1966a. Schizophrenia in twins: 16 years' consecutive admissions to a psychiatric clinic. British Journal of Psychiatry, 112, 809–818.

———, and J. SHIELDS. 1966b. Contributions of twin studies to perspectives on schizophrenia. *In* Maher, B. A. ed., Progress in Experimental Personality Research, Vol. 3, 1–84, New York, Academic Press.

———, and J. SHIELDS. 1966c. In pursuit of the schizophrenic genotype. Paper presented at the 2nd Conference on Human Behavior Genetics, Louisville, Ky.

GREGORY, I. 1960. Genetic factors in schizophrenia. Amer. J. Psychiat., 116:961–972.

GRÜNEBERG, H. 1952. Genetic studies on the skeleton of the mouse. IV. Quasicontinuous variations. J. Genet., 51:95–114.

GUILFORD, J. P. 1954. Psychometric Methods, 2nd ed. New York, McGraw-Hill.

———. 1959. Personality, New York, McGraw-Hill.

———. 1966. Intelligence: 1965 model. Amer. Psychol., 21:20–26.

HAGGARD, E. A. 1958. The Intraclass Correlation Coefficient, New York, Dryden.

HALDANE, J. B. S. 1961. Natural selection in man. *In* Steinberg, A. G. ed., Progress in Medical Genetics, Vol. I, 27–37, New York, Grune and Stratton.

HALLER, M. H. 1963. Eugenics: Hereditarian Attitudes in American Thought, Brunswick, Rutgers Univ. Press.

HARRIS, D. L. 1965. Biometrical genetics in man. *In* Vandenberg, S. G. ed., Methods and Goals in Human Behavior Genetics, 81–94, New York, Academic Press.

HESTON, L. L. 1966. Psychiatric disorders in foster home reared children of schizophrenic mothers. Brit. J. Psychiat., 112:819–825.

HIGGINS, J., ELIZABETH W. REED, and S. REED. 1962. Intelligence and family size: a paradox resolved. Eugen. Quart., 9:84–90.

HIMWICH, H. E., S. S. KETY, and J. R. SMYTHIES, eds. 1966. Amines and Schizophrenia, Oxford, Pergamon Press.

HIRSCH, J. 1964. Breeding analysis of natural units in behavior genetics. Amer. Zool., 4:139–145.

———. ed. 1967. Behavior-genetic Analysis. New York, McGraw-Hill.

———, and L. ERLENMEYER-KIMLING. 1962. Studies in experimental behavior genetics. IV. Chromosome analyses for geotaxis. J. Comp. Physiol. Psychol., 55:732–739.

HOLLINGSHEAD, A. B., and F. C. REDLICH. 1958. Social Class and Mental Illness. New York, Wiley.

HOLZINGER, K. J. 1929. The relative effect of nature and nurture influences on twin differences. J. Educ. Psychol., 20:241–248.

HUIZINGA, J., and J. A. VAN DER HEIDEN. 1957. The percentages of concordance in twins and mode of inheritance. Acta Genet. Med. (Roma), 6:437–450.

HURST, C. C. 1932. A genetic formula for the inheritance of intelligence in man. Proc. Roy. Soc. [Biol.], 112 (Series B):80–97.

HUXLEY, J., E. MAYR, H. OSMOND, and A. HOFFER. 1964. Schizophrenia as a genetic morphism. Nature (London) 204:220–221.

INOUYE, E. 1961. Similarity and dissimilarity of schizophrenia in twins. Proc. Third World Congress Psychiat., vol. I, 524–530, Montreal, Univ. Toronto Press.

JUEL-NIELSEN, N. 1965. Individual and environment, a psychiatric-psychological investigation of monozygotic twins reared apart. Acta Psychiat. Scand., Suppl. 183.

KALLMANN, F. J. 1938. The genetics of schizophrenia. New York, Augustin.

———. 1946. The genetic theory of schizophrenia: an analysis of 691 schizophrenic twin index families. Amer. J. Psychiat., 103:309–322.

———. ed. 1962. Expanding Goals of Genetics in Psychiatry. New York, Grune and Stratton.

———, and G. S. BAROFF. 1955. Abnormalities of behavior (in the light of psychogenetic studies). Ann. Rev. Psychol., 6:297–326.

———, A. FALEK, M. HURZELER, and L. ERLENMEYER-KIMLING. 1964. The developmental aspects of children with two schizophrenic parents. Psychiat. Res. Rep. Amer. Psychiat. Ass., 19:136–145.

KALOW, W. 1962. Pharmacogenetics, heredity and the response to drugs. Philadelphia, W. B. Saunders.

KEMPTHORNE, O., and R. H. OSBORNE. 1961. The interpretation of twin data. Amer. J. Hum. Genet., 13:320–339.

KETY, S. S. 1965. Biochemical theories of schizophrenia. Int. J. Psychiat., 1:409–430.

KISHIMOTO, K. 1957. A study on the population genetics of schizophrenia. In Proceedings Second International Congress for Psychiatry, Vol. 2, 20–28, Zurich, Orell Füssli Arts Graphiques.

LAWRENCE, E. M. 1931. An investigation into the relation between intelligence and inheritance. Brit. J. Psychol. Monogr., Suppl. 16.

LEAHY, A. M. 1935. Nature-nurture and intelligence. Genet. Psychol. Monogr., 17:236–308.

LEJEUNE, J. 1964. Trisomy—current stage of chromosomal research. In Steinberg, A. G., and A. G. Bearn eds., Progress in Medical Genetics. Vol. III, 144–177, New York, Grune and Stratton.

LERNER, I. M. 1958. The Genetic Basis of Selection, New York, Wiley.

LEWIS, A. 1958. Fertility and mental illness. Eugen. Rev., 50:91–106.

LEWONTIN, R. C., and J. L. HUBBY. 1966. A molecular approach to the study of genetic heterozygosity in natural populations. II. Amount of variation and degree of heterozygosity in natural populations of Drosophila pseudoobscura. Genetics, 54:595–609.

LIDZ, T., A. R. CORNELISON, D. TERRY, and S. FLECK. 1958. Intrafamilial environment of the schizophrenic patient. VI. The transmission of irrationality. Arch. Neurol. Psychiat., 79:305–316.

LOEHLIN, J. C. 1965. A heredity-environment analysis of personality inventory data.

In VANDENBERG, S., ed., Methods and Goals in Human Behavior Genetics, 163–170, New York, Academic Press.

————, and S. G. VANDENBERG. 1966. Genetic and environmental components in the covariation of cognitive abilities: an additive model. Paper presented at the 2nd Conference on Human Behavior Genetics, Louisville, Ky., April, 1966.

LOEVINGER, JANE. 1957. Objective tests as instruments of psychological theory. Psychol. Rep., 3:635–694.

LUSH, J. L. 1945. Animal Breeding Plans, Ames, Iowa State College Press.

LUXENBURGER, H. 1928. Vorläufiger Bericht über psychiatrische Serienuntersuchungen an Zwillingen. Z ges. Neurol. Psychiat., 116:297–326.

MACSORLEY, K. 1964. An investigation into the fertility rates of mentally ill patients. Ann. Hum. Genet., 27:247–256.

MAHER, B. A. 1963. Intelligence and brain damage. *In* Ellis, N. ed., Handbook of Mental Deficiency: Psychological Theory and Research, 224–252, New York, McGraw-Hill.

MAXWELL, J. 1954. Intelligence, fertility and the future. Eugen. Quart., 1:244–274.

MAYR, E. 1963. Animal species and evolution. Cambridge, Harvard Univ. Press.

MCNEMAR, Q. 1942. The Revision of the Stanford-Binet Scale, Boston: Houghton Mifflin.

MEDNICK, S. A., and F. SCHULSINGER. 1965. A longitudinal study of children with a high risk for schizophrenia: a preliminary report. *In* Vandenberg, S. ed., Methods and goals in human behavior genetics, 255–295, New York, Academic Press.

MEEHL, P. E. 1962. Schizotaxia, schizotypy, schizophrenia. Amer. Psychol., 17:827–838.

MITRA, S. 1966. Occupation and fertility in the United States. Eugen. Quart., 13:141–146.

MORAN, P. A. P. 1965. Class migration and the schizophrenic polymorphism. Ann. Hum. Genet., 28:261–268.

MORTON, N. E. 1967. The detection of major genes under additive continuous variation. Amer. J. Hum. Genet., 19:23–34.

NEEL, J. V., S. S. FAJANS, J. W. CONN, and RUTH DAVIDSON. 1965. Diabetes mellitus. *In* Neel, J. V., Margery Shaw, and W. J. Schull eds., Genetics and the epidemiology of chronic diseases. 105–132. Public Health Service Publication No. 1163.

————, and W. J. SCHULL. 1954. Human Heredity. Chicago, Univ. Chicago Press.

NEWMAN, H. H., F. N. FREEMAN, and K. J. HOLZINGER. 1937. Twins: A Study of Heredity and Environment. Chicago, Univ. Chicago Press.

NICHOLS, R. C. 1965. The inheritance of general and specific ability. Research Report No. 1, Evanston, Illinois, National Merit Scholarship Corporation.

NISSEN, H. W. 1958. Axes of behavioral comparison. *In* Roe, Anne, and G. G. Simpson, eds., Behavior and Evolution, 183–205, New Haven, Yale Univ. Press.

ODEGAARD, O. 1959. Discussion of families with manic-depressive psychosis. Eugen. Quart., 6:137–141.

PAP, A. 1958. Semantics and Necessary Truth, New Haven, Yale Univ. Press.

PEARSON, K. 1904. On the laws of inheritance in man. II. On the inheritance of the mental and moral characters in man, and its comparison with the inheritance of the physical characteristics. Biometrika, 3:131–190.

PENROSE, L. S. 1948. The supposed threat of declining intelligence. Amer. J. Ment. Defic., 53:114–118.

————. 1953. The genetical background of common diseases. Acta Genet. (Basel) 4:257–265.

————. 1963. The Biology of Mental Defect. 3rd. ed., New York, Grune and Stratton.

PICKFORD, R. W. 1949. The genetics of intelligence. J. Psychol., 28:129–145.

PITTENDRIGH, C. S. 1958. Adaptation, natural selection, and behavior. *In* Roe, Anne, and G. G. Simpson eds., Behavior and Evolution, 390–416, New Haven, Yale Univ. Press.

PLANANSKY, K. 1966. Schizoidness in twins. Acta Genet. Med. (Roma), 15:151–166.

POLLIN, W., J. R. STABENAU, L. MOSHER, and J. TUPIN. 1966. Life history differences in identical twins discordant for schizophrenia. Amer. J. Orthopsychiat., 36:492–509.

PORTER, I. A. 1966. The genetics of drug susceptibility. Diseases of the Nervous System (Monogr. Suppl.), 27:25–36.

PRIBRAM, K. 1958. Comparative neurology and the evolution of behavior. *In* Roe, Anne, and G. Simpson eds., Behavior and Evolution, 140–164, New Haven, Yale Univ. Press.

RACE, R. R., and R. SANGER. 1962. Blood Groups in Man, 4th ed., Oxford, Blackwell.

RADO, S. 1962. Theory and therapy: the theory of schizotypal organization and its application to the treatment of decompensated schizotypal behavior. *In* Rado, S., Psychoanalysis of Behavior, Vol. 2, 127–140, New York, Grune and Stratton.

REED, S. C. 1965. The evolution of human intelligence. Amer. Sci., 53:317–326.

REED, ELIZABETH W., and S. C. REED. 1965. Mental Retardation: A Family Study. Philadelphia, Saunders.

ROBERTS, J. A. F. 1952. The genetics of mental deficiency. Eugen. Rev., 44:71–83.

ROE, ANNE, and G. G. SIMPSON, eds. 1958. Behavior and Evolution, New Haven, Yale Univ. Press.

ROSANOFF, A. J., L. M. HANDY, I. R. PLESSET, and S. BRUSH. 1934. The etiology of so-called schizophrenic psychoses with special reference to their occurrence in twins. Amer. J. Psychiat., 91:247–286.

―――, L. M. HANDY, and I. R. PLESSET. 1937. The etiology of mental deficiency with special reference to its occurrence in twins. Psychol. Monogr., 216:1–137.

ROSENTHAL, D. ed., and others. 1963. The Genain Quadruplets. New York, Basic Books.

―――. 1966. The offspring of schizophrenic couples. J. Psychiat. Res., 4:169–188.

ROYCE, J. R. 1957. Factor theory and genetics. Educational and Psychological Measurement, 17:361–376.

SCARR, S. 1964. Genetics and human motivation. Unpublished Doctoral Dissertation, Harvard Univ.

SHIELDS, J. 1962. Monozygotic twins brought up apart and brought up together. London, Oxford Univ. Press.

―――, and E. SLATER. 1961. Heredity and psychological abnormality. *In* Eysenck, H. J., ed., Handbook of Abnormal Psychology, 298–343, New York, Basic Books.

―――, and E. SLATER. 1967. Genetic aspects of schizophrenia. Hospital Medicine. 1:579–584.

SIMPSON, G. G. 1958. The study of evolution: methods and present status of theory. *In* Roe, Anne and G. G. Simpson eds., Behavior and Evolution, 7–26, New Haven, Yale Univ. Press.

SLATER, E. 1953. Psychotic and Neurotic Illnesses in Twins. London, Her Majesty's Stationery Office.

―――. 1958. The monogenic theory of schizophrenia. Acta Genet. (Basel), 8:50–56.

―――. 1963. Diagnosis of zygosity by finger prints. Acta Psychiat. Scand., 39:78–84.

SMITH, S. M., and L. S. PENROSE. 1955. Monozygotic and dizygotic twin diagnosis. Ann. Hum. Genet., 19:273–289.

STAFFORD, R. E. 1965. Nonparametric analysis of twin data with the Mann-Whitney U-test. Amer. Psychol., 20:584 (abstr.)

STERN, C. 1960. Principles of Human Genetics, 2nd ed., San Francisco, Freeman.

―――. 1966. Genes and people. Public lecture sponsored by the National Foundation, at the 3rd International Congress of Human Genetics. Chicago, September 9.

TAX, S. ed. 1966. Behavioral Consequences of Genetic Differences in Man, New York, Viking Fund.

TERMAN, L. M., and MAUDE A. MERRILL. 1937. Measuring Intelligence, Boston, Houghton Mifflin.

THOMPSON, W. R. 1957. Traits, factors, and genes. Eugen. Quart., 4:8–16.

———. 1966. Multivariate experiment in behavior genetics. *In* Cattell, R. B. ed., Handbook of Multivariate Experimental Psychology, 711–731, Chicago, Rand McNally.

THURSTONE, L. L. 1938. Primary mental abilities. Psychometric Monogr., 1, Chicago, Univ. Chicago Press.

TRYON, R. C. 1958. Cumulative community cluster analysis. Educational and Psychological Measurement, 18:3–35.

VANDENBERG, S. G., ed. 1965. Methods and Goals in Human Behavior Genetics, New York, Academic Press.

———. 1966. Contributions of twin research to psychology. Psychol. Bull., 66:327–352.

———. 1967a. The nature and nurture of intelligence. *In* Glass, D. C. ed., Biology and Behavior: Genetics, New York, Russell Sage Foundation. In press.

———. ed. 1967b. Proceedings Second Invitational Conference on Human Behavioral Genetics. (In preparation.)

VERNON, P. E. 1961. The Structure of Human Abilities, 2nd ed., London, Methuen.

WILLOUGHBY, R. R., and MIRANDI GOODRIE. 1927. Neglected factors in the differential birth rate problem. J. Genet. Psychol., 34:373–393.

ZERBIN-RÜDIN, E. 1967. Endogene Psychosen. *In* Becker, P. ed., Humangenetik, Vol. 2. Stuttgart, Thieme. In press.

8

Suture-Zones of Hybrid Interaction Between Recently Joined Biotas

CHARLES L. REMINGTON

Department of Biology, Yale University, New Haven, Connecticut

Introduction 322
Tabulation of Suture-Zone Examples 323
The Individual Nearctic Zones 329
     I. Northeastern–Central Suture-Zone 331
    II. Northern Florida Suture-Zone 339
    III. Central Texas Suture-Zone 342
    IV. Rocky Mountain–Eastern Suture-Zone 348
    V. Northern–Rocky Mountain Suture-Zone 352
    VI. Pacific–Rocky Mountain Suture-Zone 356
    VII. Minor or Little-Known Zones 358
        a. Northern–Cascade Suture-Zone 360
        b. Cascade–Sierran Suture-Zone 360
        c. California Desert–Pacific Slope Suture-Zone 360
        d. Rocky Mountain–Sonoran Suture-Zone 362
        e. Southwestern New Mexico Suture-Zone 364
        f. Louisiana–East Texas Suture-Zone 364
        g. Southern Appalachian–Ozark–Ouachita Suture-Zone 366
        h. Other Minor Nearctic Zones 366
    VIII. Arctic Treeline Contact Line 368
Suture-Zones Outside North America 368
Stabilization After Early Suturing 371
Hybridization Outside Suture-Zones 372

Discussion 374
    Models of Speciation and of its Sequelae 374
    Early Events Following Geographic Contact 375
        General Theory 375
        Evidence for Initial Hybridizing 380
        Evidence for Initial Competition 381
        Suturing and Mimicry 382
        Suturing and Host Specificity 383
Acknowledgments 383
Tables 385
Appendix 411
References 415

Introduction

From a study of the geographic occurrences of contemporary hybridization among North American animals, it has become apparent that most of the hybrids are produced in a few relatively localized zones, with little hybridizing in the vast areas between these zones of mixing. The hybrids tend to be at least moderately fertile and therefore to be a source of significant gene exchange between the typically allopatric pairs of species and semispecies. There is a wide variety of consequences from this introgression, with greater or lesser influence on the parental populations, and large portions of the fauna and probably flora are involved. An appropriate term for such a belt of interfaunal and interfloral linkage is **suture-zone.**[1]

Taxonomists have been recording individual pairs of hybridizing Nearctic animal species for more than a century, but these reports have been sporadic and casual for the most part until recent years. Lately, detailed analyses have appeared for some birds and amphibians due to the work of C. G. Sibley, the Blairs, and their associates. No general geographic pattern of this hybridization has been described for the continent as a whole. Since existing published work has tended to deal with a single genus, and only rarely with even a single order, the failure to recognize a general pattern is not surprising. The present paper will show that in the suture-zones numerous crosses are occurring among Lepidoptera and other insects, mammals, birds, reptiles, amphibians, and spermato-

---

[1] From SUTURE, a stitching-together or firm line of union between formerly separate entities. The term **suture-zone** is introduced to denote a band, whether narrow or broad, of geographic overlap between major biotic assemblages, including some pairs of species or semispecies which hybridize in the zone. Within a **suture-zone** are the individual **hybrid-zones** of the various pairs of taxa which are interbreeding there. Mayr (1963) has recently used the terms "hybrid belts" and "hybrid zones" interchangeably, in one of the few published discussions touching on the suture-zone phenomenon.

phytes. It is probable that such hybridizing is present in all or most groups of terrestrial animals and plants which have mainly allopatric near relatives on the two sides of each zone. The implications of the large amount of natural hybridizing and its localization in suture-zones will be discussed after a review of some of the evidence.

There appear to be six major suture-zones in North America (see numbered zones in Figs. 1 and 2). A possible seventh, in the arctic, can only be suggested due to inadequate sampling. There are several lesser zones of hybridization, each involving genera in more than one order, class, or phylum (see lettered zones in Figs. 1 and 2). Roman numerals are preferred as the designations of the major zones, but they are too bulky for the map symbols and are represented as Arabic numerals on the figures. For convenience in distinguishing what I regard as major from minor zones, the sequence of majors goes basically east to west and of minors from west to east. Since the suture-zone concept is based on the analysis of hybridization of many kinds of animals and plants, I am illustrating this paper with figures of actual specimens of suturing taxa for each zone. A majority of these are Lepidoptera on which I have been working, but various other groups are represented if specimens or suitable published figures were easily available. Most of the specimens represent entities never before illustrated or are otherwise significant for research reference, so the full data for each specimen are given in an appendix.

Tabulation of Suture-Zone Examples

This paper is intended to be a general discussion of a phenomenon for which the documentation is, in part, conclusive but for every zone still small. Tables I–X were compiled to show specific examples of pairs of closely related taxa which are known or suspected to meet in each suture-zone. This being a synoptic paper, no attempt will be made here to describe the hybridization phenomena for the individual pairs. To simplify the tables, authors of Latin names are omitted; most can be found in easily accessible taxonomic works.

In have tried to use the contemporary names for the organisms listed in the tables, but have not made strenuous efforts to ferret out all the latest taxonomic and nomenclatural views. Most subject to obsolescence are the hierarchical levels of the species and subspecies; successive recent papers

---

Fig. 1. *Suture-zones of Western North America. Figures 1 and 2 `show the general positions of the six major zones (Arabic numerals) and seven minor zones (letters).*

Fig. 2. *Suture-zones of Eastern North America. Numbering of the major zones is the same as in the discussions in the text and the tables. Minor zones A–G of Figures 1 and 2 are numbered VIIa–VIIg in the text and tables.*

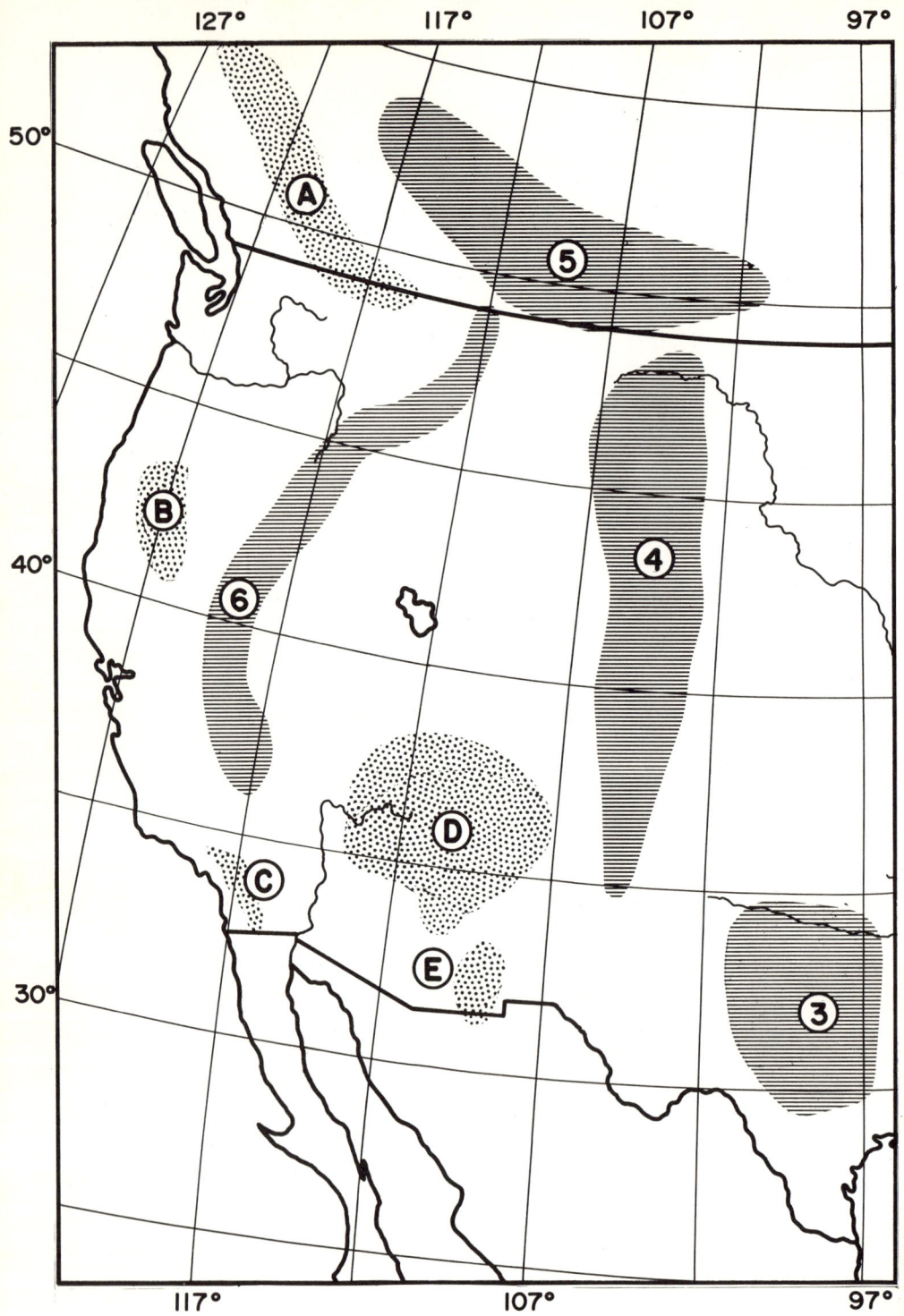

Fig. 1. (Legend on page 323)

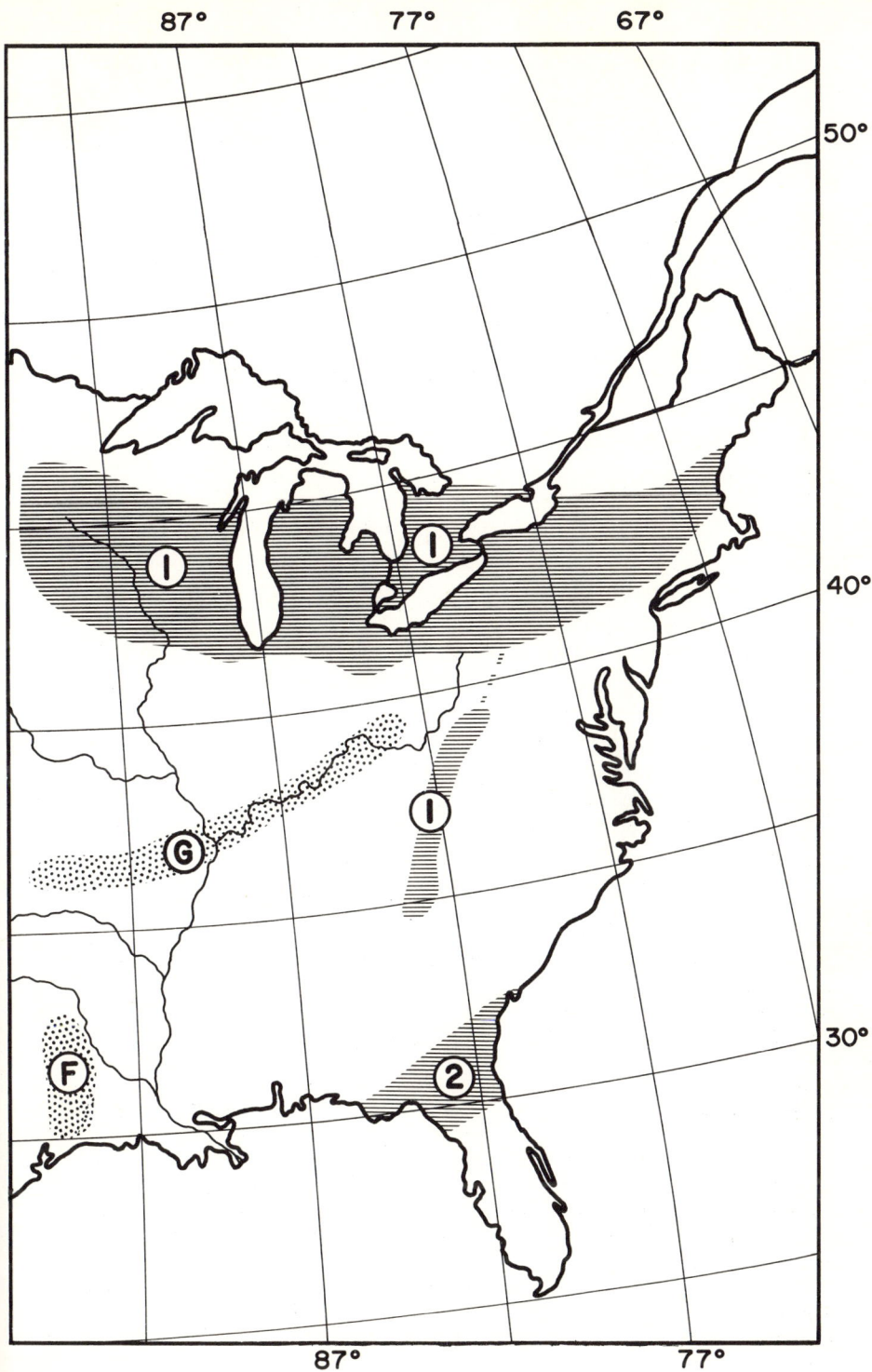

Fig. 2. (Legend on page 323)

by different leading authorities have confidently come to opposite conclusions for the same taxa. Much of this up-and-down manipulation is unjustified until adequate compatibility data are obtained by experimental hybridization. (See discussion of the theoretical interpretation of natural hybridizing later in this paper.) To an outside observer of recent research on vertebrate hybridization, the best supported conclusions on hierarchical status are those on Anura, the least supported are those on Aves. That there is a debate over the hierarchical status of so many pairs of the hybridizers in the suture-zone suggests that the breeding relationships in nature are now in a period of rapid evolution. The present paper is of course largely concerned with this problem.

In the "A" portion of each table, pairs are listed for which wild hybrids have already been reported; under "Known Areas" are listed only the places from which hybrids are definitely known, and where these known localities are very local, it is likely that there is actually much wider hybridizing along the suture-zone. The column headed "×ing" (=crossing) gives an estimate, often necessarily uncertain, of whether the amount of interbreeding is intensive, medium, or slight. I did not attempt to investigate every published record of Nearctic suture-zone hybridizing, but aside from the plants, undoubtedly the majority of known cases are listed here in the "A" tables.

In the "B" portion of each table are listed some pairs of taxa which are known to have partial sympatry or whose recorded distribution suggests that more detailed collecting will reveal geographic contact; these pairs are actually or potentially in breeding contact but have not yet been investigated in the field sufficiently to show whether they are now hybridizing in

---

Fig. 3. *Diagrammatic Model of Hypothetical Sequence of Events Producing a Suture-zone by Removal of a Physical Barrier such as a Band of Forest. At time A, pre-glacial, there is a single stable community symbolized by one species, each, of butterfly, toad, snake, bird, and tree. At time B, with a glacial maximum pushing the whole community southward and fragmenting it into refugia, each species is divided into two isolates without gene exchange. At time C, with glacial recession, the isolates move to new ranges with some fusion, and large genetic differences have accumulated. At time D, range expansions have resulted in two new stable communities separated by a physical barrier which prevents gene exchange between the northern and southern sister species of butterflies, of birds, etc. At time E, the barrier has been breached and both the northern and the southern communities expand into the former barrier region, meeting each other for the first time and tending to hybridize and to compete in the new suture-zone. At time F hybridization has ceased for the butterflies, toads, birds, and trees, evidence that their speciation had gone to completion by time D. However, the snakes have formed a clinal series from north to south, evidence that full speciation had not been attained. Competitively, the toads, birds, and trees have evolved niche separation and wide sympatry, whereas the southern butterfly species has excluded the northern relative from much of its former range and these two are allopatric. Compare with Figure 18.*

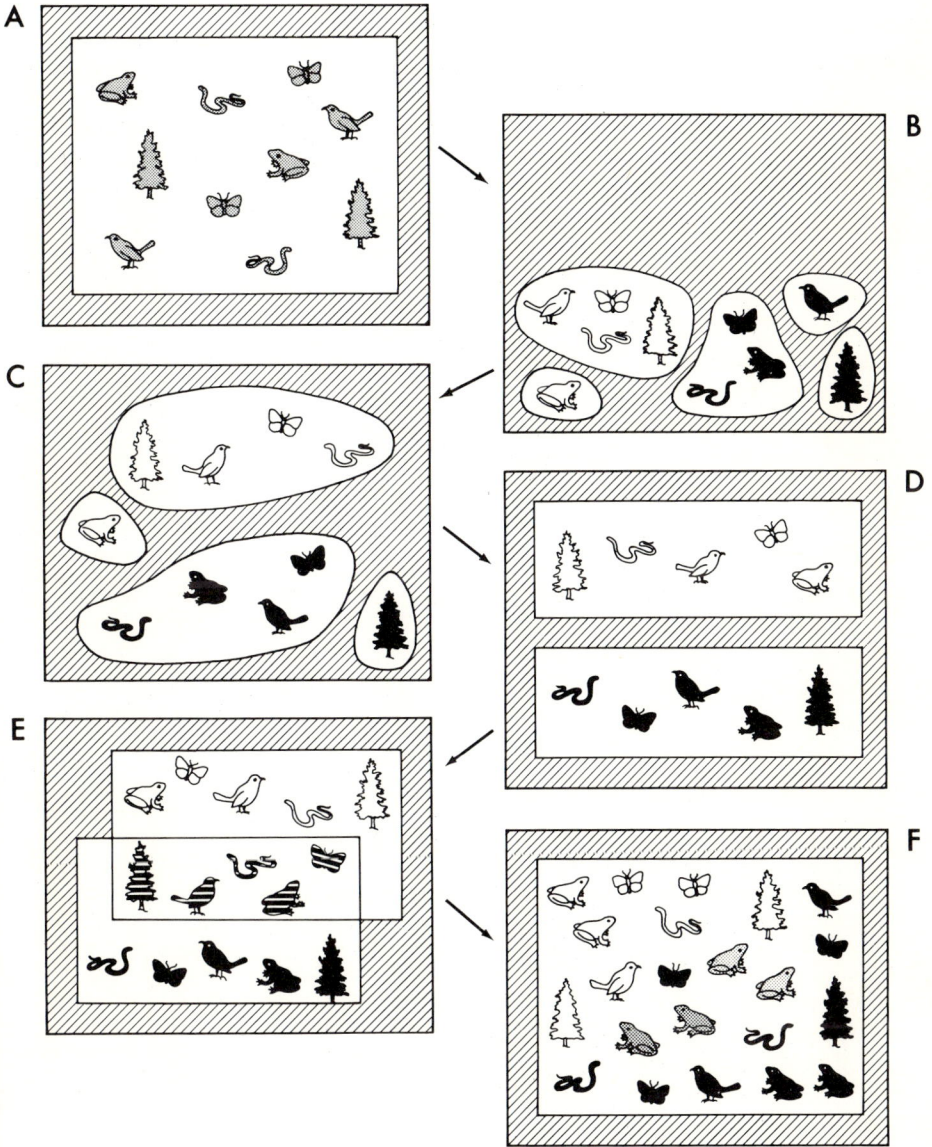

Fig. 3. (Legend on page 326)

some part of the zone. The column headed "Contact Areas" therefore includes conjectural as well as proven apposition. Even a casual inspection of present geographic ranges of most groups of terrestrial animals and plants with biparental reproduction should allow the "B" lists to be enormously enlarged. Future research will allow the transfer of many examples from the "B" to the "A" lists as well as the entire deletion of others.

Although I have personal field and museum experience with many of the vertebrate pairs which are known or suspected to be hybridizing, in compiling the "B" lists I have taken advantage of the fact that the geographic distribution of the North American tetrapods is now well known and has been conveniently mapped in contemporary manuals. Especially useful were those by Stebbins (1954) and Blair et al. (1957) for Amphibia and Reptilia, and the two elaborate volumes on the Mammalia by Hall and Kelson (1959). The many Lepidoptera examples are largely from my own experience.

Unlike ornithologists, herpetologists, and lepidopterists, the mammalogists have as yet concerned themselves relatively little with the existence of natural interspecific hybridization and have even tended to reject the possibility of such hybridization (e.g., Durrant, 1959, p. 18). Workers in groups of terrestrial invertebrates, other than butterflies and a few families of moths, have looked for hybridization even less. It is my hope that this paper will encourage systematists working with these presently neglected groups to scrutinize their material for natural hybridization. The suture-zones outlined below are obvious areas in which to begin the search for hybrids.

With few exceptions, freshwater fishes, which are well known to have extensive interspecific hybridization (e.g., see Hubbs, 1955), are unlikely to share with the terrestrial fauna the suture-zones that are the principal subject of this paper. The geological processes and human activities which influence stream systems, ponds, and lake basins are apparently the most usual factors controlling the isolation and subsequent sympatry of purely aquatic animals such as freshwater fishes and most pelecypods, gastropods, leeches, crustaceans, and others. In a special sense, suturing of freshwater biotas surely exists but is not being examined in this paper.

It seems likely that organisms living in the shallower parts of the oceans and very large lakes will be found to have suture-zone phenomena identical to those of the terrestrial biota. The proposed sea-level canal connecting the Atlantic and Pacific Oceans in Central America should produce a magnificent new suture-zone, and it would be a great loss to evolutionary biology if the events were not monitored by continuous or frequent periodic sampling on the two coasts.

The Individual Nearctic Zones

Most of the taxa now hybridizing in the Nearctic suture-zones are closely related. Thus the isolation that brought about their genetic differentiation need not have started earlier than the beginning of the Pleistocene. It is widely presumed that the movement and fragmentation of geographic ranges that produced this isolation were influenced in all the Nearctic biotas, by the drastic environmental events of Pleistocene glacial retreats and advances. Data-based biogeographic speculations on these events have been elaborated in recent years by Woodson (1947a), Rand (1948), Deevey (1949), Braun (1950), Smith (1957), Moreau (1963), Mengel (1964), Blair (1965), Ross (1965), Howden (1966), and many others. Did most of these taxa differentiate during the Pleistocene, or did many become separated by climatic and geophysical changes long after the beginning of the retreat of the last Wisconsin glacier (Valders maximum, dated about 11,500 years B.P. by Broecker and Farrand, 1963)? Either may be true, but it is not a purpose of this paper to examine the evidence for biogeography of the initial isolation or of the successive range changes prior to the presence of the barriers whose disintegration produced the present suture-zones. The barriers and their removal are the essential factors for suture-zone analysis.

The suture-zone hypothesis presumes that the suturing biotas moved together when a former barrier disappeared. This is not "hybridization of the habitat" (Anderson, 1948), and suture-zones are here conceived as entirely different from Anderson's disturbed and intermediate environments in which the hybrids are physiologically superior to the parental species. On the contrary, it appears to be certain that in suture-zones the hybrid animals (perhaps not always the plants) are typically selected against, and hybridization will ultimately cease and the hybrids will disappear. The character of the barriers and the time of their removal are important elements of the analysis of suture-zone phenomena. The removal of the physical barriers might have been linked to the immediate post-Pleistocene glacial ebb, and my initial working hypothesis, on which I first lectured a few years ago, had been that the time of removal of the barrier in each zone was in the range of 8,000 to 11,000 years B.P. This general view has probably been suggested by several writers independently over the years (see e.g. Mayr, 1963, pp. 371–2). However, a search of the recent literature, mainly palynological, failed to produce persuasive evidence of the existence and disappearance of the necessary barriers during this period (although I have no doubt that various major suture-zones did appear from time to time, and the suturing then proceeded to completion). Furthermore, the data in the vast original literature on animal hybridization suggests that extensive interspecific hybridization in animals normally accompanies the **beginning** of sympatry, and rather rapidly declines due to the establishment of anti-hybridization mechanisms (see fuller discussion below). The persistence of the present suture-zones since the beginning of postglacial time would appear therefore to be much too long. Therefore, each zone has been investigated for the disappearance of physical barriers within the last two or three millenia.

The palynologists have usually concentrated their studies on the glacial, interglacial, and immediate postglacial periods; thus, the published palynological record is skimpiest for the last 2,000 years. This is unfavorable for the present analysis, because the interpretation of the geography of today's vegetation depends most on the most recent past, the period within which the present suture-zones were probably being created by pregnation of the older barriers. The palynological evidence, like most fossil records, has serious deficiencies of geographic coverage, chronological correlation, pollen dispersal distance, and not least, the lack of a representative sampling of the complete flora due to gross underrepresentation of insect-pollinated species (see discussions by Curtis, 1959, pp. 440–445; Potter and Rowley, 1960; Davis and Goodlett, 1960; Davis, 1963; and others). The shortage of published palynological data for the top of the Quaternary is partly offset by the many descriptive accounts of vegetation by early travelers. Evidence from the earliest written history is necessarily spotty and anecdotal, but it is supplemented by detailed official records of the more recent presence and quality of timber. Thus, each of the many pieces of evidence, taken alone, has a low certainty value. But the directional consistency of this large body of data is so strong, that inferences can be made with the confidence that future research will only confirm and elaborate them.

Inspection of the suture-zone lists (Tables I to VII) shows that the hybridizing taxa can be pooled in terms of their present environmental parameters. Thereby the nature of the recent barriers which had kept them apart can be inferred. Probably the kinds of barriers and the causes of their removal differ for various major suture-zones. Human cultural practices during the last one or more centuries are the likely cause of the present suturing in some zones, and recent climatic changes in others. Definite conclusions about environmental changes in North America during the Quaternary, which can be reached through palynology, glaciology, and historical accounts of early European travelers and settlers, will be summarized for each zone. For some zones the evidence is so voluminous as to be almost unmanageable. For others it is still too scanty to allow more than cautious speculation.

I. Northeastern–Central Suture-Zone

A band approximately 100 miles wide, centered on a line roughly from Cape Cod to central Minnesota, is a major zone of hybridization between northern and southern relatives of several species pairs (see Fig. 2). There is a narrow, interrupted outlier extending southward down the Appalachians, and for some pairs this suture-zone continues northwestward into north-central British Columbia. Many known and possibly hybridizing pairs are listed in Table I, and several are illustrated in Figures 4 through 8.

The climate and physiography on the two sides of the zone tend to be so similar that preadaptation to different physical niches prior to sympatry must have been less than in Zones III and IV, and competition is probably severe between most of the hybridizing species. Nevertheless, the known major niche differences between members of several pairs in Zone I are of the sort that may be expected to arise by chance during isolation (Remington, 1958) and then be fortified rapidly during sympatry. For example: *Limenitis astyanax* feeds principally on Rosaceae, and *L. arthemis* on *Betula lutea; Papilio g. glaucus* feeds mainly on *Prunus* and *Fraxinus, Liriodendron* and *P. g. canadensis* on Salicaceae; *Erynnis baptisiae* feeds on *Baptisia,* and *E. lucilius* on *Aquilegia; Alypia octomaculata* feeds on Vitaceae, and *A. langtoni* on *Epilobium; Hyalophora cecropia* feeds on *Prunus, Acer,* and several other woody angiosperms, and *H. columbia* on *Larix;* and the breeding season of *Bufo americanus* is earlier than that of *B. fowleri* in any given locality.

As the theory would predict for all suture-zones, some species have moved into Zone I with the northern or southern community but are not hybridizing because there had been no close relative in the opposite community. Examples from the south are: *Anthocaris genutia, Papilio troilus, Asterocampa celtis* and *clyton, Deidamia inscriptum* (all Lepidoptera); *Didelphis marsupialis, Scalopus aquaticus, Icteria virens, Crotalus horridus* (all vertebrates). Examples from the north are: *Polygonia faunus, Strymon acadica, Colias interior* (all Lepidoptera); *Martes pennanti,* and *americana, Mustela erminea, Erethizon dorsatum, Lepus americanus,*

Sphyrapicus varius, Junco hiemalis, Zonotrichia albicollis (all vertebrates). Similar lists could be prepared easily for all the zones. As shown elsewhere, such species have a role in the other suturing processes, but the problems of hybridization are not relevant to them.

In transects from north of Zone I to south of it, the absence of large physical changes suggests that the former barrier to sympatry was biotic. From an inspection of Table IA, it is clear that most of the heavy hybridizers are adapted to shrubby field-edges or open woods. A broad band of largely unbroken forest (or possibly of treeless grassland) would have kept these field-edge organisms isolated into northern and southern segregates. This would have resulted in the accumulation of genetic differences leading to speciation or semispeciation. It would also have piled up on each side of the barrier whole communities, including members of pairs which had speciated elsewhere and then gradually converged in their ranges until stopped by the barrier. Destruction of the continuity of such a forested region during the last two or more centuries would then have brought together the two communities and inevitably produced the kind of suturing which exists today.

Figure 3 shows diagrammatically the hypothetical steps which could have produced Zone I: first genetic differentiation, then a barrier on either side of which the differentiates accumulated, then the break-down of the barrier to create a suture-zone, and finally the stabilization of reproductive and competitive relationships in the formerly suturing taxa.

What is the evidence for such a sequence of events in the present suture-zone? The many pollen profiles published in recent years for localities scattered over the region show a reasonably consistent record of the vegetational, and by inference climatic, postglacial history of the zone. Davis (1965) recently summarized the phases of this record as follows:

1. For over 2,000 years following the last (Wisconsin) glaciation, the vegetation of southern New England was similar to that of the present subarctic regions, i.e., low and treeless.
2. About 11,500 years ago the forests began to develop, they were predominantly of spruce until about 9,500 B.P.

Examples of Pairs of Known Hybridizers in Zone I (see Table I).
Fig. 4. (a) Papilio glaucus canadensis; (b) P. g. glaucus. *Fig. 5.* (a) Limenitis arthemis; (b) *wild hybrid* L. astyanax × arthemis; (c) L. astyanax. *Fig. 6.* (a) Hyalophora columbia; (b) H. cecropia. *Fig. 7.* (a) Vermivora chrysoptera; (b) *wild hybrid* V. chrysoptera × pinus; (c) V. pinus. *Fig. 8.* (a) Populus tremuloides; (b) P. grandidentata.

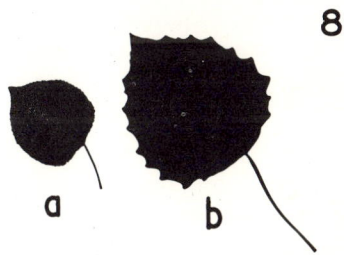

Figs. 4, 5, 6, 7, 8. (Legends on page 332)

3. Gradually, spruce was supplanted by balsam fir and sugar maple and then by pines and birch.
4. About 8,000 B.P. temperate forests developed, dominated by oaks, as well as white pine and then hemlock.
5. About 5,000 B.P. hemlock declined markedly, perhaps due to increased warmth and aridity; oaks, hickories, and chestnut became abundant.
6. Finally, during the last 2,000 years or less there has been a major change in the vegetation, demonstrated in pollen profiles by increases in pines, red maple, spruce, fir, and especially in weedy herbs such as *Ambrosia,* and recently *Plantago.*

Heavy forestation is indicated for phases 4 and 5 by the prevalence of pollen of birches, beech, oaks, hickories, pines, and hemlock. This general pattern has been reported widely, for example from southwestern Vermont (Whitehead and Bentley, 1963); Cape Cod (Butler, 1959) and Martha's Vineyard (Ogden, 1961); central (Davis, 1958) and western (Deevey, 1943) Massachusetts; Connecticut (Deevey, 1939, 1943; Sears, 1963); southern New York (Cox, 1959); northern New Jersey (Potzger, 1946; Niering, 1953; Heusser, 1963); Ohio (Sears, 1942; Potter, 1947); Indiana (Potzger, 1946); southern Michigan (Sears, 1942; Potzger, 1946, 1948); southeastern Wisconsin (West, 1961); and southern (Potzger, 1946; Wright et al., 1963) and northern (McAndrews, 1966) Minnesota. It is remarkable that the pollen diagrams from all of these localities in the present Suture-Zone I show a similar large increase in *Quercus* and decrease in *Pinus* in pollen zone C-1 (terminology of Deevey, 1939), the oaks staying at a high level through C-2 and early C-3. Diagrams for all localities north of Zone I tend to lack the *Quercus* surge: for example, northern Vermont (Davis, 1965), most of New York (Cox, 1959; Cox and Lewis, 1965), northern Michigan and Wisconsin (Potzger, 1946), and northeastern Minnesota (Fries, 1962), as contrasted to southern Vermont, southern New York, southern Michigan and Wisconsin, southern and northwestern Minnesota (all referred to above). Other typical northern pollen profiles are from coastal Maine (Potzger and Friesner, 1948) and central Ontario (Potzger and Courtemanche, 1956). An exception is Cape Breton Island, which appears to be intermediate between Zone I and more northern typical profiles (Livingstone and Livingstone, 1958).

South of Zone I there are too few available pollen diagrams of the upper strata to allow comparisons with the Zone I records. A profile from the high glades of West Virginia (Darlington, 1943) shows surprising uniformity throughout the column after the early decline of spruce and fir. No major increase of oak pollen or decrease of pine pollen is reported. The diagram for the Dismal Swamp, Virginia (Whitehead, 1965), generally resembles the Zone I profiles, but *Betula* and *Tsuga* are much scarcer than in southern New England during the last 3,500 years, and the abundant *Taxodium* and *Liquidambar* do not occur in the New England samples.

In summary, it is clear that profound changes occurred about five or six thousand years ago in the forests of the area now occupied by Suture-Zone I. Oaks became prevalent, while pines and hemlock declined, probably due to a warmer, drier climate that continued until very recent times. Possibly the oak-dominated woodlands were relatively open, and the aridity probably allowed greatly increased patchy deforestation by natural fires.

The "potential natural vegetation" (i.e., the presumed vegetation that would develop if man were removed entirely and no climatic change occurred) has been mapped in detail by Küchler (1964). This map depicts the vegetation that would have existed until a few centuries ago, prior to extensive agriculture and forest-burning by Indians and then European colonists. In the present Suture-Zone I, Küchler showed all or parts of seven vegetation types: (1) transition between Appalachian Oak Forest and Northern Hardwoods, (2) Beech-Maple Forest, (3) Elm-Ash Forest, (4) Oak-Hickory Forest (in Michigan only), (5) Maple-Basswood Forest, (6) Oak Savanna, and (7) northern portions of the Appalachian Oak Forest. Braun (1950, 1955) classified the eastern deciduous forest zones somewhat similarly. Weaver and Clements (1938) distinguished fewer subdivisions, and their Beech-Maple, Deciduous Forest, lying along the southern edge of the Lake Forest, seems to fit Zone I best of all. Livingston and Shreve (1921) included Zone I in the northern fraction of their Deciduous Forest, and they made an interesting observation, variously echoed by other writers: "There is not one of the vegetational areas of the United States that has been more completely and profoundly altered by man. . . . The virgin stands of deciduous forest were made up solely of deciduous broad-leaved trees over extensive areas, and these forests were both dense and of a stature as great as 100 feet, or even more" (p. 38). It seems to be generally agreed that the precolonial vegetation of the present suture-zone was principally a solid, continuous hardwood forest. Gleason and Cronquist (1964, p. 307) use a similar forest classification, and their statement is: "The deciduous forest once covered hundreds of thousands of square miles in the eastern half of the present United States. . . . Most of this virgin forest has now been destroyed."

To fulfill the formula (see above) for a barrier whose breakup would produce a suture-zone, evidence must be examined on whether the previous vegetation in Zone I was a broad band of continuous, dense, large forest. Braun (1950) demonstrated that these were the characteristics of the Beech-Maple Region of southeastern Wisconsin, the southern half of

Michigan, southern Ontario, western Pennsylvania, and northern Ohio and Indiana. She placed the more eastern portion of Zone I in the Glaciated Section of the Oak-Chestnut Region, although here is also an abundance of beech and hard maple, as well as oaks, chestnut, hemlock, hickories, and white pine. This, too, was a relatively solid, continuous forest in its virgin state.

My formula also presumes that the vegetation to the north and south was more open. The entire northern border of the Zone I region is a transition to the Hemlock-White Pine-Northern Hardwoods Region. The evidence is plentiful that the latter had a mixed, semi-open vegetation. Braun noted: "The coniferous communities which occur at intervals almost throughout the region are of two general types: (1) those of more or less dry sand plains and ridges where white pine, red or Norway pine, and jack pine prevail; and (2) those of poorly drained areas, bogs or muskegs, where black spruce, arbor vitae (northern white cedar), and larch prevail" (p. 337). The southern border blends into several forest types in which *Quercus* has long been a major member (Niering, 1953), and oak woods tend to be much drier and more open than beech-maple-hemlock forests. In the Middle West the southern edge was bounded by a somewhat interrupted vegetational type which was treeless, with low herbs: the Prairie Peninsula. On the west, it became bur oak—savanna (Curtis, 1959). Thus there is at least a general indication that the forest barrier existed and that much more open woods, perhaps very old, extended north and south of the broad forested band.

Early European travelers not only observed these woodlands but also recorded their decimation by the settlers. Fascinatingly, the heavy virgin forests deduced by phytogeographers may have already lost their density in many places by the time European colonists arrived in northeastern North America. Nichols (1913, pp. 199–200) concluded in his study of virgin forests in Connecticut: "At the time of its settlement, early in the seventeenth century, practically the entire state of Connecticut was densely wooded. [However], even before the advent of civilization there were doubtless considerable tracts, at least in the lowland, which in a more or less primitive way had been brought under cultivation." Bromley (1935, p. 64) reviewed a number of seventeenth century eyewitness accounts:

> On one subject, all are in accord and that is the observation that the original forest was, in most places, extremely open and parklike, due to the universal factor of fire, fostered by the original inhabitants to facilitate travel and hunting. . . . The burning of the forests and grasslands, it must be remembered, was a universal custom among other regions of the world as well.

Curtis (1959) similarly noted for Wisconsin that nomadic hunting Indians routinely started huge forest fires, and he went on to note that these nomadic tribes were present throughout Wisconsin during the entire post-glacial period. Unfortunately, the best proof cited by Curtis of frequent burning by Indians is for the grasslands; evidence for major burning of the precolonial forests of Wisconsin seems to be equivocal. Bromley's evidence now appears even more tenuous. Raup (1937, p. 111) reexamined Bromley's sources, and not only did he doubt that fire was a major forest regulator in the inland woods, but he even questioned the supposed openness of the northeastern forests, coastal or inland. Raup does agree with the view that "a warmer and drier climate existed in southern New England within the past 3,000 years," followed by a cooling trend within comparatively recent time.

Clearing by settlers soon breached the forests even where Indians had not done so. Bromley (1935, p. 65) wrote:

> Many of the old records state that the colonists continued the burning of the woods after the expulsion of the Indians to maintain grass for pasturage in the areas not completely deforested. Progressive clearing of the land for agriculture continued until the early part of the nineteenth century. Between 1820 and 1850 the area of cleared land attained its maximum amounting to 75 or 80 percent of the total in many southern New England counties.

Raup and Carlson (1941) estimated that 85 percent of the landscape in central Massachusetts was cleared by the middle of the 19th century, after which the westward migration allowed much of the farmland to revert to woodland. Davis (1965, p. 397) added that "the New England landscape showed a far greater influence of man 150 years ago than it does now." Huge tracts of forested land were purchased by land-speculators and then divided into small parcels and sold to settlers. Extracts from a guide written around 1807 for potential settlers in the western counties of New York gave detailed instructions on wise forest removal:

> Where the basswood, the butter-nut, the sugar-maple, white-ash, elm, and tall redbeech is the prevailing timber, you may be certain of a good soil both for grain and grass. If it is interspersed with hemlock, it is not the worse. The black-walnut is never found but in strong and durable ground. The large topped, short mossy-beech denotes ungenerous land. The poplar in our climate promises good wheat. The pitch-pine uniformly bespeaks a thin sandy land. . . . Those lands which produce spontaneously the birch and the spruce are the last taken up (Cooper, 1810, p. 34).

There is likewise abundant evidence from the top of pollen profiles, showing the beginning of the influence of man in dissecting the forests. The presence of the European weeds, *Plantago major, P. lanceolata,* and *Rumex acetosella,* marks the beginning of European agricultural clearing. Other nonarboreal pollens of this age, especially weeds like *Ambrosia, Chenopodium,* and *Amaranthus,* sometimes accompanied by pollen of maize or charcoal and ash, are also indicators of clearing and agriculture by Indians or European settlers. Of course, the pollen record is often directly associated with human cultural objects, especially by P. S. Martin and his associates in the Southwest. In most of the earliest palynological papers, the top of the column was ignored or only casually noted, but recently there has been a lively interest in it. Examples of locations where clear evidence of human disturbance is known include: Nova Scotia (Livingstone and Livingstone, 1958); coastal Maine (Potzger and Friesner, 1948); central Massachusetts (Davis, 1958); Martha's Vineyard (Ogden, 1961); Cape Cod (Butler, 1959); Connecticut (Sears, 1963); northern New Jersey (Heusser, 1963); Ontario (Potzger and Courtemanche, 1956); eastern Wisconsin (West, 1961); southeastern (Wright, et al., 1963), northeastern (Fries, 1962), and northwestern (McAndrews, 1966) Minnesota.

Hence there is no doubt that the heavy forests of Zone I were extensively opened up within the last few centuries by man: Indians in many places more than 300 years ago, white settlers virtually everywhere else, from 300 years ago (east coast) to 100 years ago in Wisconsin (Curtis, 1959). Thus, requisite conditions have existed for establishment of very recent suturing here.

It must be recalled, however, that there was a major replacement of pine by oak pollen in the northeastern records at the beginning of C-1 time, with oaks remaining abundant until very recently. As has been noted above, oaks are associated with drier, much more open forests than are hemlock, beech, pines, or yellow birch. Thus, there may have been a change in the forest about 5,000 years ago that could have created the suture-zone. Nichols (1913) and Bromley (1935) had favored Indian burning as the mechanism that opened up the forests before and during colonization. Raup (1937) preferred the view that the altithermal period beginning "about 3,000 years ago" was responsible.

There is little evidence for the dating of Suture-Zone I; suturing could have begun a few centuries or a few millenia ago. The lack of reproductive isolation in several species pairs makes me favor the younger rather than the older date.

A second possibility concerning the Zone I barrier is that it was a treeless sector separating the northern and southern woodland biotas. The Prairie Peninsula, which extended as a tongue of the central grassland far eastward across Illinois and Indiana into Ohio, might have been such a barrier during the xerothermic maximum. However, there is no evidence to support any hypothesis that this grassland was sufficiently broad and unbroken to serve as the barrier (Cushing, 1965), and this possibility must be ruled slight.

II. Northern Florida Suture-Zone

This major zone of hybridization between peninsular Floridian species, semispecies, and distinctive races and their continental near relatives is centered almost exactly along the northern border of Florida. For some paris the zone now extends northeastward along the Atlantic coast into South Carolina and/or westward along the Gulf coast into Alabama or Mississippi. Table II lists some known and suspected hybridizing pairs; Figures 9 to 11 show specimens of some of these pairs.

Although most of the Floridian stocks are now classified as only sub-specifically separable from the mainland stocks, they tend to be pheno-typically very distinct, with little or no smoothly clinal trend from one into the other. Some of the so-called subspecies will probably prove to have strong sterility and/or inviability barriers when tested by experimental hybridization. The suture-zone is so narrow and the hybridizing so localized, that it is possible that the area has only recently become habitable to many pairs of related taxa. As with Zone I, a transect from north of the zone to south of it shows little physiographic and climatic change. This is very different from the tropical-temperate contact in southern Florida, discussed below. The former barrier in Zone II there-fore appears most likely to have been vegetational rather than physical, and the northern and southern taxa may have evolved only modest physiological differences prior to their meeting in Zone II. The model in figure 3 shows the hypothetical sequence of events leading to suturing in Zone III.

What is known of the postglacial history of the vicinity of this suture-zone? Regrettably little compared to the great mass of literature for the Northeast. Details have been mapped of the Pleistocene and later shore-lines of Florida at times when the sea level was higher than it is today (Cooke, 1945; MacNeil, 1950). Much less is known of the shorelines when the sea level was lowest (as during glacial maxima), but fortunately this is of little importance in interpreting the environmental history of the terrestrial animals and plants now mixing in Zone II. Peninsular Florida is extremely flat and has long been so. This means that substantial rises in sea level would have inundated much of the present land, and this did occur during each interglacial. According to Cooke, the earliest Pleistocene in-undation left only a few small islands above water, in the area of present-day central and north-central Florida (Sunderland Shore, believed to be Yarmouth in age). The subsequent glacial advance to the north (the Illinoian) may have dropped the sea 300 feet below its present level in Florida, exposing as a terrestrial habitat all of present-day Florida plus a wide area on the Gulf and northeastern sides. This was followed by an inundation, correlated with the Sangamon interglacial by MacNeil, which covered southernmost Florida and much of the east and west coasts, leaving a large island and several smaller ones. If the marine barrier lasted for a sufficiently long time, it must have produced speciation in some organisms during their insular separation from the mainland. However, these two large rises in sea level ended much too long ago to account for the present suturing (more than 70,000 B.P. is suggested for the Sangamonian Stage; see Frye et al., 1965, and Richards, 1965). Anderson (1948) emphasized the great importance of northern Florida for plant hybridization at present, but his presumption, and that of Mayr (1963), that this hybridization is the consequence of the reuniting of the "Tertiary" island with the mainland appears to be unfounded.

The only supposed postglacial rise was to the "Silver Bluff" shoreline of Cooke, which was eight feet higher than today's level. MacNeil reasonably associated the Silver Bluff with the "thermal maximum" of the palynologists, perhaps 6,000 to 4,000 B.P., although Puri and Vernon (1964) and Richards (1965) consider the Silver Bluff rise more likely to be older, "late Wisconsin Interglacial." While this small rise in shoreline would have made no significant difference in topography for terrestrial organisms, two other factors would have had important effects on the vegetation. First, the modest rise in sea level would also have raised the ground water in much of lowland Florida, producing wetter conditions for tree growth. Second, it would have been associated with a rise in temperature. The only pollen diagram so far reported for Florida is from Mud Lake, Marion County, on the southern edge of Suture-Zone II (W. A. Watts, unpublished; see Whitehead, 1965). It shows a sharp rise in *Pinus* pollen and a decline in *Quercus, Myrica,* and *Chenopodium* at about 5,200 B.P., followed soon after by a rise in *Taxodium*. This must indicate a change to a cool, moist period. The pollen frequencies of these five indicator genera then remained fairly constant up to the present. Thus, there is evidence that northern Florida entered a period of moist, dense forestation about 3,000 to 4,000 years ago. Nothing is known about postglacial vegetational history in Georgia or southern Florida, bordering the suture-zone on the north and south.

The presently suturing species pairs listed in Table II whose biotopes are known to me are preponderantly animals and plants of semiwooded, scrubby habitats. *Limenitis, Papilio, Automeris, Terrapene, Colinus, Pipilo, Juniperus* and *Asclepias* are paradigmatic of this environment today.

Two lines of evidence indicate a relatively recent breakup of the forests that would have allowed occupancy by biotas specialized for semi-wooded biotopes: from archeology and from accounts by early travelers.

First, archeological evidence from near Malabar in northeastern Florida suggests that "about 700 A.D. or earlier" (Goggin, 1948) the prairie was completely dry and the sea level lower than now. The same region has a higher water table in modern times, probably produced by a higher sea level and/or greater precipitation. Also, around "1,000 to 1,500 A.D." the mollusc *Livona pica* was abundant in Florida coastal water, whereas it now lives only to the south of Florida. These archeological estimates and similar dates suggested by Rouse (1951) appear to be too young, but the climatic sequence seems to be well supported. In the last two centuries the subtropical Royal Palm has disappeared from northeastern Florida, where it was seen by William Bartram (1791). Florida's climate has been growing gradually colder during the last several centuries, after a warming trend. What effect this would have on the forest is not certain in the absence of dated details of the precipitation trends, but near the top of the Mud Lake pollen column there is a moderate increase in *Quercus* and *Myrica* and a similar decrease in *Pinus* and *Carya,* suggesting a very recent decline in forestation.

Of course, a disintegration of forest continuity at such a recent date immediately suggests human influence. The sketchy historical descriptions of the precolonial forests of the Georgia Piedmont allude to pure pine stands in the drier, more western section and pure hardwood on the moister, more eastern "red lands" (Nelson, 1957). As in the Northeast, frequent burning by precolonial Indians and then early European farmers has affected the solidity of the Southeastern forests for centuries (Bartram, 1791; Quaterman and Keever, 1962; Braun, 1950; Gleason and Cronquist, 1964). It is not unlikely that human disturbance was responsible for creating Suture-Zone II.

Representative Floridian species which have moved into the suture-zone but are not hybridizing because there is no near relative in the continental community that entered the zone from the north include: *Heliconius charitonius, Danaus gilippus* (Lepidoptera); *Rhineura floridana, Liodytes alleni* (Reptilia); *Speotyto cunicularia, Aphelocoma coerulescens* (Aves); *Neofiber alleni* (Mammalia). Northern counterparts with no near relative from the south include: *Lethe* spp., *Incisalia* spp., *Sphinx chersis, Paonias* spp., *Hyalophora cecropia* (Lepidoptera); *Haldea* spp., *Macroclemys temmincki* (Reptilia); *Reithrodontomys humulis, Microtus* (*Pitymys*) *pinetorum* (Mammalia).

Peninsular Florida has two exceptionally different biotas: (1) subtropical and (2) temperate. The southern tip has many subtropical residents, such as the crocodile, scarlet ibis, roseate spoonbill, flamingo, papaya, gumbo-limbo, strangler figs, several palms, mahogany, lignumvitae, and many butterflies (*Phocides batabano, Polygonus lividus, Ephyriades brunnea, Asbolis capucinus, Appias drusilla, Phoebis statira, Papilio aristodemus, Dryas julia, Anartia jatrophae, Eunica tatila, Marpesia petreus, Eumaeus atala, Strymon acis, Strymon martialis, Chlorostrymon maesites, Leptotes cassius* are the most typical). The tropical character of the climate is underlined by the large number of tropical species that have recently been introduced into southern Florida by man and have then become successful residents (see Neill, 1957, for many examples). These tropical residents tend to be not only **species** distinct from all Nearctic relatives, but most of them are in discrete **genera** which are widespread and highly characteristic of the entire Neotropical region. The only area in the United States with a comparable array of tropical forms is the extreme southern end of the Rio Grande Valley, in Texas. The temperate biota of Florida is the antithesis, the Floridian representatives tending to be close relatives of typically Nearctic groups that are present in much of North America.

Since the major interfaunal (and floral) boundary in Florida is far to the south at about 27° S latitude, why is the only prominent suture-zone far to the north, at about 30° N? One answer must be that there is no recently-removed barrier to breeding contact at 27°, so that any potential interbreeding should have been halted by the evolution of anti-hybridization mechanisms. A second is that few, if any, tropical species have close relatives among their temperate neighbors. And a third is that the tropical limit is determined by one or more physical factors, such as the limit of frost freedom, that prevents the spread of tropical forms beyond the firm climatic barrier.

There are, however, two fascinating Lepidoptera pairs which are hybridizing in extreme southern Florida, each having one tropical member cf clear Antillean affinities and one warm-temperate member occurring over much of the southern half of the eastern and central United States. The tropical species (or semispecies) are *Junonia evarete* and *Utetheisa ornatrix;* the United States relatives are *Junonia coenia* and *Utetheisa bella.* In the southern tip of Florida hybrids are at a high frequency, but the four parental types seem to be maintaining their discreteness. This hybridization pattern is strong evidence of new sympatry, possibly due to the arrival of *J. evarete* and *U. ornatrix* within the last several decades through some unknown human transport mechanism. The *Utetheisa* case has been worked out in great detail by Pease (1964a, 1964b). There is a preliminary paper on *Junonia* by Forbes (1928) and much more is known from recent unpublished observations. Figures 12 and 13 show specimens of these hybridizing pairs.

III. Central Texas Suture-Zone

The breadth, length, and shape of this contact line in Texas and Oklahoma (and probably northeastern Mexico) have not been determined in detail, but various genera have western species hybridizing with eastern relatives along the eastern and southern edge of the Edwards Plateau and in some cases to the north along the adjoining Comanche Plateau and the

Fig. 9. *The mimicry complex involving:* (a) *the Monarch,* Danaus plexippus; (b) *the Queen,* D. gilippus; (c) *the northern Viceroy,* Limenitis a. archippus, *mimicking the Monarch;* (d) *the Florida Viceroy,* L. a. floridensis, *mimicking the Queen; and* (e)-(h) *wild hybrids of* L. a. archippus × a. floridensis *from Zone II.*

somewhat separated low mountains of south-central Oklahoma. Anderson (1948, p. 7) was enthusiastic about the Edwards Plateau as an "area in which introgression may have operated on a grand scale" and for plant hybridization considered it "one of the outstanding centers east of the Rockies." But his suggestion that this hybridizing is linked to the time "when the Edwards Plateau came into being [and] united older land masses in Mexico and in the United States" is obviously far from the mark chronologically. Blair (1958b) has also commented on the biogeographic significance of the Edwards Plateau for plants, as well as for vertebrates.

Table III lists some species pairs hybridizing or believed to be meeting here. This suture-zone, unlike Zones I and II, is along an abrupt geomorphic boundary. Western species reach their southeastern limit on the rugged plateaus of west-central Texas and Oklahoma, and some of them now meet and hybridize with near relatives of the east whose southwestern limits are in the lowlands or up on the highlands. A few of the western members of hybridizing pairs are essentially west-Texan endemics, but most are weakly differentiated subpopulations of species which extend, with some gaps, hundreds of miles westward through the highlands of the southwestern states or turn farther northward in the Rocky Mountains. None of the eastern hybridizers is endemic to Texas, and most of them continue far to the east. The distributions of most animals and plants are so incompletely known that distributional gaps have tended to be attributed to inadequate collecting. But there is a warning in the interesting work of Blair (1958b) in Texas, showing that various widespread eastern species occurring now on the Edwards Plateau do not have continuous distributions but are actually disjuncts isolated on the Plateau. Figures 14 to 17 show some representative pairs in Zone III.

A comparison of Tables III and IV will show that some of the same species pairs are in contact in both zones. The two are quite different, however, Zone III being a contact area for Southwestern and Eastern biotas and Zone IV for Rocky Mountain and Eastern biotas.

There is probably a secondary concentration of hybridization along the western and northwestern edges of the Edwards, Stockton, and Comanche Plateaus, where eastern taxa and Plateau endemics (Hall, et al., 1961) meet relatives that do not extend as far east and south as the Balcones Escarpment (geomorphic terminology follows Fenneman, 1931, 1938). Hybrids between *Agathymus mariae* and *carlsbadensis* (Lepidoptera) have

---

Examples of Pairs of Known Hybridizers in Zone II (see Table II).
Fig. 10. (a) Papilio t. troilus; (b) P. t. ilioneus. *Fig. 11.* (a) Automeris i. io; (b) *wild hybrid* A. i. io × i. lilith; (c) A. i. lilith.
Examples of Pairs of Known Hybridizers in Southern Florida.
Fig. 12. (a) Utetheisa bella; (b) *laboratory hybrid* U. bella × ornatrix; (c) U. ornatrix.
Fig. 13. (a) Junonia coenia; (b) J. evarete.

Figs. 10, 11, 12, 13. (Legends on page 344)

been reported northwest of the Stockton Plateau (Stallings et al., 1960). Tucker (1961) showed the eastern *Quercus muhlenbergii* hybridizing with the more western *Q. mohriana* in this region. Peterson (1963) appropriately refers to a "twilight zone" between eastern and western birds in Texas, the eastern limit being the present Zone III, and the western edge being the Pecos River valley.

What is known of the post-Pleistocene history of central Texas that might explain today's suturing? During Pleistocene glacial maxima the presently arid Southwest underwent pluvial periods which were cooler and wetter than the present one (see discussion for Zone VIIIe.) Schoenwetter (1962) has published evidence that in Arizona and New Mexico there has been little change in total annual rainfall during the last 3,500 years, but the seasonal intensity has varied. Periods of heavy summer rain have produced xerophytic vegetation, whereas periods of more evenly distributed rainfall have produced a more mesophytic flora with a more arboreal character. The latter condition lasted for a few centuries just prior to about 1000 A.D. This could have been a period when there was a broad continuous connection between the biota of the Texas plateaus and the main body of Rocky Mountain and Southwestern highlands biota. With the commencement, around 1000 A.D., of localization of the rainfall into an intense summer period, the lusher vegetation would have disappeared, and the gene-exchange corridor between Texas and the Southwest would have been interrupted. This isolation in the aridity-intolerant species has continued to the present. The arrival of some of the eastern biota at the eastern side of the plateaus may have been an ancient event, perhaps enhanced somewhat in the last cenutry by the planting and irrigation of treeless areas by early white settlers and subsequent residents. Muller (1952) showed that the eastern oak, *Quercus stellata,* must have ranged at least 100 miles farther west than it now does. There are hybrid clones of *Q. stellata* × *Q. havardi* at the western edge of the Texas Panhandle, but pure *Q. stellata* is now extinct there. Although a substantial part of the lowlands near the Balcones and adjoining escarpments is treeless, it is crossed by forested bands leading to the woods to the northeast, the "cross-timbers", and has many local patches of woodlands. The stability of this vegetation was convincingly demonstrated by Graham and Heimsch (1960). Their analysis of the pollen column from Soefje Bog, Gonzales County, about 50 miles east of the Edwards Plateau, showed that the relative abundance of pollen of *Carya, Quercus, Pinus, Ulmus,* and Compositae

---

Examples of Pairs of Known and Suspected Hybridizers in Zone III (see Table III).
Fig. 14. (a) Parus atricristatus; (b) P. bicolor. Fig. 15. (a) Limenitis astyanax arizonensis; (b) L. a. astyanax. Fig. 16. (a) Papilio multicaudatus; (b) P. glaucus. Fig. 17. (a) Limenitis archippus obsoleta; (b) L. a. archippus.

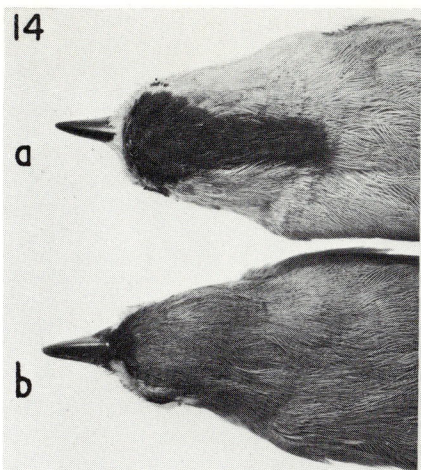

Figs. 14, 15, 16, 17. (Legends on page 346)

did not vary significantly from bottom to top of the core. A carbon-14 date for peat at the bottom of the sample was 7,820 ± 350 years.

These fragmentary indications, coupled with the low frequency of hybridizing of species pairs such as *Scaphiopus couchii* and *hurteri, Bufo valliceps* and *woodhousei,* and the apparent lack of crossing between *Centurus carolinus* and *aurifrons* (Selander and Giller, 1963), appear to mean that Zone III is a mature zone that has by now evolved partial stability. The continuation of some of the hybridization may be explained by the rather sharp boundary between the two biotic regions, which would slow the evolution of a highly efficient antihybridization genome (see Discussion, below).

IV. Rocky Mountain–Eastern Suture-Zone

The western edge of the Great Plains, the northern Plains Ranges, and the eastern foothills of the Rocky Mountains comprise the suture-zone that is the best known, due to the many hybrid birds found there. In the southern Rockies (southern Wyoming to northern New Mexico) the contact between the high mountains and plains is very abrupt. There the western mountain-dwelling member of each pair tends to have only scant opportunity to meet its eastern relative, usually where the latter reaches the edge of the mountains in river valleys, and more recently in land where man has planted trees or agricultural crops in the previously treeless plains. Hybridization here seems to be rarer than to the north, where the Plains Ranges provide extensive areas of low montane environment inhabitable both by Rocky Mountains elements and by their eastern relatives. Principally, the Plains Ranges are the Bighorn, Pryor, Big Snowy, Little Snowy, Crazy, Castle, Little Belt, Highwood, Bearpaw, and Judith Mountains of Montana, and farther to the east the Black Hills, Dakota Badlands, and Killdeer Mountains. Due to the very large number of pairs of cogeneric taxa on the two sides of this zone, this may be the most active Nearctic suture-zone. See Table IV for known hybridizers and other pairs in contact here. Representative specimens are shown in Figures 19 to 23.

It is of interest that the eastern member of each pair is not primarily a plains species, but rather a typical inhabitant of eastern wooded environments. The eastern members apparently are spreading westward slowly or not at all after they meet the eastern slopes of the mountains. By contrast, several of the Rocky Mountain taxa (e.g., *Passerina amoena, Pheucticus melanocephalus, Piranga olivacea, Colaptes cafer,* and *Icterus bullockii*) are spreading eastward, far beyond the mountains. Perhaps *Sturnella neglecta* and *Colias eurytheme* also belong in this category, but they have long been abundant residents of the Plains and much of the Midwest. *C. eurytheme,* at least, has definitely spread through the entire East from the West within the last century. *Juniperus scopulorum, Populus angusti-folia, Speyeria lurana, Polygonia satyrus,* and *Papilio rutulus* may also be examples of western types that have rather recently spread eastward.

Assuming for the present that the current spread from the eastern Rocky Mountains is largely or wholly this one-way movement eastward of the montane biota, how might it be explained? The following conjectures may provide an answer: (1) The cold climate that continued to character-ize the Rocky Mountains long after the retreat of the Wisconsin glacier to the east would have kept all the temperately adapted animals and plants at low altitudes on the two sides of the mountains, completely isolated genetically. (2) The gradual warming, accelerated during the hypsithermal (= altithermal) interval from about 6,000 to 3,000 B.P. (Pennak, 1963) or 6,500 to 4,000 B.P. (Richmond, 1965), would have gradually moved these organisms up the mountains, until they finally reached passes through which they could spill over onto the other side and occupy the whole climatic belt to which they were adapted on their ancestral side. (3) The shakiest point in this hypothesis then supposes that the physiological (ecotypic) tolerance of the western populations would have included the climates of wooded patches on the High Plains just to the east of the Front Range, whereas the High Plains taxa, faced with the relatively abrupt eastern slope, would never have had a large mountain slope population in which they could have evolved plasticity including adaptation to montane biotopes. (4) Thus, eastern types could have spread westward along the very gradually up-sloping Great Plains region but would be unable to expand up into the mountains and thence westward through the passes. The only sectors of the Rocky Mountain chain that have low altitudes for wide segments are in Wyoming, between the Laramie Range and the Wind River Mountains, and in New Mexico everywhere south of the Sangre de Cristo range limit near Santa Fe. But these are probably the only two extremely arid regions in the whole cordillera, and populations of both eastern and western members of suturing pairs must be rare or absent in these areas. Figure 18 shows diagrammatically the hypothetical steps that could have produced Zone III.

Some intriguing complications have recently become known concerning the vegetation and climates of the western Plains and the eastern Rockies. These suggest the possibility that suture-zone conditions nearly came into being many centuries ago, but optimum conditions for western occurrence of the eastern woodland biota were not synchronous with optimum condi-tions for trans-montane spillage of the western biota. Wells (1965) has reviewed many occurrences of present-day non-riparian woodlands on escarpments and other rugged topography all over the Great Plains. He suggests that the relative treelessness of this region today may be a recent development, due perhaps primarily to the addition of frequent burning (man having been there "for at least the last 11,000 years") to an already "hazardous environment for woody plants in a region of droughty climate and strong winds" (p. 249). Sears (1961a), in a preliminary sampling of lakes in the Nebraska sandhills, showed moderately high levels of pine pollen at about 5,000 B.P. and again fairly near the surface, about halfway down to a level dated at 1,110 ± 75 B.P. Intermediate levels, including the latter dated one, showed low pine counts and high grass counts. The post-hypsithermal period, beginning around 4,000 B.P., was cool and presumably moist. In the Rocky Mountains it has been termed the Neoglaciation. Included were two periods of substantial growth of montane glaciers, one C-14 dated about 3,100 to 1,000 B.P., the other somewhat later, after an interstadial of melting, and ending about one century ago (Richmond, 1965).

If the treelessness goes back to Well's implied time of 11,000 or more B.P., the former woodlands probably have little relevance to the suturing. However, the rather recent expansion of the local forested sites suggested by the Sears pollen profile and the Neoglaciation would have allowed an effective westward extension of the eastern species now suturing in Zone IV. The disjunct presence of quite surprising eastern types near the eastern base of the Colorado Rockies (e.g., the marsh butterflies, *Lethe eurydice* and *Speyeria idalia*) may be remnants of this recent more wooded period (perhaps with cooler, wetter summers than now). Such a cool interval, while moving outliers of the eastern woodland biota westward, would also have pushed montane species to lower elevations, perhaps too low to permit their reaching passes through which they could have spilled into the newly available environment to the east. So suturing may have been narrowly missed several centuries ago.

There are today some large patches of disjunct "montane" forest to the east of the main body of Rocky Mountain forest. The Black Forest is the most notable of the several of these in central Colorado. It is about 50 miles long by 25 miles wide and is surrounded by mixed prairie vegetation. Livingston (1949) showed that the timber of this tract is mainly *Pinus ponderosa,* but some *Pseudotsuga, Populus tremuloides,* and other mountain species are also present. He noted that the forest is flourishing on a coarse-textured soil, which holds water well, and that it is abruptly bounded by the nonarboreal vegetation wherever the soil becomes fine-textured, even though the precipitation, temperature, and altitude are identical over the two soil types. He considered it probably a remnant of a former typical montane forest.

It seems likely that the warming trend since 1850, following the end of the Neoglaciation, has brought the mid-altitude biota up the mountains and very recently allowed western-slope organisms to move across to the eastern slope. Their establishment in the east would have been facilitated

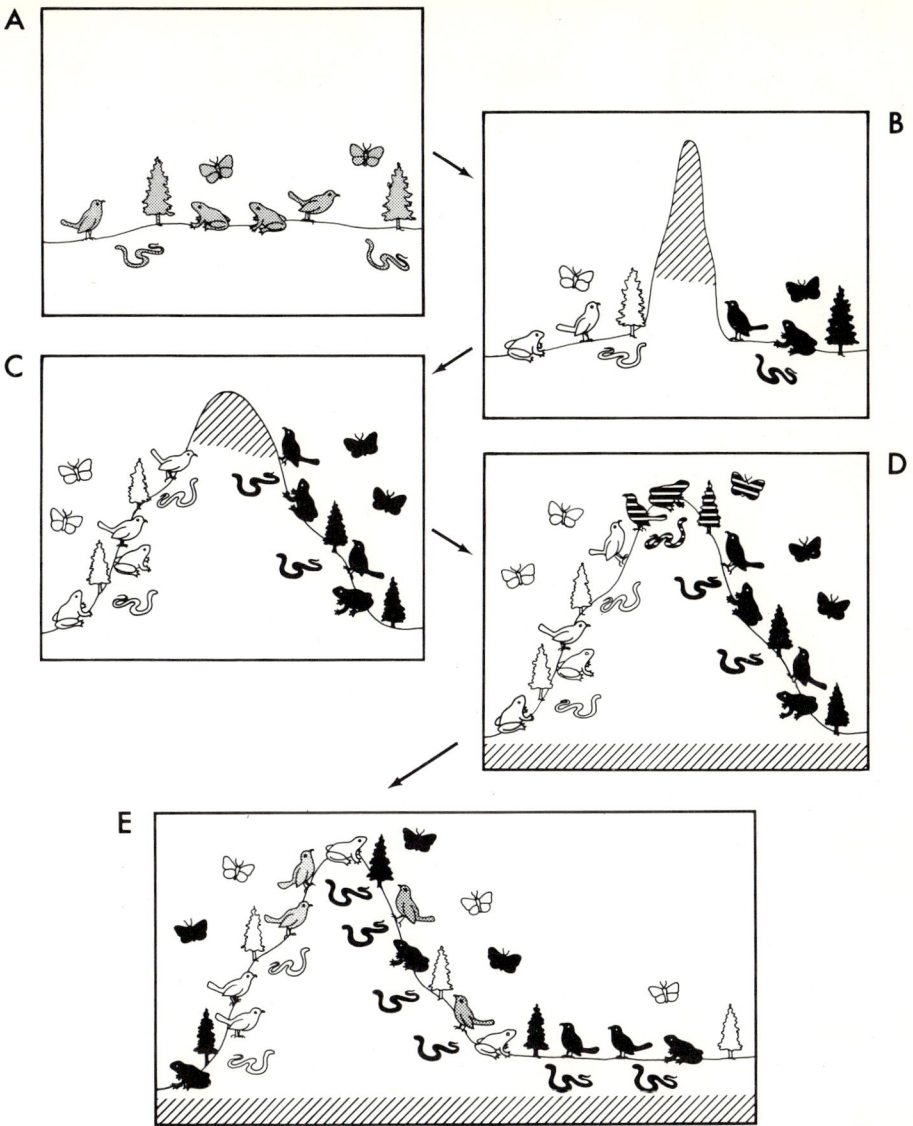

Fig. 18. *Diagrammatic Model of Hypothetical Sequence of Events Producing a Suture-zone by Climatic Amelioration in a Former Barrier of Mountains. At time A, prior to montane uplift, there is a single stable community symbolized by one species, each, of butterfly, toad, snake, bird, and tree. At time B, this community is divided by the cold climate of the higher elevations into two isolates with no gene exchange. At time C, with gradual climatic warming, the two communities move gradually up the mountains, and large genetic differences have accumulated. At time D, the inhabitable area includes the mountain top or at least the passes, and the two communities spill over into each other's ranges. The sister species tend to hybridize and compete in the new suture-zone. At time E, hybridization has ceased for the butterflies, toads, snakes, and trees, evidence that their speciation had gone to completion by time C. However, the birds have formed a clinal series from west to east, evidence that full speciation had not been attained. Competitively, the butterflies, toads, and trees have evolved niche separation and wide sympatry, where the western snake species has excluded the eastern relative from much of its former range and these two are allopatric. Compare with Figure 3.*

and expanded by the extensive planting of trees in towns, homesteads, shelterbelts, etc. on the western Plains. Virtually all of the celebrated hybridizers from the west in Zone IV could reasonably be accounted for by this sequence. *Papilio gothica, P. rutulus, Colias alexandra, Hyalophora gloveri, Speyeria lurana* (essentially *hesperis* of the eastern Rockies), *Colaptes cafer, Piranga olivacea, Icterus bullockii, Dendroica auduboni, Pheucticus melanocephalus, Passerina amoena,* and *Pipilo maculatus* in the suture-zone are inseparable (noohybrid individuals, of course) from members of their "pure" populations on the western side of the Continental Divide, which I consider strong evidence of their recent arrival in the radically different environment of the High Plains.

Westward occupation of the plains by most of the eastern members may also be due to human introduction of woody plants and perhaps also to recent absolute control of prairie burning. Gates (1940) reported that the forests have been moving slowly westward from Manhattan, Kansas, since the ending of the "great prairie fires." He regarded this as "a reoccupation rather than a migration," a view similar to that of Wells (1965, see above). Gates found no evidence of the Rocky Mountain flora extending eastward into Kansas, a rather surprising conclusion in view of the present occurrence in western Kansas of the birds *Colaptes cafer, Icterus bullockii, Pheucticus melanocephalus,* and *Passerina amoena* (Johnston, 1964) and the butterflies *Asterocampa celtis montis, Phyciodes campestris, Plebejus acmon,* and *Papilio bairdii brucei* (Field, 1940). Perhaps many plants spread less rapidly than winged animals when a suitable habitat becomes newly available.

Various species pairs on the two sides of Zone IV still have a geographic gap between them but appear to be approaching contact. New hybridizers are therefore likely, as is probably usual for a young suture-zone.

V. Northern–Rocky Mountain Suture-Zone

This zone in part resembles Zone IV, but the major difference is that the Rocky Mountain members of hybridizing pairs are crossing with species whose distribution tends to be northeast, north, and northwest of the zone. Another basic difference from Zone IV is that the environments of the two sides of Zone V tend to be much less different. Northern low-

Examples of Pairs of Known Hybridizers in Zone IV (see Table IV).
Fig. 19. (ventral sides): (a) Papilio rutulus; (b) laboratory hybrid P. rutulus × glaucus; (c) P. g. glaucus. Fig. 20. (a) Hyalophora gloveri; (b) laboratory hybrid H. gloveri × cecropia; (c) H. cecropia. Fig. 21. (after Brayshaw, 1965: p. 30): (a) Populus angustifolia; (b) wild hybrid P. angustifolia × deltoides occidentalis; (c) P. d. occidentalis; (d) wild hybrid P. d. occidentalis × tremuloides; (e) P. tremuloides. Fig. 22. (a) Papilio gothica; (b) laboratory hybrid P. gothica × polyxenes; (c) P. polyxenes. Fig. 23. (a) Pheucticus melanocephalus; (b) P. ludovicianus.

Figs. 19, 20, 21, 22, 23. (Legends on page 352)

land environments resemble southern montane ones. This being a region where glaciation is very late in receding, much of the potential region of geographic contact has only recently become habitable or has not yet progressed that far. Species pairs in various tree genera typify this condition. Raup (1946) found in north-central Alberta that *Picea glauca glauca* (northeastern) and *P. g. albertiana* (Rocky Mountains) have met and are hybridizing, and the same is true of the northeastern *Pinus banksiana* and the Rocky Mountain *P. contorta,* but *Abies balsamea* (eastern) and *A. lasiocarpa* (Rocky Mountain) were near but not yet in contact. Moss (1959) has since mentioned probable mixing of the two *Abies,* but it is not clear that he had actual evidence.

Zone V is almost entirely in Canada, along the east slope of the northern Rockies, but it extends slightly into northern Montana for various genera. Known and suspected hybridizers are listed in Table 5. As with Zone IV, there are semi-isolated outliers to the east, on the Sweetgrass Hills, Cypress Hills, Turtle Mountains, and Riding Mountain. For many groups, collecting in northern Canada and southeastern Alaska is so incomplete, that distributional details are unknown for the crucial contact areas, and much probable hybridization has not yet been detected.

These Canadian Plains Ranges present a mixture of Zone IV and Zone V suturing and even some unique hybridizing of their own. (The same is true of the northernmost Plains Ranges listed in Zone IV.) One of the uniques is the hybridizing on Riding Mountain, and probably in other Plains Ranges, of a complex *Papilio* pair. On Riding Mountain is the yellow-banded *P. avinoffi,* which may be a race or a sibling of *P. hudsonianus* (the latter has been considered on inconclusive grounds as a race of the Holarctic *P. machaon*). Farther west, in the foothill country of central Alberta, is a quite pure population of the *avinoffi* type, uncomplicated by any near relative. But on Riding Mountain there is also a black form, presently nameless, which appears to be a long-resident northern Plains offshoot of the widespread eastern *P. polyxenes.* The sexes of the Manitoba black entity are alike and resemble males of *P. polyxenes,* whereas *P. polyxenes* has evolved black and blue females mimicking the unpalatable *Battus philenor.* On Riding Mountain the more easterly dark swallowtail hybridizes freely with the more westerly yellow-banded species (Remington, 1956, 1958). There is evidence that this sympatry is extremely recent and possibly linked with human interference.

Examples of Pairs of Known Hybridizers in Zone V (see Table V).
Fig. 24 (ventral sides). (a) Limenitis arthemis rubrofasciata; (b) *wild hybrid* L. a. rubrofasciata × weidemeyerii; (c) L. weidemeyerii. Fig. 25. (a) Betula papyrifera; (b) *wild hybrid* B. papyrifera × fontinalis; (c) B. fontinalis; (d) *wild hybrid* B. fontinalis × glandulosa; (e) B. glandulosa. Fig. 26 (after Brayshaw, 1965: p. 30). (a) Populus balsamifera; (b) *wild hybrid* P. balsamifera × angustifolia; (c) P. angustifolia; (d) *wild hybrid* P. angustifolia × tremuloides; (e) P. tremuloides.

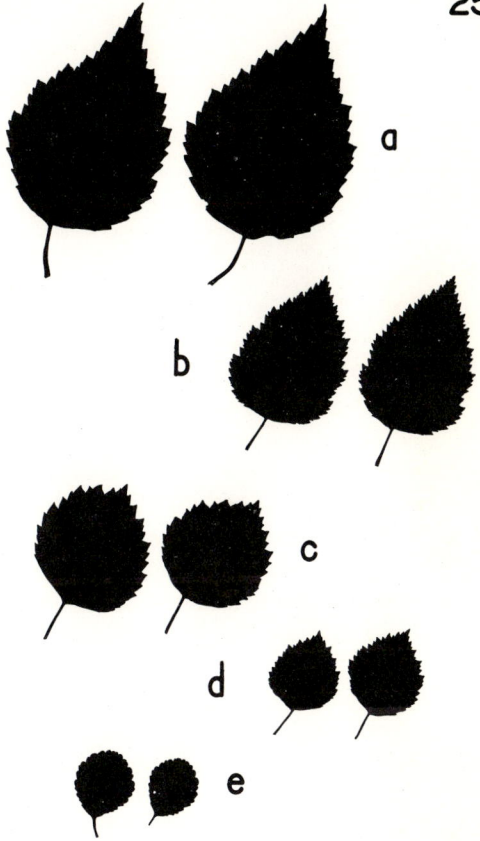

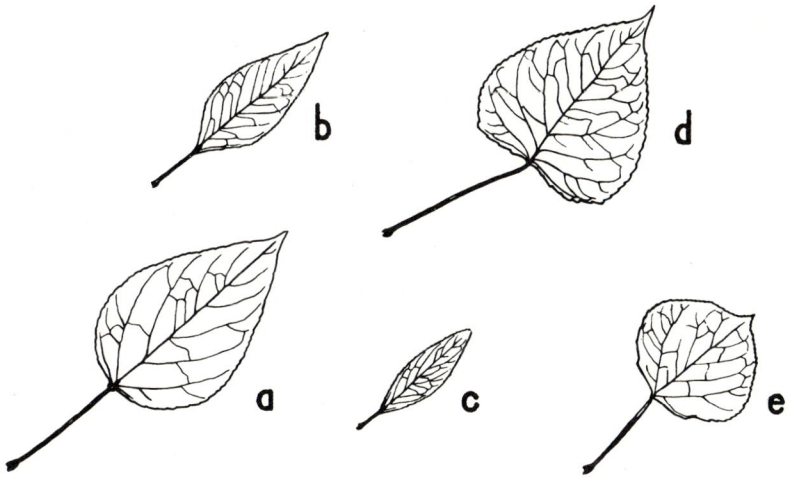

Figs. 24, 25, 26. (Legends on page 354)

In her comprehensive survey of the postglacial development of the Manitoba flora, Löve (1959, p. 579) stated that "the most intensive mixing [of eastern and western boreal forest] has taken place as late as from the hypsithermal to the present time, i.e., during the last 4,000 years," especially north of the Manitoba lakes. This would imply a suture-zone between eastern and western members of northern species pairs. For animals, at least, the taxonomic distinction is usually between the northern members of Rocky Mountain taxa and western members of more northern but transcontinental taxa. There tends to be a moderate step-cline between eastern and western populations of the pan-boreal species (e.g., *Limenitis a. arthemis* of the northeast and *L. a. rubrofasciata* of the northwest; note, also, the two races of *Picea glauca* already mentioned). But the morphological differences tend to be so modest and the intergradation so marked, that there is little inclination for classical taxonomists to see species differences. Löve reports a climatic sequence south of the Manitoba lakes that could account for the distinct step that I have noted in the western Ontario–Manitoba region for species like *L. arthemis*. Linked with the post-hypsithermal cooling trend, the last 2,000 years have seen steady cooling, with a resultant spread of aspen, then spruce, over the prairie that had been formed during the warmer, drier period. Ritchie (1964) analyzed three unusually complete postglacial pollen sections from small lakes on Riding Mountain. This presently forested plateau was

> occupied by an initial closed spruce forest, followed by a more or less treeless episode characterized by grassland vegetation, replaced by a deciduous forest episode with birch, poplar, and oak, culminating in a mixed spruce —deciduous forest closely similar to the vegetation of today . . . a climatic sequence not incompatible with the familiar Anathermal—Hypsithermal—Hypothermal sequence.

Thus, some of the more southern species of the "Northern" group now hybridizing with Rocky Mountain relatives in Alberta have apparently been able to cross a former plains barrier only recently and make contact with the biota of the Rockies. But this bridge of woodland that includes Riding Mountain is surely older than the connection across the western Plains in Zone IV.

VI. Pacific–Rocky Mountain Suture-Zone

In a very narrow band, extending northward from central-eastern California along the eastern edge of the Sierras and southern Cascades and then northeastward to western Montana, is a major zone of hybridization between far western species and their relatives of the Great Basin ranges and Rocky Mountains. The portion of this zone from northeastern

California to western Idaho is largely unstudied, but the similarity between hybridization patterns of the California–Nevada contact and the Idaho–Montana area makes it likely that the hybrid line continues in eastern Oregon and perhaps southeastern Washington and southwestern Idaho. Table VI lists some known and potentially hybridizing pairs here, and representative specimens are shown in Figures 27 to 29.

This is the narrowest and one of the youngest of the six major zones. Everywhere the suturing tends to be concentrated in the vicinity of the mountain passes. In California there is little or no hybridizing on the Pacific slope of the high Sierras, but extensive suturing on the Basin slope. In Montana the hybridization is partly in the passes and in part extends much lower. As with Zones IV and V it is striking how unidirectionally the suture-zone extends from the montane divide. Not that in Zone IV the band is on the east slope and is expanding eastward. In Zone V the eastern foothills contain most of the hybrids. Identically in Zone VI, the hybridizing is on the east side of the Sierras and the east slope of the Continental Divide in Montana and extreme southern Canada.

Limenitis lorquini and *L. weidemeyerii* are the prototype pair for this zone. Everywhere that they meet they hybridize freely, although there is evidence that the interspecific matings have a lower probability of grandchildren than pure *l* × *l* or *w* × *w* matings. The implication is, therefore, that this is a very recent sympatry. Both tend to be abundant throughout their range. *L. lorquini* is found throughout California, Oregon, and Washington except in the desert regions of southeastern California, eastern Oregon, and southeastern Washington. It is present in southwestern British Columbia and Vancouver Island, in most of the mountains of Idaho, and in all of Montana west of the Divide. *L. w. weidemeyerii* is in all the mountains of the Great Basin states, and is abundant on both slopes of the Rockies in New Mexico, Colorado, and Wyoming, on the east slope only in Montana, and on all the Plains Ranges of the United States. It has two other distinctive races—one in the Spring Mountains of southwestern Nevada and the other in most of the forested mountains of Arizona and New Mexico. The distributional details are unknown in eastern Oregon and southern Idaho. I have found *L. weidemeyerii* in extreme northeastern Oregon, but *L. lorquini* did not seem to be there. The two species hybridize where they both occur at Mono Lake and Bridgeport, California, in the border ranges just northward in Nevada, and probably in every major pass across the Continental Divide in Montana and just north of the Canadian border. At several places in Montana hybrids of various degrees of backcrossing extend many miles to the east (or north or south in the direction of the Atlantic watershed in the sectors where the Divide happens to go east-west), but in California and most of Montana the hybrid zone is very

narrow. At no locality yet discovered is there a substantial number of
hybrids extending into the normal territory of *L. lorquini* on the Pacific
slope. The explanation is unknown for the distribution of *L. weidemeyerii*
at least 600 miles to the west of the Continental Divide of Colorado and
Wyoming, but its complete absence west of the Divide from Yellowstone
Park northward. But so many species of animals and plants have the same
distribution that there must be a fundamental cause in the geological and
climatic history of the region. See Figure 19 for a hypothetical model of
suture-zone events in Zone VI.

The cause of suturing in Zone VI seems clearly to be the recent warming
trend, which has allowed low montane taxa to move gradually up the
mountains until the spillover effect developed when low points in the
mountain barricade were reached.

The recent warming tendency is indicated by the pollen records (Pennak,
1963; Sears, 1961a) and glacial sequences (Richmond, 1965) discussed
above. Hansen (1955) found evidence of temperature rise in the abrupt
reappearance of grasses, chenopods, and composites at the top of his pollen
records from the southern part of the intermountain region of British
Columbia. Dightman and Beatty (1952, p. 89) reported a period of "40
years or more of recession" up to about 1950 in the glaciers of Glacier
National Park. The same recent trend is shown from a different data source
—averages of annual and seasonal sunshine hours for eleven weather stations in British Columbia. Powell (1965, p. 95) concluded: "In comparison with the period 1921–40, sunshine hours have been less in all
seasons at most stations for the periods 1901–20 and 1941–60." Pollen
records for the immediate vicinity of Zone VI seem to be nonexistent so
far, but to the extent that extension from the above more distant sites is
valid, there is evidence of substantial warming (1) during the altithermal
about 4,000 years B.P., probably too old to account for the present hybridizing, (2) for about the past 100 years, which is much more promising, and
(3) for the period 1920 to 1940, which seems too brief and too recent
to explain the very lengthy suture-zone.

Miller (1951, p. 618), in his survey of California bird distributions,
concluded that "Connections between the lowland coastal areas and the
Great Basin are probably greater now than during much of the Pleistocene." Farther north, Dalquest (1948), in his faunistic study of the
mammals of Washington, found that the eastern part of the state is the
meeting ground of many strongly differentiated Rocky Mountain and
Pacific coast populations.

Examples of Pairs of Known and Suspected Hybridizers in Zone VI (see Table VI).
Fig. 27. (a) Limenitis lorquini; (b) *wild hybrid* L. lorquini × weidemeyerii; (c) L. weidemeyerii. *Fig. 28.* (a) Hyalophora euryalus; (b) H. gloveri. *Fig. 29.* (a) Sphyrapicus
ruber; (b) S. nuchalis.

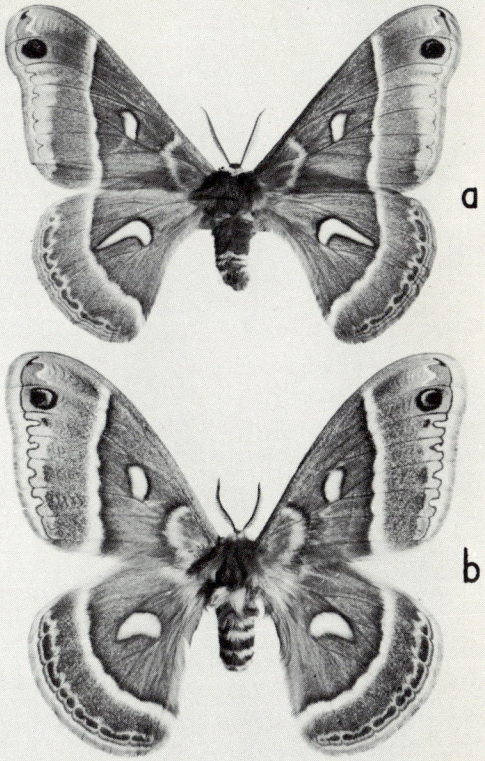

Figs. 27, 28, 29. (Legends on page 358)

VII. Minor or Little-Known Zones

The remaining Nearctic zones will not be discussed at length in this paper. Hopefully they will receive individual attention in future papers, most appropriately by genetical biogeographers residing for some years near the respective zones. Brief comments follow on these minor or little-known zones. Known and suspected pairs are listed in Tables VIIa–VIIg, and representative specimens are illustrated in Figures 30–41.

a. *Northern–Cascade Suture-Zone*. This is the area of contact between the biotas of the vast northern transcontinental woodland and the northern extension of the far western mountain chain. These central valleys of British Columbia perhaps have long had Cascade residents, but arrival of the northern biota from the north may have been impossible until very recently due to extensive barricades of severe glaciation. It is remarkable how closely the "northern" members of species pairs listed in Table VIIa resemble populations as far away as southeastern Canada and northern New England. Either rapid spread northwestward, which seems most likely, or intense panmixia would be needed to produce this condition.

b. *Cascade–Sierran Suture-Zone*. A jog in the mountain chain between the more Coastal Cascades and the inland Sierras. There appears to have been a barrier to contact across the bridge section, and the present suturing may be due to recent disappearance of the barrier in southwestern Oregon and/or northwestern California. Most of the pairs are so similar that they are appropriately considered distinctive subspecies; see Table VIIb. Peabody and Savage (1958) referred to the area as a corridor along which reptiles and amphibians dispersed from north and south in the late Cenozoic. The products of this dispersal might then have differentiated and recently resumed contact, forming the suture-zone.

c. *California Desert—Pacific Slope Suture-Zone*. This is a series of relatively low, inland mountain ranges oriented parallel to the southern California coast. In sequence, from west to east, these mountains are the boundary between extraordinarily different climatic regions, and the line of demarcation separating the xerophytic desert slope from the mesophytic Pacific slope is one of the most abrupt abutments in North America. In spite of this extreme difference, many species of animals, and some of plants, occur in both environments and have of course evolved conspicuously different subspecies. The contact sites tend to be localized at passes in the mountains and at the gaps between successive ranges. Some of the

---

Examples of Pairs of Known Hybridizers in Zones A and B (see Tables VIIa, VIIb).
Fig. 30. (a) Limenitis arthemis rubrofasciata; (b) wild hybrid L. a. rubrofasciata × L. lorquini; (c) L. lorquini. Fig. 31 (ventral sides). (a) Papilio glaucus canadensis; (b) P. rutulus. Fig. 32 (after Van Gelder, 1959: pp. 328, 336). (a) Spilogale putorius latifrons; (b) S. pu. phenax.

Figs. 30, 31, 32. (Legends on page 360)

pairs meeting here include one member extending with little variation hundreds of miles to the east, and the other member widespread in the coastal portion of California. The hybridization between some genetically incompatible species could be due to recent sympatry or to the sharp abutment phenomenon described in the Discussion below. G. L. Stebbins has suggested to me that this zone may prove to continue much farther northwestward from the San Gabriels, in the inner South Coast Ranges (see, e.g., Tucker, 1952, 1953a, 1953b). Various mainly allopatric plant pairs are known to overlap and hybridize there, but the animal interactions have not yet been examined.

d. *Rocky Mountain–Sonoran Suture-Zone.* As with the previous zone, little documentation is available for this contact zone. It somewhat resembles the situation in the Plains Ranges of Zone IV in having montane elements in the higher elevations of the ranges (such as the San Francisco, Hualpai, White, and Chiricahua Mountains of Arizona, the Mogollons, Gallos, and others of New Mexico, and the ranges of southwestern Utah) hybridizing in the middle elevations with relatives of more Sonoran environments. The shading in Figure 1 does not imply that hybridization is occurring over that whole region, but rather that it is occurring locally at mid-altitudes in many places scattered over the region. The higher species of each pair typically ranges much farther to the north into various component parts of the Rocky Mountain system of Nevada, Utah, and Colorado. This is a region that has been affected strongly by pluvial variations. For example, the remarkable relationship between *Quercus gambelii,* a typical Rocky Mountain oak, and *Q. turbinella,* a distinctly Sonoran species, was unravelled by Cottam et al. (1959). The oaks hybridized in the recent past in northern Utah, during a warmer period that permitted *Q. turbinella* to occur so far north. When the climate turned cooler and perhaps wetter, *turbinella* became extinct in the north. But undoubted hybrid clones still survive a long distance beyond the present range of *turbinella,* presumably because they have enough of the *gambelii* cold-tolerant genome to hang on even though not reproducing. This is similar to the persistence in western Texas of hybrids of *Q. stellata,* although pure *stellata* has become extinct there due to increased aridity (Muller, 1952).

When more is known of the suturing details, it should be possible to interpret the history with the many appropriate pollen analyses now available for the Southwest (e.g., Smiley, 1958; Sears, 1961b; Martin and Gray,

---

Fig. 33. Zerene Hybridization in Zone C (see Table VIIc). (a) Z. eurydice ♂; (b) Z. eurydice ♀; (c) Z. cesonia ♂; (d) Z. cesonia ♀; (e, f) wild hybrid ♂ ♂; (g, h) wild hybrid ♀ ♀.

Fig. 33. (Legend on page 362)

1962; Schoenwetter, 1962; Martin, 1963; Malde, 1964; Martin and Mehringer, 1965; et al.).

e. *Southwestern New Mexico Suture-Zone.* The faunal contacts are complicated in this region. Probably there are several linear desert ranges that have allopatric near-relatives on the two sides, with the possibility of meeting and mixing where low passes allow new contacts, as in the key instance of hybridization in this region, *Cnemidophorus tigris gracilis* × *C. t. marmoratus,* described originally by Zweifel (1962). The two lizard "subspecies" are very different, at least to the eye of a nonherpetologist, and their present taxonomic status is suggestive of the early treatment of various other organisms in which experimental studies subsequently demonstrated considerable hybrid incompatibility. It is quite possible that these are actually "biological species" whose active hybridizing in this localized region is due to recency of sympatry rather than genetic closeness. However, the hierarchical decision is relatively trivial compared to the interest of this case of two previously isolated populations now in hybridizing contact. Martin and Mehringer (1965, p. 444) reasonably concluded that this hybridization "almost surely represents post-glacial contact along the Arizona-New Mexico boundary. The two populations were widely separated during the full-glacial. It is possible that the hybrids developed only since the arroyo-cutting, overgrazing, and climatic change of the last 70 to 80 years." See Figure 37 for photographs of specimens from Zweifel's original series.

Lowe (1955, p. 345) regarded this place as the juncture between the Sonoran Desert ecologic formation and the Chihuahuan Desert. "In the general vicinity of this limit, the biota of the Sonoran Desert is in complex ecotone with more eastern biotas." He noted that the Gila Monster (*Heloderma suspectum*), Sonoran Coral Snake (*Micruroides euryxanthus*), and the lizards *Callisaurus draconoides ventralis* and *Coleonyx variegatus bogerti* have their eastern limit there. None of these four has a near relative on the eastern side, so they are not hybridizing. See Table VIIe for lists of taxa in this zone.

f. *Louisiana–East Texas Suture-Zone.* This is the contact between the Mississippi Embayment region of Louisiana and the western coastal plain of southeastern Texas. Probably there are many species or lower taxa which have evolved distinctive types in Louisiana or which are widespread in the southeast but reach their western limit here. Table VIIf lists a few

---

Examples of Pairs of Known and Suspected Hybridizers in Zone D (see Table VIId).
Fig. 34 (ventral sides). (a) *Quercus gambelii;* (b) *wild hybrids* Q. gambelii × turbinella; (c) Q. turbinella. *Fig. 35.* (a) Limenitis weidemeyerii; (b) *wild hybrid* L. weidemeyerii × astyanax arizonensis; (c) L. a. arizonensis. *Fig. 36.* (a) Papilio brucei; (b) *wild hybrid* P. brucei × bairdii; (c) P. bairdii.

Figs. 34, 35, 36. (Legends on page 364)

of the pairs. Among others which are approaching each other in this region are *Gopherus berlandieri* of southern Texas, which is said by Gunter (1945) to be extending its range northward and eastward, and its near relative *G. polyphemus,* whose western limit is now in extreme southeastern Texas.

g. *Southern Appalachian–Ozark–Ouachita Suture-Zone.* This may prove to be a major zone, perhaps more important for plants than animals. Anderson (1953, p. 297) stated: "all the widespread eastern species which have so far been analyzed exhibit introgression along the Ozark-Texas axis, if they have a related taxon in that direction." It is not certain whether this is one nearly continuous suture-zone, or whether there are two or more unconnected lines along which different species pairs are hybridizing. The recent barrier that is to be predicted by the many hybridizers listed in Table VIIg seems to have been a hiatus in the heavy forest. Beilmann and Brenner (1951) reported that records of early travelers show that vegetationally the Ozarks were prairie and open parkland with widely spaced trees. Settlers then achieved the reduction of prairie fires around 1820 to 1850, and this brought about the restoration of the Ozark forests. These were again reduced temporarily by intensive lumbering, which reached its peak around 1900 but is now negligible.

h. *Other Minor Nearctic Zones.* There are indications of a few other localized concentrations of biotic intermingling. It may become appropriate to define these in due course. For example, the eastern edge of the Mississippi Embayment in southeastern Louisiana has local hybridization between the salamanders *Amphiuma tridactylum* of Louisiana and *A. means* of the Southeast (Baker, 1947; Hill, 1954). *Limenitis a. archippus* and *L. a watsoni* and some turtles mapped by Tinkle, 1958, are hybridizing in the same area. Several other pairs of animals are in contact there. A second additional zone may be the salt-marsh coastal area of the Canadian Maritime Provinces, where there are relict populations of organisms that are believed to have evolved on the extensive land that was exposed during lowered sea level at glacial maxima. These are now hybridizing with nearrelatives which had been isolated far to the south but have recently reoccupied the northern territory. An example is the *Papilio brevicauda–P. polyxenes* pair. Third, there may be a significant suture-zone of mixing between the prairie biota and the eastern deciduous forest forms along the eastern edges of the Plains. And finally, there are suggestions in Wyoming of a zone of interaction between the northern and southern Rocky Mountain

Fig. 37. Cnemidophorus Hybridization in Zone E (see Table VIIe). (a) C. tigris gracilis (dorsal); (b) C. t. gracilis (ventral); (c) wild hybrid (dorsal); (d) wild hybrid (ventral); (e) C. t. marmoratus (dorsal); (f) C. t. marmoratus (ventral).

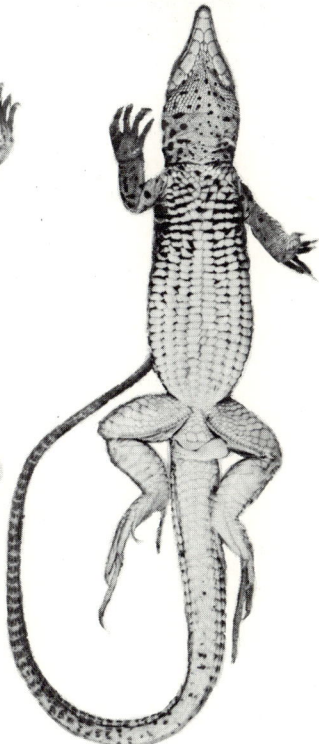

a

c

e

b

d

f

Fig. 37. (Legend on page 366)

biotas, comparable to the situation where the Cascades and Sierras meet in the Pacific states (Zone VIIb, above).

VIII. Arctic Treeline Contact Line

Many genera have one species broadly distributed in the tundra region north of treeline and a widespread relative in the forested area to the south, with a slender but very long line of contact along the treeline. In some regions, north of the main treeline there are islands of woods surrounded by treeless tundra. Hybridization between these pairs of relatives is to be expected but is almost entirely unstudied. Table VIII lists some pairs of taxa that meet at treeline. Specimens of several of the butterflies are shown in Figure 42.

Suture-Zones Outside North America

The present study has been focused on the Nearctic region. A rapid preliminary check of the hybridization literature indicates that suturing is a cosmopolitan phenomenon. A few promising places will be discussed briefly, in support of the view that suturing is widespread.

For example, there are hints of some of the suture-zones which are to be expected in Mexico. The detailed analyses by Sibley and his associates (1950 and later) of hybridization in the towhees, *Pipilo ocai* and *erythrophthalmus,* show that crossing is significant in a long band from west of Guadalajara eastward to the region of Mount Orizaba. Sampling of other animal groups has not been nearly as intensive as for the towhees, but there are fragmentary records of hybrids of other animals from the towhee mixture zone, and various principally allopatric species-pairs potentially overlap and hybridize there. Some are listed in Table IX. Another possible area of general hybridizing, of arid-land animals, is the vicinity of Culiacán, Sinaloa, where Cohn (1965) found *Neobarrettia sinaloae* and *hakippah* (Orthoptera) crossing, and a hybrid *Coleonyx variegatus sonoriensis* × *C. fasciata* (Reptilia) has been taken (Conant, 1965); possibly meeting and hybridizing in the same area are the bat species-pairs, *Macrotus mexicanus* and *californicus* and *Myotis fortidens* and *occultus.* The discovery of a hybrid between *Calocitta formosa,* a jay of the arid tropical forest, and *Psilorhinus mexicanus,* of the tall deciduous forest, in western Chiapas at one of the points where cultivation has al-

Examples of Pairs of Known Hybridizers in Zone F (see Table VIIf).
Fig. 38. (a) Limenitis a. archippus; (b) *wild hybrid* L. a. archippus × a. watsoni; (c) L. a. watsoni. *Fig. 39.* (a) Erynnis funeralis; (b) E. zarucco.
Examples of Pairs of Known and Suspected Hybridizers in Zone G (see Table VIIg).
Fig. 40. (a) Synchlora aerata; (b) *wild hybrid* S. aerata × denticularia; (c) S. denticularia.
Fig. 41. (a) Magicicada septendecim; (b) M. tredecim; (c) M. cassinii; (d) M. tredecassinii.

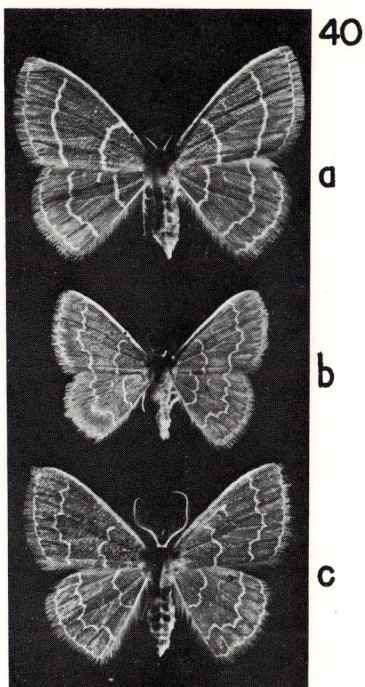

Figs. 38, 39, 40, 41. (Legends on page 368)

lowed the arid forest to invade the tall deciduous forest (Pitelka et al., 1956), points to the likelihood that other pairs of near relatives living in the two formerly separated habitats are also in early sympatry and are hybridizing; potential mammal pairs are *Reithrodontomys gracilis* and *R. mexicanus,* and *Sciurus socialis* and *S. griseoflavus.* Elsewhere, D. B. Stallings and I have for some years been examining a complex hybridizing array of the *Agathymus micheneri* group (Lepidoptera), in northeastern Mexico, where we suspect that a suture-zone exists. The *Agathymus* mixing is probably centered on a roughly north-south line passing about 70 km west of Saltillo. Substantial interspecific hybridization is indicated by evidence from the chromosomes, wing markings, genitalia, and larval biology. The gopher snakes (*Pituophis deppei* and *P. catenifer affinis*) appear to be hybridizing in the *Agathymus* zone (Conant, 1965). Other potential hybridizers there include the bats, *Pipistrellus subflavus* and its western relative, *P. hesperus.*

South America is largely unknown in the present context. It is suggestive that the swallowtail butterflies, *Papilio melasina* and *americus* (presently supposed to be dimorphic forms of the South American "race" of *P. polyxenes;* see Fig. 43), appear to be crossing in the same region in western Colombia where hybrids between the tanagers, *Ramphocelus flammigerus* and *R. icteronotus,* have long been known (see Sibley, 1958). Somewhat to the northeast, hybridization in the pokeweeds, between the higher elevaion *Phytolacca rugosa* and the lower *P. rivinoides,* is extensive in intermediate altitudes with human disturbance (Fassett & Sauer, 1950). Crossing between these two species extends northward far into Central America and may also be occurring southward in the same areas as the hybrid tanagers and swallowtails.

There appears to be a major suture-zone in Australia in the general vicinity of northeastern New South Wales and southeastern Queensland. This is the only major overlap site between the floristic Temperate and Tropical zones of Burbidge (1960), and is so remarkable that she coined a special term ("MacPherson-Macleay Overlap") for it. It is also the contact line between two of Paramonov's (1959) four main zoogeographic divisions of Australia, "Bororientalis" and "Merorientalis" (also similarly demarcated by Key, 1959). Table X lists pairs of northeastern and southeastern taxa meeting here. Burbidge discussed the role of post-Pleistocene climatic changes in shifting plant boundaries, with the present arid phase causing tropical plants to retreat northward and temperate plants to expand northward. Undoubtedly the factors controlling plant distribution similarly affect the animals.

Suture-zones in much of the Palearctic region and Africa may be blurred due to the antiquity of extensive human cultivation and chronic deforestation, but there are undoubtedly several typical suture-zones. Loukashkin

(1943) recorded introgression of *Lepus tolai* with *L. timidus* in a changing environment in northern Manchuria that seems to be ideal for suturing. Johansen (1955) has described mass-hybridization of eastern and western birds in the Yenisey region. In the Urals there appear to be eastern and western relatives which have recently met and are hybridizing freely (e.g., *Martes zibellina* and *M. martes*—see Ognev, 1931), much like the pairs in my Nearctic Zone VI. Several species pairs of birds have well-documented hybrid zones in central Europe (summarized by Mayr, 1963).

Stabilization after Early Suturing

It must also be presumed that there are former suture-zones in which by now: (1) hybridization among the animals, but not necessarily the plants, has essentially ceased; (2) some species pairs are widely sympatric; (3) some species pairs are allopatric, either because they are mutually exclusive or one excludes the other by competitive dominance; or (4) some former species have been driven to total extinction following initial sympatry. Some of these consequences are symbolized in the final stage of the models shown in Figures 3 and 19.

Likely examples of (1) plus (2) include the Virginia and Mule Deer (*Dama virginiana* and *hemionus*) in the West, the widespread Chipping and Field Sparrows (*Spizella passerina* and *pusilla*) and Downy and Hairy Woodpeckers (*Dendrocopos pubescens* and *villosus*), the tanagers (*Piranga rubra* and *erythromelas*) and orioles (*Icterus galbula* and *spurius*) in the eastern states, the eastern frogs (*Rana pipiens* and *palustris*), the tiger swallowtails of the West (*Papilio rutulus, eurymedon, multicaudatus*), the sympatric silkmoths of the East (*Callosamia promethea* and *angulifera*), the southeastern toads (*Bufo quercicus* and *terrestris*), and the three widespread fritillary butterflies (*Speyeria cybele, atlantis,* and *aphrodite*).

Possible examples of (1) plus (3) include Chipping and Tree Sparrows (*Spizella passerina* and *arborea*), the White-breasted and Red-breasted Nuthatches (*Sitta carolinensis* and *canadensis*), the Black-capped and Brown-capped Chickadees (*Parus atricapillus* and *hudsonicus*), the Eye-spot Hawk-moths (*Smerinthus geminatus* and *cerisyi*), the contemporary status of *Pieris napi oleracea* and the introduced *P. rapae,* and the Bobcat and Lynx (*Lynx rufus* and *L. canadensis*). Several pairs of taxa in my "B" tables probably will also prove to have this status, such as *Zapus* in Zones IV, V, and VI (see Krutsch, 1954).

Examples of (4) are extremely difficult to get, because the record is usually erased quickly. Assuming they hybridized early in their sympatry, the hares (*Lepus europeus* and *timidus*) in Great Britain are a partial example of (4); Barrett-Hamilton (1912) (see also Lack, 1945) has shown that the present geography and the abundant Pleistocene and post-Pleistocene fossils indicate that *L. timidus* was a highly successful breeding resident of Great Britain (as it still is in Ireland) until *L. europeus* reached the island, after which *timidus* disappeared everywhere but at certain high elevations in Scotland. The present distribution of ptarmigan species in northwestern North America could be explained by the British *Lepus* model. In northern British Columbia and Yukon Territory, *Lagopus leucurus* occupies the higher mountains and *L. rupestris* (= *L. mutus*) the barren, rocky areas at lower altitudes; but where *leucurus* is absent, *rupestris* is in the mountains as well as lowlands. There is evidence that the Blue-winged warbler (*Vermivora pinus*) may be replacing the Golden-wing (*V. chrysoptera*) at present (Berger, 1958, and others). Denisov (1961) has reported that in the Lower Volga region of Russia a southern ground squirrel, *Spermophilus pygmaeus,* has recently been extending its range northward, and in the contact line with *S. suslica* the latter is being replaced by *S. pygmaeus;* like the warblers, these ground squirrels hybridize freely where sympatric.

Hybridization Outside Suture-Zones

While the actuality and the rationale of the suture-zones are clear, it would be ridiculous to suppose that all natural hybridizing is restricted to these zones. It is to be assumed that special local events are bringing together pairs of closely related species from time to time, with some hybridization resulting. With plants, the events may be simply a local disturbance of a previously stable habitat in which crossing had been slight or absent between a pair of widely sympatric species (Anderson, 1949, and Riley, 1938, among numerous carefully analyzed instances; similar results have been reported for hylid frogs (Mecham, 1960) and pond turtles (Crenshaw, 1965)). Others among several examples of pairs that may be hybridizing outside suture-zones are *Crotalus pricei* and *C. tigris* in Arizona (Klauber, 1956), *Microdipodops pallidus* and *M. megacephalus* in the Penoyer Valley of Nevada (Hall, 1941), various hummingbirds near San Francisco Bay (Banks & Johnson, 1961), and the Bronzed and Purple Grackles (*Quiscalus versicolor* and *Q. quiscula*), whose present zone of massive hybridizing is mainly near a line from Cape Cod to New Orleans (Huntington, 1952). Huntington considered the grackles to be a single species due to the intensity of crossing, but as I have shown below (see Discussion) large-scale hybridizing is to be expected at the begin-

Fig. 42. *Examples of Pairs of Potentially Hybridizing Species Meeting at Treeline in the Arctic (see Table VIII) (left row—below treeline; right row—north of treeline). (a) Oeneis jutta; (b) O. melissa; (c) Erebia discoidalis; (d) E. rossii; (e) Erebia disa (ventral); (f) E. fasciata (ventral); (g) Colias pelidne; (h) C. hecla; (i) C. nastes.*

Fig. 42. (Legend on page 372)

ning of sympatry, even between taxa which will ultimately evolve isolating mechanisms that will halt all crossing. Few or no significant data have yet been published on the fertility and viability of hybrids as compared to non-hybrid individuals of these grackles, nor of the flickers (Short, 1963), the Mallard and Black Ducks (Sibley, 1938; Johnsgard, 1960), the sulphur butterflies—*Colias eurytheme* and *philodice* (Hovanitz, 1944), admiral butterflies—*Limenitis astyanax* and *arthemis* (Chermock, 1950; Klots, 1951), and other hybridizing pairs which are claimed to be conspecific. There are models that could be used for the objective analysis of genetic relationships in hybridizing complexes such as the flickers, ducks, and grackles and of the status in a biological species concept of allopatric pairs such as many listed in my "B" tables. Lorković (see 1958 for a summary), by hybridization experiments, chromosome counts, genitalic comparisons, and intensive geographic sampling, has analyzed the *Erebia tyndarus* group of European alpine butterflies. He concluded: "Six of the seven geographic isolates which have been discussed up to now have reached a degree of reproductive isolation that justifies regarding them as full species or at least as semispecies. This contradicts the usual treatment of such similar allopatric populations as subspecies" (p. 324). My own as yet unpublished research, on genetic relations in *Limenitis* and the *Papilio polyxenes—machaon* complex of North American butterflies, is even more germane to an interpretation of the natural hybridizing of birds *in the same* geographic areas (Suture-Zones I and IV). It has demonstrated that freely hybridizing pairs may in fact have profound incompatibility, as shown by low hybrid fertility and/or viability and by ecological inadaptiveness. A similar result has emerged from the work of W. F. Blair and his associates with Bufonidae and Hylidae, much of it still in progress (see e.g., Blair, 1959). It may be that the hybridizing birds and most allopatric putative "subspecies" have little or none of this incompatibility, as is claimed by recent authors, but the evidence is lacking at present for making any confident judgement.

Discussion

Models of Speciation and of Its Sequelae

The theoretical model of speciation events for organisms with the breeding patterns characteristic of all of the suture-zone hybrids can be summarized as follows:

1. A single parental population becomes divided into two subpopulations by barriers to breeding contact (usually, but not necessarily, absolute physical isolation; see Voipio, 1952, for discussion of speciation with "isolation by semi-independence").

2. The separation persists long enough for an accumulation of hereditary differences sufficient to constitute a barrier of some hybrid sterility and/or invariability should the two daughter populations hybridize; thus, they reach the level of separate biological species even while still entirely allopatric. Hybrids between separate biological species tend to have an average fitness lower than that of either non-hybrid parental type. However, in the environment where sympatry occurs, some hybrids between "good" species may have a higher extrinsic viability than either parental type, but hybrid sterility or intrinsic inviability (e.g., embryonic lethality) may be so severe that the hybrid pairings have a lower probability of grandchildren than the non-hybrid matings. If so, event (4), below, is to be expected. There are special circumstances in which the super-viable hybrid type might evolve full fertility by reorganization of the genome to eliminate sterility factors, but at present this appears to be unusual.

3. When they become sympatric they will at first be likely: (a) to hybridize freely and (b) to be in ecological competition. For sympatry to be maintained, two speciation sequelae must evolve, namely:

4. Any genetic tendencies toward intraspecific breeding in the presence of potential mate-choice will be favored, and this will tend to fix **anti-hybridization** mechanisms in due course; and

5. Any genetic tendencies toward niche specialization in the presence of potential interspecific competition will be favored, with **anti-competition** mechanisms becoming fixed in due course.

Early Events Following Geographic Contact

General Theory. Among the many implications of this model is the reasonable certainty that any two sister species will tend to hybridize when they first meet, often mating at random with their own and the other species. Further, **the amount of natural hybridizing is commonly more a measure of recency of first sympatry than a measure of the amount of genetic dissimilarity** making up the interspecific barrier of hybrid sterility and/or inviability. **Chance** development of anti-hybridization mechanisms while the parental species are allopatric will usually be trivial compared to the rapid **selection** for an anti-hybridization genome in each parental species when they are sympatric.

There is, however, a mechanism which should be expected to preadapt a species with anti-hybridization responses in the presence of a close relative never before encountered. Suppose Species A is sympatric with Species B, and hybrids between the two are intersterile. In due course, A will evolve a powerful anti-hybridization genome which causes it to choose with precision only individuals of the opposite sex having characters unique to A, such as the specific odor, call, behavior, form, color, and pattern by which A differs from B. Then suppose that Species A becomes sympatric with Species C; the self-recognition adaptations, functioning for the rejection of B at mating time, could also be pre-adaptations for rejecting C. Some instances interpreted as accidental pre-adaptations for anti-hybridization may be due to this form of selection; furthermore, some others may be due to former sympatry between A and B, followed by complete geographic separation, finally followed by the resumption of sympatry. These two possibilities need to be considered when mate-preference mechanisms are discovered between presently allopatric populations. Suggestive cases include the call differences and mate choice in the presently allopatric chorus frogs, *Pseudacris ornata* and *streckeri* (Blair and Littlejohn, 1960) and the strong sexual isolation between the Wyoming and Michigan strains of *Drosophila athabasca* studied by Miller (1958b). Examples of developmental inviability in hybrids (as with the *Bufo* studied by Volpe, 1955) seem to be irrelevant to this question of anti-hybridization sequelae as a function of sympatry; it is unclear why selection should favor increased hybrid inviability (the "intrinsic isolating genes" of Volpe), although it seems obvious that selection would favor anti-hybridization genes in sympatric species with inviable hybrids (Dobzhansky and Koller, 1938).

The special interest that attaches to suture-zones is that extensive hybridization among several different groups of organisms in a geographically limited area indicates: A) that species pairs have so recently met that there has not yet been time for the establishment of the anti-hybridization mechanisms that will later terminate all or most crossing, and B) that since many diverse species pairs are hybridizing in the same zone, this is strong evidence that there was a barrier to gene exchange so exclusive that two large faunal (and probably floral) communities were kept out of contact with each other until recently.

There are several different possible consequences to the beginning of sympatry, as in a suture-zone:

1. The anti-hybridization and anti-competition sequelae may develop perfectly, and one or both species will be able to extend its range far into the range of the other species;

2. the competitively inferior species may be driven to extinction by the superior species;

3. one species may remain inferior and be extirpated wherever the superior species can survive the physical environment, but it may persist permanently in regions where climate or other physical factors prevent the survival of the competitively superior species;

4. each species may be competitively superior to the other in its own pre-suturing environmental combination, and both may survive allopatrically because each excludes the other from otherwise inhabitable regions.

Obviously, only the first of these four consequences allows permanent sympatry. An implication of this argument is that **allopatry is by no means a necessary indicator of subspecific status** for the allopatric taxa.

The rates at which the sequelae develop may differ very greatly in different organisms. Likewise, their rates of range expansion may differ. For these reasons we should expect that the hybridization zone for any one species pair may soon become wider or lie in a somewhat different line from that of another species pair in the same suture-zone. This could account for the difference in details of the hybridization zones of *Vermivora pinus* and *chrysoptera* (Aves), *Bufo fowleri* and *americanus* (Amphibia), and *Limenitis astyanax* and *arthemis* (Lepidoptera), although all are meeting in the major Northeastern–Central Suture-Zone. Grinnell and Swarth (1913, p. 394), in a thoughtful early commentary on this problem, noted that "there is *no exact concordance* in the distributional behavior of all the animals of a region . . . where two faunas meet."

An extreme example of rapid extension of sympatry following initial suturing is the *Colias eurytheme–philodice* pair (Lepidoptera). It appears that with the westward spread of European agricultural practices in the eighteenth and nineteenth centuries, suitable larval foodplants became distributed so that the southwestern *C. eurytheme* was able to spread into the Great Plains and eventually to make contact with eastern populations of *C. philodice philodice* in the Midwest. *C. eurytheme* then continued to spread eastward until it reached the Atlantic coast early in the twentieth century. At the same time *C. p. philodice* was probably expanding westward across the Great Plains into areas occupied by few, if any, *Colias* in precolonial times. *C. eurytheme* and *C. p. philodice* are now sympatric breeding residents, from the western plains to the Atlantic coast and from about 34° to 44° N. latitude (there are some local climatically and topographically controlled deviations from this gross geographic outline). Throughout this vast new zone of sympatry the two species interbreed freely at presen, in eastern areas apparently at random.

The rates of expansion of sympatry may differ for a given species pair at different times. There is some indication (Lanyon, 1966) that the Western Meadowlark (*Sturnella neglecta*) has recently begun to spread, much more rapidly than before, eastward into the range of *S. magna*. The geography and interactions between the meadowlarks may soon resemble closely those of the two *Colias* but surely with much less hybridizing. The increase in rate of expansion may have been made possible by the development of a new genotype in *S. neglecta* populations.

The Australian tephritid fruit flies, *Dacus tryoni* and *D. neohumeralis*, can be also compared with *Colias eurytheme* and *C. philodice,* respectively. The pre-agricultural ranges of these two *Dacus* are not known, but at present they are hybridizing extensively, and this probably indicates recency of sympatry. Their most recent movements and biology have been closely observed (see Birch, 1961 and 1965). Both have moved over to cultivated fruits from their ancestral wild hosts, and *D. tryoni* in particular has spread rapidly southward throughout eastern Australia. M. A. Bateman has shown that major genetic differences already exist between various recently established populations of *D. tryoni.*

The several Nearctic zones of active suturing considered in Tables I to VII all appear to have enough current hybridization to indicate that they are in the earliest stages of the development of sequelae to sympatry. They are thus ideal sites for the study of evolutionary events. Some research of this sort has already been carried out, perhaps most notably with the Blue-winged and Golden-winged warblers (recently summarized by Short, 1963; see also Gill & Lanyon, 1964). They are also favorable loci for the study of the community integration that must follow the merging of two previously isolated communities. This would include the competitive role of species in one ancestral community that have no near relative in the suturing complex, i.e., those that must evolve anti-competition but not anti-hybridization mechanisms.

One of the most important aspects of hybridization is its role in providing potentially increased fitness to both parental species due to the transfer, by introgression, of blocks of genetic material from one species into the other (Anderson, 1949). In animals (and perhaps to a greater extent in plants than is generally assumed), selection usually fixes anti-hybridization mechanisms rapidly when two species meet and at first interbreed at random. The result is that many species pairs rarely or never hybridize in nature now, although they may show high interfertility when tested experimentally.

In suture-zone phenomena, there is one interesting mechanism which can be expected to allow permanent hybridization between animals with even low interfertility. This is the situation in the Colorado portion of the Rocky Mountains, mentioned for Zone IV, above. Here the Front Range rises abruptly out of the level plains, within a few map miles going from 5,500 to 9,000 ft in elevation and from essentially treeless plain to lush montane forest. The species adapted for mountains, such as *Papilio gothica* (Lepidoptera), meet their lowland relatives, such as *Papilio polyxenes,* only in an extremely limited environment of the canyon mouths. This *Papilio* pair hybridizes freely and the hybrids are moderately fertile, although normally only one sex survives. Since these canyon mouths

---

Examples of Pairs of Known Hybridizers from Non-Nearctic Zones.
Fig. 43. (a) Papilio polyxenes melasina; (b) P. p. americus. Fig. 44 (after Selander, 1964: pl. 5). (a) Campylorhynchus rufinucha nigricaudatus; (b) wild hybrids C. r. nigricaudatus × r. humilis; (c) C. r. humilis. Fig. 45. (a) Tisiphone abeona morrisi; (b) wild hybrid T. ab. morrisi × ab. aurelia; (c) T. ab. aurelia (dorsal side of each specimen at left, ventral side of same three specimens at right).

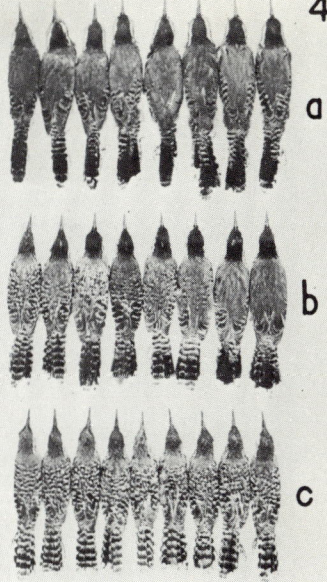

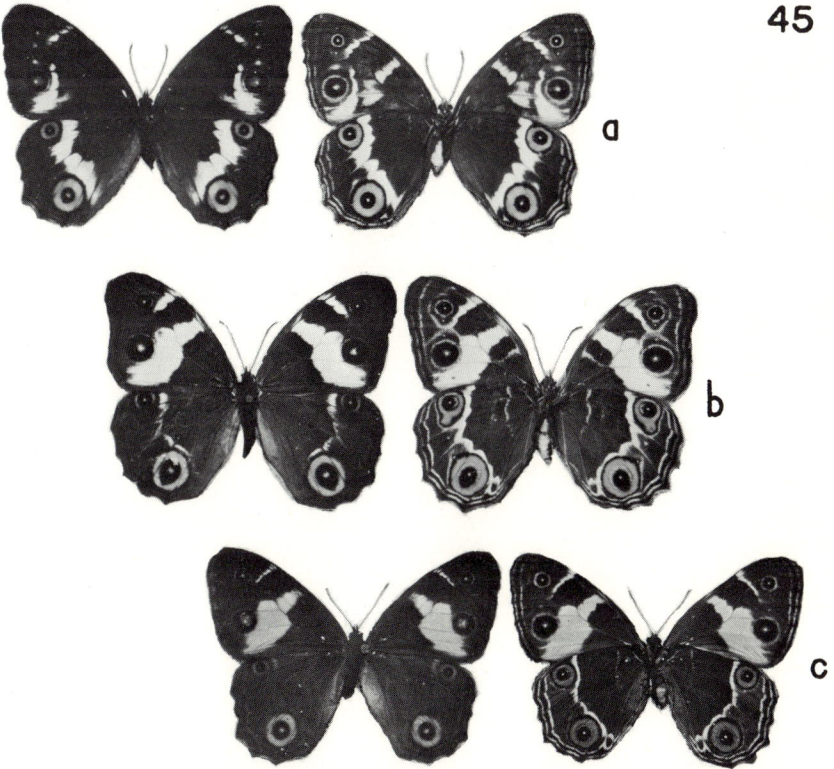

Figs. 43, 44, 45. (Legends on page 378)

provide only a knife-edge of contact between two vast, exceptionally pan-
mictic populations, the offspring of any individual favored by selection due
to an allele for intraspecific mate preference are likely to breed some distance from the contact line and the allele will promptly lose its adaptive
advantage. Under these environmental conditions, it is unlikely that anti-
hybridization mechanisms will evolve in Colorado in either species. [A
similarly local zone of contact appears to connect the two wren populations
in the Ocuilapa Valley on the coast of Chiapas (Selander, 1964)]. In
contrast to the most usual situation along the Colorado Front Range, in
the Black Hills and other Plains ranges to the north the two *Papilio* species
are in large-scale contact over a broad area, and all the theoretical requirements are present for rapid evolution of anti-hybridization mechanisms and
the end of introgression.

Evidence for Initial Hybridizing. It has been emphasized above that any
two moderately or closely related species will tend to hybridize when they
first meet. Some samples of the extensive but scattered data bearing on
this contention will be briefly reviewed.

The best test would of course be the direct observation for a few years
of the initial contact of pairs of species, one or both of which had been
extending their native range. Such research opportunities must be very
infrequent, and of the few that develop, most are probably missed by
relevant specialists until too late. However, almost as suitable are the
interactions between native animals and plants with exotic relatives which
are introduced into their home ranges by man.

Introductions of animals, followed by hybridization, have been less well
documented than those of plants, but a few examples can be mentioned.
In New Zealand, Red Deer (*Cervus elephas*) have been introduced from
Europe and Wapiti (*C. canadensis*) from North America; both are well
established, and they apparently are hybridizing "freely" (Howard, 1965),
although the fate of the hybrids does not seem to have been studied. The
hare, *Lepus europeus,* was introduced into southern Sweden for sport, and
it then hybridized commonly with the native *L. timidus;* "as they become
accustomed to each other, hybridism decreases" (Barrett-Hamilton, 1912,
p. 238). The interactions of the introduced European Hare with the
American Varying Hare seem not to have been examined. The eastern
Bullfrog, *Rana catesbiana,* has been successfully introduced into several
places west of the Rocky Mountains, and in Oregon it has been recorded
in amplexus with the very different native frog, *R. aurora* (Storm, 1952).
Phillips (1923) noted that the Mallard Duck (*Anas platyrhynchos*) in
North America hybridizes freely in the Lake States and Ontario with the
more eastern native Black Duck (*A. rubripes*); he believed that the hy-

bridizing Mallards are mostly domesticated or semidomesticated call-ducks, rather than the native Mallards. Johnsgard (1965) similarly referred to the extensive hybridization in New Zealand and Australia between the recently introduced Mallard and the native *Anas poecilorhyncha,* although the latter does not hybridize with Mallards in India or Burma, where the two ducks are naturally sympatric.

Among insects, three examples of Lepidoptera are probably representative. For many years the late Otto H. Schroeter, an amateur lepidopterist of New London, Connecticut, imported large numbers of cocoons of the Asiatic Luna Moth (*Actias selene*) for the purpose of making hybrids for his own amusement by crossing it with the distinctively different American species, *A. luna.* His technique was to tether several virgin females of *A. selene* overnight in a grove of trees known to have a resident population of the native *A. luna.* He reported that the next morning he would find most, or all, of the females *in copulo* with wild males. *A. luna* has no relatives in North America and has no doubt been long isolated from its Asiatic congeners. A celebrated and similar instance was the overnight exposure in New York City of a virgin female of a sphinx moth, *Smerinthus ocellatus,* from Europe; the next morning she was *in copulo* with a wild male of the American *Paonias astylus,* and ultimately a substantial brood of male intergeneric hybrids was reared (Neumögen, 1894). Sexual attraction in these two families is by wind-borne pheromones, and these widely allopatric species have not evolved pheromones which the males of the "wrong" species ignore. There is also evidence that the Cabbage Butterfly, *Pieris rapae,* introduced from Europe about a century ago, at first hybridized rather frequently with *P. napi oleracea* in New England and even copulated with the distantly related species, *P. protodice.* Hybridization now seems to be rare or entirely ended.

It is well known that introduced outcrossing plants tend to hybridize freely with native relatives. Smith (1943) recorded several *Populus* combinations and the frequent crossing of the English Oak, *Quercus robur,* with native American white oaks such as *Q. prinus* and *Q. alba.* Mulligan (1965) recently summarized this phenomenon for various herbaceous plants.

This is a fragment of the large and consistent body of conclusive evidence that allopatric related species are likely to hybridize when they first meet.

Evidence for Initial Competition. As with anti-hybridization mechanisms, anti-competition systems in newly sympatric species are rarely accessible for study under natural conditions. There is, however, an indisputable corpus of experimental evidence for the universality of niche competition between related species. A number of recent ecological treatises have dealt with this subject thoroughly. Most competition is trophic, and closely related, long-sympatric species show elaborate food specializations that take them almost entirely out of interspecific competition. Consistent with the corresponding prediction of competition theory, allopatric near relatives typically resemble each other closely in the kind, size, and site of their food.

Field data bearing on the initial competition between newly sympatric relatives come mainly from work or newly introduced species. The New Zealand deer introductions are again relevant. Howard (1965) summarizes previous work showing that in one region, at least, the Sika Deer (*Cervus nippon*) is outcompeting the Red Deer (*C. elephas*) and replacing it. Various other examples have been mentioned above, such as the current spread of the Blue-winged Warbler at the expense of the Golden-wing and of the Russian *Spermophilus pygmaeus* at the expense of *S. suslica,* the history of hares in Great Britain, and the probable history of the ptarmigan species of northwestern Canada.

Suturing and Mimicry. There are several instances in which movement into suture-zones has affected the mimicking species. In Zone II the dark Floridian Viceroy, *Limenitis archippus floridensis,* is now in breeding contact with the much lighter *L. a. archippus* of the north. Each is an effective mimic of a different species of *Danaus,* but the hybrids now being produced in the suture-zone are poor mimics of both distasteful models. The hybridization zone of *L. archippus* appears to be much narrower than for other animals and plants crossing there, and it is likely that selection so strongly favors the two mimetic phenotypes that the introgressants are rapidly eliminated. A similar problem confronts the white-spotted black pyraustid moth, *Ennychia funebris,* in Zone I, where it has a northern model, *Alypia langtoni,* and a slightly different southern model, *A. octomaculata.* But much worse is the problem of *Limenitis astyanax* in Zone I. *L. astyanax* is an effective mimic of the unpalatable swallowtail, *Battus philenor,* which has no relative to the north and does not quite reach the suture-zone, due to the restricted range of its foodplant. *L. astyanax,* however, has probably spread rapidly northward into the suture-zone due to abundance of its foodplants there, and it is now hybridizing with a northern relative, *L. arthemis,* which is not a mimetic butterfly. The hybrids would be rather poor mimics (Remington & Remington, 1957), and at present they rarely reach the range of the model, *B. philenor.* These and other suturing mimics have all the problems of sequelae that confront any hybridizer in the zone, plus the disruptive influence on the perfection of their mimetic genotype.

Suturing and Host Specificity. For host-specific parasites and phytophages, suturing adds extra selective pressures beyond those of euryophagous or photosynthetic organisms. In the new territory they are likely to meet hosts to which they are not physiologically well adjusted, but they may not have the genome that will produce a response-system causing them to reject the unsuitable hosts. For example, Straatman (1962) has described his research in Australia on four species of butterflies which freely oviposit on introduced relatives of their native foodplants, but the larvae died on the introduced plants. Host specificity is thus another of the several environmental parameters which tend to produce adaptational stress in suture-zones, because the suturing species are confronted suddenly with a complex array of organisms to which they have not been directly adapted.

Acknowledgements

In the preparation of this paper I have been aided at every step by my wife, Jeanne E. Remington. She did much of the general literature search, and her editorial criticism unearthed many unclear and misleading passages in the text. My son, Sheldon T. C. Remington, worked long hours checking the reference list and parts of the typescript. Thomas A. Brown made the photographs, several of them difficult because conspicuous color differences had to be brought out in black and white. William Vars drew figures 1 and 2 and assembled the plates, and Ward Whittington drew figures 3 and 18. Janet R. Dugle generously presented me with leaf specimens of her hybridizing birches, shown in figure 25. Richard G. Zweifel kindly loaned the original negatives of the photographs of his remarkable *Cnemidophorus* and hybrids, for use in my figure 37, and R. S. L. Davis (via Thomas W. Davies) loaned me color transparencies of the *Tisiphone* shown in figure 45. Robert K. Selander, Richard G. Van Gelder, and T. C. Brayshaw permitted the reproduction from their published work of my figures 21, 26, 32, and 44. As I was working out the suture-zone concept, I benefitted very greatly from the discussions that arose repeatedly with my colleagues and students in the Graduate Seminar on Evolution at Yale. Very many others were helpful in conversation, but I want especially to acknowledge the valuable comments and suggestions of Janet R. Dugle, Peter H. Raven, Frank C. Vasek, Edward S. Deevey, Douglas C. Ferguson, and Don B. Stallings, and the encouragement of Theodosius Dobzhansky. Various parts of the research mentioned here were supported by National Science Foundation Grants G-23781 and GB-03763. To all I am grateful.

TABLES

Abbreviations for Tables I–X

Ala.——Alabama
Alta.——Alberta
Ariz.——Arizona
Ark.——Arkansas
B.C.——British Columbia
C——Central
Calif.——California
Colo.——Colorado
Conn.——Connecticut
Fla.——Florida
Ga.——Georgia
Ida.——Idaho
Ill.——Illinois
Ind.——Indiana
Kans.——Kansas
Ky.——Kentucky
La.——Louisiana
Labr.——Labrador
Man.——Manitoba
Mass.——Massachusetts
Md.——Maryland
Me.——Maine
Mich.——Michigan
Minn.——Minnesota
Miss.——Mississippi
Mo.——Missouri
Mont.——Montana
Mt.——Mount

Mts.——Mountains
N.C.——North Carolina
N.E.——New England
Nebr.——Nebraska
N.J.——New Jersey
N.M.——New Mexico
N.S.——Nova Scotia
N.S.W.——New South Wales
N.W.T.——North West Territories
N.Y.——New York
Okla.——Oklahoma
Ont.——Ontario
Oreg.——Oregon
Pa.——Pennsylvania
Phila.——Philadelphia
Que.——Quebec
Queens.——Queensland
Sask.——Saskatchewan
S.C.——South Carolina
S.D.——South Dakota
Tenn.——Tennessee
Tex.——Texas
U.S.——United States
Va.——Virginia
Vt.——Vermont
Wash.——Washington
Wisc.——Wisconsin
Wyo.——Wyoming

TABLE I.   Northeastern–Central Suture-Zone

A. KNOWN HYBRIDIZING PAIRS

Northern	Central	Known Areas	Xing*	References
Stenonema interpunctatum canadense (E)†	*S. i. heterotarsale*	St. Lawrence valley	M?	Spieth (1947)
Allocapnia minima? (Plecoptera)	*A. maria*?	Sawyerville, Que.	M?	Hanson (1960)
Limenitis a. arthemis (L)	*L. astyanax*	N & S of Hartford to Minneapolis line	I	Remington (unpubl.)
Cercyonis pegala nephele group (L)	*C. p. alope* group	Same as *Limenitis*	I	Remington (unpubl.)
Papilio glaucus canadensis (L)	*P. g. glaucus*	N N. Engl.; SE Canada	I?	Remington (unpubl.)
Erynnis lucilius (L)	*E. baptisiae*	Conn.	S?	Maeki & Remington (1960), Burns (1964)
Alypia langtoni (L)	*A. octomaculata*	S Ont.	M?	Forbes (1960)
Hyalophora columbia (L)	*H. cecropia*	Me. to Ont.	S	Sweadner (1937)
Nemoria bistriaria rubromarginaria (L)	*N. b. bistriaria*	S Conn.; S N.Y.; C N.J.; SE & C-W Pa.	I?	Ferguson (in press)
Synchlora albolineata (L)	*S. aerata*	N Conn.; C N.Y.; S Ont.; C Mich.; S Wisc.		Ferguson (in press)
Rhyacophila banksi (T)	*R. parantra*	Maynooth, Ont.	M?	Ross (1965)
Vespula vulgaris (Hy)	*V. maculifrons*	Ottawa, Ont., et al.	S?	Miller (1961)
Ambystoma laterale (Am)	*A. jeffersonianum*	N Ind. NE to Me., Canada (triploids)	M?	Uzzell (1964)
Bufo americanus (Am)	*B. fowleri*	wide	I	Blair (1941), Volpe (1952), Cory & Manion (1955), et al.
Thamnophis sauritus septentrionalis (R)	*T. s. sauritus*	S Me. to E Ill.	I?	Rossman (1963)
Bonasa umbellus togata (A)	*B. u. umbellus*	N N. Engl. to C Minn.	I?	Pitelka (1941)
Vermivora chrysoptera (A)	*V. pinus*	S Minn. to N Ohio; now widely in Northeast	I	Faxon (1913), Parkes (1951), Berger (1958), Short (1963)
Spizella pallida (A)	*S. passerina*	N Mich.; Ithaca, N.Y.	S	Storer (1954), McIlroy (1961)
Peromyscus maniculatus gracilis (M)	*P. m. bairdii*	N Mich.	S?	Blair (1958b), Hall & Kelson (1959)

*Symbols for amount of crossing in Tables I-X, are: I = intense, M = moderate, S = slight.
† Symbols in parentheses after first species name in each pair refer to orders of insects or classes of vertebrates: (A) = Aves, (Am) = Amphibia, (C) = Coleoptera, (D) = Diptera, (E) = Ephemeroptera, (H) = Hemiptera, (Hy) = Hymenoptera, (L) = Lepidoptera, (M) = Mammalia, (O) = Odonata, (R) = Reptilia, (S) = Siphonaptera, (T) = Trichoptera. (P) = plants.

TABLE I (Cont.)

Northern	Central	Known Areas	Xing	References
Juniperus horizontalis (P)	*J. virginiana*	N. Engl. to C Minn.	I	Fassett (1945b), Hall (1952, 1961)
Populus tremuloides (P)	*P. grandidentata*	Que. to N. Engl. W to Ohio	S?	Smith (1943)
Populus balsamifera (P)	*P. deltoides*	S Que.; Ont.; Vt.; N.Y.; SE & S-C Alta.	M?	Smith (1943), Brayshaw (1966)
Populus tremuloides (P)	*P. deltoides*	C Alta.	M?	Brayshaw (1966)

B. SUSPECTED HYBRIDIZERS

Northern	Central	Contact Areas	References
Enallagma vernale (O)	*E. geminatum*	SW Que.; S Ont.; Mich.	Walker (1953)
Lestes d. disjunctus (O)	*L. d. australis*	S N. Engl.; N.Y.; Pa.; N Ohio; N Ind.; Iowa	Walker (1953)
Gomphus brevis (O)	*G. viridifrons*	S Ont.; N.Y.; Pa.	Walker (1958)
Gomphus notatus (O)	*G. plagiatus*	Pa. to Mo.	Walker (1958)
Celithemis monomelaena (O)	*C. fasciata*	?	Needham & Westfall (1955)
Ladona julia (O)	*L. deplanata*	?	Needham & Westfall (1955)
Sphinx borealis (L)	*S. gordius*	SE Canada; N N. Engl.	J. C. Riotte (unpublished)
Pieris napi oleracea (L)	*P. virginiensis*	W Mass.; S Vt.; NW Mich.	Reinthal (1956), Voss & Wagner (1956)
Anopheles "maculipennis" (earlei?) (D)	*A. quadrimaculatus*	Leverett, Mass.	Freeborn (1923)
Vespula acadica (Hy)	*V. vidua*	N N. Engl.; N Lake States	Miller (1961)
Empidonax traillii (A)	*E. brewsteri*	E Pa. to S Wisc. to C B.C.	Stein (1963)
Spinus pinus (A)	*S. tristis*	N N. Engl.; SE Canada; Appalachian Mts.	Am. Orn. Un. (1957
Glaucomys sabrinus (M)	*G. volans*	Va.; N.Y.	Remington (unpublished)
Peromyscus maniculatus gracilis (M)	*P. leucopus*	S N.S.; S Ont.; N N. Engl.; N Lake States	Hall & Kelson (1959)
Populus balsamifera (P)	*P. grandidentata*	NE U.S.	Preston (1961)

TABLE II.   Northern Florida Suture-Zone

A. KNOWN HYBRIDIZING PAIRS

Northern	Floridian	Known Areas	Xing	References
Hexagenia munda sspp. (E)	*H. m. orlando*	N Fla.	M?	Spieth (1941)
Limenitis a. archippus (L)	*L. a. floridensis*	Fla.-Ga. boundary	M	Remington (1958)
Papilio t. troilus (L)	*P. t. ilioneus*	N Fla.	I	Remington (unpublished)
Automeris i. io (L)	*A. io lilith*	N Fla. to E S.C.	M	Kimball (1965), Remington (unpublished)
Limnophila m. macrocera (D)	*L. m. suffusa*	N Fla.	M?	Rogers (1933)
Hydroporus cimicoides (C)	*H. hebes*	Marion Co., Fla.	M?	Young (1954)
Hydroporus lobatus (C)	*H. hebes*	S Ga.	I?	Young (1954)
Coptomus i. interrogatus (C)	*C. i. obscurus*	uplands of N Fla.	M?	Young (1954)
Bidessus p. pullus (C)	*B. p. floridanus*	N Fla., esp. Walton Co.	M?	Young (1954)
Tropisternus collaris striolatus (C)	*T. c. viridis*	NW Fla.; SW & S Ga.; SE S.C.	I?	Young (1961 et seq.)
Acris g. gryllus (Am)	*A. g. dorsalis*	N Fla.; SE Ga.	?	Carr & Goin (1955)
Kinosternon s. subrubrum (R)	*K. subrubrum steindachneri*	N Fla.	I?	Carr & Goin (1955)
Sternothaerus m. peltifer (R)	*S. m. minor*	NW Fla.; SW Ala.	I	Tinkle (1958)
Terrapene c. carolina (R)	*T. c. bauri*	NE Fla.; SE Ga.	I?	Carr & Goin (1955)
Thamnophis s. sauritus (R)	*T. sauritus sackenii*	SE Ga.; S S.C.	M?	Rossman (1963), Neill (1957)
Tantilla c. coronata (R)	*T. c. wagneri*	N Fla.; S Ga.	?	Carr & Goin (1955)
Sistrurus m. miliarius (R)	*S. m. barbouri*	S Ala.; S Ga.; S S.C.	?	Klauber (1956)
Colinus v. virginianus (A)	*C. v. floridanus*	N Fla.	I?	Sprunt (1954)
Meleagris gallopavo silvestris (A)	*M. g. osceola*	N Fla.	S?	Sprunt (1954)
Sitta p. pusilla (A)	*S. p. caniceps*	NW Fla.; SE Ga.	M?	Sprunt (1954), Burleigh (1958)
Thryothorus l. ludovicianus (A)	*T. l. miamensis*	NW Fla.; SE Ga.	I?	Sprunt (1954) Burleigh (1958)
Pipilo erythrophthalmus rileyi (A)	*P. e. alleni*	N Fla.	I?	Dickinson (1952)
Aimophila a. aestivalis (A)	*A. a. bachmani*	N Fla.; S Ga.	I?	Am. Orn. Un. (1957)
Sciurus n. niger (M)	*S. n. shermani*	NW Fla.; S Ga.	I?	Moore (1956)
Spilogale p. putorius (M)	*S. p. ambarvalis*	N Fla.; S Ga.	I?	Van Gelder (1959)
Pinus e. elliottii (P)	*P. e. densa*	Central E & W coasts of Fla.; coast of Ala. & Miss.	M?	Mergen (1958)
Juniperus virginiana (P)	*J. barbadensis*	N Fla.; S Ga.; S Ala.	I?	Hall (1952)
Asclepias tuberosa (P)	*A. rolfsii*	Fla.-Ga. boundary	M?	Woodson (1947b, 1964)

TABLE II (Cont.)

B. SUSPECTED HYBRIDIZERS

Northern	Floridian	Contact Areas	References
Perithemis tenera (O)	*P. seminole*	N Fla.	Needham & Westfall (1955)
Celithemis leonora (O)	*C. bertha*	N Fla.; S Ga.	Needham & Westfall (1955)
Progomphus obscurus (O)	*P. alachuensis*	N Fla.	Byers (1940)
Battus philenor (L)	*B. polydamas*	N Fla.	Remington (unpublished)
Erynnis b. brizo (L)	*E. b. somnus*	extreme N Fla.	Burns (1964)
Synanthedon decipiens (L)	*S. sapygaeformis*	N Fla.	Engelhardt (1946)
Sylvora a. acerni (L)	*S. a. buscki*	N Fla.; S Ga.	Engelhardt (1946)
Pilaria recondita (D)	*P. arguta*	N Fla.	Rogers (1933)
Nephrotoma ferruginea (D)	*N. suturalis*	N Fla.; S Ga.	Rogers (1933)
Rhagoletis mendax (no. form) (D)	*R. mendax* (Fla. form)	N Fla.; S Ga.	Bush (1966)
Rhagoletis cornivora (no. form) (D)	*R. cornivora* (Fla. form)	N Fla.; S Ga.	Bush (1966)
Latrodectus m. mactans (Arachnida)	*L. m. bishopi*	N Fla.	Kaston (1938)
Pseudemys scripta (R)	*P. nelsoni*	N Fla.	Carr & Goin (1955)
Storeria dekayi wrightorum (R)	*S. d. victa*	N Fla.; S Ga.	Carr & Goin (1955)
Cryptotis p. parva (M)	*C. p. floridana*	N Fla.; S Ga.	Hall & Kelson (1959)
Eptesicus f. fuscus, etc. (M)	*E. f. osceola*	N Fla.	Hall & Kelson (1959)
Sylvilagus palustris (M)	*S. aquaticus*	SW Ga. to S Miss.	Hall & Kelson (1959)
Glaucomys volans saturatus (M)	*G. v. querceti*	NE Fla.; SE Ga.	Hall & Kelson (1959)
Peromyscus nuttalli (M)	*P. floridanus*	N-C Fla.	Hall & Kelson (1959)
Microtus (Pitymys) pinetorum (M)	*M. (P.) parvulus*	N Fla.; SW Ga.; SE Ala.	Hall & Kelson (1959)
Ursus a. americanus (M)	*U. a. floridanus*	N Fla.	Hall & Kelson (1959)
Procyon lotor varius (M)	*P. l. elucus*	N Fla.	Hall & Kelson (1959)

TABLE III.   Central Texas Suture-Zone

A. KNOWN HYBRIDIZING PAIRS

Eastern	Western	Known Areas	Xing	References
Limenitis a. astyanax (L)	*L. a. arizonensis*	C-W Tex.?	?	Remington (unpublished)
Limenitis a. archippus (L)	*L. a. obsoletus*	C-W Tex.?	?	Remington (unpublished)
Erynnis br. brizo (L)	*E. br. burgessi*	C Tex. (esp. Kerr Co.)	M	Burns (1964)
Tropisternus collaris striolatus (C)	*T. c. mexicanus*	C Tex.; E Okla.; W Ark.; E Kans.	M?	Young (1961 et seq.)
Trox s. spinulosus (C)	*T. s. dentibius*	C Tex.	I?	Vaurie (1955)
Scaphiopus holbrooki hurteri (Am)	*S. couchii*	C Tex.	S	Wasserman (1957)
Bufo fowleri (Am)	*B. woodhousei*	C Tex.	I	Blair (1956, 1958b)
Bufo valliceps (Am)	*B. woodhousei*	C Tex.	S	Thornton (1955)
Gastrophryne carolinensis (Am)	*G. olivacea*	C Tex.; E Okla.	S	Hecht & Matalas (1946), Blair (1958b)
Thamnophis p. proximus (R)	*T. p. rubrilineatus*	Balcones Escarpment	M?	Rossman (1963)
Elaphe obsoleta lindheimeri (R)	*E. o. bairdi*	Edwards Plateau	I?	Blair (1958b)
Colinus virginianus (A)	*Callipepla squamata*	Concho Co., Tex.	?	McCabe (1954)
Parus bicolor (A)	*P. atricristatus*	C Tex. (esp. near Austin)	M	Dixon (1955)
Passerina ciris (A)	*P. versicolor*	Los Fresnos, Tex.	M?	Storer (1961)
Perognathus h. hispidus (M)	*P. h. paradoxus*	C Tex.	?	Glass (1947)
Juniperus virginiana (P)	*J. ashei*	C Tex.; E Okla.; S Mo.	I	Hall (1952, 1961), Hall & Carr (1964)
Quercus stellata (P)	*Q. havardi*	C Tex.	M	Muller (1952)
Quercus stellata (P)	*Q. mohriana*	C Tex.	S	Muller (1952)
Prunus s. serotina (P)	*P. s. eximia*	C Tex.	?	McVaugh (1951, 1952)
Bumelia lanuginosa rigida (P)	*B. l. texana*	C Tex.	?	McVaugh (1952)

B. SUSPECTED HYBRIDIZERS

Eastern	Western	Contact Areas	References
Erythrodiplax umbrata (O)	*E. funerea*	Tex.	Needham & Westfall (1955)
Mitoura gryneus (L)	*M. siva*	W-C Tex.	R. O. Kendall (unpublished)
Papilio glaucus (L)	*P. multicaudatus*	Tex. (S. & N. Dak ?, Mont. ?)	Remington (unpublished)
Erynnis juvenalis (L)	*E. meridianus*	W Tex.	Burns (1964)

TABLE III (Cont.)

Eastern	Western	Contact Areas	References
Erynnis horatius (L)	*E. tristis*	W Tex.	Burns (1964)
Erynnis juvenalis (L)	*E. tristis*	W Tex.	Burns (1964)
Drosophila affinis & algon-quin (D)	*D. pseudoobscura & azteca*	C Tex.	Patterson & Stone (1952), Miller (1958a)
Lytta lecontei (C)	*L. fulvipennis*	C Tex.	Selander (1960)
Bufo valliceps (Am)	*B. debilis*	C Tex.	Blair et al. (1957)
Bufo fowleri (Am)	*B. debilis*	C Tex.	Blair et al. (1957)
Pseudacris triseriata (Am)	*P. clarkii*	C Tex.; C Okla.; C Kans.	Mecham (1961), Blair (1958b)
Cnemidophorus sexlineatus (R)	*C. inornatus*	C Tex.	Stebbins (1954)
Cnemidophorus sexlineatus (R)	*C. sacki*	C Tex.; SW Okla.	Stebbins (1954)
Pituophis melanoleucus (R)	*P. catenifer*	C Tex.	Smith & Kennedy (1951), Blair (1958b)
Crotalus horridus (R)	*C. viridis*	C Tex.	Klauber (1956)
Crotalus horridus (R)	*C. atrox*	C Tex.	Klauber (1956)
Archilochus colubris (A)	*A. alexandri*	Edwards Plateau	Simmons (1925), Peterson (1963)
Centurus carolinus (A)	*C. aurifrons*	E Edwards Plateau	Selander & Giller (1963)
Dendrocopos pubescens (A)	*D. scalaris*	a band from Coleman & Austin to Dallas & Mexia	Peterson (1963)
Contopus virens (A)	*C. sordidulus*	Edwards Plateau	Am. Orn. Un. (1957)
Sayornis phoebe (A)	*S. nigricans*	Edwards Plateau	Peterson (1963)
Myiarchus crinitus (A)	*M. cinerascens*	Edwards Plateau	Peterson (1963)
Sialia sialis (A)	*S. mexicana*	C-W Tex.	Am. Orn. Un. (1957)
Piranga rubra (A)	*P. flava*	Edwards Plateau	Peterson (1963)
Icterus spurius (A)	*I. bullockii*	C Tex.	Simmons (1925)
Sylvilagus aquaticus (M)	*S. auduboni*	C Tex.	Davis (1960)
Spermophilus tridecemlinea-tus (M)	*S. mexicanus*	E,N,W margins Edwards Plateau	Hall & Kelson (1959), Davis (1960)
Spermophilus tridecemlinea-tus (M)	*S. spilosoma*	SE & NW of Edwards Plateau & panhandle	Davis (1960)
Reithrodontomys fulvescens (M)	*R. montanus*	C Tex.	Davis (1960)
Neotoma floridana (M)	*N. micropus*	C Tex.; W Okla.; W Kans.	Hall & Kelson (1959) Davis (1960)
Canis niger (M)	*C. latrans*	C & E Tex.	Davis (1960)
Canis niger (M)	*C. lupus*	formerly C Tex.; C Okla.	Hall & Kelson (1959)
Spilogale putorius interrupta (M)	*S. p. leucoparia*	C Tex.	Van Gelder (1959)
Morus rubra (P)	*M. microphylla*	Balcones Escarpment	McVaugh (1952)

TABLE IV. Rocky Mountain–Eastern Suture-Zone

A. KNOWN HYBRIDIZING PAIRS

Western	Eastern	Known Areas	Xing	References
Speyeria lurana (L)	*S. atlantis*	Black Hills	I to S	Grey, Moeck, Evans (1963)
Papilio gothica (L)	*P. polyxenes*	Colo.; Mont.	S	Remington (1958 & unpublished)
Papilio rutulus (L)	*P. glaucus*	S.D.; Mont.		Brower (1959), Remington (unpublished)
Colias alexandra (L)	*C. christina*	Mont.; Man.; Alta.		Remington (unpublished)
Hesperia pawnee (L)	*H. leonardus*	Minnesota	M?	J. S. Nordin (in litt.)
Hyalophora gloveri (L)	*H. cecropia*	C Colo.; W N.D.	M to S	Sweadner (1937), Remington (unpublished)
Podosesia fraxini (L)	*P. syringae*	St. Paul, Minn.	M?	Engelhardt (1946)
Diplotaxis obscura (C)	*D. tristis*	Manitoba	M?	Howden (1966)
Bufo woodhousei (Am)	*B. americanus*	Tulsa, Okla.	M	Blair (1941)
Bufo woodhousei (Am)	*B. fowleri*	100 mi. strip: E Okla., W Mo. to E Nebr.	I	Blair (1941)
Thamnophis elegans terrestris (R)	*T. marcianus*	SW Kans.	S?	Smith (1946)
Tympanuchus phasianellus (A)	*T. cupido*	Man.; Sask.; Alta.; E Colo.	S	Rowan (1926), Evans (1966)
Sphyrapicus nuchalis (A)	*S. varius*	Nebr.; W Alta.; Tex.; Jalisco	M?	Howell (1952)
Colaptes cafer (A)	*C. auratus*	W Okla. to Mont. to C B.C.	I	Short (1965)
Dendrocopos pubescens leucurus (A)	*D. p. medianus*	Man. to Kans.	M?	Pitelka (1941)
Contopus sordidulus (A)	*C. virens*	W Nebr.; W S.D.	M?	Short (1961)
Piranga olivacea (A)	*P. ludoviciana*	W Nebr.; S.D.	S?	Short (1961)
Icterus bullockii (A)	*I. galbula*	strongest in C Daks.; W Nebr.; W Kans.	I?	Sibley & Short (1964)
Dendroica auduboni (A)	*D. coronata*	E Colo.; Ariz.; B.C.; Alaska	M	Alexander (1945), Mengel (1964), Monson & Phillips (1964)
Pheucticus melanocephalus (A)	*P. ludovicianus*	S.D.; W Nebr.; W Kans.	S?	Short (1961), Johnston (1964)
Passerina amoena (A)	*P. cyanea*	E Mont. to W Kans.; N Ariz.		Sibley & Short (1959), Monson & Phillips (1964)
Junco oreganus (A)	*J. aitkeni*	Bighorn Valley of S Mont.	5/73	Miller (1955)
Junco oreganus (A)	*J. hyemalis*	Jasper, etc., Alta.	9/44	Miller (1955)
Pipilo maculatus arcticus (A)	*P. e. erythrophthalmus*	rivers in NW Plains	I	Sibley & West (1959)
Juniperus scopulorum (P)	*J. virginiana*	SW N.D.; C & W S.D.; W Nebr.; N Tex.	I?	Fassett (1944), Hall (1952, 1961)

TABLE IV (Cont.)

Western	Eastern	Known Areas	Xing	References
Populus tremuloides (P)	*P. deltoides*	C Alta.	M?	Brayshaw (1966)
Populus angustifolia (P)	*P. deltoides*	S Alta.; Mont.; Wyo.; S.D.; Nebr.; Colo.; N.M.; W Tex.	M?	Smith (1943), Vines (1960), Brayshaw (1966)
Quercus breviloba (P)	*Q. stellata*	E edge Plains	I	Hall & Carr (1964)
Acer grandidentata (P)	*A. saccharum*	E edge Plains	I	Hall & Carr (1964)

B. SUSPECTED HYBRIDIZERS

Western	Eastern	Contact Areas	References
Amphiagrion abbreviatum (O)	*A. saucium*	Minn.; Man.; Kansas	Walker (1953)
Ischnura perparva & *cervula* (O)	*I. verticalis* & *posita*	Man.; etc.?	Walker (1953)
Progomphus borealis (O)	*P. obscurus*	W Plains	Byers (1940)
Aeschna multicolor (O)	*A. mutata*	Plains	Walker (1958)
Limenitis w. weidemeyerii (L)	*L. a. astyanax*	W Plains	Remington (unpublished)
Polygonia satyrus (L)	*P. comma*	W Plains	Remington (unpublished)
Everes amyntula (L)	*E. comyntas*	Minn.; Man.; Mont.; Colo.	Clench (1958), J. S. Nordin (in litt.)
Incisalia iroides (L)	*I. augustinus*	Colo. to Alta.	Clench (1958)
Mitoura siva (L)	*M. gryneus*	C & W S.D.; W Nebr.	Remington (unpublished)
Erynnis telemachus (L)	*E. juvenalis*	E Colo.; E N.M.	Burns (1964)
Erynnis afranius (L)	*E. lucilius*	W Minn.; Nebr.	Burns (1964)
Erynnis afranius (L)	*E. baptisiae*	E Colo.; E N.M.	Burns (1964)
Erynnis pacuvius (L)	*E. martialis*	Colo. Front Range	Burns (1964)
Hesperia h. harpalus (L)	*H. h. ochracea*	W Plains	MacNeill (1964)
Hesperia p. pahaska (L)	*H. pawnee*	W Plains	MacNeill (1964)
Synchlora liquoraria (L)	*S. aerata*	Plains?	Ferguson (in press)
Drosophila pseudoobscura (D)	*D. affinis* & *algonquin*		Patterson & Stone (1952)
Rhagoletis zephyria (D)	*R. pomonella*	Minn.; N.D.; southward	Bush (1966)
Rhagoletis suavis (D)	*R. completa*	Iowa; Kans.; W Plains	Bush (1966)
Exema mormona (C)	*E. canadensis*	W Plains; C Tex.	Karren (1966)
Exema conspersa (C)	*E. dispar*	W Plains; C Tex.	Karren (1966)
Vespula pennsylvanica (Hy)	*V. maculifrons*	W Plains (E Mont. to C N.M.)	Miller (1961)
Vespula atropilosa (Hy)	*V. vidua*	Plains	Miller (1961)
Bufo cognatus (Am)	*B. americanus*	Plains	Blair et al. (1957)
Bufo cognatus (Am)	*B. fowleri*	S Plains	Blair et al. (1957)
Bufo hemiophrys (Am)	*B. americanus*	N Plains	Blair et al. (1957)
Bufo debilis (Am)	*B. fowleri*	S Plains	Blair et al. (1957)
Thamnophis elegans terrestris (R)	*T. radix*	W Plains	Stebbins (1954)

TABLE IV (Cont.)

Western	Eastern	Contact Areas	References
Archilochus alexandri (A)	*A. colubris*	W Alta.; C Mont.	Bent (1940)
Selasphorus platycercus (A)	*A. colubris*	W S.D.; W Nebr.; W Kans.	Bent (1940)
Selasphorus rufus (A)	*A. colubris*	SW Alta.; C Mont.	Bent (1940)
Stellula calliope (A)	*A. colubris*	SW Alta.; C Mont.; C Wyo.	Bent (1940)
Chaetura vauxi (A)	*C. pelagica*	S Alta.; N-C Mont.	Bent (1940)
Cyanocitta stelleri (A)	*C. cristata*	W edge of Plains	Blair et al. (1957)
Sialia mexicana (A)	*S. sialis*	C Mont.; E Colo.	Am. Orn. Un. (1957)
Sialia currucoides (A)	*S. sialis*	W Plains	Am. Orn. Un. (1957)
Myotis thysanodes (M)	*M. keenii*	Black Hills; S Wash.	Hall & Kelson (1959)
Sylvilagus nuttallii (M)	*S. floridanus*	W Plains	Hall & Kelson (1959)
Sylvilagus audubonii (M)	*S. floridanus*	W Plains	Hall & Kelson (1959)
Tamiasciurus fremonti (M)	*T. hudsonicus*	SE Wyo.; N-C Colo.	Hall & Kelson (1959)
Perognathus fasciatus (M)	*P. flavescens*	C Dakotas; NW Nebr.; NE Colo.	Hall & Kelson (1959)
Zapus princeps (M)	*Z. hudsonius*	E Man.; Sask.; N.D.; N S.D.; E Mont.; S Alta.; C & N B.C.	Krutsch (1954)
Spilogale putorius gracilis (M)	*S. interrupta*	W S.D.; W Nebr.; E Wyo.; E Colo.; W Okla.	Van Gelder (1959)

TABLE V.   Northern–Rocky Mountain Suture-Zone

A. KNOWN HYBRIDIZING PAIRS

Northern	Rocky Mountain	Known Areas	Xing	References
Limenitis arthemis (L)	*L. weidemeyerii*	Cariboo region, B.C.; N Mont.	M	Remington (unpublished
Papilio hudsonianus (L)	*P. gothica*	Mont.?; Man.?; Sask.?; B.C.?	M?	Remington (unpublished)
Hyalophora nokomis (L)	*H. gloveri*	Man.; Alta.	M	Sweadner (1937)
Lytta viridana (C)	*L. nuttalli*	Alta.	S?	Selander (1960)
Canachites canadensis (A)	*C. franklini*	C-W Alta.; C B.C.	S	Rand (1948)
Perisoreus c. canadensis (A)	*P. can. capitalis*	S B.C. mts.	M?	Rand (1948)
Junco hiemalis (A)	*J. oreganus*	E foothills Alta. mts.; C B.C.	M?	Miller (1941)
Zonotrichia leucophrys gambelii (A)	*Z. l. oriantha*	SW Alta.; SE B.C.	I?	Rand (1948)
Pinus banksiana (P)	*P. contorta*	Edmonton to Peace River, Alta.	I?	Moss (1955)
Picea glauca albertiana (P)	*P. engelmannii*	W Alta. (mid elev.)	I?	Raup (1946), Moss (1959)
Abies balsamea (P)	*A. lasiocarpa*	Lesser Slave Lake, Athabasca River, Alta.	I?	Moss (1959)
Juniperus horizontalis (P)	*J. scopulorum*	SW Alta.; W N.D.; NE Wyo.	M?	Fassett (1945a), Stevens (1950), Moss (1959)
Populus balsamifera (P)	*P. angustifolia*	S-C & SW Alta.	M?	Brayshaw (1966)
Populus tremuloides (P)	*P. angustifolia*	Bow R., Alta.	S?	Brayshaw (1966)
Betula fontinalis (P)	*B. glandulosa*	N.W.T.; Yukon; Alta.	I	Dugle (1966)
Betula fontinalis (P)	*B. papyrifera*	Yukon; E B.C.; Alta.; Colo.; Ida.; Mont.; Utah	I	Dugle (1966)

B. SUSPECTED HYBRIDIZERS

Northern	Rocky Mountain	Contact Areas	References
Hesperia manitoba (L)	*H. h. harpalus*	S-C B.C.	MacNeill (1964)
Lytta viridana (C)	*L. cyanipennis*	W Alta.; E B.C.	Selander (1960)
Anoplodera mutabilis (C)	*A. aspera*	NW Alta.; NE B.C.	Howden (1966)
Oporornis agilis (A)	*O. tolmiei*	Alta.; B.C.	Mengel (1964)
Oporornis philadelphia (A)	*O. tolmiei*	Alta. (no contact?)	Mengel (1964)
Marmota caligata (M)	*M. flaviventris*	Mont.; B.C.; C Wash.	R. S. Hoffman (unpublished), Carl et al. (1951), Dalquest (1948)

TABLE V (Cont.)

Northern	Rocky Mountain	Contact Areas	References
Marmota caligata (M)	*M. monax*	Banff to Jasper, Alta.	Soper (1964)
Spermophilus undulatus (M)	*S. columbianus*	N B.C. (no contact now)	Hall & Kelson (1959)
Oropsylla id. bertholfi (S) (on *S. undulatus*)	*O. i. idahoensis* (on *S. columbianus*)	N B.C. (no contact now)	Holland (1963)
Zapus hudsonius (M)	*Z. princeps*	C & N B.C.	Krutsch (1954)
Ovis dalli (M)	*O. canadensis*	Peace River district, B.C. (no contact now)	Cowan (1940)
Betula resinifera (P)	*B. fontinalis*	C Sask.; W Alta.; N.W. T.; etc.	Dugle (1966)
Betula glandulifera (P)	*B. fontinalis*	Man.; Sask., Alta., E B.C. W N.W.T.	Dugle (1966)

TABLE VI.   Pacific–Rocky Mountain Suture-Zone

A. KNOWN HYBRIDIZING PAIRS

Pacific	Rocky Mountain	Known Areas	Xing	References
Limenitis lorquini (L)	*L. weidemeyerii*	Mono Lake, Calif.; E Ida.; W Mont.	I	Brown (1934), Remington (unpublished)
Hesperia h. yosemite (L)	*H. h. harpalus*	E Sierran foothills	I?	MacNeill (1964)
Hyalophora euryalus (L)	*H. gloveri*	E B.C.; N Ida.; W Mont. (? Ariz. & Calif.?)		Sweadner (1937)
Chilocorus tricyclus (C)	*C. hexacyclus*	Crowsnest Pass, Alta.	M	Smith (1963)
Sceloporus graciosus gracilis (R)	*S. g. graciosus*	W slope of Sierra; SE Oreg.	I?	Stebbins (1954)
Thamnophis e. elegans (R)	*T. e. terrestris*	E Calif.; W Nev.; S-C Oreg.	I?	Stebbins (1954)
Crotalus viridis oreganus (R)	*C. v. viridis* + *v. lutosus*	Mono Co., Calif.; SE Oreg.; W Ida.	M?	Klauber (1956)
Bonasa umbellus brunnescens (A)	*B. u. umbelloides*	C B.C.; C Wash.; C Oreg.	I?	Pitelka (1941)
Sphyrapicus daggetti (A)	*S. nuchalis*	Mono Co., Modoc Co., Calif.	S	Howell (1952)
Sphyrapicus ruber (A)	*S. nuchalis*	Manning Park & Cariboo, B.C. region; Yakima Co., Wash.; near Sisters, Oreg.	M	Howell (1952)
Dendrocopos pubescens turatii + *gairdneri* (A)	*D. p. leucurus*	E Wash.; E Oreg.; NE Calif.	I?	Pitelka (1941)
Perisoreus canadensis obscurus group (A)	*P. c. canadensis* group	N-C Wash.; S-C B.C.	I?	Miller (1943), Aldrich (1943)
Junco oreganus thurberi (A)	*J. c. caniceps*	W Nev.; Mono Co., Calif.	M?	Miller (1941)
Junco oreganus mearnsi (A)	*J. c. caniceps*	SE Ida.; NW Nev.; N Utah; SW Wyo.	I	Miller (1941)
Martes caurina (M)	*M. americana*	N Ida.; W Mont.	I	Wright (1953)
Spilogale putorius phenax (M)	*S. p. gracilis*	NE & E-C Calif	?	Van Gelder (1959)
Ovis can. californiana (M)	*O. c. canadensis*	SE Wash.; N Ida.; W Nev.	?	Cowan (1940)
Populus trichocarpa (P)	*P. balsamifera*	C & E B.C.; W Alta.; (? W Mont., Ida., E Wash., E Oreg.?)	I	Brayshaw (1965, 1966)

B. SUSPECTED HYBRIDIZERS

Pacific	Rocky Mountain	Contact Areas	References
Sympetrum occidentale (O)	*S. fasciatum*	Alta.	Needham & Westfall (1955)
Papilio zelicaon (L)	*P. gothica*	NE & C Calif.	Remington (unpublished)

TABLE VI (Cont.)

Pacific	Rocky Mountain	Contact Areas	References
Erynnis propertius (L)	*E. meridianus*	NW Nev. mts.	Burns (1964)
Hesperia nevada (W. type) (L)	*H. nevada* (Rocky Mt. type)	Alta.	MacNeill (1964)
Rhyacophila tucala (T)	*R. alberta*	?	Ross (1965)
Drosophila occidentalis (D)	*D. suboccidentalis*	E Calif.; W Nev.; Ida?	Sears (1947), Patterson & Stone (1952)
Tenthredo rhammisia (Hy)	*T. stricklandi*	?	Ross (1965)
Masticophis lateralis (R)	*M. taeniatus*	E side of Sierra mts.	Stebbins (1954)
Dendragapus fuliginosus (A)	*D. obscurus*	E-C B.C.; C Wash.	Pitelka (1941)
Parus rufescens (A)	*P. atricapillus*	N B.C.; W Mont.; E Oreg.; etc.	Swarth (1922), Peterson (1941)
Vermivora ruficapilla ridgwayi (A)	*V. virginiae*	SE Calif.; SW Nev.; S Ida.	Linsdale (1936), Mengel (1964)
Sorex preblei (M)	*S. cinereus*	E Oreg.; W Ida.	Hall & Kelson (1959)
Sorex v. vagrans group (M)	*S. v. obscurus* group	SE B.C.; W Mont.; W Ida.	Findley (1955)
Sorex trowbridgii (M)	*S. merriami*	C Wash.; C Oreg.; NE Calif.	Hall & Kelson (1959)
Eutamias townsendii (M)	*E. minimus*	C Wash.; C Oreg.; NE Calif.	Hall & Kelson (1959)
Tamiasciurus douglasii (M)	*T. hudsonicus*	S-C B.C.; N-C Wash.; C Oreg.	Hall & Kelson (1959), Dalquest (1948)
Monopsyllus ciliatus (S) (on *T. douglasii* mainly)	*M. vison* (on *T. hudsonicus*)	SW Alaska; S-C B.C.?	Holland (1963)
Clethrionomys occidentalis (M)	*C. gapperi*	C B.C.; C Wash.	Hall & Kelson (1959)
Zapus trinotatus (M)	*Z. princeps*	SW B.C.	Krutsch (1954)
Juniperus occidentalis (P)	*J. osteosperma*	SW Ida.; E slope Sierras; W Nev.	Munz & Keck (1959)
Juniperus occidentalis (P)	*J. scopulorum*	E Wash.; E Oreg.; W Ida.	Davis (1952), Peck (1961)
Populus trichocarpa (P)	*P. angustifolia*	SW B.C.; C Wash.; Ida.; W Nev.; Inyo Co., Calif.	Preston (1961), Munz & Keck (1959)
Populus fremontii (P)	*P. angustifolia*	Inyo Co., Calif.	Munz & Keck (1959)

TABLE VIIa. Northern–Cascade Suture-Zone

A. KNOWN HYBRIDIZING PAIRS

Northern	Cascade	Known Areas	Xing	References
Limenitis arthemis (L)	*L. lorquini*	C B.C.; NW Mont	I	Remington (unpublished)
Papilio glaucus canadensis (L)	*P. rutulus*	Central valleys, B.C.	I	Brower (1959)
Sphyrapicus varius (A)	*S. ruber*	Telegraph Cr., B.C.	M?	Swarth (1922), Howell (1952)
Colaptes auratus (A)	*C. cafer*	Hazelton area, B.C.	M?	Swarth (1924)
Dendroica townsendii (A)	*D. occidentalis*	C Wash.; N Oreg.	M?	Jewett (1944)
Junco hiemalis connectens (A)	*J. o. oreganus*	Flood Glacier, N B.C.	M?	Swarth (1922)

B. SUSPECTED HYBRIDIZERS

Northern	Cascade	Contact Areas	
Hylocichla ustulata swainsoni (A)	*H. u. ustulata*	Coast range passes, B. C.	Swarth (1922)
Marmota monax (M)	*M. caligata*	Hazelton, B.C. region	Swarth (1924)
Tamiasciurus hudsonicus (M)	*T. douglasi*	Interior B.C.; E Wash.	Swarth (1922)
Clethrionomys dawsoni (M)	*C. wrangeli*	Stikine valley, B.C.	Swarth (1922)
Zapus hudsonius (M)	*Z. trinotatus*	Okanagan area, B.C.	Cowan & Guiguet (1956)
Populus balsamifera (P)	*P. trichocarpa*	S Yukon; N B.C. [see Table VIa]	Preston (1961), [Brayshaw, 1965]

TABLE VIIb.   Cascade–Sierran Suture-Zone

A. KNOWN HYBRIDIZING PAIRS

Northern (Cascade)	Sierran	Known Areas	Xing	References
Drosophila athabasca (D)	*D. azteca*	Coos Co., Oreg.; Del Norte Co., Calif.	M?	Sulerud & Miller (1966)
Gerrhonotus multicarinatus scincicauda (R)	*G. m. multicarina-tus*	NW Calif.	I?	Stebbins (1954)
Gerrhonotus coeruleus prin-cipis (R)	*G. c. shastensis*	SW Oreg.; NW Calif.	I?	Stebbins (1954)
Sphyrapicus ruber (A)	*S. daggetti*	SW Oreg.; NW Calif.	I	Howell (1952)
Junco oreganus shufeldti (A)	*J. o. thurberi*	SW 1/4 Oreg.	I	Miller (1941)
Spilogale putorius latifrons (M)	*S. pu. phenax*	N Calif.	I?	Van Gelder (1959)

B. SUSPECTED HYBRIDIZERS

Northern (Cascade)	Sierran	Contact Areas	References
Sympetrum occidentale (O)	*S. californicum*	S Oreg.; N Calif.	Needham & Westfall (1955)
Rana cascadae (Am)	*R. boylei*	S Oreg.; N Calif.	Dumas (1966)
Thamnophis ordinoides (R)	*T. elegans terres-tris* & *T. e. hydro-phila*	SW Oreg.; NW Calif.	Stebbins (1954)
Scapanus townsendii (M)	*S. latimanus*	SW Oreg.; NW Calif.	Hall & Kelson (1959)
Scapanus orarius (M)	*S. latimanus*	C Oreg.; NW Calif.	Hall & Kelson (1959)
Aplodontia r. rufa group (M)	*A. r. californica*	N-C Calif.	Hall & Kelson (1959)

TABLE VIIc.   California Desert—Pacific Slope Suture-Zone

A. KNOWN HYBRIDIZING PAIRS

Western	Eastern	Known Areas	Xing	References
Zerene eurydice (L)	Z. cesonia	San Bernardino & San Jacinto Mts.	M?	Wright (1905), Riddell (1941), Remington (unpublished)
Hypsiglena torquata klauberi (R)	H. t. deserticola	W of S Calif. desert	M?	Tanner (1944)
Rhinocheilus lecontei clarus (R)	R. l. lecontei	W side of Colo. desert, Calif., etc.	M	Klauber (1941)
Crotalus ruber (R)	C. atrox	29 Palms, Calif., to La Bamba, Baja Calif.	S?	Klauber (1956)
Lophortyx californica (A)	L. gambelii	passes & E bases San Bernardino & San Jacinto Mts.	up to 5%	Henshaw (1885), Miller (1955)
Sylvilagus auduboni sanctidiegi (M)	S. a. arizonae	Banning, Calif.	M?	Grinnell & Swarth (1913)
Perognathus f. fallax (M)	P. f. pallidulus	Banning, Calif.	M?	Grinnell & Swarth (1913)
Dipodomys a. agilis (M)	D. a. perplexus	Cajon Pass area	M?	Vaughan (1954)
Dipodomys merriami parvus (M)	D. m. merriami	Cajon Pass, Calif.	S?	Lidicker (1960)
Peromyscus maniculatus gambeli (M)	P. m. sonoriensis	Blue Ridge, San Gabriel Mts.	M?	Vaughan (1954)
Neotoma fuscipes macrotis (M)	N. f. simplex	Blue Ridge, San Gabriel Mts.	M?	Vaughan (1954)
Lynx rufus californicus (M)	L. r. baileyi	5000 ft. elevation San Gabriel Mts.	M?	Vaughan (1954)
Quercus dumosa (P)	Q. turbinella californica	Cajon Summit	I	Tucker (1953a)

B. SUSPECTED HYBRIDIZERS

Western (Pacific slope)	Eastern (desert slope)	Contact Areas	References
Pipistrellus hesperus merriami (M)	P. h. hesperus	low passes	Vaughan (1954)
Spermophilus b. beecheyi (M)	S. b. fisheri	low passes	Vaughan (1954)
Thomomys bottae pallescens (M)	T. b. neglectus	mid elevations	Vaughan (1954)
Perognathus californicus dispar (M)	P. c. bernardinus	passes	Vaughan (1954)
Perognathus longimembris brevinasus (M)	P. l. bangsi	passes	Grinnell & Swarth (1913)
Dipodomys merriami parvus (M)	D. m. simiolus	low passes	Grinnell & Swarth (1913)
Peromyscus eremicus fraterculus (M)	P. e. eremicus	low passes	Vaughan (1954)

TABLE VIIc (Cont.)

Western (Pacific slope)	Eastern (desert slope)	Contact Areas	References
Neotoma lepida intermedia (M)	*N. l. lepida*	low passes	Vaughan (1954)
Taxidea taxus neglecta (M)	*T. t. berlandieri*	low passes	Vaughan (1954)
Canis latrans ochropus (M)	*C. l. mearnsi*	passes	Vaughan (1954)
Urocyon cinereoargentatus californicus (M)	*U. c. scotti*	?	Grinnell & Swarth (1913)
Juniperus occidentalis (P)	*J. californica*	mid elevations	Munz & Keck (1959)

TABLE VIId.   Rocky Mountain—Sonoran Suture-Zone

A. KNOWN HYBRIDIZING PAIRS

Higher (or Rocky Mtns.)	Lower (Sonoran)	Known Areas	Xing	References
Limenitis w. weidemeyerii (L)	*L. w. angustifascia*	S Utah; S-C Nev.; N N.M.; N Ariz.	I	Remington (unpublished)
Limenitis w. angustifascia (L)	*L. astyanax arizonensis*	mid altitudes, Ariz.	S	Bauer (1954), Remington (1958 and unpublished)
Papilio brucei (L)	*P. bairdii*	SW Colo.; N Ariz.	I	Remington (unpublished)
Crotalus viridis viridis + *C. v. lutosus* (R)	*C. v. nuntius* + *C. v. cerberus*	S Utah; N & E Ariz.	I?	Klauber (1956)
Callipepla squamata (A)	*Lophortyx gambelii*	S-C Ariz.; S N.M.	S	Bailey (1928), Phillips et al. (1964), Hubbard (1966)
Selasphorus platycercus (A)	*Calypte costae*	Rincon Mts., Ariz.	S	Banks & Johnson (1961)
Selasphorus platycercus (A)	*Archilochus alexandri*	(migrant) both parents may be W Rocky Mts.	S	Banks & Johnson (1961)
Eugenes fulgens (A)	*Cynanthus latirostris*	Huachuca Mts., Ariz.	S	Phillips et al. (1964)
Aphelocoma coerulescens woodhousei group (A)	*A. coe. california* group	W-C & NW Nev.	I	Pitelka (1951)
Ovis c. canadensis (M)	*O. c. mexicana*	S Nev.; Utah; N N.M.	?	Cowan (1940)
Juniperus scopulorum (P)	*J. pachyphloea*	W-C N.M.; E-C Ariz.	I?	Hall (1952)
Quercus gambelii (P)	*Q. turbinella*	N Utah (relict hybrids); SW Utah; NW Ariz.	I	Cottam et al. (1959)
Quercus gambelii (P)	*Q. arizonica*	Mid altitudes, Ariz. & Sonora	S	Tucker (1963)

B. SUSPECTED HYBRIDIZERS

Higher (or Rocky Mtns.)	Lower (Sonoran)	Contact Areas	References
Limenitis a. archippus (L)	*L. a. obsoletus*	S Nev.; S Utah; N N.M.	Remington (unpublished)
Rhagoletis boycei (D)	*R. juglandis*	N.M.; Ariz. (5000 ft. - 6000 ft.)	Bush (1966)
Drosophila montana (D)	*D. novamexicana*	N.M.; Ariz.; SW Colo.	Patterson & Stone (1952)
Drosophila athabasca (D)	*D. azteca*	C N.M.; C Ariz.	Sulerud & Miller (1966)
Crotalus viridis (R)	*C. lepidus*	S N.M.; SE Ariz.	Klauber (1956)
Glaucidium gnoma (A)	*G. brasilianum*	C Ariz.	Phillips et al. (1964)
Piranga ludoviciana (A)	*P. flava* (& *P. rubra?*)	C & S Ariz	Phillips et al. (1964)
Neotoma cinerea (M)	*N. albigula*	N N.M.; N Ariz.	Hall & Kelson (1959)

TABLE VIIf. Louisiana–East Texas Suture-Zone

A. KNOWN HYBRIDIZING PAIRS

Southeastern	Texan	Contact Areas	Xing	References
Limenitis archippus watsoni (L)	*L. a. archippus*	N La.; E Tex.	I	Remington & Manley (unpublished)
Erynnis zarucco (L)	*E. funeralis*	W La.	M?	Burns (1964)
Bufo fowleri (Am)	*B. valliceps*	Point Blank, Tex.	M?	Blair (1941)
Passerina c. ciris (A)	*P. c. pallidior*	SE Texas	I?	Storer (1951)

B. SUSPECTED HYBRIDIZERS

Southeastern	Texan	Contact Areas	References
Scaphiopus h. holbrooki (Am)	*S. ho. hurteri*	C La.	Wasserman (1958)
Rana sevosa (Am)	*R. areolata*	W La.	Blair (1958a)
Sistrurus miliarius (R)	*S. catenatus*	E Tex.	Klauber (1956)
Tadarida cynocephala (M)	*T. mexicana*	E Tex.	Davis (1960)
Reithrodontomys humulis (M)	*R. montanus*	SE Tex.	Blair (1958b), Hall & Kelson (1959), Davis (1960)

TABLE VIIe. Southwestern New Mexico Suture-Zone

A. KNOWN HYBRIDIZING PAIRS

West Side	East Side	Known Areas	Xing	References
Cnemidophorus tigris gracilis (R)	*C. t. marmoratus*	Peloncillo Mts. passes, N.M.	M	Zweifel (1962), Dessauer et al. (1962), Martin & Mehringer (1965)
Rhinocheilus l. lecontei (R)	*R. l. tessellatus*	SE Ariz.; SW N.M.	M?	Klauber (1941), Shannon & Humphrey (1963)
Hypsiglena torquata ochrogaster (R)	*H. t. texana*	W N.M.	M?	Tanner (1944)

B. SUSPECTED HYBRIDIZERS

Western	Eastern	Contact Areas	References
Erynnis juvenalis clitus (L)	*E. telemachus*	SE Ariz.	Burns (1964)
Erynnis scudderi (L)	*E. telemachus*	SE Ariz.	Burns (1964)
Erynnis scudderi (L)	*E. j. juvenalis*	SW N.M.	Burns (1964)
Gastrophryne mazatlanensis (Am)	*G. olivacea*	NW Chihuahua; NE Sonora	Stebbins (1954), Blair et al. (1957)
Coleonyx variegatus (R)	*C. brevis*	SW N.M.; SE Ariz.	Stebbins (1954)
Sceloporus jarrovi (R)	*S. poinsetti*	SW N.M.	Stebbins (1954)
Sceloporus clarki (R)	*S. poinsetti*	SW N.M.	Stebbins (1954)
Phrynosoma solare (R)	*P. modestum & cornutum*	SE Ariz.; NE Sonora	Stebbins (1954)
Lampropeltis pyromelana (R)	*L. triangulum*	SW N.M.	Stebbins (1954)
Masticophis bilineatus (R)	*M. taeniatus*	SW N.M.	Stebbins (1954)
Crotalus tigris (R)	*C. lepidus*	SE Ariz.	Klauber (1956)
Thomomys umbrinus sspp. (M)	*T. baileyi mearnsi*	SW N.M.; SE Ariz.	Hall & Kelson (1959)

TABLE VIIg.   Southern Appalachian–Ozark–Ouachita Suture-Zone

A. KNOWN HYBRIDIZING PAIRS

Northern (or highland)	Southern	Known Areas	Xing	References
Synchlora aerata (L)	*S. denticularia*	Miss.; W N.C.; E Tenn.		Ferguson (in press)
Drosophila a. americana (D)	*D. a. texana*	Morrilton, Ark. (Va. & N.C. to Ark., C Okla., C Tex.)	M?	Stone & Patterson (1947), Patterson & Stone (1952)
Lytta sayi (C)	*L. unguicularis*	?	?	Selander (1960)
Plethodon glutinosus (Am)	*P. ouachitae*	LeFlore Co., Okla.	I	Blair & Lindsay (1965)
Eurycea longicauda gutto-lineata (Am)	*E. lucifuga*	Cumberland plateau, Tenn. & Ky.	S?	Mittleman (1942)
Pseudacris triseriata (Am)	*P. brachyphona*	S Appalachians	?	Mecham (1961)
Bufo fowleri (Am)	*B. terrestris*	York & Benton, Ala.; Richmond Co., etc., Ga.	S?	Blair (1941), Neill (1949b)
Bufo americanus (Am)	*B. terrestris*	"line of juncture"	M?	Blair (1947)
Pseudemys rubriventris (R)	*P. floridana*	NE N.C.	I	Crenshaw (1965)
Terrapene carolina (R)	*T. ornata*	Wayne Co., Mo.	S?	Shannon & Smith (1949)
Crotalus horridus (R)	*C. adamanteus*	Barbour Co., Ala.	S	Klauber (1956)
Dendrocopos pubescens medianus (A)	*D. p. pubescens*	Carolinas; N Ga.; S Ind.; S Ill.; C Mo.; SE Kans.	I?	Pitelka (1941)
Parus atricapillus (A)	*P. carolina*	C N.J.; SW Ill.; SE Mo.	S?	Chapman (1924), Brewer (1963)
Peromyscus leucopus (M)	*P. gossypinus*	N-C Ala.; SE La.; E Tex.	S	McCarley (1954)
Uvularia grandiflora (P)	*U. perfoliata*	N Ga.; N Ala.; N Miss.; Tenn.	M?	Dietz (1952)
Cercis canadensis (P)	*C. reniformis*	Ozarks, Ark. & Mo.	I	Hall & Carr (1964)

B. SUSPECTED HYBRIDIZERS

Northern (or highland)	Southern	Contact Areas	References
Magicicada septendecim (H)	*M. tredecim*	S Va.; N Tenn.; Ind.; E Ill.; Iowa; W Mo.; E Kans.	Alexander & Moore (1962)
Magicicada cassini (H)	*M. tredecassini*	Tenn.; NE Ill.; S Iowa; Kans.	Alexander & Moore (1962)
Magicicada septendecula (H)	*M. tredecula*	W Ky.; C Ind.; E Ill.; W Mo.	Alexander & Moore (1962)
Lampropeltis d. triangulum (R)	*L. d. doliata*	Upper piedmont: N.C.; S.C.; NE Ga.	Conant (1943), Neill (1949a)
Sorex cinereus (M)	*S. longirostris*	Md.; S Ind.; Ill.	Hall & Kelson (1959)

TABLE VIII.   Arctic Treeline Contact Line

ALL POSSIBLY HYBRIDIZING; NONE CERTAIN

Below Treeline	Tundra	Contact Areas	References
Boloria freija tarquinius (L)	*B. f. freija*	C Keewatin; N Mackenzie; N & SW Yukon	Freeman (1958)
Boloria eunomia dawsoni (L)	*B. e. triclaris*	NW Que.; N Ont.; NE Man.; C N.W.T.; N Yukon	Freeman (1958)
Boloria frigga gibsoni (L)	*B. f. saga*	N Man.; N N.W.T.; N Yukon; NE Alaska	Freeman (1958)
Oeneis jutta (L)	*O. melissa*	N Que.; N Man.; N Mackenzie; N Yukon	Freeman (1951, 1958)
Erebia disa (L)	*E. fasciata*	Dawson, Yukon; Reindeer Depot, Coppermine River, Great Slave Lake, N.W.T.	Freeman (1951), Ehrlich (1958a,b)
Erebia discoidalis (L)	*E. rossii*	Churchill & Gillam, Man.	Freeman (1951)
Plebejus a. aquilo (L)	*P. a. lacustris*	NW Que.; N Man.; N Mackenzie; N Yukon	Freeman (1958)
Colias pelidne (L)	*C. hecla & nastes*	N Labr.; N Que.; (SW Alta.)	Freeman (1951, 1958)
Aedes fitchii, excrucians, implicatus (D)	*A. nigripes & impiger*	entire holarctic treeline	Barr & Ehrlich (1958), Stone et al. (1959)
Oropsylla idahoensis (S)	*O. alaskensis*	treeline in Alaska	Holland (1963)
Megabothris quirini (S)	*M. groenlandicus*	Ft. Yukon, Alaska; Mackenzie River mouth; S of Churchill, Man.; Richmond Gulf, Que.; NE Labr.	Holland (1958)
Vespula arctica (Hy)	*V. albida*	across entire arctic	Miller (1961)
Gavia immer (A)	*G. adamsii*	forest boundary, exc. Baffin Is.	Rand (1948)
Lepus americanus (M)	*L. arcticus & othus*	across entire arctic	Hall & Kelson (1959)

TABLE IX.   Central Mexico Suture-Zone

A. KNOWN HYBRIDIZING PAIRS

Northern & Eastern	Southern & Western	Contact Areas	Xing	References
Lytta variabilis - orange phase (C)	*L. variabilis* - black phase	Guanajuato; Querétaro; Hidalgo	?	Selander (1960)
Centurus aurifrons (A)	*C. uropygialis*	Calvillo, Aguas Calientes; Santa Ana, Jalisco	5%	Selander & Giller (1963)
Pipilo erythrophthalmus (A)	*P. ocai*	Veracruz to Jalisco		Sibley (1950)
Peromyscus maniculatus (M)	*P. melanotis*	Base of Mt. Orizaba	2♀♀	Davis (1944)
Spilogale putorius leucoparia (M)	*S. p. angustifrons*	Hidalgo; Guanajuato; N Jalisco	I?	Van Gelder (1959)

B. SUSPECTED HYBRIDIZERS

Northern & Eastern	Southern & Western	Contact Areas	References
Tropisternus collaris mexicanus (C)	*T. c.* ssp. nov.?	San Blas–Guadalajara region	Young (1961 & oral)
Crotalus lepidus (R)	*C. triseriatus*	E Nayarit; C Jalisco	Klauber (1956)
Sorex saussurei (M)	*S. oreopolus*	whole zone	Hall & Kelson (1959)
Notiosorex crawfordi (M)	*N. gigas*	Michoacan; Jalisco	Hall & Kelson (1959)
Sylvilagus audubonii (M)	*S. cunicularius*	Veracruz to Jalisco	Hall & Kelson (1959)
Lepus californicus (M)	*L. mexicanus*	Hidalgo to Jalisco	Hall & Kelson (1959)
Cratogeomys zinseri (M)	*C. gymnurus*	Jalisco	Hall & Kelson (1959)
Peromyscus pectoralis (M)	*P. hylocetes*	Mexico City to Jalisco	Hall & Kelson (1959)
Baiomys taylori (M)	*B. musculus*	Veracruz to Jalisco	Hall & Kelson (1959)
Neotoma mexicana (M)	*N. alleni*	Puebla to SW Jalisco	Hall & Kelson (1959)

TABLE X. Northern New South Wales Suture-Zone

A. KNOWN HYBRIDIZING PAIRS

Southern (N.S.W.)	Northern (Queens.)	Known Areas	Xing	References
Tisiphone abeona aurelia (L)	*T. a. morrisi*	Port Macquarie area	I	Waterhouse (1914, 1922 a,b)
Platycercus eximius (A)	*P. adscitus*	N of Port Macquarie	S	Keast (1961)
Neositta c. chrysoptera (A)	*N. c. leucocephala*	Port Macquarie area	M?	Keast (1961)

B. SUSPECTED HYBRIDIZERS

Southern (N.S.W.)	Northern (Queens.)	Contact Areas	References
Hypocysta metirius (L)	*H. irius*	S Queens.	Waterhouse (1932)
Candalides simplexa (L)	*C. erinus*	S Queens.; N N.S.W.	Common (1964)
Hypochrysops i. ignita (L)	*H. i. chrysonotus*	S Queens.; N N.S.W.	Common (1964)
Delias harpalyce (L)	*D. nigrina*	C N.S.W.	Waterhouse (1932)
Hasora khoda haslia (L)	*H. chromus contempta*	S Queens.	Waterhouse (1932)
Emydura macquarii (R)	*E. latisternum*	N N.S.W.	Worrell (1963)
Oedura robusta (R)	*O. tryoni*	S Queens.; N N.S.W.	Worrell (1963)
Phyllurus platurus (R)	*P. cornutus*	S Queens.; N N.S.W.	Worrell (1963)
Lygosoma paraeneum (R)	*L. graciloides*	SE Queens.; NE N.S.W.	Worrell (1963)
Lygosoma reticulatum (R)	*L. punctulatum*	SE Queens.; NE N.S.W.	Worrell (1963)
[& both with *L. lentiginosus* & *verreauxii* ?]			

APPENDIX

Specimen Data For Illustrations

Fig. 4.   (a) *Papilio glaucus canadensis* (♂, Mile 15, Alaska Hwy., B.C., 16 June 1964, J. D. & J. R. Eff); (b) *P. g. glaucus* (♂, West Rock, New Haven, Conn., 17 June 1954, C. L. Remington and R. W. Pease).

Fig. 5.   (a) *L. ar. arthemis* (♂, Germania, Potter Co., Penna., 7 July 1950, R. P. Seibert); (b) wild hybrid *L. ar. astyanax × ar. arthemis* (♂, Germania, Potter Co., Penna., 5 July 1951, R. P. Seibert); (c) *Limenitis as. astyanax* (♂, Pittsburgh, Penna., 20 August 1935).

Fig. 6.   (a) *Hyalophora columbia* (♂, Dorchester, N.B., bred on *Larix,* eclosed 4 April 1953, D. C. Ferguson); (b) *H. cecropia* (♂, Greenwich, Fairfield Co., Conn., 14 June 1939, P. Starrett).

Fig. 7.   (a) *Vermivora chrysoptera* (♂, Thomaston, Litchfield Co., Conn., 7 June 1963, D. H. Parsons, YPM 78268); (b) wild hybrid *V. chrysoptera × pinus* (♂, Thomaston, Litchfield Co., Conn., 25 May 1962, D. H. Parsons, YPM 78270); (c) *V. pinus* (♂, Thomaston, Litchfield Co., Conn., 25 May 1962, D. H. Parsons, YPM 78248).

Fig. 8.   (a) *Populus tremuloides;* (b) *P. grandidentata* (both a and b, Burlington, Hartford Co., Conn., 11 June 1921, G. E. Nichols).

Fig. 9.   (a) *Danaus plexippus* (♂, Greenwich, Fairfield Co., Conn., 2 August 1940, P. Starrett); (b) *D. gilippus berenice* (♂, Port Charlotte, Charlotte Co., Fla., 19 June 1964); (c) *Limenitis a. archippus* (♂, 4 mi. SW of Norfolk, Litchfield Co., Conn., 3 July 1966, C. L. & S. T. C. Remington); (d) *L. a. floridensis* (♂, Archbold Biol. Sta., Highlands Co., Fla., bred, eclosed 18 January 1958, R. W. Pease & L. J. Brass); (e-h) introgressing types, *L. a. archippus × floridensis* (♂ ♂, Greenville, Madison Co., Fla., 2 July 1965 and 30 June 1966, T. R. Manley).

Fig. 10.   (a) *Papilio t. troilus* (♂, Woodbridge, New Haven Co., Conn., 5 May 1951, R. W. Pease); (b) *P. t. ilioneus* (♂, Welaka, Putnam Co., Fla., 13 March 1962, D. C. Ferguson).

Fig. 11.   (a) *Automeris i. io* (♂, Washington, Litchfield Co., Conn., 19 June 1966, S. A. Hessel); (b) wild hybrid *A. i. io × lilith* (♂, Coosawhatchie, Jasper Co., S.C., 9 May 1966, R. B. Dominick); (c) *A. i. lilith* (♂, Archbold Biol. Sta., Highlands Co., Fla., 16 February 1965, L. J. Brass).

Fig. 12.   (a) *Utetheisa bella* (♀, Achbold Biol. Sta., Highlands Co., Fla., bred, eclosed February 1958, R. W. Pease); (b) laboratory hybrid *U. bella × ornatrix* (♀, Archbold Biol. Sta., Highlands Co., Fla., bred, eclosed February 1958, R. W. Pease); (c) *U. ornatrix* (♀, Archbold Biol. Sta., Highlands Co., Fla., bred, eclosed 14 January 1958, R. W. Pease).

Fig. 13. (a) *Junonia coenia* (♂ , Mountain Lake, Polk Co., Fla., 22 December 1940, P. Starrett); (b) *J. evarete* (♂ , Flamingo, Dade Co., Fla., 18 January 1955, C. L. & J. E. Remington).

Fig. 14. (a) *Parus atricristatus* (♂ , Travis Co., Texas, 10 May 1905, E. Perry, YPM 2313); (b) *P. bicolor* (♀ , Wooster, Wayne Co., Ohio, 5 February 1891, YPM 5235).

Fig. 15. (a) *Limenitis astyanax arizonensis* (♂ , Huachuca Canyon, Cochise Co., Ariz., 30 April 1966, R. F. Sternitzky); (b) *L. a. astyanax* (♂ , Jackson, Hinds Co., Miss., 15 June 1956, B. Mather).

Fig. 16. (a) *Papilio multicaudatus* (♂ , Ramsey Canyon, Huachuca Mts., Cochise Co., Ariz., 3 July 1965, R. F. Sternitzky); (b) *P. glaucus* (♂ , West Rock, New Haven, Conn., 17 June 1954, C. L. Remington & R. W. Pease).

Fig. 17. (a) *Limenitis archippus obsoleta* (♂ , Charleston, Chiricahua Mts., Cochise Co., Ariz., 9 September 1966, R. F. Sternitzky); (b) *L. a. archippus* (♂ , St. Louis, Mo., 4 September 1933, F. R. Arnhold).

Fig. 19. (a) *Papilio rutulus* (♂ , Los Angeles, Calif., 9 June 1930), F. R. Arnhold); (b) laboratory hybrid ♂ F_1 of ♂ *P. rutulus* (Gunnison Co., Colo., S. A. Ae) × ♀ *P. glaucus* (Willimantic, Conn., bred H. Wilhelm), eclosed 9 May 1958, bred C. L. Remington & R. W. Pease; (c) *P. glaucus* (♂ , West Rock, New Haven, Conn., 17 June 1954, C. L. Remington & R. W. Pease).

Fig. 20. (a) *Hyalophora gloveri* (♀ , Idaho Springs, Gilpin Co., Colo., 6 July 1955, E. E. Remington); (b) laboratory hybrid ♂ F_1 of ♂ *H. gloveri* (Sheridan, Wyo., bred D. Downey) × ♀ *H. cecropia* (New England), eclosed 4 May 1958; (c) *H. cecropia* (♂ , Greenwich, Fairfield Co., Conn., 14 June 1939, P. Starrett). Photographed with a filter to bring out in black-gray-white the color contrasts between inner and outer ground color of the three genotypes; see Figure 28b for an unfiltered photo of the *H. gloveri* shown in Figure 20a.

Fig. 22. (a) *Papilio gothica* (♂ , Gothic, Gunnison Co., Colo., 27 June 1956, E. E. Remington); (b) laboratory hybrid ♂ F_1 of ♂ *P. gothica* (Cumberland Pass, Gunnison Co., Colo., C. L. Remington) × ♀ *P. polyxenes* (Illinois, E. Dluhy), eclosed 14 May 1956, bred C. L. Remington; (c) *P. polyxenes* (♂ , Greenwich, Fairfield Co., Conn., 13 September 1940, P. Starrett).

Fig. 23. (a) *Pheucticus melanocephalus* (♂ , Fort Robinson, Dawes Co., Nebr., 24 June 1953, R. V. Clem, YPM 7953); (b) *P. ludovicianus* (no data, YPM 3807).

Fig. 24. (a) *Limenitis arthemis rubrofasciata* (♂ , 4 mi. N. of Kiowa, Glacier Co., Mont., 14 July 1964, C. L. Remington); (b) wild hybrid *L. a. rubrofasciata* × *L. weidemeyerii* (♂ , West Butte, Sweetgrass Hills, Toole Co., Mont., 21 July 1964, E. E. Remington & R. C. Stein); (c) *L. weidemeyerii* (♂ , West Butte, Sweetgrass Hill, Toole Co., Mont., 21 July 1964, E. E. Remington & R. C. Stein).

Fig. 25. (a) *Betula papyrifera* (Cameron Falls, Waterton, Alta., 12 August 1961, J. R. Dugle, no. 1678, chromosomes 2n = 84); (b) wild hybrid (× *utahensis*) *B. papyriferia* × *fontinalis* (Thunder Hill cpgd., Columbia Lake, B.C., 29 July 1963, J. R. Dugle, no. 2329); (c) *B. fontinalis* (1 mi. past Blakiston Falls, Waterton, Alta., 11 Aug. 1961, J. R. Dugle, no. 1649, chromosomes 2n = 28); (d) wild hybrid (×

eastwoodae) *B. fontinalis* × *glandulosa* (entrance, Jasper N. P., Alta., 22 June 1962, J. R. Dugle, no. 1906); (e) *B. glandulosa* (Bog, 23 mi. SW of Edson, Alta., 21 June 1962, J. R. Dugle, no. 1881).

Fig. 27. (a) *Limenitis l. lorquini* (♂, Iceberg Meadows, Tuolumne Co., Calif., 22 July 1953, W. A. Hammer); (b) wild hybrid *L. lorquini* × *weidemeyerii* (♂, Mono Lake, Mono Co., Calif., 17 July 1938, C. W. Kirkwood); *L. w. weidemeyerii* (♂, Mono Lake, Mono Co., Calif., 14 July 1938, C. W. Kirkwood).

Fig. 28. (a) *Hyalophora euryalus* (♂, Santa Cruz, Santa Cruz Co., Calif., bred, eclosed 17 July 1959, R. F. Sternitzky); (b) *H. gloveri* (♀, Idaho Springs, Gilpin Co., Colo., 6 July 1955, E. E. Remington).

Fig. 29. (a) *Sphyrapicus ruber* (♂, Cabin Creek, 3300'; 33 mi. W. of Yakima, Wash., 9 May 1948, J. B. Hurley, YPM 8142); (b) *S. nuchalis* (♂, 4 mi. W. of Sisters, 3100'; Deschutes Co., Oreg., 23 April 1955, J. B. Hurley, YPM 8140).

Fig. 30. (a) *Limenitis arthemis rubrofasciata* (♂, Loon Lake, Cariboo, B.C., 28 June 1957, D. L. Bauer); (b) wild hybrid *L. a. rubrofasciata* × *L. lorquini* (♂, Loon Lake, Cariboo, B.C., 28 June 1957, D. L. Bauer); (c) *L. lorquini* (♂, Brewster, Okanagan Co., Wash., 22 May 1942, J. C. Hopfinger).

Fig. 31. (a) *Papilio glaucus canadensis* (♂, Mile 15, Alaska Hwy., B.C., 16 June 1964, J. D. & J. R. Eff); (b) *P. rutulus* (♂, Salmon Meadows, 4200', Okanagan Co., Wash., 14 July 1954, J. C. Hopfinger).

Fig. 32. (a) *Spilogale putorius latifrons* (♂, Port Angeles, Clallam Co., Wash., AMNH 15722); (b) *S. pu. phenax* (♂, Half Moon Bay, San Mateo Co., Calif., AMNH 144923); both from Van Gelder, 1959.

Fig. 33. (a) *Zerene eurydice* (♂, Soboba Hot Springs, Riverside Co., Calif., 28 May 1949, C. W. Kirkwood); (b) *Z. eurydice* (♀, Irvine, Orange Co., Calif., 17 Aug. 1929); (c) *Z. cesonia* (♂, Mingus Mt., Yavapai Co., Ariz., 18 July 1952, D. L. Bauer); (d) *Z. cesonia* (♀, Pena Blanca Lake, Sta. Cruz Co., Ariz., 27 March 1964, S. A. Hessel); (e-h) wild hybrids *Z. eurydice* × *cesonia* (e, f-♂ ♂; g, h-♀ ♀) (all Upper Santa Ana River, San Bernardino Mts., Calif., C. H. Ingham). The half-tone reproduction does not bring out the conspicuous ground-color differences in ♂ ♂ of *Z. eurydice* (hindwing organge, forewing with purplish iridescence) and *Z. cesonia* (hindwing and forewing yellow); the hybrid of Figure e has the color of *cesonia* and the markings of *eurydice,* and the hybrid ♂ of Figure f has the color of *eurydice* and the markings of *cesonia.*

Fig. 34. (all 1 mi. N. of New Harmony, Washington Co., Utah, 9 Aug. 1967, C. L. & J. E. Remington): (a) *Quercus gambelii;* (b) wild hybrid *Q. gambelii* × *turbinella;* (c) *Q. turbinella.*

Fig. 35. (a) *Limenitis weidemeyerii* (♂, Kaibab Plateau, Coconino Co., Ariz., 1 July 1952, D. L. Bauer); (b) wild hybrid *L. weidemeyerii* × *astyanax arizonensis* (♂, Oak Creek Canyon, 5600', Coconino Co., Ariz., 28 Aug. 1941, C. W. Medlar, *ex* F. T. Thorne); (c) *L. astyanax arizonensis* (♂, Huachuca Canyon, Cochise Co., Ariz., 30 April 1966, R. F. Sternitsky).

Fig. 36. (a) *Papilio brucei* (♂, Glenwood Springs, Garfield Co., Colo., 3 Aug. 1960, C. L. Remington); (b) wild hybrid *P. brucei* × *bairdii* (♂, Glenwood Springs, Garfield Co., Colo., 3 Aug. 1960, C. L. Remington); (c) *P. bairdii* (♂, Mingus Mt., Yavapai Co., Ariz., 15 July 1952, D. L. Bauer).

Fig. 37. (a) *Cnemidophorus tigris gracilis* (♀, 8.6 mi. W of Animas, Hidalgo Co., N.M., AMNH 80755); (b) *C. t. gracilis* (♂, 3.4 mi. N, o.5 mi. W of Cienega Ranch, Cochise Co., Ariz., H. Pough, AMNH 84911); (c) wild hybrid *C. t. gracilis* × *t. marmoratus* (♀, Crystal Mine, 8.2 mi. S of Road Forks, Hidalgo Co., N.M., 23 July 1958, R. G. Zweifel, AMNH 80737); (d) wild hybrid *C. t. gracilis* × *t. marmoratus* (♂, Crystal Mine, 8.2 mi. S of Road Forks, Hidalgo Co., N.M., 12 July 1960, R. G. Zweifel, AMNH 84881); (e) *C. t. marmoratus* (♀, 8 mi. NW of Lordsburg, Hidalgo Co., N.M., R. G. Zweifel, AMNH 84842); (f) *C. t. marmoratus* (♂, 10.5 mi. NE of Lordsburg, Hidalgo Co., N.M., AMNH 84869).

Fig. 38. (a) *Limenitis a. archippus* (♀, Louisville Reservoir, Boulder Co., Colo., 10 Aug. 1963, J. D. & J. R. Eff); (b) wild hybrid *L. a. archippus* × *watsoni* (♀, Red River W of Plain Dealing, Bossier Par., La., 28 Sept. 1962, R. O. & C. A. Kendall); (c) *L. a. watsoni* (♀, Iberville Par., La., 12 Oct. 1936, F. R. Arnhold).

Fig. 39. (a) *Erynnis funeralis* (♀, Carlsbad Caverns N. P., N.M., 26 July 1959, E. E. Remington); (b) *E. zarucco* (♀, Augusta, Richmond Co., Ga., bred, eclosed 14 April 1941, H. W. Eustis).

Fig. 40. (a) *Synchlora aerata* (♀, Highlands, 3865', Macon Co., N.C., 6 Aug. 1958, J. G. Franclemont); (b) wild hybrid *S. aerata* × *denticularia* (♂, Highlands, 3865', Macon Co., N.C., 12 Aug. 1958, J. G. Franclemont); (c) *S. denticularia* (♀, Archbold Biol. Sta., Highlands Co., Fla., 19 Dec. 1957, R. W. Pease).

Fig. 41. (a) *Magicicada septendecim* (♀, Hamden, New Haven Co., Conn., July 1945, S. C. Ball); (b) *M. tredecim* (♂, Orange Co., N.C., 8 May 1959, Barnes); (c) *M. cassinii* (♂, Aberdeen, Brown Co., Ohio, 31 May 1957, F. W. Mead); (d) *M. tredecassini* (♀, Butler Co., Mo., 25 May 1959, T. E. Moore & R. D. Alexander).

Fig. 42. (a) *Oeneis jutta* (♀, Mile 918, Alaska Hwy., Whitehorse, Yukon, 23 June 1964, J. D. & J. R. Eff); (b) *O. melissa* (♀, Eagle Summit, Alaska, 19 June 1966, K. W. Philip); (c) *Erebia discoidalis* (♂, Beulah, Man., 14 Aug. 1905, J. Mesele); (d) *E. rossii* (♀, Lake Rosabella, Churchill, Man., 4 July 1941, J. Struthers); (e) *E. disa* (♂, Mile 918, Alaska Hwy., Whitehorse, Yukon, 23 June 1964, J. D. & J. R. Eff); (f) *E. fasciata* (♂, Mile 45, Dempster Hwy., 3800', N fork of Klondike, Yukon, 27 June 1964, J. R. Eff); (g) *Colias pelidne* (♂, Nigel Pass, Banff-Jasper Hwy., Alta., 2 Aug. 1966, J. Scott & C. Curtis); (h) *C. hecla* (♂, Driftwood, Alaska, 5 Aug. 1952, P. F. Bellinger); (i) *C. nastes* (♂, Chandler Lake, Alaska, 19 July 1952, G. W. Rawson).

Fig. 43. (a) *Papilio polyxenes melasina* ♂; (b) *P. p. americus* ♂ (both Dagua, Valle del Cauca, Colombia).

Fig. 44. (a) *Campylorhynchus rufinucha nigricaudatus* (♂ ♂, southeastern Chiapas, Mexico); (b) introgressing types, *C. r. nigricaudatus* × *humilis* (♂ ♂, Rio Dulce and Rancho Ocuilapa, Ocuilapa Valley, southwestern Chiapas); (c) *C. r. humilis* (♂ ♂, Isthmus of Tehuantepec, Oaxaca, Mexico); all from Selander, 1964: plate 5.).

Fig. 45. (a) *Tisiphone abeona morrisi* (♀, Crescent Head, N. S. W. Australia, 11 Oct. 1966, R. S. L. Davis); (b) wild hybrid (*joanna*) *T. ab. morrisi* × *ab. aurelia* (♀, Port Macquarie, N.S.W., 10 Oct. 1966, R. S. L. Davis); (c) *T. ab. aurelia* (♀, Camden Haven, N. S. W., 3 Dec. 1966, R. S. L. Davis).

References

ALDRICH, J. W. 1943. Relationships of the Canada Jays in the Northwest. Wilson Bull., 55:217–222, 1 pl., 1 fig.

ALEXANDER, G. 1945. Natural hybrids between *Dendroica coronata* and *D. auduboni*. Auk, 62:623–626.

ALEXANDER, R. D., and T. E. MOORE. 1962. The evolutionary relationships of 17-year and 13-year cicadas, and three new species (Homoptera, Cicadidae, *Magicicada*). Univ. Michigan Mus. Zool., Misc. Publ. 121:59, 1 pl., 10 figs.

AMERICAN ORNITHOLOGISTS' UNION. 1957. Check-list of North American Birds, 5th ed. Baltimore, American Ornithologists' Union. 691 pp.

ANDERSON, E. 1948. Hybridization of the habitat. Evolution, 2:1–9, 1 fig.

———. 1949. Introgressive Hybridization. New York, Wiley. 109 pp., 23 figs.

———. 1953. Introgressive hybridization. Bot. Rev., 28:280–307.

ATWOOD, W. W. 1940. The Physiographic Provinces of North America. Boston, Ginn. 536 pp., map, 5 pls., 281 figs.

BAILEY, V. 1928. A hybrid Scaled × Gambel's Quail from New Mexico. Auk, 45:210.

BAKER, C. L. 1947. The species of *Amphiumae*. J. Tennessee Acad. Sci., 22:9–21, 8 figs.

BANKS, R. C., and N. K. JOHNSON. 1961. A review of North American hybrid hummingbirds. Condor, 63:3–28, 1 pl., 4 figs. 236–312 num. figs.

BARR, A. R., and P. R. EHRLICH. 1958. Mosquito records from the Chukchi sea coast of northwestern Alaska. Mosquito News, 18:12–14.

BARRETT-HAMILTON, G. E. H. 1912. A History of British Mammals, vol. 2:236–312, num. figs. London, Gurney and Jackson.

BARTRAM, W. 1791. Travels through North and South Carolina, Georgia, East and West Florida, the Cherokee Country, the extensive territories of the Muscogulges, or Creek Confederacy, and the Country of the Chacktaws. Philadelphia. 522 pp., 8 pls.

BAUER, D. L. 1954. An apparent hybrid *Limenitis* from Arizona. Lepid. News, 8:129–130.

BEILMANN, A. P., and L. G. BRENNER. 1951. The recent intrusion of forests in the Ozarks. Ann. Missouri Bot. Garden, 38:261–282.

BENT, A. C. 1940. Life histories of North American cuckoos, goatsuckers, hummingbirds and their allies. U.S. Nat. Mus. Bull., 176:506. 73 pls.

BERGER, A. J. 1958. The Golden-winged—Blue-winged Warbler complex in Michigan and the Great Lakes area. Jack-Pine Warbler, 36:37–73.

BESSEY, C. E. 1905. Plant migration studies. Univ. Nebraska Stud., 5:11–37. 67 figs.

BIRCH, L. C. 1961. Natural selection between two species of tephritid fruit fly of the genus *Dacus*. Evolution, 15:360–374, 14 figs.

———. 1965. Evolutionary opportunity for insects and mammals in Australia. *In* Baker, H. G., and G. L. Stebbins, eds., The Genetics of Colonizing Species. 197–211, 2 figs. New York, Academic Press.

BLAIR, A. P. 1941. Variation, isolating mechanisms, and hybridization in certain toads. Genetics, 26:398–417, 6 figs.

———. 1947. Variation of two characters in *Bufo fowleri* and *Bufo americanus*. Amer. Mus. Novitates, 1343:5.

——— and H. L. LINDSAY, JR. 1965. Color pattern variation and distribution of two large *Plethodon* salamanders endemic to the Ouachita Mountains of Oklahoma and Arkansas. Copeia, 1965:331–335, 1 fig.

BLAIR, W. F. 1951. Interbreeding of natural populations of vertebrates. Amer. Nat., 85:9–30.

————. 1956. Call difference as an isolation mechanism in southwestern toads (genus *Bufo*). Texas J. Sci., 8:87–106, 5 figs.

————. 1958. Distributional patterns of vertebrates in the southern United States in relation to past and present environments. Amer. Assoc. Adv. Sci. Publ., 51: 433–468, 11 figs.

————. 1959. Genetic compatibility and species groups in U.S. toads (*Bufo*). Texas J. Sci., 11:427–453, 5 figs.

————. 1965. Amphibian speciation. *In* Wright, H. E., Jr., and D. G. Frey, eds., The Quaternary of the United States, 543–556, 4 figs. Princeton, Princeton Univ. Press.

————, A. P. BLAIR, P. BRODKORB, F. R. CAGLE, and G. A. MOORE. 1957. Vertebrates of the United States. New York, McGraw-Hill. 819 pp., 426 figs.

————, and M. J. LITTLEJOHN. 1960. Stage of speciation of two allopatric populations of chorus frogs (*Pseudacris*). Evolution, 14:82–87.

BRAUN, E. L. 1950. Deciduous Forests of Eastern North America. Philadelphia, Blakiston. 596 pp. num-figs.

————. 1955. The phytogeography of unglaciated eastern United States and its interpretation. Bot. Rev., 21:297–375.

BRAYSHAW, T. C. 1965. The status of the Black Cottonwood (*Populus trichocarpa* Torrey and Gray). Canad. Field-nat., 79:91–95, 1 fig.

————. 1966. Native poplars of southern Alberta and their hybrids. Canada Dept. Forestry Publ., 1109:40, 21 figs., 1 pl.

BREWER, R. D. 1963. Ecological and reproductive relationship of Black-capped and Carolina Chickadees. Auk. 80:9–47, 2 figs.

BROECKER, W. S., and W. R. FARRAND. 1963. Radiocarbon age of the Two Creeks forest bed, Wisconsin. Bull. Geol. Soc. Amer., 74:795–802, 3 figs.

BROMLEY, S. W. 1935. The original forest types of southern New England. Ecol. Monogr., 5:61–89, 8 figs.

BROOKS, C. E. P. 1951. Geological and historical aspects of climatic change. *In* Malone, T. F., Compendium of Meteorology. 1004–1018, 4 figs. Boston, American Meteorological Society.

BROOKS, C. F., A. J. CONNOR, et al. 1936. Climatic Maps of North America. Cambridge, Blue Hill Meteorological Observatory Harvard Univ. 1 p., 27 pls.

BROWER, L. P. 1959. Speciation in butterflies of the *Papilio glaucus* group. I. Morphological relationships and hybridization. Evolution, 13:40–63, 8 figs.

BROWN, C. 1934. Notes on *Basilarchia lorquini* Bdv., form *fridayi* Gund. (Lepid.: Nymphalidae). Entomol. News, 45:205–206.

BURBIDGE, N. T. 1960. The phytogeography of the Australian region. Australian J. Bot., 8:75–211, 12 figs.

BURLEIGH, T. D. 1958. Georgia Birds. Norman, Univ. Oklahoma Press. 746 pp., 48 pls., 15 figs.

BURNS, J. M. 1964. Evolution in skipper butterflies of the genus *Erynnis*. Berkeley, Univ. California Publ. Entomol., 37:216 pp., 1 pl., 24 figs.

BUSH, G. L. 1966. The taxonomy, cytology, and evolution of the genus *Rhagoletis* in North America (Diptera, Tephritidae). Bull. Mus. Comp. Zool., 134:431–562, 237 figs., 14 maps.

BUTLER, P. 1959. Palynological studies of the Barnstable Marsh, Cape Cod, Massachusetts. Ecology, 40:735–737, 1 fig.

BYERS, C. F. 1940. A study of the dragonflies of the genus *Progomphus* (*Gomphoides*) with a description of a new species. Proc. Florida Acad. Sci., 4:19–85, 6 pls.

CARL, C. F., C. J. GUIGUET, and G. A. HARDY. 1952. A natural history survey of the Manning Park area, British Columbia. Occ. Papers British Columbia Prov. Mus., 9:130 pp., 22 figs.

CARR, A. F., JR., and C. J. GOIN. 1955. Guide to the Reptiles, Amphibians and Fresh-water Fishes of Florida. Gainesville, Univ. Florida Press. 341 pp., 67 pls., 30 figs.

CHAPMAN, F. M. 1924. Criteria for the determination of subspecies in systematic ornithology. Auk, 41:17–29.

CHERMOCK, R. L. 1950. A generic revision of the Limenitini of the world. Amer. Midl. Nat., 43:513–569, 68 figs.

CLENCH, H. K. 1958. [Review of] Brown, F. M., et al., Colorado Butterflies. Part III. Lepid. News, 11:57–60.

COHN, T. J. 1965. The arid-land katydids of the North American genus *Neobarrettia* (Orthoptera: Tettigoniidae): their systematics and a reconstruction of their history. Univ. Michigan Mus. Zool., Misc. Publ., 126:179 pp., 1 pl., 24 figs.

COMMON, I. F. B. 1964. Australian Butterflies. Brisbane, Jacaranda Press. 131 pp., 509 figs.

CONANT, R. 1943. The Milk Snakes of the Atlantic coastal plain. Proc. New England Zool. Club, 22:3–34, pls. 2–4, 1 fig.

––––––. 1965. Miscellaneous notes and comments on toads, lizards, and snakes from Mexico. Amer. Mus. Novitates, 2205:38, 13 figs.

COOKE, C. W. 1945. Geology of Flórida. Florida Geol. Surv. Geol. Bull., 29:339, 1 pl., 47 figs.

COOPER, W. 1810. A Guide to the Wilderness. Dublin, Gilbert and Hodges. 41 pp.

CORY, L., and J. J. MANION. 1955. Ecology and hybridization in the genus *Bufo* in the Michigan-Indiana region. Evolution, 9:42–51, 1 fig.

COTTAM, W. P., J. M. TUCKER, and R. DROBNICK. 1959. Some clues to Great Basin postpluvial climates provided by oak distributions. Ecology, 40:361–377, 6 figs.

COWAN, I. M. 1940. Distribution and variation in the native sheep of North America. Amer. Midl. Nat., 24:505–580, 4 figs.

––––––, and C. J. GUIGUET. 1956. The mammals of British Columbia. British Columbia Prov. Mus. Handb., 11:413, num. figs.

COX, D. D. 1959. Some postglacial forests in central and eastern New York state as determined by the method of pollen analysis. Bull. New York State Mus. Sci. Serv., 377:52, 19 figs.

––––––, and D. M. LEWIS. 1965. Pollen studies in the Crusoe Lake area of prehistoric Indian occupation. Bull. New York State Mus. Sci. Serv., 397:29, 7 figs.

CRENSHAW, J. W. 1965. Serum protein variation in an interspecies hybrid swarm of turtles of the genus *Pseudemys*. Evolution, 19:1–15, 4 figs.

CURTIS, J. T. 1959. The Vegetation of Wisconsin. Madison, Univ. Wisconsin Press. 629 pp., 66 pls., 88 figs.

CUSHING, E. J. 1965. Problems in Quaternary phytogeography of the Great Lakes region. *In* Wright, H. E., Jr., and D. G. Frey, eds., The Quaternary of the United States: 403–416, 3 figs. Princeton, Princeton Univ. Press.

DALQUEST, W. W. 1948. Mammals of Washington. Univ. Kansas Publ. Mus. Nat. Hist., 2:1–444, 140 figs.

DARLINGTON, H. C. 1943. Vegetation and substrate of Cranberry Glades, West Virginia. Bot. Gaz., 104:371–393, 17 figs.

DAVIS, J. H., JR. 1946. The peat deposits of Florida, their occurrence, development, and uses. Florida Geol. Surv., Geol. Bull., 30:247, 1 pl., 36 figs.

DAVIS, M. B. 1958. Three pollen diagrams from central Massachusetts. Amer. J. Sci., 256:540–570, 4 pls., 8 figs.

––––––. 1961. Pollen diagrams as evidence of late-glacial climatic change in southern New England. Ann. New York Acad. Sci., 95:623–631, 2 figs.

––––––. 1963. On the theory of pollen analysis. Amer. J. Sci., 261:897–912, 1 fig.

––––––. 1965. Phytogeography and palynology of northeastern United States. *In* Wright, H. E., Jr., and D. G. Frey, eds., The Quaternary of the United States: 377–401, 4 figs. Princeton, Princeton Univ. Press.

––––––, and J. C. GOODLETT. 1960. Comparison of the present vegetation with pollen-spectra in surface samples from Brownington Pond, Vermont. Ecology, 41:346–357, 6 figs.

DAVIS, R. J. 1952. Flora of Idaho. Dubuque, W. C. Brown. 828 pp.

Davis, W. B. 1944. Notes on Mexican mammals. J. Mamm. 24:370–403, 1 fig.
———. 1960. The mammals of Texas. Texas Game Fish Comm. Bull., 41:252, 73 figs.
Day, G. M. 1953. The Indian as an ecological factor in the northeast forest. Ecology, 34:329–346.
Deevey, E. S., Jr. 1939. Studies on Connecticut lake sediments. I. A. postglacial chronology for southern New England. Amer. J. Sci., 237:691–724, 11 figs.
———. 1943. Additional pollen analyses from southern New England. Amer. J. Sci., 241:717–752, 16 figs.
———. 1949. Biogeography of the Pleistocene. Bull. Geol. Soc. Amer., 60:1315–1416, 27 figs.
Denisov, V. P. 1961. [Relationship of *Citellus pygmaeus* Pall. and *C. suslica* Guld. on the junction of their ranges.] Zool. Zhurn., 40:1079–1085, 7 figs.
Dessauer, H. C., W. Fox, and F. H. Pough. 1962. Starch-gel electrophoresis of tranferrins, esterases and other plasma proteins of hybrids between two subspecies of whiptail lizard (genus *Cnemidophorus*). Copeia, 1962: 767–774, 3 figs.
Dice, L. R. 1943. The Biotic Provinces of North America. Ann Arbor, Univ. Michigan Press. 78 pp., 1 pl.
———. 1952. Natural Communities. Ann Arbor, Univ. Michigan Press. 547 pp., 51 figs.
Dickinson, J. C., Jr. 1952. Geographic variation in the Red-eyed Towhee of the eastern United States. Bull. Mus. Comp. Zool., 107:271–352, 15 figs.
Dietz, R. A. 1952. Variation in the perfoliate Uvularias. Ann. Missouri Bot. Garden, 39:219–247, 26 figs.
Dightman, R. A., and M. E. Beatty. 1952. Recent Montana glacier and climate trends. Monthly Weather Rev., 80:77–81, 1 fig.
Dixon, K. L. 1955. Crested titmice in Texas. Univ. California Publ. Zool., 54:125–206, 3 pls., 15 figs.
Dobzhansky, T., and P. C. Koller. 1938. An experimental study of sexual isolation in *Drosophila*. Biol. Zentralblatt, 58:589–607, 3 figs.
Dorf, E. 1959. Climatic changes of the past and present. Contr. Mus. Paleont. Univ. Michigan, 13:181–210, 1 pl., 3 figs.
Dugle, J. R. 1966. A taxonomic study of western Canadian species in the genus *Betula*. Canad. J. Bot., 44:929–1007, 45 figs.
Durrant, S. D. 1959. The nature of mammalian species. J. Arizona Acad. Sci., 1:18–21.
Ehrlich, P. R. 1958a. Lepidoptera collected in the tundra-taiga ecotone at Kotzebue, Alaska. Entomol. News, 69:17–20.
———. 1958b. Problems of Arctic-Alpine insect distribution as illustrated by the butterfly genus *Erebia* (Satyridae). Proc. X. Int. Congr. Entomol., 1:683–686, 8 figs.
Englehardt, G. P. 1946. The North American clear-wing moths of the family Aegeriidae. U.S. Nat. Mus. Bull., 190:222, 32 pls.
Evans, K. 1966. Observations on a hybrid between the Sharp-tailed Grouse and the Greater Prairie Chicken. Auk, 83:128–129, 1 fig.
Fassett, N. C. 1944. *Juniperus virginiana, J. horizontalis* and *J. scopulorum*. II. Hybrid swarms of *J. virginiana* and *J. scopulorum*. Bull. Torrey Bot. Club, 71:475–483, 6 figs.
———. 1945a. idem—III. Possible hybridization of *J. horizontalis* and *J. scopulorum*. Bull. Torrey Bot. Club, 72:42–46, 4 figs.
———. 1945b. idem—IV. Hybrid swarms of *J. virginiana* and *J. horizontalis*. Bull. Torrey Bot. Club, 72:379–384.
———, and J. D. Sauer. 1950. Studies of variation in the weed genus *Phytolacca*. I. Hybridizing species in northeastern Colombia. Evolution, 4:332–339, 4 figs.
Faxon, W. 1913. Brewster's Warbler (*Helminthophila leucobronchialis*) a hybrid between the Golden-winged Warbler (*Helminthophila chysoptera*) and the Blue-winged Warbler (*Helminthophila pinus*). Mem. Mus. Comp. Zool., 40:311–316.

FENNEMAN, N. M. 1931. Physiography of the Western United States. New York, McGraw-Hill. 534 pp., 173 figs.

————. 1938. Physiography of the Eastern United States. New York, McGraw-Hill. 714 pp., 5 pls., 197 figs.

FERGUSON, D. C. 1950. Collecting a little-known *Papilio*. Lepid. News, 4:11–12.

————. A revision of the moths of the subfamily Geometrinae occurring in America north of Mexico. Bull. Peabody Mus. Nat. Hist. In press.

FIELD, W. D. 1940. A manual of the butterflies and skippers of Kansas (Lepidoptera, Rhopalocera). Bull. Univ. Kansas, 39(10):327, 2 pls.

FINDLEY, J. S. 1955. Speciation of the Wandering Shrew. Univ. Kansas Mus. Nat. Hist. Publ., 9:1–68, 18 figs.

FORBES, W. T. M. 1928. Variation in *Junonia lavinia* (Lep., Nymphalidae). Journ. New York Entomol. Soc., 36:306.

————. 1960. Lepidoptera of New York and neighboring states. Part IV. Cornell Univ. Agr. Exper. Sta. Mem., 371:188, 188 figs.

FREEBORN, S. B. 1923. The range overlapping of *Anopheles maculipennis* Meig. and *Anopheles quadrimaculatus* Say. Bull. Brooklyn Ent. Soc., 18:157–158.

FREEMAN, T. N. 1951. Northern Canada and some northern butterflies. Lepid. News, 5:41–42, 1 fig.

————. 1958. The distribution of arctic and subarctic butterflies. Proc. X. Int. Congr. Ent., 1:659–671, 27 figs.

FRIES, M. 1962. Pollen profiles of late Pleistocene and Recent sediments from Weber Lake, northeastern Minnesota. Ecology, 43:295–308, 10 figs.

FRYE, J. C., H. B. WILLMAN, and R. F. BLACK. 1965. *In* Wright, H. E., Jr., and D. G. Frey, eds., The Quaternary of the United States: 43–61, 7 figs. Princeton, Princeton Univ. Press.

GATES, F. C. 1940. Annotated List of the Plants of Kansas: Ferns and Flowering Plants. Manhattan, Kansas, Agric. Exper. Sta. 266 pp., 80 pls.

GILL, F. B., and W. E. LANYON. 1964. Experiments on species discrimination in Blue-winged Warblers. Auk, 81:53–64, 2 figs.

GLASS, B. P. 1947. Geographic variation in *Perognathus hispidus*. J. Mamm., 28:174–179, 1 fig.

GLEASON, H. A., and A. CRONQUIST. 1964. The Natural Geography of Plants. New York, Columbia Univ. Press. 420 pp. num. pls.

GOGGIN, J. M. 1948. Florida archeology and recent ecological changes. J. Washington Acad. Sci., 38:225–233, 1 fig.

GRAHAM, A., and C. HEIMSCH. 1960. Pollen studies of some Texas peat deposits. Ecology, 41:751–763, 5 figs.

GREY, L. P., A. H. MOECK, and W. H. EVANS. 1963. Notes on overlapping subspecies. II. Segregation in the *Speyeria atlantis* of the Black Hills (Nymphalidae). J. Lepid. Soc., 17:129–147, 4 figs.

GRIGGS, R. F. 1914. Observations of the behavior of some species at the edges of their ranges. Bull. Torrey Bot. Club, 41:25–49, 6 figs.

GRINNELL, J., and H. S. SWARTH. 1913. An account of the birds and mammals of the San Jacinto area of southern California with remarks on the behavior of geographic races on the margins of their habitats. Univ. California Publ. Zool., 10:197–406, pls. 6–10, 3 figs.

GUNTER, G. 1945. The northern range of Berlandier's Tortoise. Copeia, 1945:175.

HAGEN, D. W. 1967. Isolating mechanisms in Threespine Sticklebacks (*Gasterosteys*) J. Fisheries Res. Board Canada, 24:1637–1692, 18 figs.

HALL, E. R. 1941. Revisions of the rodent genus *Microdipodops*. Field Mus. Nat. Hist., Zool. Ser., 27:233–277, 8 figs.

————, and K. R. KELSON. 1959. The Mammals of North America. New York, Ronald Press. 1201 pp., 553 figs.

HALL, M. T. 1952. Variation and hybridization in *Juniperus*. Ann. Missouri Bot. Garden, 39:1–64, 1 pl., 18 figs.

————. 1961. Notes on cultivated junipers. Butler Univ. Bot. Studies, 14(1):73–90.

————, and C. J. CARR. 1964. Differential selection in juniper populations from the Baum Limestone and Trinity Sand of southern Oklahoma. Butler Univ. Bot. Studies, 14(2):21–40.

————, J. F. McCORMICK, and G. G. FOGG. 1961. Hybridization between *Juniperus ashei* Buchholz and *Juniperus pinchoti* Sudworth in southwestern Texas. Butler Univ. Bot. Studies, 14(1):9–28, 7 figs.

HANSEN, H. P. 1955. Postglacial forests in south central and central British Columbia. Amer. J. Sci., 253:640–658, 8 figs.

HANSON, J. F. 1960. A case for hybridization in Plecoptera. Bull. Brooklyn ent. Soc., 55:25–34.

HECHT, M. K., and B. L. MATALAS. 1946. A review of middle North American toads of the genus *Microhyla*. Amer. Mus. Nov., 1315:21, 12 figs.

HENSHAW, H. W. 1885. Hybrid quail (*Lophortyx gambeli* × *L. californicus*). Auk, 2:247–249.

HEUSSER, C. J. 1956. Postglacial environments in the Canadian Rocky Mountains. Ecol. Monogr., 26:263–302, 14 figs.

————. 1960. Late-Pleistocene environments of north Pacific North America. Amer. Geogr. Soc. Spec. Publ., 35:308, 25 pls., 49 figs.

————. 1963. Pollen diagrams from three former cedar bogs in the Hackensack tidal marsh, northeastern New Jersey. Bull. Torrey Bot. Club, 90:16–28, 2 figs.

HILL, I. R. 1954. The taxonomic status of the mid-Gulf Coast *Amphiuma*. Tulane Studies Zool., 1:189–215, 11 figs.

HOLLAND, G. P. 1958. Distribution patterns of northern fleas (Siphonaptera). Proc. X. Int. Congr. Entomol., 1:645–658, 6 figs.

————. 1963. Faunal affinities of the fleas (Siphonaptera) of Alaska with an annotated list of species. *In* Gressitt, J. L., ed., Pacific Basin Biogeography: 45–63, 9 figs. Honolulu, Bishop Mus. Press.

HOVANITZ, W. 1944. Genetic data on the two races of *Colias chrysotheme* in North America and on a white form occurring in each. Genetics, 29:1–30.

HOWARD, W. E. 1965. Interaction of behavior, ecology, and genetics of introduced mammals. *In* Baker, H. G., and G. L. Stebbins, eds., The Genetics of Colonizing Species: 461–480. New York, Academic Press.

HOWDEN, H. F. 1966. Some possible effects of the Pleistocene on the distributions of North American Scarabeidae (Coleoptera). Can. Entomol. 98:1177–1190, 26 figs.

HOWELL, T. R. 1952. Natural history and differentiation in the Yellow-bellied Sapsucker. Condor, 54:237–282, 7 figs.

HUBBARD, J. P. 1966. A possible back-cross hybrid involving Scaled and Gambel's Quail. Auk, 83:136–137, 1 fig.

HUBBS, C. L. 1955. Hybridization between fish species in nature. Syst. Zool., 4:1–20, 8 figs.

HUNTINGTON, C. E. 1952. Hybridization in the Purple Grackle, *Quiscalus quiscula*. Syst. Zool., 1:149–170, 6 figs.

JEWETT, S. G. 1944. Hybridization of Hermit and Townsend Warblers. Condor, 46:23–24.

JOHANSEN, H. 1955. Die Jenissei-Faunenscheide. Zool. Jahrb., 83:237–247, 1 fig.

JOHNSGARD, P. A. 1960. A quantitative study of sexual behavior of Mallards and Black Ducks. Wilson Bull., 72:133–155, 6 figs.

————. 1965. Handbook of Waterfowl Behavior. Ithaca, Cornell Univ. Press. 378 pp., 20 pls., 96 figs.

JOHNSTON, R. F. 1964. The breeding birds of Kansas. Univ. Kansas Mus. Nat. Hist. Publ., 12:575–655, 10 figs.

KARREN, J. B. 1966. A revision of the genus *Exema* of America, north of Mexico (Chrysomelidae, Coleoptera). Univ. Kansas Sci. Bull., 46:647–695, 65 figs.

KASTON, B. J. 1938. Notes on a new variety of Black Widow Spider from southern Florida. Florida Entomol., 21:60–61.

KEAST, A. 1961. Bird speciation on the Australian continent. Bull. Mus. Comp. Zool., 123:305–495, 28 figs.

KEY, K. H. L. 1959. The ecology and biogeography of Australian grasshoppers and locusts. *In* Keast, A. R. L. Crocker, and C. S. Christian, Biogeography and Ecology in Australia: 192–210, Den Haag, Junk.

KIMBALL, C. P. 1965. The Lepidoptera of Florida an Annotated Checklist. Gainesville, Florida Div. Plant Ind. 363 pp., 26 pls.

KLAUBER, L. M. 1941. The Long-nosed Snakes of the genus *Rhinocheilus*. San Diego Soc. Nat. Hist. Trans., 9:289–332, 2 pls.

———. 1956. Rattlesnakes, vol. I. Berkeley & Los Angeles, Univ. California Press. 708 pp., 1 pl., 185 figs.

KLOTS, A. B. 1951. A Field Guide to the Butterflies of North America, East of the Great Plains. Boston, Houghton Mifflin. 349 pp., 40 pls., 8 figs.

KRUTSCH, P. H. 1954. North American jumping mice (genus *Zapus*). Univ. Kansas Publ. Mus. Nat. Hist., 7:349–472, 47 figs.

KÜCHLER, A. W. 1964. Potential natural vegetation of the conterminous United States. Amer. Geog. Soc. Spec. Publ., 36:156 pp., map, 116 figs.

LACK, D. 1945. The ecology of closely related species with special reference to Cormorant (*Phalacrocorax carbo*) and Shag (*P. aristotelis*). J. Anim. Ecol., 14:12–16.

LANYON, W. E. 1966. Hybridization in meadowlarks. Bull. Amer. Mus. Nat. Hist., 134:1–26, 8 pls., 3 figs.

LIDICKER, W. Z., Jr. 1960. An analysis of intraspecific variation in the kangaroo rat *Dipodomys merriami*. Univ. California Publ. Zool., 67:125–218, pls. 9–12, 20 figs.

LINSDALE, J. M. 1936. The birds of Nevada. Pacific Coast Avifauna, 23:145 pp., 1 fig.

LIVINGSTON, B. E., and F. SHREVE. 1921. The distribution of vegetation in the United States, as related to climatic conditions. Carnegie Inst. Washington Publ., 284:590 pp., 73 pls., 74 figs.

LIVINGSTON, R. B. 1949. An ecological study of the Black Forest, Colorado. Ecol. Monogr., 19:123–144, 16 figs.

LIVINGSTONE, D. A., and B. G. R. LIVINGSTONE. 1958. Late-glacial and postglacial vegetation from Gillis Lake in Richmond County, Cape Breton Island, Nova Scotia. Amer. J. Sci., 256:341–359, 1 fig.

LÖVE, D. 1959. The postglacial development of the flora of Manitoba. Canad. J. Bot., 37:547–585, 4 figs.

LORKOVIĆ, Z. 1958. Some peculiarities of spatially and sexually restricted gene exchange in the *Erebia tyndarus* group. Symp. Quant. Biol. (Cold Spring Harbor), 23:319–325, 4 figs.

LOUKASHKIN, A. S. 1943. On the hares of northern Manchuria. J. Mamm., 24:73–81, 1 pl., 2 figs.

LOWE, C. H., JR. 1955. The eastern limit of the Sonoran Desert in the United States with additions to the known herpetofauna of New Mexico. Ecology, 36:343–345.

LUTZ, H. J. 1930. The vegetation of Heart's Content, a virgin forest in northwestern Pennsylvania. Ecology, 11:1–29, 10 figs.

MACNEIL, F. S. 1950. Pleistocene shore lines in Florida and Georgia. U.S. Geol. Surv. Prof. Paper, 221–F:95–107, pls. 19–25.

MACNEILL, C. D. 1964. The skippers of the genus *Hesperia* in western North America with special reference to California. Univ. California Publ. Ent., 35:221 pp., 8 pls., 28 figs.

MAEKI, K., and C. L. REMINGTON. 1960. Studies of the chromosomes of North American Rhopalocera. 2. Hesperiidae, Megathymidae, and Pieridae. J. Lepid. Soc., 14:37–57, 7 pls.

MALDE, H. E. 1964. Environment and man in arid America. Science, 145:123–129, 1 fig.

MARTIN, P. S. 1963. Early man in Arizona: the pollen evidence. Amer. Antiquity, 29:67–73, 1 fig.

———, and J. GRAY. 1962. Pollen analysis and the Cenozoic. Science, 137:103–111, 5 figs.

———, and P. J. MEHRINGER, JR. 1965. Pleistocene pollen analysis and biogeography of the Southwest. *In* Wright, H. E., Jr., and D. G. Frey, eds., The Quaternary of the United States: 433–451, 6 figs. Princeton, Princeton Univ. Press.

MAYR, E. 1963. Animal Species and Evolution. Cambridge, Mass., Harvard Univ. Press. 797 pp., num. figs.

———, and T. GILLIARD. 1952. Altitudinal hybridization in New Guinea honey-eaters. Condor, 54:325–337.

McANDREWS, J. H. 1966. Postglacial history of prairie, savanna, and forest in northwestern Minnesota. Mem. Torrey Bot. Club, 22(2):72 pp., 6 pls.

McCABE, R. A. 1954. Hybridization between the Bob-white and Scaled Quail. Auk, 71:293–297, 1 pl., 1 fig.

McCARLEY, W. H. 1954. Natural hybridization in the *Peromyscus leucopus* species group of mice. Evolution, 8:314–323, 3 figs.

McCULLOCH, W. F. 1939. A postglacial forest in central New York. Ecology, 20:264–271, 2 figs.

McILROY, D. W. 1961. Possible hybridization between a Clay colored Sparrow and a Chipping Sparrow at Ithaca. Kingbird, 11:7–10.

McVAUGH, R. 1951. A revision of the North American black cherries (*Prunus serotina* Ehr., and relatives). Brittonia, 7:279–315.

———. 1952. Suggested phylogeny of *Prunus serotina* and other wide-ranging phylads in North America. Brittonia, 7:317–346, 29 maps.

MECHAM, J. S. 1960. Introgressive hybridization between two southeastern tree-frogs. Evolution, 14:445–457, 3 figs.

———. 1961. Isolating mechanisms in anuran speciation, *In* Blair, W. F., ed., Vertebrate Speciation: 23–61, 2 figs. Austin, Univ. Texas Press.

MENGEL, R. M. 1964. The probable history of species formation in some northern wood warblers (Parulidae). Living Bird, 1964:9–43, 8 figs.

MERGEN, F. 1958. Genetic variation in needle characteristics of Slash Pine and in some of its hybrids. Silvae Genetica, 7:1–9, 12 figs.

MILLER, A. H. 1941. Speciation in the avian genus *Junco.* Univ. California Publ. Zool., 44:173–434, 33 figs.

———. 1943. A new race of Canada Jay from coastal British Columbia. Condor, 45:117–118.

———. 1951. An analysis of the distribution of the birds of California. Univ. California Publ. Zool., 50:531–644, pl. 30–40, 5 figs.

———. 1955. Concepts and problems of avian systematics in relation to evolutionary process. *In* Wolfson, A., ed., Recent Studies in Avian Biology: 1–22. Urbana, Univ. Illinois Press.

MILLER, C. D. F. 1961. Taxonomy and distribution of Nearctic *Vespula.* Canad. Entomol. Suppl., 22: 52 pp., 99 figs.

MILLER, D. D. 1958a. Geographical distributions of the American *Drosophila affinis* subgroup species. Amer. Midl. Nat., 60:52–70, 1 fig.

———. 1958b. Sexual isolation and variation in mating behavior within *Drosophila athabasca.* Evolution, 12:72–81.

MITTLEMAN, M. B. 1942. A new Long-tailed Eurycea from Indiana, and notes on the *longicauda* complex. Proc. New England Zool. Club, 21:101–105, pl. 20.

MOORE, J. C. 1956. Variation in the Fox Squirrel in Florida. Amer. Midl. Nat., 55:41–65, 7 figs.

MOREAU, R. E. 1963. Vicissitudes of the African biomes in the late Pleistocene. Proc. Zool. Soc. London, 141:395–421, 4 figs.

MOSS, E. H. 1955. The vegetation of Alberta. Bot. Rev., 21:493–567.

———. 1959. Flora of Alberta. Toronto, Univ. Toronto Press. 546 pp.

MULLER, C. H. 1952. Ecological control of hybridization in *Quercus:* a factor in the mechanism of evolution. Evolution, 6:147–161, 3 figs.

MULLIGAN, G. A. 1965. Recent colonization by herbaceous plants in Canada. *In* Baker, H. G., and G. L. Stebbins, eds., The Genetics of Colonizing Species: 127–143, 6 figs. New York, Academic Press.

MUNZ, P. A., and D. D. Keck. 1959. A California Flora. Berkeley and Los Angeles, Univ. California Press. 1687 pp., 134 figs.

NEEDHAM, J. G., and M. F. WESTFALL, JR. 1955. A Manual of the Dragonflies of North America (Anisoptera). Berkeley and Los Angeles, Univ. California Press. 615 pp., 314 figs., 1 pl.

NEILL, W. T. 1949a. The distribution of Milk Snakes in Georgia. Herpetologica, 5:8.

———. 1949b. Hybrid toads in Georgia. Herpetologica, 5:30–32.

———. 1957. Historical biogeography of present-day Florida. Bull. Florida State Mus. Biol. Sci., 2:175–220.

NELSON, T. C. 1957. The original forests of the Georgia Piedmont. Ecology, 38:390–397.

NEUMÖGEN, B. 1894. Notes on a remarkable "interfaunal" hybrid of *Smerinthus.* Entomol. News, 5:326–327.

NICHOLS, G. E. 1913. The vegetation of Connecticut. II. Virgin forests. Torreya, 13:199–215, 5 figs.

NIERING, W. A. 1953. The past and present vegetation of High Point State Park, New Jersey. Ecol. Monogr., 23:127–148, 23 figs.

OGDEN, J. G., III. 1961. Forest history of Martha's Vineyard, Massachusetts. I. Modern and pre-colonial forests. Amer. Midl. Nat., 66:417–430, 3 figs.

———. 1965. Pleistocene pollen records from eastern North America. Bot. Rev., 31:481–504.

OGNEV, S. I. 1931 [English transl. 1962]. Zveri vostochnoi Evropy i severnoi Azii [Mammals of eastern Europe and northern Asia], vol. 2. Moscow and Leningrad, Glavnauka-Gosudarstvennoe Izdatel'stvo. 590 pp., 202 figs.

PARAMONOV, S. J. 1959. Zoogeographical aspects of the Australian Dipterofauna. *In* Keast, A., R. L. Crocker, and C. S. Christian, Biogeography and Ecology in Australia, pp. 164–191, 1 fig. The Hague, Junk.

PARKES, K. C. 1951. The genetics of the Golden-winged × Blue-winged Warbler complex. Wilson Bull., 63:5–15, 1 pl.

PATTERSON, J. T., and W. S. STONE. 1952. Evolution in the Genus *Drosophila.* New York, Macmillan. 610 pp., 74 figs.

PEABODY, F. E., and J. M. SAVAGE. 1958. Evolution of a Coast Range corridor in California and its effect on the origin and dispersal of living amphibians and reptiles. Amer. Assoc. Adv. Sci. Publ., 51:159–186, 19 figs.

PEASE, R. W., JR. 1964a. Chromosome numbers in geographic populations of the *Utetheisa ornatrix* complex. J. Lepid. Soc., 17:231–233, 1 fig.

———. 1946b. Evolution and hybridization in the *Utetheisa ornatrix* L. complex. Yale Univ. Ph.D. thesis.

PECK, M. E. 1961. A Manual of the Higher Plants of Oregon, 2nd ed. Portland, Binfords and Minot. 936 pp., 98 figs.

PENNAK, R. W. 1963. Ecological and radiocarbon correlations in some Colorado mountain lake and bog deposits. Ecology, 44:1–15, 10 figs.

PETERSON, R. T. 1941. A Field Guide to Western Birds. Boston, Houghton Mifflin Co. 240 pp., num. pls. figs.

———. 1963. A Field Guide to the Birds of Texas and Adjacent States. Boston, Houghton Mifflin Co. 304 pp., 60 pls.

PHILLIPS, A. R., J. T. MARSHALL, JR., and G. MONSON. 1964. The Birds of Arizona. Tucson, Univ. Arizona Press. 220 pp., 45 pls.

PHILLIPS, J. C. 1923. A Natural History of the Ducks, Volume II. Boston, Houghton Mifflin Co. 409 pp., 26 pls.

PITELKA, F. A. 1941. Distribution of birds in relation to major biotic communities. Amer. Midl. Nat., 25:113–137, 11 figs.

————. 1951. Speciation and ecologic distribution in American jays of the genus *Aphelocoma*. Univ. California Publ. Zool., 50:195–464, pls. 17–30, 21 figs.

————, R. K. SELANDER, and M. ALVAREZ DEL TORO. 1956. A hybrid jay from Chiapas, Mexico. Condor, 58:98–106, 5 figs.

POTTER, L. D. 1947. Postglacial forest sequence of north-central Ohio. Ecology, 28:396–417, 17 figs.

————, and J. ROWLEY. 1960. Pollen rain and vegetation, San Augustin Plains, New Mexico. Bot. Gaz., 122:1–25, 3 figs.

POTZGER, J. E. 1946. Phytosociology of the primeval forest in central-northern Wisconsin and Upper Michigan, and a brief postglacial history of the Lake Forest formation. Ecol. Monogr., 16:211–250, 31 figs.

————. 1948. A pollen study in the tension zone of Lower Michigan. Butler Univ. Bot. Studies, 8:161–177, 3 figs.

————, and A. COURTEMANCHE. 1956. Pollen study in the Gatineau Valley, Quebec. Butler Univ. Bot. Studies, 13:12–23, 4 figs.

————, and R. C. FREISNER. 1948. Forests of the past along the coast of southern Maine. Butler Univ. Bot. Studies, 8:178–203, 3 figs.

————, M. E. POTZGER, and J. McCORMICK. 1956. The forest primeval of Indiana as recorded in the original U.S. Land Surveys and an evaluation of previous interpretations of Indiana vegetation. Butler Univ. Bot. Studies, 13:95–111, 9 figs.

POWELL, J. M. 1965. Changes in the amounts of sunshine in British Columbia, 1901–1960. Quart. J. Roy. Meteor. Soc., 91:95–98, 2 figs.

PRESTON, R. J. 1961. North American Trees, 2nd ed. Ames, Iowa State Univ. Press. 395 pp., 160 pls., 5 figs.

PURI, H. S., and R. O. VERNON. 1964. Summary of the geology of Florida and a guide book to the classic exposures, rev. ed. Florida Geol. Surv. Spec. Publ., 5:312, 37 figs.

QUARTERMAN, E., and C. KEEVER. 1962. Southern mixed hardwood forest: climax in the southeastern Coastal Plain, U.S.A. Ecol. Monogr., 32:167–185, 2 figs.

RAND, A. L. 1948. Glaciation, an isolating factor in speciation. Evolution, 2:314–321, 5 figs.

RAUP, H. M. 1937. Recent changes of climate and vegetation in southern New England and adjacent New York. J. Arnold Arb., 18:79–117.

————. 1946. Phytogeographic studies in the Athabasca—Great Slave Lake region, II. J. Arnold Arb., 27:1–85, 5 pls., 6 figs.

————, and R. E. CARLSON. 1941. The history of land use in the Harvard Forest. Harvard For. Bull., 20:64 pp.

REINTHAL, W. J. 1956. In search for *Pieris virginiensis* in Massachusetts. Lepid. News, 10:25–28.

REMINGTON, C. L. 1956. Interspecific relationships of two rare swallowtail butterflies, *Papilio nitra* and *Papilio hudsonianus,* to other members of the *Papilio machaon* complex. Yearbook Amer. Phil. Soc., 1955:142–146.

————. 1958. Genetics of populations of Lepidoptera. Proc. X. int. Congress Entomol. 2:787–805, 13 figs.

REMINGTON, J. E., and C. L. REMINGTON. 1957. Mimicry, a test of evolutionary theory. Yale Sci. Mag., 32(1):10–11, 13–14, 16–17, 19, 21, 4 figs.

RICHARDS, H. G. 1965. Pleistocene stratigraphy of the Altantic Coastal Plain. *In* Wright, H. E., Jr., and D. G. Frey, eds., The Quaternary of the United States: 129–133, 1 fig. Princeton, Princeton Univ. Press.

RICHMOND, G. M. 1954. Modification of the glacial chronology of the San Juan Mountains, Colorado. Science, 119:614–615.

————. 1965. Glaciation of the Rocky Mountains. *In* Wright, H. E., Jr., and D. G. Frey, eds., The Quaternary of the United States: 217–230, 3 figs. Princeton, Princeton Univ. Press.

RIDDELL, J. 1941. Some remarkable forms and aberrations in the subgenus *Zerene* Hübner (Lepidoptera, Pieridae). Trans. R. Entomol. Soc. London, 91:447–457, pls. 1–3.

RILEY, H. P. 1938. A character analysis of colonies of *Iris fulva, I. hexagona* var. *giganticaerulea* and natural hybrids. Amer. J. Bot., 25:727–738, 3 figs.

RITCHIE, J. C. 1964. Contributions to the Holocene paleoecology of westcentral Canada. I. The Riding Mountain area. Canad. J. Bot., 42:181–196, 11 figs.

ROGERS, J. S. 1933. The ecological distribution of the craneflies of northern Florida. Ecol. Monogr., 3:1–74, 25 figs.

ROSS, H. H. 1965. Pleistocene events and insects. *In* Wright, H. E., Jr., and D. G. Frey, eds., The Quaternary of the United States: 583–596, 5 figs., Princeton, Princeton Univ. Press.

ROSSMAN, D. A. 1963. The colubrid snake genus *Thamnophis:* a revision of the *sauritus* group. Bull. Florida State Mus., 7:99–178, 10 figs.

ROUSE, I. 1951. A survey of Indian River archeology. Yale Univ. Publ. Anthrop., 44:296 pp., 8 pls., 15 figs.

ROWAN, W. 1926. Comments on two hybrid grouse and on the occurrence of *Tympanuchus americanus americanus* in the province of Alberta. Auk, 43:333–336, 2 pls.

SCHOENWETTER, J. 1962. The pollen analysis of eighteen archeological sites in Arizona and New Mexico. Fieldiana: Anthropology, 53:168–209, 4 figs.

SEARS, J. W. 1947. Relationships within the *quinaria* species group of *Drosophila*. Univ. Texas Publ., 4720:137–156, 1 fig.

SEARS, P. B. 1942. Forest sequences in the north central states. Bot. Gaz., 103:751–761, 6 figs.

———. 1961a. A pollen profile from the grassland province. Science, 134:2038–2040, 2 figs.

———. 1961b. Palynology and the climatic record of the Southwest. Ann. New York Acad. Sci., 95:632–641, 2 figs.

———. 1963. Vegetation, climate, and coastal submergence in Connecticut. Science, 140:59–60, 1 fig.

SELANDER, R. B. 1960. Bionomics, systematics, and phylogeny of *Lytta,* a genus of blister beetles (Coleoptera, Meloidae). Illinois Biol. Monog., 28:295 pp., 1 pl., 350 figs.

SELANDER, R. K. 1964. Speciation in wrens of the genus *Campylorhynchus*. Univ. California Publ. Zool., 74:1–305, 22 pls., 36 figs.

———. 1965. Avian speciation in the Quaternary. *In* Wright, H. E., Jr., and D. G. Frey, eds., The Quaternary of the United States: 527–542, 6 figs. Princeton, Princeton Univ. Press.

———, and D. R. GILLER. 1963. Species limits in the woodpecker genus *Centurus* (Aves). Bull. Amer. Mus. Nat. Hist., 124:213–274, pls. 53–56, 17 figs.

SHANNON, F. A., and F. L. HUMPHREY. 1963. Analysis of color pattern polymorphism in the snake, *Rhinocheilus lecontei*. Herpetologica, 19:153–160, 3 figs.

———, and H. M. SMITH. 1949. Herpetological results of the University of Illinois Field Expedition, spring 1949. Trans. Kansas Acad. Sci., 52:494–509.

SHELFORD, V. E. 1963. The Ecology of North America. Urbana, Univ. Illinois Press. 610 pp., num. figs.

SHORT, L. L., JR. 1961. Notes on bird distribution in the central plains. Nebraska Bird Rev., 29:2–22.

———. 1963. Hybridization in the wood warblers *Vermivora pinus* and *V. chrysoptera*. Proc. XIII. Int. Orn. Congr.:147–160, 2 figs.

———. 1965. Hybridization in the flickers (*Colaptes*) of North America. Bull. Amer. Mus. Nat. Hist., 129:307–428, 11 figs.

SIBLEY, C. G. 1950. Species formation in the Red-eyed Towhees of Mexico. Univ. California Publ. Zool., 50:109–194, pls. 11–16, 18 figs.

———. 1958. Hybridization in some Colombian tanagers, avian genus *Ramphocelus*. Proc. Amer. Phil. Soc., 102:448–453, 1 fig.

———, and L. L. SHORT, JR. 1959. Hybridization in the buntings (*Passerina*) of the Great Plains. Auk, 76:443–463, 5 figs.

————, and L. L. Short, Jr. 1964. Hybridization in the orioles of the Great Plains. Condor, 66:130–150, 3 figs.

————, and D. A. West. 1959. Hybridization in the Rufous-sided Towhees of the Great Plains. Auk, 76:326–338, 3 figs.

Sibley, C. L. 1938. Hybrids of and with North American Anatidae. C. R. IX. Congr. Orn. Int., 327–335.

Simmons, G. F. 1925. Birds of the Austin Region. Austin, Univ. Texas Press. 387 pp., figs.

Smiley, T. L. 1958. Years, centuries, and millennia. *In* Smiley, T. L., ed., Climate and Man in the Southwest. Univ. Arizona Program in Geochron., Contrib., 6:10–18.

Smith, E. C. 1943. A study of cytology and speciation in the genus *Populus* L. J. Arnold Arb., 24:275–305, 4 pls.

Smith, H. M. 1946. Hybridization between two species of garter snakes. Univ. Kansas Mus. Nat. Hist. Publ., 1:99–100.

————, and J. P. Kennedy. 1951. *Pituophis melanoleucus ruthveni* in eastern Texas and its bearing on the status of *P. catenifer*. Herpetologica, 7:93–96, 1 fig.

Smith, P. W. 1957. An analysis of post-Wisconsin biogeography of the Prairie Peninsula region based on distributional phenomena among terrestrial vertebrate populations. Ecology, 38:205–218, 46 figs.

Smith, S. G. 1963. Natural hybrids between coccinellid species. Canada Dept. Forestry, Forest Ent. Path. Branch, Bi-monthly Prog. Rep., 19(4):2.

Soper, J. D. 1964. The Mammals of Alberta. Edmonton, Govt. of Alberta. 410 pp., 67 pls., maps, 41 figs.

Spieth, H. T. 1941. Taxonomic studies on the Ephemeroptera. II. The genus *Hexagenia*. Amer. Midl. Nat., 26:233–280, 6 pls.

————. 1947. Taxonomic studies of the Ephemeroptera. IV. The genus *Stenonema*. Annals Ent. Soc. Amer., 40:87–122, 31 figs.

Sprunt, A., Jr. 1954. Florida Bird Life. New York, Coward-McCann. 527 pp., 56 pls., 65 figs. [with 1963 "Addendum", 24 pp.]

Stallings, D. B., J. R. Turner, and V. N. Stallings. 1960. Apparent wild hybrids among the Megathymidae. J. Lepid. Soc., 13:204–206, 1 pl.

Stebbins, R. C. 1954. Amphibians and Reptiles of Western North America. New York, McGraw-Hill. 528 pp., 104 pls., 51 figs.

Stein, R. C. 1963. Isolating mechanisms between populations of Traill's Flycatchers. Proc. Amer. Phil. Soc., 107:21–50, 14 figs.

Stevens, O. A. 1950. Handbook of North Dakota Plants. Fargo, North Dakota Agric. College. 324 pp., 319 figs.

Stone, A., K. L. Knight, and H. Starcke. 1959. A synoptic catalogue of the mosquitoes of the world. Entomol. Soc. Amer. Thos. Say Found., 6:358 pp.

Stone, W. S., and J. T. Patterson. 1947. The species relationships in the *virilis* group. Univ. Texas Publ., 4720:157–160.

Storer, R. W. 1951. Variation in the Painted Bunting (*Passerina ciris*), with special reference to wintering populations. Occas. Papers Mus. Zool. Univ. Michigan, 532:12 pp.

————. 1954. A hybrid between the Chipping and Clay-colored Sparrows. Wilson Bull., 66:143–144.

————. 1961. A hybrid between Painted and Varied Buntings. Wilson Bull., 73:209.

Storm, R. M. 1952. Interspecific mating behaviour in *Rana aurora* and *Rana catesbiana*. Herpetologica, 8:108.

Straatman, R. 1962. Notes on certain Lepidoptera ovipositing on plants which are toxic to their larvae. J. Lepid. Soc., 16:99–103.

Sulerud, R. L., and D. D. Miller. 1966. A study of key characteristics for distinguishing several *Drosophila affinis* subgroup species, with a description of a new related species. Amer. Midl. Nat., 75:446–474, 11 figs.

SWARTH, H. S. 1922. Birds and mammals of the Stikine River region of northern British Columbia and southeastern Alaska. Univ. California Publ. Zool., 24:125–314.

————. 1924. Birds and mammals of the Skeena River region of northern British Columbia. Univ. California Publ. Zool., 24:315–394, pls. 9–11, 1 fig.

SWEADNER, W. R. 1937. Hybridization and the phylogeny of the genus *Platysamia*. Ann. Carnegie Mus., 25:163–241, pls. 15–19.

TANNER, W. M. 1944. A taxonomic study of the genus *Hypsiglena*. Great Basin Nat., 5:25–92, 4 pls.

THORNBURY, W. D. 1965. Regional Geomorphology of the United States. New York, John Wiley. 609 pp., 321 figs.

THORNTON, W. A. 1955. Interspecific hybridization in *Bufo woodhousei* and *Bufo valliceps*. Evolution, 9:455–468, 7 figs.

TINKLE, D. W. 1958. The systematics and ecology of the *Sternothaerus carinatus* complex. Tulane Studies Zool., 6:3–56, 57 figs.

TUCKER, J. M. 1952. Evolution of the Californian Oak *Quercus alvordiana*. Evolution, 6:162–180, 14 figs.

————. 1953a. The relationship between *Quercus dumosa* and *Quercus turbinella*. Madroño, 12:49–60, 4 figs.

————. 1953b. Two new oak hybrids from California. Madroño, 12:119–127, 1 fig.

————. 1961. Studies in the *Quercus undulata* complex. I. A preliminary statement. Amer. J. Bot., 48:202–208, 5 figs.

————. 1963. Studies in the *Quercus undulata* complex. III. The contribution of *Q. arizonica*. Amer. J. Bot., 50:699–708, 6 figs.

UZZELL, T. M., JR. 1964. Relations of the diploid and triploid species of the *Ambystoma jeffersonianum* complex (Amphibia, Caudata). Copeia, 1964:257–300, 23 figs.

VAN GELDER, R. G. 1959. A taxonomic revision of the spotted skunks (genus *Spilogale*). Bull. Amer. Mus. Nat. Hist., 117:229–392, 47 figs.

VAUGHAN, T. A. 1954. Mammals of the San Gabriel Mountains of California. Univ. Kansas Publ. Mus. Nat. Hist., 7:513–582, 4 pls., 1 fig.

VAURIE, P. 1955. A revision of the genus *Trox* in North America (Coleoptera, Scarabeidae). Bull. Amer. Mus. Nat. Hist., 106:1–90, 27 figs.

VINES, R. A. 1960. Trees, Shrubs and Woody Vines of the Southwest. Austin, Univ. Texas Press. 1104 pp., num. figs.

VOIPIO, P. 1952. Subspecific boundaries and genodynamics of populations in mammals and birds. Ann. Zool. Soc. Vanamo, 15(4):32 pp., 1 fig.

VOLPE, E. P. 1952. Physiological evidence for natural hybridization of *Bufo americanus* and *Bufo fowleri*. Evolution, 6:393–406, 7 figs.

————. 1955. Intensity of reproductive isolation between sympatric and allopatric populations of *Bufo americanus* and *Bufo fowleri*. Amer. Nat., 89:303–317, 4 figs.

VOSS, E. G., and W. H. WAGNER, JR. 1956. Notes on *Pieris virginiensis* and *Erora laeta*—two butterflies not hitherto reported from Michigan. Lepid. News, 10:18–24, 2 figs.

WALKER, E. M. 1953. The Odonata of Canada and Alaska, vol. 1. Toronto, Univ. Toronto Press. 292 pp., 44 pls.

————. 1958. The Odonata of Canada and Alaska, vol. 2. Toronto, Univ. Toronto Press. 318 pp., 64 pls.

WASSERMAN, A. O. 1957. Factors affecting interbreeding in sympatric species of spadefoots (genus *Scaphiopus*). Evolution, 11:320–338, 7 figs.

————. 1958. Relationships of allopatric populations of spadefoots (genus *Scaphiopus*). Evolution, 12:311–318, 2 figs.

WATERHOUSE, G. A. 1914. A monograph of the genus *Tisiphone* Hübner. Australian Zool., 1:15–19, 1 pl.

————. 1922a. An account of some breeding experiments with the satyrine genus *Tisiphone*. Proc. Linnean Soc. New South Wales, 47:ix–xv, 3 pls.

————. 1922b. The relation of *Tisiphone* to the physiography of eastern Australia. Proc. Linnean Soc. New South Wales, 47:xv–xvii.

————. 1932. What Butterfly is That? A Guide to the Butterflies of Australia. Sydney, Angus and Robertson. 291 pp., 34 pls., 4 figs.

WEAVER, J. E., and F. E. CLEMENTS. 1938. Plant Ecology, 2nd ed. New York, McGraw-Hill. 601 pp., 271 figs.

WELLS, B. W. 1937. Southern Appalachian grass balds. J. Elisha Mitchell Sci. Soc., 53:1–26, 5 pls., 2 figs.

WELLS, P. V. 1965. Scarp woodlands, transported grassland soils, and concept of grassland climate in the Great Plains region. Science, 148:246–249, 2 figs.

WEST, R. G. 1961. Late- and postglacial vegetational history in Wisconsin, particularly changes associated with the Valders readvance. Amer. J. Sci., 259:776–783, 6 figs.

WHITEHEAD, D. R. 1965. Palynology and Pleistocene phytogeography of unglaciated eastern North America. *In* Wright, H. E., Jr., and D. G. Frey, eds., The Quaternary of the United States. Pp. 417–432, 5 figs. Princeton, Princeton Univ. Press.

————, and D. R. BENTLEY. 1963. A post-glacial pollen diagram from southwestern Vermont. Pollen et Spores, 5:115–127, 2 figs.

WOODSON, R. E., JR. 1947a. Notes on the 'historical factor' in plant geography. Contr. Gray Herb. Harvard Univ., 165:12–25.

————. 1947b. Some dynamics of leaf variation in *Asclepias tuberosa*. Ann. Missouri Bot. Garden, 34:353–432, 15 figs.

————. 1964. The geography of flower color in Butterflyweed. Evolution, 18:143–163, 1 pl., 7 figs.

WORRELL, E. 1963. Reptiles of Australia. Sydney, Angus and Robertson, Ltd. 207 pp., 63 pls. figs.

WRIGHT, H. E., JR., T. C. WINTER, and H. L. PATTEN. 1963. Two pollen diagrams from southeastern Minnesota: problems in the regional late-glacial and postglacial vegetational history. Bull. Geol. Soc. Amer., 74:1371–1396, 3 pls., 4 figs.

WRIGHT, P. L. 1953. Intergradation between *Martes americana* and *Martes caurina* in western Montana. J. Mamm., 34:74–86, 1 pl., 1 fig.

WRIGHT, W. G. 1905. The Butterflies of the West Coast of the United States. San Francisco, Whitaker & Ray. 264 pp., 32 pls.

YAMASHINA, Y. 1948. Notes on the Marianas Mallard. Pacific Sci., 2:121–124.

YOUNG, F. N. 1954. The water beetles of Florida. Univ. Florida Studies Biol. Sci. Ser., 5(1):238 pp., 31 figs.

————. 1961. Geographical variation in the *Tropisternus mexicanus* (Castelnau) complex. Verh. XI. Int. Kongr. Ent., 1:112–116, 2 figs.

ZWEIFEL, R. G. 1962. Analysis of hybridization between two subspecies of the Desert Whiptail Lizard, *Cnemidophorus tigris*. Copeia, 1962:749–766, 6 figs.

Author Index

Complete references are given on pages indicated by italics.

Aaronson, S.   151, *153*
Abelson, P. H.   119, *153*
Aeppli, E.   88, *96*
Ahmed, I. A.   247, *274*
Aldrich, J. W.   *415*
Alexander, G.   *415*
Alexander, R. D.   *415*
Allen, G.   288–289, 295, *314*
Allen, G. M.   245, 262, *274*
Allport, G. W.   281, *314*
Alvarez del Toro, M.   *424*
Alvarez, J.   77, *96*
Amadon, D.   *32*
American Ornithologists' Union   *415*
Anastasi, A.   281, 285, 297, 301, *314*
Anderson, E.   329, 340, 344, 366, 372, 378, *415*
Anderson, V. E.   277, 294, *314*
Andrewartha, H. G.   13, 14, *32*
Asimov, I.   4, *32*
Atwood, W. W.   *415*
Atz, E. H.   77, *96*

Baccetti, B.   233, *235*
Bailey, V.   *415*
Bajema, C. J.   297, 299, *314*
Baker, C. L.   366, *413*
Baker, H. G.   *415*
Baldwin, E.   73, *96*
Banks, R. C.   372, *415*
Banta, A. M.   43, 71, 92, *96*
Barber, V. C.   222, *235*
Barghoorn, E. S.   115–116, *153, 155*
Barnes, B. G.   198, *235*
Baroff, G. S.   277, *317*
Barr, A. R.   *415*
Barr, T. C.   39, 42–43, 49–50, 51, 57, 62, 65, 67–68,' 80, 83–85, 91–92, 95, *96–97, 99*
Barrett-Hamilton, G. E. H.   371, 380, *415*
Bartram, W.   341, *415*
Bastock, M.   160, 172, 176, 185, 189, *191*
Bateman, M. A.   378

Battley, E. H.   *154*
Bauer, D. L.   *415*
Beardmore, J. A.   15, 22, *32*
Beatty, R. A.   73, *97*
Beatty, M. E.   358, *418*
Becker, H. E.   216, *235*
Beckner, M.   16, *32*
Bedel, L.   42, *97*
Bedini, C.   233, *235*
Beerstecher, E., Jr.   116, *153*
Beilmann   366, *415*
Benaim, S.   277, 289, *314*
Benirschke, K.   247, *274*
Bent, A. C.   *415*
Bentley, D. R.   334, *428*
Berger   372, *415*
Berkner, L. V.   106–107, 111–113, 151, *153*
Bernard, Claude   10
Bern, H. A.   221, 233, *239–240*
Berns, K. I.   130, *153*
Bessey, C. E.   *415*
Bien, E.   115, 118, *153*
Bijers, T. J.   *153*
Birch, L. C.   13, 14, *32, 378, 415*
Bischof, L. J.   294, *314*
Bishop, S. C.   73, *97*
Bisset, K. A.   148, *153*
Black, R. F.   *419*
Blair, A. P.   322, *415–416*
Blair, W. F.   322, 328–329, 344, 374, 376, *415–416*
Bleuler, E.   302, *314*
Blewett, D. B.   288, 294, *314*
Bloch, K.   150, *154*
Blume, J.   65, *97*
Blythe, E.   18
Bock, W.   7, *32*
Bolívar, C.   57
Bolton, E. T.   122, 129, 135, 138, 145, 147, *153, 155*
Bonet, F.   *97*
Böök, J. A.   307, 310, *314*
Bossert, W. H.   247, *274*
Bouillon, M.   73, *102*
Bowman, T. E.   85, *97*

Brace, C. L.   69, 76, 97
Bradshaw, A. D.   10, 32
Brandenburg, J.   233, 235
Brandenburger, J. L.   201, 221, 236
Braun, E. L.   329, 335–336, 341, 416
Brayshaw, T. C.   352, 416
Breder, C. M., Jr.   50, 78, 90, 95, 97, 101
Brenner, L. G.   366, 415
Bretz, J. H.   39, 41–42, 97
Brewer, R. D.   416
Brinkley, B. R.   200, 241
Britten, R. J.   153
Brodkorb, P.   416
Brodbeck, M.   286, 315
Broecker, W. S.   329, 416
Bromley, S. W.   336–338, 416
Bronson, F. H.   254, 275
Brooks, C. E. P.   416
Brooks, C. F.   416
Brower, L. P.   416
Brown, C.   416
Brown, F. A.   66–67, 97
Brown, K. T.   195, 216, 235, 241
Brown, R. G. B.   190
Brncic, D.   168, 176, 191–192
Brues, A. M.   8, 32
Brush, S.   319
Buffon, G. L. L.   255, 274
Bullock, T. H.   279–280, 314
Bünning, E.   65, 77, 97
Burbanck, M. P.   90, 97
Burbanck, W. D.   90, 97
Burbidge, N. T.   370, 416
Burch, P. R. J.   310, 314
Burfield, S. T.   212–213, 235
Burleigh, T. D.   416
Burns, J. M.   416
Burt, C.,   293, 300, 314
Bush, G. L.   414
Butler, P.   334, 338, 416
Byers, C. F.   416

Cable, R. M.   217, 240
Cagle, F. R.   416
Cairns, J.   130, 153
Caius, J.   261, 274
Caldwell, P. J.   130, 153
Calman, W. T.   83, 97
Cannon, W. B.   10, 32
Capenos, J.   232, 242

Carl, C. F.   416
Carlson, E. A.   300, 314
Carlson, R. E.   337, 424
Carr, A. F.   91, 98
Carr, C. J.   420
Carson, H. L.   187, 192
Carter, C. O.   299, 310, 314
Castro, E.   155
Cattell, R. B.   282, 292, 297, 302, 314–315
Caumartin, V.   48, 98
Cazier, M. A.   175, 192
Chermock, R. L.   374, 417
Christiansen, K. A.   60, 77, 93, 98
Chugunov, Iu. D.   196, 235
Clark, A. W.   227–229, 235
Clark, P. J.   291, 315
Clarke, B.   205, 239
Clayton, R. K.   205, 236
Clements, F. E.   335, 428
Clench, H. K.   416
Cleverdon, R. C.   155
Cloney, R. A.   222, 232, 238
Cloud, P. E.   153
Cody, M. L.   29, 32
Cohen, A. I.   216, 236
Cohen-Bazire, G.   195, 236
Cohn, T. J.   368, 417
Coiffait, H.   43, 98
Collin, J.-P.   216, 236
Comfort, A.   8, 32
Common, I. F. B.   417
Conant, R.   417
Cone, R. A.   216, 235
Configliachi, P.   73, 98
Conn, H. W.   49, 98
Conn, J. W.   302, 310, 318
Connor, A. J.   416
Conrad, H. S.   292, 300, 315
Contis, G.   232, 242
Cooke, C. W.   339, 417
Cooper, R. M.   293, 315
Cooper, W.   337, 417
Cope, E. D.   43, 71, 98
Cornelison, A. R.   317
Cory, L.   416
Cottam, W. P.   362, 417
Cournoyer, D. N.   48, 98
Courtemanche, A.   334, 338, 424
Cowan, I. M.   417
Cowan, S. T.   149, 155
Cowie, D. B.   153

Cox, D. D. 334, *417*
Crenshaw, J. W. 372, *417*
Cronbach, L. J. 285–286, 315
Cronquist, A. 335, 341, *419*
Cropley, J. B. 49, *98*
Crowe, F. W. 303, *315*
Cuenot, L. 80, *98*
Cullen, E. 181, 192
Cummins, C. S. 151, *153*
Curl, R. L. 36, *98*
Curtis, J. T. 330, 336, 338, *417*
Cushing, E. J. 338, *417*
Cvijič, J. 42, *98*

DaCunha, A. 167, *192*
Dahr, E. 249, *274*
Dalquest, W. W. 358, *417*
Danneel, R. 233, *236*
Darlington, H. C. 334, *417*
Darwin, C. 15–16, 18, 21, *32*, 38, 70–71, 73, 75–76, *98*, 245, 248, *274*
Daum, J. 94, *99*
Dauvillier, A. 106, *153*
Davenport, C. B. 80, *98*
David, P. R. 277, *315*
Davidson, R. 302, 310, *318*
Davies, W. E. 36, 41, 98
Davis, J. H., Jr. *417*
Davis, M. B. 330, 332, 334, 337–338, *417*
Davis, R. J. *417*
Davis, W. B. *418*
Davis, W. M. 39, 42, *98*
Day, G. M. *418*
Deamer, D. W. 76, *98*
DeBeer, G. 137, *153*
Deevey, E. S., Jr. 329, 334, *418*
Degerbøl, M. 249–250, *274*
Deguchi, N. 232, *242*
DeKay, J. E. 42, *98*
Deleurance-Glaçon, S. 66, 94, *98*
DeLey, J. 118–120, 123–127, 129–130, 134–136, 139–142, 144–147, 150, *153–155*
Del Solar, E. 185, 190, *192*
Denisov, V. P. 372, *418*
Dennis, E. S. 130, *154*
Dennis, M. J. 221, *237*
Descartes, R. 2
Desikachary, T. V. 121, *154*
Dessauer, H. C. *418*

Dethier, V. G. 171, *192*
Dhainaut-Courtois, N. 222, *236*
Dice, L. R. *418*
Dickinson, J. C., Jr. *418*
Dietz, R. A. *418*
Dightman, R. A. 358, *418*
Dilly, N. 215, *236*
Dixon, K. L. *418*
Dobzhansky, Th. 7, 10, 14, 15, 17, 22, *32*, 69, 79, *98*, 167, 176, 184, 187, 190–191, *192*, 277–278, 286, 293, 306, 309, *315*, 376, *418*
Dodt, E. 196, *236*
Dole, M. 106, *154*
Dombrowski, H. 117–119, *154*
Dorf, E. *418*
Doty, P. *155*
Dresco-Derouet, L. 82, 90, *98*
Drobnick, R. *417*
Dubos, R. 2, 8, 10, *32*
Duc-Nguyen, H. 136, *154*
Dudich, E. 48, *98*
Dugle, J. R. *418*
Dunn, L. C. 23 *32*, 277, *315*
Durand, J. P. 73, *98*
Durrant, S. D. 328, 418

Eakin, R. M. 195–196, 198–201, 207, 210–211, 213–214, 216, 221–222, 228, 232–233, *236–237, 241*
Easler, C. 278, *316*
Eberly, W. *98*
Echlin, P. 119, *154*
Edwards, D. A. 233, *241*
Edwards, J. H. 310, *315*
Edwards, J. P. 90, *97*
Eguchi, E. 203, 232–233, *237*
Ehrlich, P. R. *415, 418*
Ehrman, L. 185, *192*
Eigenmann, C. H. 43, 49, 69, 71, 94, *98*
Eiseley, L. 18, *33*
Elliott, P. O. *32*
Elsässer, G. 303, *315*
Emerson, A. E. 77, 98
Engelhardt, G. P. *418*
Erlenmeyer-Kimling, L. 277, 293, 308–309, *315, 317*
Erwin, J. 150, *154*
Essen-Möller, E. 287, 308, 311, *315*
Evans, E. 222, *235*
Evans, K. *418*

Evans, W. H.   *419*
Ewing, A. W.   172, *192*
Eysenck, H. J.   282, *315*

Fahrenbach, W. H.   232, *237*
Fajans, S. S.   302, 310, *318*
Falconer, D. S.   11, *33,* 278, 290–291, 300, 302, 310–311, *315*
Farrand, W. R.   329, *416*
Fassett, N. C.   370, *418*
Fauré-Fremiet, E.   205, *237, 240*
Faxon, W.   *418*
Feigl, H.   286, *315*
Fenneman, N. M.   344, *419*
Fentress, J. C.   255, *274*
Ferguson, D. C.   *419*
Fernández-Morán, H.   232–233, *237*
Field, W. D.   352, *419*
Findley, J. S.   *419*
Fischer, A.   222, *237*
Fisher, R. A.   13, *33,* 300, *315*
Fleck, S.   *317*
Flügge, C.   172, *192*
Fogg, G. E.   144, *154*
Fogg, G. G.   *420*
Forbes, W. T. M.   342, *419*
Fosberg, F. R.   26, *33*
Fox, W.   *418*
Franz, V.   214, *237*
Freeborn, S. B.   *419*
Freedman, D. G.   277, 294, *315*
Freeman, F. N.   293, *318*
Freeman, T. N.   *419*
Freeze, E.   131, *154*
Frenkel, A.   112, *154*
Frey, D. G.   *417*
Friedman, S.   140, *153–154*
Fries, M.   334, *419*
Friesner, R. C.   334, 338, *424*
Fritsch, F. E.   121–122, *154*
Frye, J. C.   340, *419*
Fujimoto, K.   228, *242*
Fuller, J. L.   246–247, 266–267, 270, *274–275,* 277–279, 283–284, 292, 294, 300, *316*
Futch, D.   184, *192*

Gaffron, H.   *154*
Gallik, G. J.   232, *242*
Galton, Sir Francis, 287, 292

Gardner, J. H.   42, *99*
Garrone, G.   310, *316*
Gates, F. C.   352, *419*
Gause, G. F.   136, *154*
Gesell, A.   289, *316*
Gibbons, I. R.   195, 198, *237*
Gibbs, S. P.   195, 205, *237*
Gibson, B.   20, *34*
Gill, F. B.   378, *419*
Giller, D. R.   348, *425*
Gilmore, V. H.   *275*
Ginet, R.   66, 91, 95, *99*
Ginsburg, B. E.   255, *275,* 277, 280, *316*
Giumarro, C.   *155*
Glaessner, M. F.   114, *154*
Glass, B. P.   *419*
Glass, D. C.   277, *316*
Gleason, H. A.   335, 341, *419*
Goggin, J. M.   340, *419*
Goldman, E. A.   256, *275*
Goldsmith, T. H.   233, *237*
Goodland, H.   233, *240*
Goodlett, J. C.   330, *417*
Goodrie, M.   297, *320*
Gordon, M. S.   89, *99*
Gordon, R. O.   *155*
Gottesman, I. I.   277, 287–290, 292–293, 302–307, 309, 311–312, *316*
Goudge, Th.   16, *33*
Gounot, A. M.   48, *99*
Graham, A.   346, *419*
Graham, P. H.   149, *154*
Granick, S.   195, *237*
Grant, V.   6, 10, 24, *33*
Gray, J.   362, *422*
Gregory, I.   303, *316*
Grenacher, H.   201, *237*
Grene, M.   16, *33*
Greuet, C.   217, *237*
Grey, L. P.   *419*
Griggs, R. F.   *419*
Grimstone, A. V.   215, *237*
Grinnell, J.   377, *419*
Grossfield, J.   172–173, *192*
Gruhl, K.   175, *192*
Grüneberg, H.   278, 309, *316*
Guest, W. C.   *193*
Guiguet, C. J.   *414–415*
Guild, W. R.   *155*
Guilford, J. P.   281–283, *316*
Gunter, G.   366, *419*
Günzler, E.   65, *97*

Gupta, P. D.   233, *242*
Gurnee, R. H.   50, *99*
Gwilliam, G. F.   232, *238*

Haag, W. G.   251, *274*
Hadži, J.   217, *238*
Hagadorn, I. R.   221, *239*
Hagen, D. W.   *419*
Haggard, E. A.   290, *316*
Hahn, M.   201, *241*
Haldane, J. B. S.   *193,* 313, *316*
Hale, E. B.   252, *274*
Hall, E. R.   328, 372, *419*
Hall, J. S.   53, *100*
Hall, M. T.   344, *419–420*
Haller, M. H.   277, *316*
Hama, K.   232, *238*
Hamann, O.   42, *99*
Hamilton, W. D.   269, *274*
Handy, L. M.   295, *319*
Hansen, H. J.   94, *99*
Hansen, H. P.   358, *420*
Hansen, K.   224, *238*
Hanson, J. F.   *420*
Hardy, D. E.   26, *159,* 162, 170, 175, *192*
Hardy, G. A.   *416*
Harris, D. L.   292, *316*
Harris, S. J.   65, 67, 90, *101*
Hartline, H. K.   209, 222, *238*
Hartman, O.   *238*
Harvald, B.   289, *314*
Hatt, R. T.   261, *274*
Hawes, R. S.   50, 95, *99*
Heberlein   140–141, 152
Hecht, M. K.   *420*
Heerd, E.   196, *236, 238*
Heiden, J. A. van den   306, *317*
Heimsch, C.   346, *419*
Henshaw, H. W.   *420*
Hermann, H.   232, *238*
Hermans, C. O.   222, 232, *238*
Hersher, L.   253, *274*
Hesse, R.   212, *238*
Heston, L. L.   302, 305, *316*
Heusser, C. J.   338, *420*
Heuts, M. J.   70, 78
Higgins, J.   297, 299, *316*
Hill, I. R.   *420*
Himwich, H. E.   303, *316*
Hinshelwood, C.   130, *153*

Hirsch, J.   277, 280, *317*
Hobbs, H. H., Jr.   64, 85, *99*
Hodgson, E. S.   171, *192*
Hoffer, A.   309, *317*
Holdhaus, K.   84, *99*
Holland, G. P.   *420*
Holland, H. D.   106, 113, *154*
Hollingshead, A. B.   303, 317
Holsinger, J. R.   60, *99*
Holzinger, K. J.   290, 293, *317–318*
Horne, R. W.   215, *237*
Horridge, G. A.   209, 215, 221, 232, *238, 241*
Horowitz, N. H.   130, *154*
Horstman, E.   207, *238*
Hovanitz, W.   374, *420*
Howard, W. E.   380, 382, *420*
Howden, H. F.   329, *420*
Howell, T. R.   *420*
Hoyer, B.   *153*
Hubbard, J. P.   *420*
Hubbs, C. L.   76–77, *99,* 328, *420*
Hubby, J. L.   302, *317*
Huizinga, J.   306, *317*
Humphrey, F. L.   *425*
Huntington, C. E.   372, *420*
Hurst, C. C.   300, *317*
Hurzeler, M.   *317*
Husson, R.   94, *99*
Hutchinson, G. E.   105, *154*
Hunter, S. H.   151, *153*
Huxley, J. S.   30, *33,* 277, 309–310, *317*
Hyman, L. H.   209, 213, 217, 223, *238*

Ikeda, Y.   140–141, *156*
Inger, R. F.   65, *102*
Innes, W. T.   77, *99*
Inouye, E.   288, *317*
Isseroff, H.   217, *238*
Ivanov, M. V.   *155*

Janzer, W.   75, 79, *99*
Janzer, W. L.   75, 79, *99*
Jarvik, L. F.   293, *315*
Jeannel, R.   38, 42–43, 51, 57, 62, 70–71, 80, 82–84, 91, *97, 99*
Jegla, T. C.   53, 95, *100*
Jellison, W. L.   51, *100*
Jewett, S. G.   *420*
Johansen, H.   371, *420*

Johnsgard, P. A.   374, 381, *420*
Johnson, F.   250, *274*
Johnson, N. K.   372, *415*
Johnston, R. F.   352, *420*
Jolicoeur, P.   258, *274*
Jones, H. E.   292, 300, *315*
Jones, T. C.   *275*
Juel-Nielsen, N.   289, 307, *317*
Jung, D.   224, *238*

Kaissling, K. E.   233, 238, *241*
Kallmann, F. J.   277, 295, 303–310, *314–315, 317*
Kalwo, W.   277, *317*
Kamishima, Y.   210, *238*
Kammerer, P.   73, *100*
Karaman, S.   43, *100*
Karren, J. B.   *420*
Kaston, B. J.   *420*
Kato, B. J.   *153*
Kaul, D.   185, *192*
Kawaguti, S.   210, *238*
Keast, A.   *421*
Keever, C.   314, *424*
Kelly, A. P.   136, *154*
Kelly, D. E.   196, 203, 216, *238*
Kelson, K. R.   328, *419*
Kempthorne, O.   292, *317*
Kennedy, J. P.   *426*
Kernéis, A.   224, *239*
Kersters, K.   118–119, 150, *153*
Kety, S. S.   303, *316–317*
Key, K. H. L.   370, *421*
Khalaf, K. T.   233, *239*
Kimball, C. P.   *421*
Kirschstein, H.   216, *240*
Kishimoto, K.   303, *317*
Klauber, L. M.   *421*
Klots, A. B.   375, *421*
Kluyver, A. J.   148–149, *154*
Knight, B. C.   148, *154*
Knight, K. L.   *426*
Kofoid, C. A.   *100*
Kohls, G. M.   51, *100*
Koller, P. C.   191, *192,* 376, *418*
Koopman, K. F.   190, *192*
Kornberg, A.   129, *154*
Kosswig, C.   73, 76–77, 88, *100*
Kosswig, L.   88, *100*
Krasne, F. B.   224, 227, *239*
Krekeler, C. H.   77, *100*

Krutsch, P. H.   371, *421*
Küchler, A. W.   335, *421*
Kuehne, R. A.   49–50, 57, *97*
Kümmel, G.   217, *239*
Kunisawa, R.   195, *236*
Kuwabara, M.   203, 233, *237*
Kuznetsov, S. I.   116, *155*

Lack, D.   29, *33,* 370, *421*
Lamarck, J. B.   69–70, *100*
Land, M.   222, *235, 239*
Langer, B.   185, *193*
Lankester, E. R.   70, 75, *100*
Lanyon, W. E.   377–378, *419, 421*
Lattin, G. de   73, 76, 83, 88, *100*
Lawrence, B.   244, 246, 250–252, 262, *274*
Lawrence, E. M.   292, *317*
Lawrence, P. A.   224, 227, *239*
Leahy, A. M.   293, *317*
Leedale, G. F.   205, *239*
Lejeune, J.   277, 294, *317*
Leleup, N.   83, *100*
Lerner, I. M.   11, *33,* 268, *274,* 300, 309, *317*
Levene, H.   10, *32*
Levine, J.   139, *155*
Levins, R.   12, *33*
Lewin, R. A.   112, *155*
Lewis, A.   308, *317*
Lewis, D. M.   334, *417*
Lewontin, R. C.   12, 14, 15, 22–23, *32–33,* 302, *317*
Lickfeld, K. G.   201, *241*
Lidicker, W. Z., Jr.   *421*
Lidz, R.   302, *317*
Linsdale, J. M.   *421*
Littlejohn, M. J.   376, *416*
Livingston, B. E.   *421*
Livingston, R.   335, 350, *421*
Livingstone, B. G. R.   334, 338, *421*
Livingstone, D. A.   334, 338, *421*
Loehlin, J. C.   292, 302, *317*
Löve, D.   356, *421*
Loeringer, J.   285, *318*
Lorenz, K.   245, 252, 269, *274–275*
Lorković, Z.   374, *421*
Lotka, A. J.   13, *33*
Loukashkin, A. S.   370, *421*
Low, R. J.   247, *274*
Lowe, C. H., Jr.   364, *421*

Ludwig, W. 70, 75–76, 78–79, *100*
Lush, J. L. 290, *318*
Lutz, H. J. *421*
Luxenburger, H. *318*
Lwoff, A. 148, *155*
LyaLikowa, N. N. *155*
Lykken, D. T. 313
Lysenko, O. 149, *155*

MacArthur, R. H. 62, *100*
MacLeod, H. 130, *154*
MacNeil, C. D. *421*
MacNeil, F. S. 339–340, *421*
MacRae, E. K. 216, *239*
Macsorley, K. 308, *318*
Maeki, K. *421*
Maguire, B., Jr. 73, 78, *100*
Maher, B. A. 282, *318*
Malde, H. E. 364, *421*
Malott, C. A. 39, 42, *100*
Mangold, O. 215, *239*
Manier, E. 16, 28, *33*
Manigault, P. 136
Manion, J. J. *417*
Manning, A. 159–160, 162, 171–172, 176–177, 182, 186, 189–190, *191–192*
Manton, I. 205, *239*
Marmur, J. 139, *155*
Marshall, J. T., Jr. *423*
Marshall, L. C. 106–107, 111–113, 151, *153*
Marshall, N. B. 91, *100*
Martin, P. S. 388, 363, 364, *422*
Maser, M. D. *155*
Massie, H. R. 130, *155*
Matalas, B. L. *420*
Matthew, W. D. 245, *275*
Matthey, R. 247, *275*
Maxwell, J. 296, *318*
Maynard Smith, J. *192*
Mayr, E. 6, *33*, 39, 83, 86, 181, 190, *192*, 277, 286, 296, 302, 306, 309, *317–318*, 322, 329, 340, 371, *422*
Mayrat, A. 232, *239*
McAndrews, J. H. 334, 338, *422*
McCabe, R. A. *422*
McCarley, W. H. *422*
McCarthy, B. J. 138, 147, *153, 155*
McClary, D. O. 130, *155*
McCormick, J. F. *420*
McCready, R. G. L. 115–116, 118, *155*

McCulloch, W. F. *422*
McIlroy, D. W. *422*
McLaughlin, W. S. 277, *316*
McNemar, O. 301, *318*
McVaugh, R. *422*
Mecham, J. S. 372, *422*
Mednick, S. A. 302, *318*
Meehl, P. E. 285–286, 289, 304, *315*, *318*
Meeuse, B. J. D. 205, *239*
Mehringer, P. J., Jr. 364, *422*
Melamed, J. 233, *239*
Mellon, A. D. 232, *242*
Menaker, M. 65, *100*
Mengel, R. M. 329, *422*
Mergen, F. *422*
Merrill, M. A. 300, *319*
Mettler, L. E. 191, *192*
Milani, R. 174, *192*
Miller, A. H. 358, *422*
Miller, C. D. F. *422*
Miller, D. D. 376, *422, 426*
Miller, S. L. 106, *155*
Miller, W. H. 196, 214, 221, 233, *239*
Minckler, S. 130, *155*
Miranda, M. *153*
Missotten, L. 216, *239*
Mitchell, R. W. 53, *100*
Mitra, S. 308, *318*
Mittleman, M. B. *422*
Moeck, A. H. *419*
Mohrbacher, R. J. 30, *33*
Monson, G. *423*
Moody, M. F. 201, 216, 221, *239*
Moore, G. A. *416*
Moore, G. W. 42, *100*
Moore, J. C. *422*
Moore, T. E. *415*
Moran, P. A. P. 303, 309–310, *318*
Moreau, R. E. 329, *422*
Morowitz, H. J. 130, *155*
Morris, I. 119, *154*
Morton, N. E. 311–312, *318*
Mosher, L. *319*
Moss, E. H. 354, *422*
Motas, C. 43, *100*
Motschulsky, T. V. von 42, *100*
Muller, C. H. 362, *423*
Muller, H. J. 19, *33*, 190, *193*
Mulligan, G. A. *423*
Munz, P. A. *423*

Murie, A.   257, *275*
Myers, G. E.   115–116, 118, *155*

Nagel, E.   2, 3, *33*
Nagle, J. J.   191, *192*
Naka, K.   203, 233, *237*
Nakamura, H.   112, *155*
Nakao, T.   214, *239*
Needham, J. G.   *423*
Neel, J. V.   290, 302–303, 310, *315, 318*
Neill, W. T.   342, *423*
Nelson, T. C.   341, *423*
Neumögen, B.   381, *423*
Newman, H. H.   293, *318*
Nicholas, G.   50, *99*
Nichols, G. E.   336, 338, *423*
Nichols, R. C.   294, *318*
Niering, W. A.   336, *423*
Nishimura, T.   228, *242*
Nishioka, R. S.   221, 233, *239–240*
Nissen, H. W.   280, *318*
Noble, G. K.   73, *100*
Norman, W. W.   91, *100*
North, W. J.   215, *239*

Odbert, H. S.   281, *314*
Odegaard, O.   308, *318*
Ogden, J. G.   334, 338, *423*
Ognev, S. I.   371, *423*
Ogur, M.   130, *155*
Ohlba, S.   14, 15, *33*
Ohtsuki, H.   207, 210, *242*
Oksche, A.   196, 216, *239*
Olsson, R.   214, *240*
Osborne, R. H.   292, *317*
Osmond, H.   309, *317*
Oura, A.   233, *241*

Packard, A. S.   42, 71, 73, 83, *100*
Palmer, J. D.   51, 65, *101*
Pantin, C. F. A.   215, 237, *239*
Pap, A.   286, *318*
Paradowski, W.   277, 309, *315*
Paramonov, S. J.   370, *423*
Park, I. W.   118–119, 130, 134, 139, 142, 144–147, *153–155*
Park, O.   51, 65, 66–67, 75, 90, *101*
Parker, G. H.   214, *240*
Parkes, K. C.   *423*

Parsons, P. A.   185, *192*
Patterson, J. T.   164, 166, 183, *193, 423, 426*
Pavlovsky, O.   14, 15, 22, *32*
Peabody, F. E.   360, *423*
Pearson, K.   292, *318*
Pease, R. W., Jr.   342, *423*
Peck, M. E.   *423*
Pedler, C.   233, *240*
Pennak, R. W.   349, 358, *423*
Penrose, L. S.   277, 287, 294, 297, 310, *318*
Peterson, R. T.   346, *423*
Petit, C.   172, 185, *193*
Philip, U.   172, *193*
Phillips, A. R.   380, *423*
Phillips, J. C.   *423*
Philpott, D. E.   233, *237*
Pickford, R. W.   300, *318*
Pipa, R. L.   233, *240*
Pitelka, D. R.   205, *240*
Pitelka, F. A.   370, *423*
Pittendrigh, C. S.   280, *318*
Planansky, K.   311, *319*
Plesset, I. R.   295, *319*
Pollin, W.   289, *319*
Pond, G. G.   217, *240*
Pontecorvo, G.   130, *155*
Poos, G. I.   30, *33*
Pope, C. H.   73, *100*
Porter, I. A.   277, *319*
Potter, L. D.   330, 334, *424*
Potzger, J. E.   334, 338, *424*
Pough, F. H.   *418*
Poulson, T. L.   39, 51, 65, 67, 73, 77, 79–81, 89–92, 94–95, *101*
Powell, J. M.   358, *424*
Press, N.   217, *240*
Preston, R. J.   *424*
Primbram, K.   280, *319*
Pringsheim, E. G.   205, *239–240*
Prout, T.   69, *101*
Puri, H. S.   340, *424*

Quarterman, E.   341, *424*
Quay, W. B.   196, *237*

Race, R. R.   287, *319*
Racovitza, E. G.   42, 43, 70–71, *101*
Rado, S.   311, *319*

Rand, A. L.   329, *424*
Rasquin, P.   77, *101*
Raup, H. M.   337–338, 354, *424*
Reddell, J. R.   51, 89–91, *97*
Redlich, F. C.   303, *317*
Reed, C. A.   251, *275*
Reed, E. W.   295–298, *316, 319*
Reed, S. C.   295–299, 313, *316, 319*
Reinthal, W. J.   *424*
Reiter, M. B.   247, *275*
Remington, C. L.   382, *424*
Remington, J. E.   331, 354, 382, *424*
Rendel, J. M.   *193*
Reichle, D. E.   51, 65, *101*
Rensch, B.   25, 30–31, *33,* 73
Richards, H. G.   340, *424*
Richmond, G. M.   349–350, 358, *424*
Riddell, J.   *424*
Riley, H. P.   372, *425*
Ritchie, J. C.   356, *425*
Roberts, J. A. F.   294, *319*
Roberts, R. B.   *153*
Roberts, T. W.   65, 67, 90, *101*
Robinson, J. T.   27, *33*
Robson, E. A.   215, 232, *237, 240*
Roe, A.   277, 280, *319*
Rogers, J. S.   *425*
Röhlich, P.   201, 217, 224, *240*
Roper, J. A.   130, *155*
Rosanoff, A. J.   295, 306, *319*
Rosen, D. E.   81, 89, *99, 101*
Rosenbloom, L.   77, *101*
Rosenthal, D.   289, 303, 305, 311, *319*
Ross, H. H.   329, *425*
Rossman, D. A.   *425*
Roszkowski, A. P.   30, *33*
Rouiller, C. H.   205, *237, 240*
Rouse, I.   341, *425*
Roux, W.   73, 76, *101*
Rowan, W.   *425*
Rowley, J.   *424*
Royce, J. R.   283, *319*
Ruck, P.   233, *241*
Rüdeberg, C.   216, *241*
Rusconi, M.   72, *98*
Rutherford, D. J.   232, *241*
Rutten, M. G.   106, 113, 114, *155*

Sacher, G. H.   8, *32*
Sadoğlu, P.   77, *101*
Saito, H.   140–141, *156*

Sakamoto, S.   221, *242*
Sanger, R.   287, *319*
Sauer, J. D.   370, *418*
Savage, J. M.   360, *423*
Scarr, S.   307, *319*
Schildkraut, C. L.   139, *155*
Schiner, J. R.   42–43, *101*
Schiödte, J. C.   42, *101*
Schlagel, S. R.   90, *101*
Schmalhausen, L. I.   30, *33*
Schmidt, F. J.   42, *101*
Schmidt, V. A.   48
Schneider, D.   233, *241*
Schoenwetter, J.   346, 364, *425*
Schopf, J. W.   116, *155*
Schulsinger, F.   302, *318*
Schwalbach, G.   201, *241*
Schwartz, W.   115, 118, *153*
Scott, J. P.   246–247, 254, 257, 265–266, 269–270, *275*
Scott, W.   *101*
Scriven, M.   16, 32
Seaman, E.   139, *155*
Sears, P. B.   334, 338, 350, 358, 362, *424*
Sekhon, S. S.   232, *241*
Selander, R. B.   *425*
Selander, R. K.   348, 379–380, *414, 424–425*
Serenkov, G. P.   122, *155*
Shannon, F. A.   *425*
Shelford, V. E.   *425*
Shields, J.   277, 288–290, 293, 295, 302–307, 309, 312–313, *314, 316, 319*
Short, L. L., Jr.   374, 378, *425–426*
Shreve, F.   335, *421*
Shull, W. J.   290, 303, *315*
Sibley, C. L.   179, *193,* 322, 368, 370, 374, *425–426*
Siegel, S. M.   *155*
Simmons, G. F.   *426*
Simon, E.   42, *97*
Simpson, G. G.   2, 6, 25–26, 28, 30–31, *33,* 81, *101,* 277, 280, *319*
Sinnott, E. W.   4, 33
Sistrom, W. R.   195, *237*
Sjöstrand, F. S.   216, *241*
Slater, E.   287–288, 295, 304, 306, 310–311, *319*
Slifer, E. H.   232, *241*
Slobodkin, L. B.   12, *33*
Smart, J. C.   16, *33*

Smelser, G. K.   216, *241*
Smiley, T. L.   362, *426*
Smith, E. C.   381, *426*
Smith, H. M.   *425–426*
Smith, M. E.   278, *315*
Smith, P. W.   329, *426*
Smith, S. G.   *426*
Smith, S. M.   287, *319*
Smith, T. G., Jr.   203, *241*
Smythies, J. R.   303, *316*
Sneath, P. H. A.   118, 149, *155*
Snyder, L. H.   277, *315*
Soper, J. D.   *426*
Sotavalta, O.   233, *241*
Sotelo, J. R.   200, *241*
Spassky, B.   277, *315*
Spencer, H.   75, *101*
Spickett, S. G.   20, *34*
Spiess, E. B.   185, *193*
Spieth, H. T.   160, 165, 167–169, 173, 180, 183–184, 186–187, *193, 426*
Sprunt, A., Jr.   *426*
Spurway, H.   *193*
Srb, A.   6, *34*
Stabenau, J. R.   *319*
Stafford, R. E.   287, *319*
Stager, K. E.   51, *101*
Stallings, D. B.   344, 370, *426*
Stanier, R. Y.   120, *155*
Starcke, H.   *426*
Stark, L.   232, *238*
Stebbins, G. L.   2, *34*, 362, *415*
Stebbins, R. C.   196, *236, 241*, 328, *426*
Stein, R. C.   *426*
Stern, C.   278, 302, *319*
Stevens, O. A.   *426*
Steyn, W.   196, 216, *241*
Stone, A.   *426*
Stone, L. S.   73, 80, *101*
Stone, W. S.   164, 166, 183, *193, 423, 426*
Storer, R. W.   *426*
Storm, R. M.   380, *426*
Straatman, R.   383, *426*
Streisinger, G.   174, 184, *192–193*
Stubblefield, E.   200, *241*
Sturm, J. H.   42, *102*
Sturtevant, A. H.   159, 166, 171–172, *193*
Sueoka, N.   128, 131, *155*
Sugawara, Y.   *242*
Sulerud, R. L.   *426*

Swarth, H. S.   377, *419, 427*
Sweadner, W. R.   *427*
Swinnerton, A. C.   39, 41, 42, *102*
Szafranski, P.   *153*

Takahashi, H.   140–141, *156*
Taliaferro, W. H.   217, *241*
Tanasachi, J.   43, *100*
Tanner, W. M.   *427*
Tasaki, K.   221, *242*
Tasnádi-Kubacska, A.   137, *156*
Tax, S.   277, *319*
Teilhard De Chardin, P.   24, 28, *34*
Tellkampf, T. G.   42, *102*
Terman, L. M.   300, *319*
Terry, D.   *317*
Thinés, G. L.   91, *100*
Thoday, J. M.   20, 28, 30, *34*
Thomas, C. A.   130, *153*
Thomas, H. T.   175, 177, *193*
Thompson, H.   289, *316*
Thompson, W. R.   277–279, 283–284, 292, 294, 300, *316, 320*
Thornbury, W. D.   42, *102, 427*
Thornton, W. A.   *427*
Thrailkill, J. V.   50, *99*
Thurm, U.   232, *241*
Thurstone, L. L.   282, *320*
Tinbergen, N.   158, 177, *193*
Tinkle, D. W.   366, *427*
Tijtgat, R.   *154*
Tonosaki, A.   221, *241–242*
Török, L. J.   201, 217, 224, *240*
Tourtelotte, M. E.   *155*
Throckmorton, L. H.   164, *193*
Trujillo-Cenóz, O.   200, 233, *239, 241*
Tryon, R. C.   281, *320*
Tucker, J. M.   346, 362, *417, 425*
Tupin, J.   *319*
Turano, A.   232, *242*
Turner, J. R.   *426*
Tuurala, O.   233, *241*
Tyler, S. A.   115–116, *153*

Uchizono, K.   232, *241*
Uzzell, T. M., Jr.   *427*

Vaissière, R.   232, *241*
Vallentyne, J. R.   119, *156*

Valvasor, J. W.   42, *102*
Vandel, A.   38–39, 42–43, 47, 50–51, 70, 72–73, 77, 80, 82, 84, 86–87, 90, 94, *102*
Vandenberg, S. G.   277, 291–292, 294, *318, 320*
Van der Merwe, N. J.   248, *275*
Van Ermengem, J.   *154*
Van Gelder, R. G.   361, *427*
Van Niel, C. B.   119, 148, *154–155*
Vaughan, T. A.   *427*
Vaupel-von Harnack, M.   196, 210, 216, *240–241*
Vaurie, P.   *427*
Vernon, P. E.   282, *320*
Vernon, R. O.   340, *424*
Vines, R. A.   *427*
Viré, A.   42, *102*
Voipio, P.   374, *427*
Volpe, E. P.   376, *427*
Von Wahlert, G.   7, *32*
Voss, E. G.   *427*

Waddington, C. H.   203, 232, *241*
Wager, H.   205, *241*
Wagner, H. G.   209, *238*
Wagner, K.   246, *275*
Wagner, W. H., Jr.   *427*
Wake, R. G.   130, *153*
Wald, G.   195, *241*
Walker, E. M.   *427*
Wallace, B.   6, 10, 23, *34*
Waring, M. J.   *153*
Wasserman, A. O.   *427*
Waterhouse, G. A.   *427*
Watts, W. A.   340
Weaver, J. E.   335, *428*
Weaver, W.   2, *34*
Weed, L. L.   136, *154, 156*
Weismann, A.   69, 73, 76, *120*
Wells, B. W.   *428*
Wells, P. V.   349, 353, *428*
Werth, E.   262, *275*
West, D. A.   *426*
West, R. G.   334, 338, *428*

Westcott, R. L.   *81*
Westfall, J. A.   196, 198–199, 207, 210–211, 213–214, 221–222, 228, 232, *236–237*
Westfall, M. F.   *423*
Wetzel, B. K.   227, *241*
Wheeler, M.   175–176, *193*
Williams, G. C.   9, *34*
Willoughby, R. R.   297, *320*
Wilson, F. D.   *193*
Wingstrand, K. G.   215, *240*
Whitehead, D. R.   334, 340, *428*
Willman, H. B.   *419*
Woese, C.   *155*
Wolken, J. J.   195, 217, 221, 232–233, *242*
Woods, L. P.   65, *102*
Woodson, R. E., Jr.   329, *428*
Woodward, H. P.   39, *102*
Woolpy, J. H.   255, *275*
Worden, P. B.   195, *237*
Worrell, E.   *428*
Wright, H. E., Jr.   334, 338, *417, 428*
Wright, P. L.   *428*
Wright, S.   78, *102, 259, 268, 275*
Wright, W. G.   *428*

Yamada, E.   216, *242*
Yamamoto, T.   221, *242*
Yamashina, Y.   *428*
Yanase, T.   221, 224, *242*
Yasuzumi, G.   232, *242*
Yoshida, M.   207, 210, *242*
Young, F. N.   *428*
Young, S. P.   256, *275*

Zerbin-Rudin, E.   303–304, 310, *320*
Zeutzschel, B.   233, *236*
Zimm, B. H.   130, *155*
Zimmerman, E. C.   26, *34*
Zonana, H. V.   221, *242*
Zubek, J. P.   294, *315*
Zweifel, R. G.   364, *414, 428*

Subject Index

Abies balsamea   354
  lasiocarpa   354
Acer   331
Acetobacter   128, 147
  aceti (*mesoxydans*)   140
"*Achromobacter*"   125
Actias
  luna   381
  selene   381
Actinomycetes   151–152
Adaptability   9–11
  genetic   9
  physiological   9
Adaptation   5–6, 29
Adaptedness   5–6, 21–23, 29
  of genotype   9, 11
  quantification of   12–14
  to survive and reproduce   7
Adaptive radiation   26
Aerobacter   128
  aerogenes   130
Aeromonas   128, 150
Agathymus
  carlbadensis   344
  mariae   344
  micheneri   370
Agrobacterium   128, 140–141, 143–144, 149–150, 152
  tumefaciens   136, 147
"*Alcaligenes*"   125–126, 128
Allelomimetic behavior   269
Allometry, negative   78
"Altruistic" behavior   269
Alypia
  langtoni   382
  octomaculata   382
Amaranthus   337
Amblyopsis   91
  rosae   81, 89
  spelaea   54, 65, 92, 94–95, 98
Ambrosia   334
Ambystoma punctatum   80
Ameroduvialis   60
  jeanneli   92
Amoebaleria defessa   53

Amphipods   60, 82
Amphiuma
  means   366
  tridactylum   366
Anagenesis   25, 27
Anartia jatrophae   341
Anas
  platyrhynchos   380
  poecilorhyncha   381
  rubripes   380
Animikiea   115
Anoptichthys   77–78, 89, 91, 97
  jordani   77, 90–91
Anthocaris genutia   331
Anthrobia monmouthia   53
Anti-competition   375, 378
Anti-hybridization   375–376, 378, 380
  genome   348
Antopocerus   170
Antriadesmus fragilis   53
Aphaenops   50, 94
Aphanizomenon flos-aquae   122
Aphelocoma coerulescens
Appias drusilla   341
Aquilegia   331
Archaeorestis   115
Arctic treeline contact line   368
Arianops   75
Armandia brevis   222–223
Arthrobacter   128
Asbolis capucinus   341
Asclepias   340
Asellus   76, 91
  aquaticus   75, 77
  aquaticus cavernicolus   77, 88
  stygius   60, 92
Asterocampa
  celtis   331
  celtis montis   352
  clyton   331
Astyanax   77
  jordani   89–90
  mexicanus   77, 89, 96
Athiorhodaceae   109
Atyidae   85

Australopithecus   27
Automeris   340
  i. io   345, 411
  i. io × *lilith*   345, 411
  i. lilith   345
Azomonas   144, 147, 150
Azotobacter   144
Azotococcus   150

Bacillus   128, 141, 147
  natto   139–140
  subtilis   130, 136, 139–140, 149
  subtilis var. *aterrimus*   139
Bacteria
  chemiautotrophic   109–110, 119, 151
  fossil   116–117
  heterotrophs   110–112, 119, 151
  "living fossils"   117–119
  photoautotrophic   109–110, 119, 151
  possible dissimilarity between ancient
    and modern bacteria   135–137
  relationship with bluegreen algae   119–
    121
Bacteriochlorophyll   195
Bacterium paricoli   136
Balanced polymorphism   19–20
  heterotic   18
Baptisia   331
Batrisodes
  henroti   53, 74
  pannosus   75
  spp.   73
  valentinei   66
Bats   48
Battus philenor   354, 382
Beggiatoa   116
Behavior
  evolution of, in dogs and wolves, 264–
    265
  types of   158–159
Beijerinckia   144
Betula   335
  fontinalis   355, 412
  fontinalis × *glandulosa*   355, 412
  glandulosa   355, 412
  lutea   331
  papyrifera   412
  papyrifera × *fontinalis*   412
  papyrifera × *utahensis*   412
Bifidobacterium   128

Biospeleology, development of   42–43
Bluegreen algae
  fossil   121
  phylogeny   121–122
Bombyx mori,
  rhabdomere, development of   203
Bradysia sp.   53
Branchiostoma californiense   214
Brathinidae   81
Brathinus   81
  nitidus   81
Bufo   376
  americanus   331, 377
  fowleri   331
  quercicus   371
  terrestris   371
  valliceps   348
  woodhousei   348
Bugula avicularia, photosensitivity   233

Caecobarbus   78
  geertsi   78
Caecosphaeroma   82
  burgundum   90, 94, 98
Calandra   14
California Desert-Pacific Slope Suture-
  Zone   360
Callisaurus draconoides ventralis   364
Callosamia
  angulifera   371
  promethea   371
Calocitta formosa   368
Cambarus   72
  rusticus   90
  setosus   90
  tenebrosus   60, 90
Cambrian   25, 29, 105, 109, 112, 151
Campylorhynchus
  rufinucha humilis   379, 414
  rufinucha nigricaudatus   379, 414
  rufinucha nigricaudatus × *r. humilis*
    379, 414
Canis
  aureus   245
  dingo   245
  familiaris   244
  latrans   244
  lupus   244
    arabs   251
    pallipes   251
  mesomelas   244

Canis (*cont.*)
  niger   251
  taxonomy and distribution   244–245
Carabidae   81
Carbon cycle, possible evolution   108
Carya   341, 346
Carychium stygium   53, 55
Cascade-Sierran Suture-Zone   360
Catopidae   70, 81
Cave Biotope, extent of   36–39
Cave ecosystem
  circadian rhythms in cavernicoles, 65–67
  competition   60–65
  effects of geologic structure   67–69
  energy sources and food webs   51–60
  physical parameters   47–51
Caves
  animal life, in   43
    classification   43–45
    regional development of faunas, 45–47
    taxonomic distribution   47
  biological significance   38–39
  colonization of   75, 82–86
  origin   39–42
Central Texas Suture-Zone   342–348
Centurus
  aurifrons   348
  carolinus   348
Cervus canadensis   380
  elephas   380, 382
  nippon   382
Ceuthophilus
  conicaudux   53
  cunicularis   53
Chamaesiphon   115
Chenopodium   337, 340
Chlorobacteriaceae   109
Chlorostrymon maesites   341
Chologaster
  agassizi   65, 77, 81, 89–90, 95
  cornutus   77, 81
Chromobacterium   150
Chromosome number, *Canis*   247
Chromulina psammobia   205
Chymomyza   175, 178
Ciliary line of evolution
  Cephalochordata   214
  Chaetognatha   210–213
  Coelenterata   207–210

Ciliary line of evolution (*cont.*)
  Ctenophora   209
  Echinodermata   209–210
  Protista   205–207
  Vertebrata   216
Ciliary photoreceptors   196–200
  Bombyx   224
  Branchiomma vesiculosum   224–225
Cirolanidae   85
Cladogenesis   25
Clostridium   128
Cnemidophorus
  tigris gracilis   367, 413
  tigris marmoratus   367, 414
  tigris gracilis × *t. marmoratus*   364, 367, 413
Colaptes cafer   348, 352
Coleonyx fasciata   368
  variegatus bogerti   364
  variegatus sonoriensis   368
Colias
  alexandra   352
  eurytheme   348, 374, 377
  hecla   373, 414
  interior   331
  nastes   373, 414
  pelidne   373, 414
  philodice   374, 377
Colinus   340
Collembolans   60
Comamonas   150
Comparative intermediary metabolism, 149–150, 152
Compositae   346
Compositionism   3–5
Corynebacterium   128
Crangonyx   91
Crenothrix   115
Crotalus
  horridus   331
  pricei   372
  tigris   372
Cuon   245
Cyanophyceae   122
Cyanophyta   120
Cytophaga   123

Dacus
  neohumeralis   377
  tryoni   377

Dama
  hemionus   371
  virginiana   371
Danaus   382
  gillipus   341, 343, 409
  plexippus   343, 409
Daphnia   90
Darlingtonea kentuckensis   92, 95
Darwin finches   26
Darwinian fitness   15–19, 29
  and adaptedness   21–24
Deidamia inscriptum   331
Dendrocoelum
  infernale   96
  lacteum   96
Dendrocopos
  pubescens   371
  villosus   371
Dendroica auduboni   352
Deoxyribonucleic acid   107
  evolutionary changes in bacteria   128
  homology   137–148
  nucleotide composition of micro-organ-
    isms   122–128
      Gram negative bacteria   124–126
      Gram positive bacteria,   126–127
Desulfovibrio   107, 117, 128
Didelphis marsupialis   331
Difflugia   60
Diptera   26
DNA; *see* Deoxyribonucleic acid
Dolichopodidae   26
Down's syndrome   295
Drepaniid birds   26
Drosophila   7, 10, 13–15, 19–20, 22–23,
  157ff.
  adiastola   165
  a. americana   184
  ananassae   183, 185, 189
  arizonae   191
  athabasca   376
  auraria   173–174, 176, 190
  birchii   14
  comatifemora   180
  crucigera   180
  flavopilosa   168
  funebris   166
  grimshawi   180
  lebanensis   165
  melanogaster   159–164, 173, 176, 178,
    185–186, 189–191

Drosophila (*cont.*)
  miranda   191
  mojavensis   191
  paulistorum   185
  persimilis   185–187, 190–191
  petalopeza   165
  picticornis   160, 162, 173
  prosaltans   184
  pseudoobscura   14–15, 22, 174, 185–
    187
  quinaria   170
  serrata   14
  simulans   186
  victoria   165
  virilis   160, 164, 173–174, 184
  willistoni   186
Drosophila
  in Hawaii   26, 168, 178–181
  "sex ratio" variant   22–23
  species, numbers of   26, 159
Drosophila (*Sophophora*)   164
Drosophilidae   26
  Scaptomyza   26
  Titanochaeta   26
Dryas julia   341
Dugesia tigrina eyecup sensory cell   219

Early events following geographic contact,
  375–380
Edaphobites   43
Energy sources and food webs   51
Ennychia funebris   382
Enterobacteriaceae   138, 141, 147, 149
Entosphaeroides   115
Eoastrion   115
Eocene   29
Eosphaera   115
Ephyriades brunnea   341
Erebia
  disa   373, 414
  discoidalis   373, 414
  fasciata   373, 414
  rossii   373, 414
  tyndarus   374
Erethizon dorsatum   331
Erynnis baptisiae   331
  funeralis   369, 414
  lucilius   331
  zarucco   369, 414
Escherichia   128, 147, 150

Escherichia (*cont.*)
  coli 130–131, 139, 141, 145, 147
Euglena, photoreceptors 205–207
Eumaeus atala 341
Eunica tatila 341
Eurycea
  (= *Typhlomolge*) *rathbuni* 45, 91–92
  lucifuga 53
Evidence for initial competition 381–382
Evidence for initial hybridizing 380–381
Evolution
  of behavior, dogs and wolves 264–265
  of dog 260–264
  of micro-organisms in Precambrian 105
    end of Precambrian 112–113
  of wolf 259–260
Evolutionary changes in bacterial DNA 128
  Drastic mutational changes in DNA base composition 134
  Evolutionary drift in percent GC 129–134
  possible dissimilarity between ancient and modern bacteria 135–137
Evolutionary opportunism and adaptive radiation 23–27
  processes
    anagenesis 25
    cladogenesis 25
Evolutionary plasticity 11–12, 23, 28–29
Eye
  Arachnida 233
  Asplanchna brightwelli 221
  Haliotis 221
  Hermissenda crassicornis 221
  insect 232–233
  leeches 224, 227–228
  Nereis limnicola 222
    vexillosa 222–223
  Onychophora 228–232
  Pecten 221–222
  Platyneresis dumerilii 222
  Sagitta scrippsae 212
  Viviparus maleatus 221
Eyespot
  Asterias 209–210
  Ciona intestinalis 215
  Henricia leviuscula 210–211
  Leptasterias pusilla 210
  Patiria miniata 210

Feeding habits, *Drosophila* 166–168
Felis
  canadensis 371
  rufus 371
Fermentation 112
"Fitness" 28–29
Flavobacterium 123, 127
Fraxinus 331
Fucus, flagellum 205

Galápagos Islands 26
Gallionella 117
Gammarus pulex 90
Genetic communality 284
Genetic drift 78
Genetics and evolution, human social organization 272–273
Gluconobacter 128
Gopherus
  berlandieri 366
  polyphemus 366
Group survival and individual, survival, 269–270
Guano 53
  bats 51
  cave cricket 51
Gunflintia 115
Gyrinophilus palleucus 90–91

Hadenoecus 53
  subterraneus 51, 53, 65
Haemophilus influenzae 130
Haideotriton wallacei 91
Haldea spp. 341
Halobacterium 128
Halteria 60
Hardy and Weinberg rule 17
Heliconius charitonius 341
Helix aspera 201–202, 204
Heloderma suspectum 364
Heteromysis cotti 82
Hierarchical organization of traits 283
Homeostasis 10, 29
Homo
  africanus 27
  erectus 27
  habilis 27
  "Neanderthal" 27

Homo (*cont.*)
  robustus   27
  sapiens   27, 30
House mouse, gene alleles, *t* locus   23
Human evolution   27
Huroniospora   115
Hyalophora
  cecropia   331–333, 341, 353, 409–410
  columbia   331–333, 409
  euryalus   359
  gloveri   352–353, 359, 410
  gloveri × *cecropia*   353
Hybridization outside Suture-Zones   372–374
Hyla regilla, third eye receptor   198–200

Icteria virens   331
Icterus
  bullocki   348, 352
  galbula   371
  spurius   371
Idiomyia planitibia   179
Incisalia spp.   341
Individual Nearctic Zones   329–331
Ingolfiella   82
Intellectual abilities   292–294
Intelligence and fitness   296–299
Intraspecific variations, mating behavior, 182–185
Iridophores, *Diadema setosum*   210
Isopods   60

Junco hiemalis   332
Juniperus   340
  scopulorum   349
Junonia
  coenia   342, 345, 412
  evarete   342, 345, 412
Jurassic   29

Kakabekia   115
  -like organisms   116
Klebsiella   128, 150
Kleptochthonius cerberus   53, 58

Lagopus
  leucurus   372
  mutus   372
  rupestris   372

Lamarckism   70
Larix   331
Lek behavior   178–181
Leptotes cassius   341
Lepus
  americanus   331
  europeus   371–372, 380
  timidus   371–372, 380
  tolai   371
Lethe spp.   341
  eurydice   350
Life in anoxygenic period   106–110
Limenitis   340, 374
  archippus archippus   343, 347, 366, 369, 382, 411, 412, 414
  archippus archippus × *archippus floridensis*   343, 411
  archippus archippus × *archippus watsoni*   369, 414
  archippus floridensis   343, 382, 411
  archippus obsoleta   347, 412
  archippus watsoni   366, 369, 414
  arthemis   331–333, 374, 377, 382
  arthemis arthemis   356, 411
  arthemis rubrofasciata   355–356, 361, 412, 413
  arthemis rubrofasciata × *L. lorguini,*   361, 413
  arthemis rubrofasciata × *L. weidemeyerii*   355, 412
  astyanax   331–332, 374, 377, 382
  astyanax × *arthemis*   332–333, 411
  astyanax astyanax   346, 411–412
  astyanax arizonensis   347, 365, 412–413
  lorquini   357, 359, 361, 413
  lorquini × *weidemeyerii*   359, 413
  weidemeyerii   355, 357, 359, 365, 412–413
  weidemeyerii × *astyanax arizonensis*   365, 413
Limulus   233
Linopodes   92
Liodytes alleni   341
Liquidambar   335
Liriodendron   331
Livona pica   341
Louisiana-East Texas Suture-Zone   364
Lucifuga   94
Lucilia   117
Lycaon   245

Lyngba 115
  estuarii 122

Macpherson-Macleay Overlap 370
Macroclemys termincki 341
Macrotus
  californicus 368
  mexicanus 368
Magicicada
  cassinii 369, 414
  septendecim 369, 414
  tredecassinii 369, 414
  tredecim 369, 414
Marpesia petreus 341
Martes
  americana 331
  martes 371
  pennanti 331
  zibellina 371
Material compensation 73
Mating behavior
  Drosophila
    adaptive nature 165
    courtship patterns 160–165
    nature and function of stimuli 171–174
    ontogeny 160, 188
    significance of Drosophilid mating behavior 176–178
    specialized behavior, Hawaiian species 168–170
    summary of mating behavior 170–171
    interspecific behavior 185–189
    intraspecific variations 182–185
    other Diptera 175–176
Mechanism 3–5
Mental retardation 294–296
Meta menardii 45, 53
Metalogenium 117
Methanobacterium 107
Methanosarcina 107
Metridium senile, gastrodermal vesicles 215
Microbial fossils 114
  indirect evidence 114
  Precambrian algae-like fossils 115–116
Micrococcus denitrifans 107

Microdipodops
  megacephalus 372
  pallidus 372
Microtus (Pitymys) pinetorum 341
Micruroides euryxanthus 364
Miktoniscus mammothensis 53
Minor or little-known Zones 358–368
Models of speciation and of its sequelae 374–375
Munidopsis polymorpha 83, 97
Mustela erminea 331
Mycobacterium 128
Mycoplasma 128
  gallisepticum 130
Myotis
  fortidens 368
  occultus 368
Myotis sodalis 54
Myrica 340–341

Natural selection 15–16, 29–30
  directional 21
  "disruptive" 20
  diversifying 20–21
  heterotic balancing 18–21
  normalizing 18–20
  varieties of 18–21
Nature of psychological traits 278–286
Neaphaenops tellkampfii 62–63, 92, 95
Nelsonites 85, 92, 94
Nelumbo 118
Neobarrettia
  hakippah 368
  sinaloae 368
Neolamarckism 71
Neurospora crassa 130
Niphargus 82, 99
  orcinus virei 66, 90–91, 95, *98*
  puteanus 66
Nitrococcus 107
Nitrosomonas 107
Northeastern-Central Suture-Zone 331–338, 377
Northern-Cascade Suture-Zone 360
Northern Florida Suture-Zone 339–342
Northern-Rocky Mountain Suture-Zone 352–356
Nostoc punctiforme 122
Notoplana acticola 218

Numerical taxonomy   148
Nyctereutes   245

Oceanic islands   26
Ochromonas malthamensis   112
Oeneis
  jutta   373, 414
  melissa   373, 414
Onchidium verruculatum   221–222
Orconectes   72, 85, 90, 95
  inermis   90
  limosus   85
  pellucidus   60, 65–67, 91
Organism   70, 72
Origin of the dog   245–246
Origin of variation in dog   270–271
Orthoptera   48
Oscillatoria   112, 115
Other minor Nearctic Zones   366–368
Oxygenic atmosphere
  beginning of   110–112, 151
    geochemical evidence   113–114, 151

Pacific-Rocky   Mountain   Suture-Zone
  356–358
Palaemonias
  alabamae   85
  ganteri   60–61, 85, 95
Palynology   330
Paonias spp.   341
  astylus   381
Papilio   340, 354, 380
  americus   370
  aristodemus   341
  avinoffi   354
  bairdii   365, 413
  bairdii brucei   352
  brucei   365, 413
  brucei × *bairdii*   365, 413
  brevicauda   366
  eurymedon   371
  glaucus glaucus   331–333, 353, 411–
    413
  glaucus canadensis   331–333, 361, 411
  gothica   352–353, 378, 412
  gothica × *polyxenes*   353, 412
  hudsonius   354
  machaon   354, 374
  melasina   370
  multicaudatus   347, 371, 412

Papilio (cont.)
  polyxenes   353–354, 366, 370, 374,
    378, 412
  polyxenes americus   378, 414
  polyxenes melasina   379, 414
  rutulus   349, 352–353, 361, 371, 412–
    413
  rutulus × *glaucus*   353
  troilus   331
  troilus troilus   345, 411
  troilus ileoneus   344, 411
Paramecium   60
Paranthropus   27
  robustus   27
Parietal eye   196–197
Parus
  atricapillus   371
  atricristatus   347, 412
  bicolor   347, 412
  hudsonicus   371
Passerina amoena   348, 352
Pathways between genes and traits   273–
  275
Peranema   60
Peripatus   228–229
Permian   135, 137
Phacus   60
Phalangodes   92
  armata   53, 59
Phanetta subterranea   53
Phenotypic aspects of bacterial evolution
  148–151
Phenylketonuria   295
Pheromones   172
Pheucticus
  ludovicianus   353, 412
    melanocephalus   348, 352–353, 412
Phocides batabano   341
Phoebis statira   341
Phreatobite   43
Phyciodes campestris   352
Phyletic evolution   25–26
Phytolacca
  rivinoides   370
  rugosa   370
Picea
  glauca   356
  glauca albertiana   354
  glauca glauca   354
Pieris
  napi oleracea   371, 381
  protodice   381

Pieris (cont.)
  rapae 371, 381
Pinus 334, 340–341, 346
  banksiana 354
  contorta 354
  ponderosa 350
Pipilo 340
  erthrophthalmus 368
  maculatus 352
  ocai 368
Pipistrellus
  hesperus 370
  subflavus 46, 370
Piranga
  erthromelas 371
  olivacea 348, 352
  rubra 371
Pituophis
  catenifer affinis 370
  deppei deppei 370
Plantago 334
  lanceolata 337
  major 337
Plebejus acmon 352
Pleiotropy, indirect effect of 77–79
Pleistocene 27, 42, 76, 84, 86, 329, 339, 346, 372
Pleiston 148
Pleurobrachia pileus, apical organ 209
Pliocene 42, 84
Plusiocampa cookei 53, 56
Poecilia sphenops 89
Polygenic model for transmission of intelligence 299–302
Polygonia
  faunus 331
  satyrus 349
Polygonus lividus 341
Polyorchis, eyecup 207–208
Populus 381
  angustifolia 349, 353, 355
  angustifolia × *deltoides occidentalis* 353
  angustifolia × *tremuloides* 333, 353, 355, 411
  balsamifera 355
  balsamifera × *angustifolia* 355
  deltoides occidentalis 353
  d-occidentalis × *tremuloides* 353
  grandidentata 333, 411
  tremuloides 333, 353, 355, 411

Precambrian 105–107, 114, 135, 137, 151
Prespelaea copelandi 81
Primitive earth and its atmosphere 105
Procambarus 72, 85
  lucifugus 65
Process of domestication, *Canis*
  mutually understandable behavior patterns, 257
  polymorphism in wolf 257–258
  probable method of domestication 258–259
  process of primary socialization in dog 253–255
  similar ecological niches of man and wolf 256–257
  socialization in wolf 255
Progressive evolution 27–32
Proteus (Bacteria)
  morgani 141, 145
  vulgaris 141, 145
Proteus (Urodela) 73, 77
  anguinus 42, 73, 98
Protostomes versus Deuterostomes 234
Prototheca zopfi 111
Providence 141, 145
Prunus 331
Pselaphidae 81
Pseudacris
  ornata 376
  streckeri 376
Pseudanophthalmus 60, 64, 84–85
  audax 62, 64
  beaklei 83
  ciliaris 42
  farreli 84
  grandis 92
  hubbardi superspecies 83
  inexpectatus 62, 64
  intermedius group 84
  menetriesii 53, 62
  orlindae 42
  pubescens 62
  robustus 84–85
  striatus 62
Pseudomonas 124, 127–128, 135, 140–141, 144–145, 147, 149–150, 152
  aeruginosa 119
  campestris var. *pelargonii* 130, 134
  fluorescens 130, 134, 146
  putida 130, 134, 146

"Pseudomonas halocranaea"   118–119
Pseudosinella hirsuta   77
Pseudotsuga   350
Pseudozona mirabilis   53
Psilorhinus mexicanus   368
Psychological tests as signs and samples of behavioral traits   285–286
Ptinidae   14
Ptomaphagus   72
   hirtus   53

Quaternary   330
Quercus   334, 336, 340–341, 346
   alba   381
   gambellii   362, 365, 411
   gambellii × *turbinella*   365, 411
   mohriana   346
   muhlenbergi   346
   prinus   381
   stellata   346
   stellata × *havardi*   346
   turbinella   362, 365, 411
Quiscalus
   quiscula   372
   versicolor   372

Ramphocelus
   flammigerus   370
   icteronotus   370
Rana
   aurora   380
   catesbeiana   380
   palustris   371
   pipiens   371
Rattus norvegicus   30
Regressive evolution in cave animals 69–80
   Darwinian theories   72–79
   Lamarckian theories   69–71
   Orthogenetic theories   71–72
Reithrodontomys
   gracilis   341
   humilis   370
   mexicanus   370
Rhabdomeric line of evolution
   Annelida   222–224
   Arthropoda   232–233

Rhabdomeric line of evolution (*cont.*)
   Mollusca   217–222
   Onychophora   224–231
   Platyhelminthes   217
   Rotifera   217
Rhadine   81
   subterranea   53
Rhineura floridana   341
Rhizobium   128, 137, 141, 143–144, 147, 150, 152
Rhizoperta   14
Ribonucleic acid   107
RNA; *see* Ribonucleic acid
Rocky Mountain-Eastern Suture-Zone 348–356
Rocky Mountain-Sonoran Suture-Zone 362, 364
Rosaceae   331
Rumex acetosella   337

Saccharomyces cerevisiae   130
Salmonella   141
Sarcophaga   175, 177–178
Scalopus aquaticus   331
Scaphiopus
   couchii   348
   hurteri   348
Scaptomyza   26, 159, 175–177
Sceloporus occidentalis   196–197
Schizophrenia   302–313
   family and twin studies   303–307
   frequency   304
   natural selection   307–309
   transmission   309–313
Sciurus
   griseoflavus   370
   socialia   370
Scoterpes copei   53, 57
Serratia   128, 141
Sexual dimorphism in *Drosophila*   169– 170, 178–179
Sexual isolation in *Drosophila*   181– 182, 188–189
   origin of   190–191
Sexual selection   188–189
Shigella   141
Sickle-cell anemia   20
Siderocapsa   117
Silurian   151

Sitta
  canadensis 371
  carolinensis 371
Size of bacterial genome 130
Smerinthus
  cerisyi 371
  geminatus 371
  ocellatus 381
Social selection in dog and wolf 265–268
Southwestern New Mexico Suture-Zone 364–366
Species
  numbers of 24–25
  allochronic 27
Spelechus 75
Speleobama vana 81
Speonomus diecki 66
Speotyto 81
  cunicularia 341
Spermophilus
  pygmaeus 372, 382
  suslica 372, 382
Speyeria
  aphrodite 371
  atlantis 371
  cybele 371
  hesperis 352
  idalia 350
  lurana 349, 352
Sphaeroma serratum 90, 94
Sphaerotilus 117
  natans 117
Sphinx chersis 341
Sphyrapicus
  nuchalis 359, 413
  ruber 359, 413
  varius 332
Spilogale
  phenax phenax 361, 413
  putorius latifrons 361, 413
Spirillum 127–128
Spizella
  arborea 371
  passerina 371
  pusilla 371
Stabilization after early suturing 371–372
Staphylococcus aureus 136
Streptosporangium 141

Strymon
  acadica 331
  acis 341
  martialis 341
Sturnella
  magna 377
  neglecta 348, 377
Stygicola 94
Stygobromus 91
  exilis 60
Suture zone examples, tabulation 323–328
Suture-Zones outside North America 368–371
Suturing and host specificity 383
Suturing and mimicry 382
Synchlora
  aerata 369, 414
  aerata × *denticularia* 414
  denticularia 369, 414
Synechococcus 112
Synergistic relationship between genetic variation and social organization complexity 271–272

Tadarida brasiliensis mexicana 51, 53
Taxodium 335, 340
Tetrapods, mesozoic 25
Texamaurops reddelli 93
Thermomonospora 141
Thiorhodaceae 109–110
Time of domestication, dog 249–250
Tisiphone
  abeona aurelia 379, 414
  abeona morrisi 379, 414
  abeona morrisi × *aurelia* 379, 414
Titanochaeta 26
Tomocerus flavescens americanus 53
Trait approach 281, 283
Trechinae 71, 82, 83
Trechus 82
  hydropicus 84
  hydropicus beutenmulleri 84
  schwarzi 84
  schwarzi cumberlandus 84–85
Troglobites
  adaptation
    reproductive 94–96
    to cave environment 89–93
  definition 43

Troglobites (*cont.*)
   speciation   80–89
      genetic changes   86–89
      preadaptation   80–82
      taxonomic distribution   47
Troglocambarus   64
   maclanei   64
Troglophiles   43, 73
Trogloxenes   43, 65, 73
Tsuga   335
Twin method   287
   strategic implications   288–289
   Twin data statistics   289–292
Typhlicthys   60, 65, 91
   subterraneus   54, 65, 81
Typhlogarra widdowsoni   91
Typholotriton   73
   spelaeus   80

Ulmus   352
Ulotrichaceae   115
Ultraviolet irradiation   107, 109, 112–
   113, 134–135, 151
Ursus horribilis   30
Utetheisa
   bella   342, 345, 411
   bella × *ornatrix*   345, 411
   ornatrix   342, 345, 411

Variation and selection   268–269
Vaucheriaceae   115
Vermivora
   chrysoptera   333, 372, 377, 409
   chrysoptera × *pinus*   333, 411
   pinus   333, 372, 377, 411
Vibrio   129
Vitalism   3–5
Vulpes   245

Wolf ancestor to dog, further evidence
   admixture of wild genes   249
   chromosomes   247
   comparative anatomy   246
   from behavior   247–249

Xanthomonads   139, 145–146, 149, 152
Xanthomonas pelargonii   139–140

Zapus   371
Zerene
   cesonia   363, 413
   eurydice   363, 413
   eurydice × *cesonia*   413
Zonotrichia albicollis   332